LIBRARY
UNIV OF MAINE

LIBRARY
UNIV OF MAINE

THE CHEMICAL FORMULARY

LIBRARY
UNIV OF MAINE

LIBRARY
UNIV OF MAINE

The Chemical Formulary

*A Collection of Valuable, Timely, Practical,
Commercial Formulae and Recipes for
Making Thousands of Products in
Many Fields of Industry*

VOLUME VIII

Editor-in-Chief
H. BENNETT

1948
CHEMICAL PUBLISHING CO., INC.
BROOKLYN, N. Y., U. S. A.

COPYRIGHT, 1948
BY H. BENNETT

PRINTED IN THE UNITED STATES OF AMERICA

EDITOR-IN-CHIEF

H. BENNETT

BOARD OF EDITORS

Ajemian, D. H.	International Plastic Corp.
Albert, John R.	Pittsburgh Testing Laboratory
Aries, Robert S.	Polytechnic Inst. of Brooklyn
Allen, Austin O.	Vita-Var Corp.
Bender, H.	Tung Sol Lamp Works
Benjamin, Lionel	Consulting Chemical Engir
Beretras, H. S.	Nox-Rust Chemical Corp.
Berlow, Charles	Berlow & Schlosser
Blades, A. O.	Blades Engineering Co
Blumenthal, Saul	Food Consultant
Bowman, Clell E.	Industrial Chemist
Carpenter, S. C.	J. Laskin & Sons Corp.
Castro, L. M.	Propan Technical Service
Degering, Ed. F.	Purdue University
Ehrlich, S. D.	Atlas Powder Co.
Frediani, Harold A.	Eimer & Amend
Gidvani, B. S.	London Shellac Research Bureau
Gutkin, S. S.	Falk & Co.
Harriman, Benj. R.	General Aniline & Film Corp.
Heiberger, P.	Arrow Lacquer Co.
Hofmann, M. P.	C. O. Bartlett & Snow Co.
Hyman, S. C.	Industrial Chemist
Jacobson, W.	Antioch College
Jones, Hilton Ira	Hizone Laboratories
Karch, Herbert S.	C. P. Hall Co.
Kellogg, H. D.	Research Consultant
Krassner, Frederick	U. S. Naval Clothing Depot
Kwaselow, J.	Hudson Motor Car Co.
Lampo, Ralph	Arvey Corp.
Leistner, W. E.	Duralene Processed Fabrics
Levey, Harold A.	Consulting Chemical Engineer
Levitt, Benjamin	Consulting Chemist
Lincoln, Bert. H.	Continental Oil Co.
Maglio, M. Martin	Advance Solvents & Chem. Corp.
Martin, Louis E.	Protected Steel Products Co.

Masurovsky, B. I.	Consulting Chemist
Mathers, F. C.	Indiana University
McAnulty, John F.	Blaw-Knox Co.
McCutcheon, John W.	Consulting Chemist
Metro, F. G.	Consulting Chemist
Meyer, Martin	Brooklyn College
Miley, W. M.	Gerber Products Co.
Morris, John T.	Penn. Alcohol & Chem. Corp.
Nugey, A. L.	Consulting Chemical Engineer
O'Brien Jr., Harold C.	University of Pittsburgh
Palese, Richard A. M.	Engineering Associates
Pickard, M. H.	Pacific Coast Borax Co.
Pinnock, D. R.	Wood & Selick
Rasch, Carl H.	Riverside Chemical Co.
Robertson, G. Ross	University of California
Sagady, William	Consulting Chemist
Schuler, George H.	E. I. du Pont de Nemours & Co.
Seymour, Raymond B.	Industrial Research Institute
Shoub, H. L.	Consulting Chemist
Sklarew, S.	Ralph L. Evans Associates
Tonn Jr., W. H.	University of Oklahoma
Ward, George W.	Midwest Research Institute
Webster, J. Robert	Lindsay Ripe Olive Co.
Wendt, Arthur S.	Fred Fear & Co.
Werner, Jesse	General Aniline & Film Corp.
Whitcomb, H. G.	Sherwin-Williams Co.
Zimmerman, B. G.	Zimmerman Associates
Zunick, M. J.	General Electric X-Ray Corp.

PREFACE

Chemistry as taught in our schools and colleges is confined to synthesis, analysis and engineering—and properly so. It is part of the proper foundation for the education of the chemist.

Many a chemist on entering an industry soon finds that the bulk of the products manufactured by his concern are not synthetic or definite chemical compounds but are mixtures, blends or highly complex compounds of which he knows little or nothing. The literature in this field, if any, may be meagre, scattered or antiquated.

Even chemists, with years of experience in one or more industries, spend considerable time and effort in acquainting themselves on entering a new field. Consulting chemists, similarly, have problems brought to them from industries foreign to them. A definite need has existed for an up-to-date compilation of formulae for chemical compounding and treatment. Since the fields to be covered are many and varied, an editorial board was formed, composed of chemists and engineers in many industries.

Many publications, laboratories, manufacturing companies and individuals have been drawn upon to obtain the latest and best information. It is felt that the formulae given in this volume will save chemists and allied workers much time and effort.

Manufacturers and sellers of chemicals will find in these formulae new uses for their products. Non-chemical executives, professional men and others, who may be interested, will gain from this volume a "speaking acquaintance" with products which they may be using, trying, or with which they are in contact.

It often happens that two individuals using the same ingredients in the same formula get different results. This may be due to slight deviations or unfamiliarity with the intricacies of a new technique. Accordingly, repeated experiments may be necessary to get the best results. Although many of the formulae given are being used commercially many have been taken from patent specifications and the literature. Since these sources are often subject to various errors and omissions, due regard must be given to this factor. Wherever possible it is advisable to consult with other chemists or technical workers regarding commercial production. This will save time and money and avoid "headaches."

It is seldom that any formula will give exactly the results which one requires. Formulae are useful as starting points from which to work out one's own ideas. Formulae very often give us ideas which may help us in our specific problems. In a compilation of this kind errors of omission, commission and printing may occur. We shall be glad to receive any constructive criticism.

H. BENNETT

PREFACE TO VOLUME VIII

Additional new formulae have been gathered to compile an eighth volume of the *Chemical Formulary*—an addition which will broaden and bring up-to-date the contents of volumes I, II, III, IV, V, VI and VII. Because the board of editors feels that information of this nature, to be most helpful, should be released as soon as possible and since we have had hundreds of inquiries as to when Volume VIII would be ready, an early publication date was decided upon.

Many German formulae received from Allied Intelligence Groups, and just released, have been included.

Schools and colleges in increasing numbers seem to find it advisable to use the *Chemical Formulary* as an aid in promoting a practical interest in chemistry. By its use, students learn to make cosmetics, inks, polishes, insecticides, paints and countless other products. The result is that chemistry becomes an extremely interesting practical and useful subject. This interest often continues even when the students reach the theoretical or more difficult phases of this subject.

Since some mature users of this book have not had the good fortune to have had previous training or experience in the art of chemical compounding, the simple introductory chapter of directions and advice has been repeated. This chapter should be studied carefully by all beginners (and some more experienced workers) and some of the preparations given therein should be made before attempting to duplicate the more complex formulae in the succeeding chapters.

An enlarged directory of sources of chemicals and supplies has been added. This should prove useful in locating new as well as old materials and products.

It is a sincere pleasure to acknowledge the valuable assistance of the members of the board of editors and others who have given of their time and knowledge in contributing the special formulae which have made this volume possible.

H. BENNETT

NOTE

All the formulae in volumes I, II, III, IV, V, VI and VII (except in the introduction) are different. Thus, if you do not find what you are looking for in this volume, you may find it in one of the others.

CONTENTS

	PAGE
PREFACE	vii
PREFACE TO VOLUME VIII	ix
ABBREVIATIONS	xxv
I. INTRODUCTION	1
II. ADHESIVES	16

    Glue Solution; Gum Arabic Mucilage; Casein Mucilage; High Strength Starch; Label; Waxed Paper; Paper to Tin; Paper and Metal Foil; Paper Pad End Binding Cement; Cellophane; Thermoplastic Paper Coating; Tape Coating; Pressure-Sensitive; Thermoplastic Adhesive Tape; Anti-Stick Coating for Adhesive Rolls; Asphalt; Plywood; Rubber Latex; Cellulose Ether; Tire-Leak Seal; Gelatin Seal; Automobile Radiator Leak Seal; Stopping Boiler Leaks; Sealing Glass into Brass Fittings; Bonding Glass to Aluminum; Sealer for Lenses; Metal to Glass; Furnace; Coke Oven; Dry Furnace; Acidproof; Acidproof for Sulfite Digesters; Acidproof Digester Lining; Synthetic Resin; Pipe Joints; Threaded Joint Seal; Closing Leaks in Iron Pipes; Caulking Composition; Aquarium Sealing or Plugging; Tin; Can Seam Sealing Compound; Cement; Spark Plug; Thermoplastic; Mechanical Packing Paste; Plastic Putty; Emulsion for Vinylite Sheet; Vinylite Coated Cloth; Cellulose Acetate; Lucite and Plexiglas; Thermosetting Plastics; Laminating Glassine or Cellophane; Laminating and Waterproofing Paper; Leather; Sealing Corks; Peelable Label; Heat or Pressure Sensitive; Laboratory; Self-Curing Rubber; Glass Seal for High-Temperature Vacuum Work; Joining Glass, Ceramic and Metals; Borating Solution.

III. COSMETICS	32

    Cosmetic Formulation. Cosmetic Emulsions: Cold Creams; Absorption Base Creams; Liquid Creams. *Greaseless Creams:* Vanishing Creams; Finishing Cream; Camphor Peppermint Cream; All Purpose Cream; Sports Cream; Nourishing Cream. *Cleansing Creams. Foundation Creams:* Powder; Tropical Powder; Liquid

Foundation Make-Up; Theatrical Grease Paint; Hand; Face Lotion; Cosmetic Mucilage; Beauty Milk; Skin Cleaner and Softener; Hand Lotions; Glycerin Hand Jelly. *Sun Protection Cosmetics:* Sun Tan Lotion; Sun Tan Lotion Base; Sun Tan Oils; Sunscreen Cream; Sun Tan Cream; Sun Protective Cream; Sunburn Preventive Ointment; Cheap Sunburn Ointment; Sun Tan Compact Powder. *Leg Make-Up:* Liquid Stocking; Paste Leg Make-Up. *Lip Preparations:* Lipsticks; Liquid Lip Stain; Anti-Chap. *Hair Preparations:* Hair Tonics or Lotions; Anti-Dandruff; Balsamic; Anti-Dandruff Ointment; Cold Permanent Wave Lotion; Permanent Wave Cream; Hair Setting Mix; Chemical Heating Composition for Waving; Curly Hair Straightener; Cream; Pomade; Liquid Dressing; Brilliantine; Bleaching Paste; Shampoo; Coconut Oil Shampoo; Coconut Oil Potash Soap; Transparent Shampoo Jelly; Golden Shampoo; Silver Shampoo; Cream Shampoo; Soapless Shampoo; Dandruff Removing Shampoo. *Bath Preparations:* Bubble Bath Powder. *Depilatories:* Cream. *Shaving Preparations:* Brushless Cream; Brushless Stick; Lather Cream; Lather Shaving Cream Improver; Pre-Shave Cream; After-Shave Lotion. *Nail Preparations:* Laquer Remover; Polish Drier; Cream for Brittle Nails. *Cosmetics for the Eyes:* Mascara; Eye Shadow; Antiseptic Ointment; Drops; Eyelid Scale or Crust Remover. *Anti-Perspirants and Deodorants:* Creams; Liquid. *Face Powder and Compacts:* Rouge; Tropical Face Powder. *Foot Preparations:* Powder; Balm. *Dental Preparations:* Cleanser and Gum Hardener; Denture Cleaner; Denture Adhesive; Tooth Cavity Filling; Toothache Oil; Tooth Paste; Tooth Paste Flavors; Tooth Powder; Anti-Pyorrhea Tooth Powder; Liquid Dentifrice. *Coloring and Perfuming:* Formulae for Popular Water Soluble Colors; Oil and Wax Soluble Dye Blends; Fixing Powder of Aromatics and Essential Oils; Perfume Fixing Agents; Rose Concentrate for Face Powder; Jasmine Perfume Base; Lilac Perfume Base; Violet Perfume Base; Bay Rum.

IV. DRUG PRODUCTS 92

Aluminum Hydroxide Gel; Amino Acid; Elixir; Anti-Acid; Antiseptic Baby Oil; Antiseptic Ear Powder; Asthma Inhalant; Atabrine Tablets; Athlete's Foot Preparations. *Burn Treatments:* Flash Burn Preventive Cream; Burn Film with Sulfagel; Burn Ointment; Burn

Sprays; Burn Treating Emulsions; Triple Dye-Soap Mixtures for Burns; Calamine Lotion; Camphor Solution; Crab Lice Treatment; Enteric Coating; Impetigo Lotion; Itch Remedy; Liniments; Nasal Preparations; Plaster Dressing Deodorizer; Pill Excipient; Penicillin Pastilles; Prickly Heat Preparations; Rash Lotion; Scabies Lotion; Sulfadiazine Tablets; Sulfathiazole Dusting Powder; Sulfonamide Chewing Wafer; Styptic Cotton; Wart Remover; Vaginal Douche. *Vaginal Suppositories:* Penicillin Suppository; Sulfa Suppository; Venereal Prophylactic. *Medicated Creams and Ointments:* Creams; Base for Cream; Ointment Base; Vanishing Ointment Base; White Ointment Base; Water-In-Oil Ointment Base; Oil-In-Water Ointment Base; Absorption Base; Washable Ointment Base; Antiseptic Cream; Anti-Vesicant Ointment; Coal Tar Ointment; Penicillin Cream; Penicillin-Glycerin Jelly; Sulfa Drug Ointments; Sulfathiazole Creams; Sulfonamide Ointments; Tropical Ulcer and Wound Ointment; Vitamin A and D Ointment; Wound Healing Ointment; Medicinal Lubricating Jelly. *Industrial Protective Preparations:* Creams; Dermatitis Treatments; Dermatitis Preventives. *Antiseptics and Germicides:* Skin Lotion; Surgical Skin Antiseptic; Sticks; Surgical Instruments; Starch Sponge; Hydrogen Peroxide Stabilizer; Fumigating Sickrooms; Fumigating Cone.

V. EMULSIONS AND DISPERSIONS 111

Hydrocarbon Solvent; Chlorinated Solvent; Oil; Cocoa Butter; Benzyl Benzoate; Methyl Salicylate; Sulfathiazole; Glycol Ester Resin; DDT; Viscous Wax and Grease; Low Viscosity Wax; Irish Peat Wax; Acrawax; Acrawax C; Fatty Acid Amide; Soluble Mineral Oil; Soluble Vegetable and Animal Oil; Soluble Solvent; Industrial Mineral Oil; Cutting Oil; Soluble Oils; Soluble Neatsfoot-Mineral Oil; Soluble Pine Oils; Pectin; Zein Dispersion.

VI. FARM AND GARDEN PREPARATIONS 121

Seedless Tomato Culture; Improving Tomato Yield; Preventing Sprouting of Stored Potatoes; Plant Growth Regulator; Hydroponic Plant Food; Cleaning and Disinfecting Fruits; Tree Wound Dressing; Stimulating Gum Rosin Yields; Cut Flower Preservative; Preservative for Cut Orchids. *Preventing Seed Damping-Off:*

Standard Drench Method; Improved Method; Post-Seedling Method. *Treatment of Vegetable Seeds:* Beets-Leaf Spot; Celery and Celeriac-Bacterial Blight; Onion-Smut; Sweet Potatoes; Rhubarb-Rot; Mushroom Disease Control; Tobacco; Ornamentals; Narcissus Bulb. *Seed Potato Treatment:* Surface Borne Diseases; Preparation of the Solution; Tank and Equipment; Special Precautions; Heating Solution. *Ring Rot:* Disinfecting Equipment. *Prevention of Storage Rot and Infections:* Wet Method; Dip Method; Fumigation Method. *Grain Smut Control:* Grain Seed to Be Treated; Spray Method; Sprinkle Method; Dip Method; Testing Seed for Germination. *Dairy Cattle Feeds:* For Cows on Poor Pasture; for Cows on Good Pasture; for Cows on Fair Pasture; Molasses Cattle Feed; Improved Sweet Sorghum Silage. *Horse and Mule Feed:* Maintaining Idle Stock; Rations for Light Work; Rations for Heavy Work. *Hog Feed:* Mixtures for Sows; Mixture for Growing Pigs. *Lamb Feed:* Average Ration. *Chicken Feed. Turkey Feed:* Increase of Weight of Poultry; Worm Control in Sheep; Cattle Wound Antiseptic Powder; Cattle Wound Dressing; Bull Semen Preserving Medium. *Anthelmintic for Turkeys:* Chick Coccidiosis Control; Poultry Inhalant; Defeathering Compound; Egg Production Increaser; Egg Preserving Coating; Rabbit Deterrent; Pigeon Repellent; Dog Chaser; Shampoo for Puppies; Dog Shampoo. *Small Scale Casein Manufacture.*

VII. FOOD PRODUCTS 147

Baking Powder. Pie Fillings: Canned Apple Pie; Fresh Apple Pie; Blueberry Pie; Lemon Chiffon Pie; Canned Cherry Pie; Frozen Cherry Pie; Peach Chiffon Pie; Frozen Peach Pie; Fresh Strawberry Pie; Lemon Pie; Pie Meringue; Strawberry Whip Pie; Cherry Filling with Apple Sauce; Method for Milk Fillings; Butterscotch Pie; Golden Vanilla Cream; Chef's Chocolate Cream; Light Vanilla Cream. *Baked Goods:* Lemon Chiffon Pie; Sugarless Lemon Pie; Pie Crust Mix; Basic Sweet Dough with Soya Flour; Commercial Danish Dough; Rich Coffee Cake; Baking Powder Biscuit; Honey Muffins; Crisp Pecan Wafers; Cocoanut Bars; Banana Muffins; Almond Macaroons; Layer Cake; Pound Cake; White Cake with Soya Flour; Yellow Layer Cake; Fruit Cakes; Light Fruit Cake; Devil's Food Cake; French Cheese Cake; Doughnuts; Chocolate Doughnuts; Doughnut Glaze;

Doughnut Honey Glaze; Doughnut Icing Glaze; Doughnut Icing; Lemon Powder for Pies; Vanilla Powder; Pie Filling Powder Base; Cream Filling for Pies; Cream Filling for Ice Cream Wafers; Cream Icing and Filling; Improved Cake Icing; Cream Puff Paste; Boiled Meringue Icing; Light Meringue Icing; Sugarless Meringue; Billowy Marshmallow; Marshmallow Fluff; Egg White Substitute; Yeast Cake Improver. *Improving the Quality of Baked Goods:* Cakes; Baking Test for Goods Containing Glyceryl Monostearate; Bread; Icings; White Bread Containing Soybean Flour; Bread Improver; Irish Soda Bread; Vitamin Mixture for Flour Enrichment; Noodles; Wheat Cake; Bakers' Pan Grease; Yeast Fermentation Defoamer; Fudge Bars; Nougat Fruit Bar; Santo Domingo Nougat; Short Nougat; Fruit Bars; Candied Fruit Peel; Butterscotch Toppings; Cream Fondant for Chocolate Dipping; Candy Coatings; Sugarless Chocolate Bars; Sugarless Mapleine Syrup; Compound Table Syrups; Improved Shortening; Stabilized Whipped Cream; Imitation Whipped Cream; Whipped Cream; Frosted Chocolate Malted; Soybean Milk; Moldproofing Cheese; Custard Powder; Coconut Flavor Pudding Powder. *Imitation Flavors:* Apricot; Banana; Rum Essence; Raspberry; Peach; Pineapple; Golden Ginger Ale; Rock and Rye Whiskey Essence; Artificial Cinnamon Oil; Imitation Clove Oil; Vanilla Concentrate; Vanilla Sugar. *Flavor Emulsions:* Almond. *Spicy Flavors:* Meat Extract; Fish Extract; Sausage Extract; Smoked Meat Extract; Curry; Tarragon; Mixed Pickle; Mustard Pickle. *Spice Vinegar Flavors:* Tarragon; Fruit; Mustard; Wine; Malt; Universal Vinegar Essence; Celery; Artificial Lemon Syrup. *Frozen Desserts:* Improved Ice Cream Manufacture; Powdered Ice Cream Mix Specifications; Ice Cream Mix; Ice Milk; Rhubarb Ice; Ice Cream Stabilizer. *Low Cost Pectin Jellies:* Manufacture of Liquid Pectin; Acid Coagulant for Jellies; Apple-Raspberry; Raspberry; Imitation; Pure Orange; Cranberry; Artificial Honey. *Salad Dressing:* Russian. Sauerkraut-Tomato Juice Cocktail. *Peanut Butter Spreads:* Orange; Chocolate; Raisin; Cherry; Pickle; Olive-Pimiento. Ripe Olive Sandwich Spreads. Table Mustard; French Mustards. *Sauces:* Seafood; Mustard; Chili; Barbecue. Pepper-Onion Relish; Vinegar Tablets; Pickling Vinegar; Manufacture of Olive Oil; Beef-Soya Sausage; Pork-Soya Sausage; Meat and Vegetable Dehydration;

Stuffing for Fowls; Detecting Cold Storage Eggs; Preventing Mold Growth in Cold Storage Rooms. *Treating and Handling Fruits, Flavors, Nuts, and Colors.* Revivifying Dry Popcorn; Rhubarb Wine.

VIII. INK AND ALLIED PRODUCTS 199

Waterproof Ruling; Safety Paper; Ink From Old Mimeograph Paper; Spirit; Hectograph; Typewriter and Rubber-Stamp; Indestructible; Glass Etching; Ceramic Stencilling; Metal Marking; Bakelite Stamping; Marking; Thermoplastic Solid Printing; Thermofluid Printing; Steam-Setting Printing; Dampener for Lithographic Printing Plates; Ink Eradicator; Secret Writing Detector; Cleaner for Ruling Pens.

IX. INSECTICIDES, FUNGICIDES AND WEED KILLERS 203

Household Insecticides: German Cockroach Poison; Control of Bedbugs; Control of Ants; Control of Silverfish; Body Louse; Sticky Fly Paper; Mothproofing. *DDT Preparations:* Non-Penetrating Spray; Housefly Spray; Emulsion Screen Coating; Insecticide Emulsion Concentrate; Aerosol Insecticide; Delousing Preparations; Quick Acting Insecticide; Airplane Insecticide; Wettable Powder. *Mosquito and General Insect Repellent:* Mosquito Attractants; Insect Bite Lotion; Midge Repellent; Stable Bordeaux Mixture; Fatted Calcium Arsenate; Rotenone Insecticide; Naphthalene Plant Bactericide; Velsicol Insecticide; Derris Diluents; Nicotine Spray; Lethane Insecticide Emulsion; Cammexane Insecticide. *Fungicides:* Mildew Control; Fungus Control; Spray for Brown Rot; Apple; Cedar Rust Control; Spreader Sticker. *Insecticides Used on Animals and Plants:* Cattle Lice; Cattle Grub Control; DDT Cattle Spray; Cattle Ticks; Goat Louse Dip; Dog Flea Powder; Chicken Louse Powder; Chicken Roost Spray; Killing Insects on Bulbs and Corms; Insecticide Paste; Termite Control; Cabbage Maggots and Chinch Bugs; Potato Psyllid and Flea Beetle Control; Corn Earworm Insecticides; Corn Borer Dust; Pea Aphid Spray; Mexican Bean Beetle Control; Grasshopper Control; Cotton Flea Hopper; Grape-Bud Beetle; Coddling Moth; Peach Tree Borer Spray; Blueberry Thrips Control; Vine Moth Control; Sprays for Ornamental Plants; Horticultural Lice Spray; Protecting Stored Seed from Insects; Colorado Beetle Spray Base; Lucerne Snout Beetle Control;

Slug Killer; Pine Sawfly Control; Sandfly Control; Chiggers; Japanese Beetle Lure; Fire Ant Spray; Mound Ant Control; Treatment of Insect Infested Wheat. *Fumigants and Disinfectants:* Stored Products; Shelled Corn Crib; Narcissus Bulb Fly Fumigation; Dairies and Factories; Citrus Fruit Disinfectant; Insect; Tomato Seed; Wheat Seed; Soil; Tomato Leaf Mold Control. *Herbicides:* Poison Oak and Poison Ivy Eradicator; Weed Killer; Drain Plant Root Destroyed. *Rat Control.*

X. LEATHER TREATING PREPARATIONS 219

Hide Depilatory; Oropon Type Bate; Chrometan; Vegetable Tan; Leather Dubbing; Mildew Preventive for Leather Book Bindings; Leather Fungicide Protection; Increasing Durability of Leather Soles; Cleaning and Dressing Leather Belts; Anti-Tack Solution for Artificial Leather; Artificial Leather Dope; Leather Conditioning Agents; Preparation of Sulfonated Castor Oil; Dyeing Gloves; Wool Sheepskin Preservative.

XI. LUBRICANTS AND OILS 223

Tire Rim; Pipe Thread Seal and Lubricant; Valve; Wire and Bolt Drawing; Pressure Lubricant for Cold Rolling; Soluble Oil; Cutting Oil; Glass Grinding Fluid; Dental Grinding Coolant; Drill Lubricant; Lubricant for Split Dies; Parting Medium for Molds; Water-Soluble; Gasoline-Insoluble Plug Valve; Solvent-Insoluble; Nitration-Resistant; Packing; Rubber; Mold Lubricant for Rubber Goods; Slide Rule; Optical Instrument; Fine Instrument; Clock; Drawer, Window and Door; Lubricating Mastic; Goldbeaters' Lubricant; Wire Rope Lubricant and Preservative; Waterproof Rope; Water Removable Textile; Rayon; Belt Dressing Compound; Hydraulic Pressure Fluid; Hydraulic Brake Fluid; Fat and Oil Oxidation Inhibitor; Anti-Foaming Oil Mixture; Lubricating Oil Filter; Salt Water Protective Grease.

XII. CONSTRUCTION MATERIALS 228

Wood Preservatives; Soluble Dyes for Wood Impregnation with Dimethylol Urea; Detecting Heartwood and Sapwood in Douglas Fir Lumber; Cork Substitute; Building Brick; Dark Gray Concrete; Light Weight Cement; Weatherproof Cement; Cement Improver; Floor Composition and Road Marker; Sanitary Floor Surface; Concrete Floor Hardener; Interior Wall Plaster; Building Board; Coating Asbestos Cement Sheet; Crinkle-Finish Enamel; Preventing Scumming in Antimony-Free Enamels; Opaque Titanium Enamel; Synthetic Mica; Refractory Lining for Foundry Crucibles; Cement Coating for Steel Reaction Chambers; Refractory Lining for Melting Furnaces; Protective Coating for Refractories; High Dielectric Ceramic; Tungsten Cement; High Temperature Insulating Coating; Thermal Insulation; Fireproof Building Insulation; Insulating Mortar; Ceramic Tile and Insulation; Heat Insulating Compound; Mold Soapstone.

XIII. METALS AND THEIR TREATMENT 235
Coloring Anodized Aluminum. Bluing Steel and Iron; Black Finish on Steel; Rustproof Blackening of Stainless Steel; Black Finish for Silverware; Pretreatment of Stainless Steel for Electroplating; Fast Nickel Plating; Silver Plating Steel; Black Nickel Plating; Indium Plating; High-Speed Copper Plating; Cadmium Plating Solution; Lead Plating; Tin Plating; High-Speed Zinc Plating; Plating on Plastics; Metallographic Etchants for Magnesium; Solutions for Polishing Stainless Steel Specimens; Strip Coating; Stripping Copper, Brass or Zinc from Iron; Stripping Copper from Iron or Zinc Alloys; Stripping Metallic Coatings; Stripping Oxide Films from Aluminum; Cleaning Magnesium Welds; Aluminum Cleaner; Pickling (Rust Removing) Solution; Pickling Bath Inhibitor; Cleaning Steel Tools and Equipment; Metal Cleaner; Radiator Rust Remover; Rust Remover for Iron and Steel; Rustproofing Steel; Rust Inhibitors; Slushing (Metal Protective) Grease; Protective Treatment For Tinplate; Corrosion Protection of Magnesium Alloys; Chrome-Pickle Treatment; Sealed Chrome-Pickle Treatment; Dichromate Treatment; Galvanic Anodizing; Chrome-Sulfate Treatment; Chrome-Alum Treatment; Modified Alkali-Chromate Treatment; Borax Treatment; Caustic-Pressure Treatment; Tarnishproof Silver; Dental Amalgam Alloy; Anti-Friction Aluminum Alloy; Non-Oxidizing Bursting Disc Alloy; Printing Type Alloy; Non-Poisonous Buckshot; Galvanizing Alloy; Porous Oil-Absorbing Metal; Platinum Substitute; Light Porous Metal-Like Composition; Cigarette Lighter Flint; Spherical Metal Powders; Examples; Tempering Steel (Very Hard); Anti-Carburizing Composition; Foundry Core Wash; Casting Foundry Core; Foundry-

Core Binder; Foundry Core Crack Filler; Wetting Agent for Dental Castings; Magnesium Molding Sand Inhibitors; Foundry Parting Compound; Non-Corrosive Soldering Flux; General Soldering Flux; Stainless Steel Soldering Flux; Soft Soldering Flux; Silver Soldering Flux; Welding Flux for Magnesium Alloys; Aluminum Welding Flux; Arc-Welding Flux; Electric Welding Flux; Aluminum Melting and Degassing Flux; Flux for Melting Tin Bronze; White Metal (Pewter) Welding Composition; Aluminum Solder; Aluminum Foil Solder; Wiping Solder; Brazing Solder; Solder for Zinc; Low Melting Solder; Soft (Berzelit) Solder; Silver Solder; Bonding Cast Aluminum to Steel; Porous Metal Filters; Removing Carbide Tips from Tool Shanks; Chemical Sharpening of Files; Flotation Reagent (for Molybdenum Ore).

XIV. PAINT, VARNISH, LACQUER AND OTHER COATINGS . . . 258

Exterior House Paints: White; Paint Base; White Primer-Undercoat; Brick and Stucco; Trim and Trellis; General Utility. *Wagon, Tractor and Implement Paints:* Rust-Inhibiting Synthetic Primer; Enamel Gloss White; Enamel Gray; Light Gray Machinery Enamel; Olive Drab Lusterless; Olive Drab Gloss; Black Lusterless. *Exterior Metal Paints:* Synthetic Primers for Ferrous Metals; Rust-Inhibiting Primer; Multiple Pigment Red Lead Paints; Rustproofing Coating. *Marine Paints:* Anti-Corrosive Shipbottom; Anti-Fouling Shipbottom; Anti-Fouling for Wooden Vessels; Black Hull; Light-Gray Hull; Fire-Resisting Preservative; Interior Flat; Interior Enamel; Light Green Fire-Retardant; Waterproof Awning; Moistureproof Coating. *Emulsion Paints:* Gray; Brown; Cream-Colored; Interior; Varnish Enamel; Polyvinyl Acetate Emulsion; Paint Base; Oleoresinous Water Paint; Water Emulsion Paint. *Exterior Varnishes:* Water-Resisting Spar; Glyceryl-Phthalate Spar; Spar Clear; Chemical-Resistant Spar. *Flat Wall Finishes:* Sealer; Undercoater; Wall Paint. *Interior Semi-Gloss Paint. Interior Gloss Paints. Quick-Dry Enamels:* Interior; Non-Yellowing White; Colored; Camouflage Enamel. *Floor Enamels:* Clear Liquid Wood Sealer; Gray; Light Oak; Black Oil-Resisting; Vehicle. *Interior Varnishes:* Floor and Trim; Mixing; Quick-Dry; Short Oil Blending; Tall Oil; Electrical Insulating; Fiber and Pulp Impregnation; Phenolic Mixing; Ready-Mixed Aluminum Vehicles. *Lacquers:* Wood; Gray Finish Coat for Wood and Metal; Red Primer; Metal; Clear Polishing Furniture; Clear; Cloth Waterproofing; Gray Engine; White Cellulose Acetate; Medium Green; Olive Drab. *Cellulose Ether Coatings:* Cellolyn Lacquer Base; Nitrocellulose Furniture. *Cellulose Acetate Lacquers:* Clear Lacquer Base Compositions; Lacquer Thinners; Nitrocellulose Lacquer Thinners. *Paint and Varnish Removers. Special Paints, Coatings and Compounds:* Waterproof Emulsion Wax Coating; Gasolineproof Coating; Waterproof Label Glaze; White Stencil Paint; Acidproof Tank Lining; Asphalt Clear Sealer; Knot Sealer; Mosquitoproof Paint; Protective Coating for Methyl Methacrylate Sheet; Welding Spatter and Cleaning Shield Mixture; Thermocolor Paints; Wallpaper Remover; X-Ray Protective Coating; Coating for Inside of Petroleum Tanks; Gasolineproof Coating for Concrete; Coatings for Plastics; Plastic Coating for Iron Nails; Chemical-Resistant Paints; Caulking Compound; Saran Coating; Black Wrinkle Finish; Yellow Fluorescent Pigment; Fungus Treating Coatings; Fungusproof Wax Coating; Red Barn Paint; Whitewash; Powdered Cold Water Paint; Kalsomine; White Pigment Composition for Fire-Retardant Paint; Concrete Floor Treatment; Antiseptic Barnyard Whitewash; Acidproofing Laboratory Tables; Quick-Drying Red Primer for Wood; Improving Lithographic Varnish; Crankcase Insulating Coating; Black Shingle Stain and Preservative; Walnut Furniture Stain. *Wood Stains—Fast to Light:* Maple; Cherry; Light Oak; Dark Oak; Red Mahogany; Brown Mahogany; Light Walnut; Dark Walnut.

XV. PAPER 289
Paper Finish; Sizing; Transparentizing; Improved Drawing Surface; Cap Die Cutting Lubricant; Greaseproofing Composition; Waterproof Coating; Waterproof Coating for Cartons; Destroying Paper Mill Foam; Inhibiting Bacterial Slime.

XVI. PHOTOGRAPHY 292
Metol-Pyro Developer; Developers for Low-Temperature Processing; Fine-Grain Developer; Development of Brown or Blue-Black Tone; Developing Old Printing Paper; Succinic Acid in Photography; Recovering Silver from Photographic Film; Photographic Filter.

XVII. PLASTICS, RUBBER, RESINS AND WAXES 295
Identification Tests for Plastic Raw Materials: Identification of Raw Material By Heating Tests. *Plastic Compositions:* Black Polyvinyl Plastic; Molding Composition; Transparent Moistureproof Sheeting; Molded Wood Composition; Wood and Crack Filler; Dental Impression Plaster; Denture Mold Plaster; Linseed Fatty Acid Alkyd Resin; Synthetic Resin and Varnish; Flame-Resistant Resin; Electric Cable Coating; Electrical Insulating Compound; Insulating Tape Impregnant; Thermoplastic Shoe Stiffener; Thermoplastic Box Toe Stiffener; Shoe-Filler Composition; Rubberless Eraser; Artificial Straw; Artificial Bristles; Catalysts for Hardening Urea-Formaldehyde Resins; Plasticizer for Polyvinyl Acetate. *Plastic Film Coating:* Styrene Organosols; Vinyl Organosol. *Plating With Plastics:* Vinylite; Vinyl Chloride. *Surface Dyeing of Plastics. Fluorescent Coatings and Plastics:* Formulations for Fluorescent Enamels. *Hot-Melt Coatings:* Plastic Peel; Hot-Melt Coating. *Aminoplast Resins:* Impregnation with Aminoplast Resins; Preparation and Lubrication of Molds; Mixing Formulas; Formulas and Tests for Specific Industrial Application. *Plastic Packings:* Water-Resistant Plastic Packing; Plastic Mixture With Lead; Oil-Resistant Plastic Packing; Gasket Composition; Molded Friction Material; Silastic Gaskets. *Compressed Asbestos Sheet:* Water-Resistant; Oil-Resistant. *Foamed Plastics. Rubber Compositions:* Synthetic Sponge Rubber; Windshield Wiper Compounds; Synthetic Rubber Thread; Eraser Stock; Plastic; Rubber-to-Brass Valve Stem Adhesion; Softening Rubber Articles; to Reclaim Rubber; Anti-Oxidant Film for Ultra-Violet Protection; Mold Wash; Boring Holes in Rubber Stoppers; Anti-Tack Coating for Asphalt Tiles. *Making Flexible Molds from Neoprene Latex:* Master Dispersions. *Wax Compositions:* Beeswax Substitute; "Lost" Wax; Hard Wax Composition; Condenser Impregnating Wax; Ski Wax; Stop-Off Wax; Wetting Agent for Wax Molds. *Solvents:* Vinylite Laminating Thinners; Softening Baths for Celluloid and Cellulose Acetate Plastic.

XVIII. POLISHES 324
Automobile Polish; Auto Polishing Cloth; Metal Cleaning and Polishing Cloth; Metal Polish; Furniture Polish; Furniture Cleaner and Polish; Floor Polish; Dance Floor

Wax; Wood Laboratory Table Polish; Oil Polishes; Wax Polishes; Wax Paste Polish; Liquid Cream Wax Polish; Automobile Polish. *Rubless Polishes:* Rubless Wax Floor Polish; Triethanolamine Water-Resistant Rubless Polish; Morpholine Water-Resistant Rubless Polish; *Natural Resin Dispersions:* Shellac Dispersion; Casein Dispersion; Polishing Paste; Diamond Dust Abrasive; Abrasive Cleaner; Lens Polishing Powder; Polishing Powder; Smoothing Compound for Lucite and Plexiglas; Polishing Compound for Lucite and Plexiglas; Tumbling Barrel Polish; French Polish Base; Black Paste Shoe Polish; Soft Leather Polish; Rubber Footwear Polish.

XIX. PYROTECHNICS AND EXPLOSIVES 336

Waterproof Matches. *Smokes:* White; Colored; Smoke Composition; Explosive Type Colored Smoke Bursts; Dyes for Colored Smokes; Colored Military Flame; Ammunition Primer; Rifle Cartridge Primer; Priming Mixture for Shot-Shell Ammunition; Priming Mixture for Rimfire Ammunition; Priming Mixture for Centerfire Ammunition; Friction Primer for Hand Grenades; Electric Blasting Cap Ignition Mixture; Gasless Delay Fuze Powder; Moldable Seal for Explosive Igniter Wires.

XX. SOAPS AND CLEANERS 340

Laundry; Salt Water; Glossy Soap Finish; Floating Castile; Liquid Soap Extender. *Perfuming of Soap:* Soap Perfumes; Soap Perfuming; Perfume for Tallow Soap; Perfume for Laundry Soap; Perfume for Cold Made Soap; Toilet Soap Perfume; Perfume for Cold Made Castile Soap; Color and Perfume Stabilization in Laundry Soap. *Powdered Soap:* Household; Dishwashing Powder; Powdered Cleaner; Industrial. *Industrial Hand Cleaners and Soap:* Mechanics' Soap; Waterless Hand Cleaner; Water Softener and Cleanser; Saddle Soap; Washing Fluid; Laundry Bluing; Laundry Sour; Acid Resistant Wetting Agent; Glove Cleaner; Dry Cleaning; Rug Cleaning; Textile Tar Spot Remover; Upholstery Cleaner; Cleaning Fluid; Disinfecting Dry Cleaning Solvent; Dry Cleaning Fluid; Blending Soap with Organic Solvents; Carpet; Paint and Tar Solvent; Paint Spot Remover; Grease and Paint Remover; Wall-Paper; Wall; Oil and Grease Spot Remover for Floors; Marble; Telephone Mouthpiece; Celluloid and Fabrikoid Cleaner; Jewelry; Cigarette Stain Remover; Auto; Wash for Printing Rollers; Denture Paint Brush; Industrial; Solvent Emulsion; Cleaning Non-Ferrous Tanks; Lead-Lined Tanks; Metal; Metal Parts; Aluminum; Stainless Steel; Surgical Instrument Detergent; Piston Gum and Carbon Remover; Airplane Body and Engine; Machinery; Motor Carbon Remover; Cleaning Microscope Slides; Removing Carbon Deposits; Removing Brown Stains From Burettes; Cleaning Fermentation Tubes and Glassware; Removing Films of Silicone Lubricants; Window; Automotive Glass; Laboratory Glass Cleaning Solution; Clearing Stopped Drains; Cleaning Locomotive Boilers; Removal of Oils and Greases; Cleaning Auto Radiators; Cleaning Tarnish From Silverware; Dust Cloth Emulsion.

XXI. TEXTILES 355

Bleaching Cotton Goods; Non-Settling Bleach Suspension; Textile Waterproofing; Package Dyeing of Rayon; Dyeing of Nylon Hose; Snagproofing Nylon Hose; Nylon Oxford Dyeing; Dyeing Wool with Phosphoric Acid; Printing on Woolen Fabrics; Screen Printing on Textiles; Self-Emulsifying Base for Textile Printing; Natural Dye. *Textile Sizing:* Washable Textile Sizing; Organdie Finish for Textiles; Nylon Yarn Sizing; Linen Laundry Stiffening; Emulsion for Textile Sizes. *Fungus- and Mildewproofing:* Rotproofing; Rayon Tire Cord Treatment; Delusterant for Textiles. *Flameproofing:* Water-Soluble Flameproofing; Flameproofing Welding Curtains; Water-Resistant Flameproofing; Resilient Rayon Batting; Coir Packing Pads; Airplane Crash Pad; Salvaging Old Jute Fibers; Synthetic Textile Fiber; Asbestos Fiber Suspension; Wool Pulling Compound.

XXII. MISCELLANEOUS 380

Fire Extinguishing Compound; Fire Extinguishing Powder; Foam Fire Extinguisher; Extinguishing Phosphorus Fires; Chimney Soot Remover; Slack Coal Briquettes; Temperature Indicating Cement; Cold Producing Powder; Windshield Anti-Fog; Anti-Freeze for Radiators; Anti-Freeze Corrosion Inhibitor; Aircraft De-icing Compound; Prevention of Ice Formation in Gasoline; Gasoline and Hydrocarbon Liquid Thickener; Boiler Water Compounds; Boiler Scale Inhibitor; Bubble Fluid; Foam Powder; Dyeing of Feathers for Use as Fish Lures; Dyeing Method; Dyeing Vegetable Ivory Buttons; Solubilinzing True Gums; Gum Tragacanth Substitute; Electrical Resistance; X-Ray Contrast Composition; X-Ray Opaque Cream; Non-Foaming Drilling Mud Additive; Sealing Porous Formations in Oil Wells; Heat Absorbing Glass Batch; Infra-Red Crystal Lens; Synthetic Perspiration; Sea Water Imitation; Standard Soil Mixture (Testing); Hydrogen Sulfide Generation; Nitrogen Gas Generating Composition; Doctor (Plumbite) Solution; Non-Gelling Starch; Household Deodorant Spray; Factory Deodorant Spray; Preservation of Gross Specimens; Killing and Preserving; Tissue Embedding and Sectioning Compositions; Dialyzers; Bacterial Culture Media; Treatment of Glass Wool Air Filters; Glycerin Substitute.

TABLES 391

REFERENCES AND ACKNOWLEDGMENTS 398

TRADE NAME CHEMICALS 399

CHEMICALS AND SUPPLIES: WHERE TO BUY THEM 400

SELLERS OF CHEMICALS AND SUPPLIES 418

INDEX 429

ABBREVIATIONS

amp.	ampere
amp./dm^2	amperes per square decimeter
amp./sq. ft.	amperes per square foot
anhydr.	anhydrous
avoir.	avoirdupois
Bé.	Baumé
b.p.	boiling point
C.	Centigrade
°C.	degrees Centigrade
cc.	cubic centimeter
c.d.	current density
cm.	centimeter
cm^3	cubic centimeter
conc.	concentrated
c.p.	chemically pure
cps.	centipoises
cu. ft.	cubic foot
cu. in.	cubic inch
cwt.	hundredweight
d.	density
dil.	dilute
dm.	decimeter
dm^2	square decimeter
dr.	dram
E.	Engler
F.	Fahrenheit
°F.	degrees Fahrenheit
f.f.c.	free from chlorine
f.f.p.a.	free from prussic acid
fl. dr.	fluid dram
fl. oz.	fluid ounce
f.p.	freezing point
ft.	foot
ft.^{2}	square foot
g.	gram
gal.	gallon
gr.	grain
hl.	hectoliter
hr.	hour
in.	inch
kg.	kilogram
l.	liter
lb.	pound
liq.	liquid
m.	meter
min.	minim, minute
ml.	milliliter—cubic centimeter
mm.	millimeter
m.p.	melting point
N.	normal
N.F.	National Formulary
oz.	ounce
pH	hydrogen-ion concentration
p.p.m.	parts per million
pt.	pint
pwt.	pennyweight
q.s.	a quantity sufficient to make
qt.	quart
r.p.m.	revolutions per minute

sec.	second
sp.	spirits
sp. gr.	specific gravity
sq. dm.	square decimeter
tech.	technical
tinc.	tincture
tr.	tincture
Tw.	Twaddell
U.S.P.	United States Pharmacopeia
v.	volt
visc.	viscosity
vol.	volume
wt.	weight
x.	extra

Chapter I

INTRODUCTION

At the suggestion of a number of teachers of chemistry and home economics the following introductory matter has been included.

This section is written in a simple way so that anyone, with or without a technical education or experience, can start making simple products without any complicated or expensive machinery. For commercial productions, however, suitable equipment is necessary.

Chemical specialties en masse are composed of pigments, gums, resins, solvents, oils, greases, fats, waxes, emulsifying agents, water, chemicals of great diversity, dyestuffs, and perfumes. To compound certain of these with some of the others requires certain definite and well-studied procedures, any departure from which will inevitably result in failure. The steps for a successful compounding are given with the formulas. Follow them explicitly. If the directions require that A should be added to B, carry this out literally, and not in reverse fashion. In making an emulsion, the job is often quite as tricky as the making of mayonnaise. In making mayonnaise, you add the oil to the egg, *slowly*, with constant and even and regular stirring. If you do it correctly, you get mayonnaise. If you depart from any of these details: if you add the egg to the oil, or pour the oil in too quickly, or fail to stir regularly, the result is a complete disappointment. The same disappointment might be expected if the prescribed procedure of any other formula is violated.

The next point in importance is the scrupulous use of the proper ingredients. Substitutions are sure to result in inferior quality, if not in complete failure. Use what the formula calls for. If a cheaper product is desired, do not obtain it by substituting a cheaper material for the one prescribed: resort to a different formula. Not infrequently a formula will call for an ingredient which is difficult to obtain: in such cases, either reject the formula or substitute a similar material only after a preliminary experiment demonstrates its usability. There is a limit to which this rule may reasonably be extended. In some instances the substitution of an equivalent ingredient may legitimately be made. For example: when the formula calls for *white wax* (beeswax), yellow wax can be used, if the color of the finished product is a matter of secondary importance. Yellow beeswax can often replace white beeswax, making due allowance for color: but paraffin will *not* replace beeswax, even though its light color recommends it above yellow beeswax.

And this leads to the third point: the use of good quality ingredients, and ingredients of the correct quality. Ordinary lanolin is not the same thing as *anhydrous* lanolin: the replacement of one for the other, weight for weight, will give discouragingly different results. Use exactly what the formula calls for: if you are unacquainted with the material and a doubt arises as to just what is meant, discard the formula and use one that you understand. Buy your materials from reliable sources. Many ingredients are obtainable in a number of different grades: if the formula does not designate the grade, it is understood that the best grade is to be used. Remember that a formula and the directions can tell you only a part of the story. Some skill is often required to attain success. Practice with a small batch in such cases until you are sure of your technique. Many instances can be cited. If the formula calls for steeping quince seed for 30 minutes in cold water, your duplication of this procedure may produce a mucilage of too thin a consistency. The originator of the formula may have used a fresher grade of seed, or his conception of what "cold" water means may be different from yours. You should have a feeling for the right degree of mucilaginousness, and if steeping the seed for 30 minutes fails to produce it, steep them longer until you get the right kind of mucilage. If you do not know what the right kind is, you will have to experiment until you find out. Hence the recommendation to make small experimental batches until successful results are obtained. Another case is the use of dyestuffs for coloring lotions, and the like. Dyes vary in strength; they are all very powerful in tinting value; it is not always easy to state in quantitative terms how much to use. You must establish the quantity by carefully adding minute quantities until you have the desired tint. Gum tragacanth is one of those products which can give much trouble. It varies widely in solubility and bodying power; the quantity prescribed in the formula may be entirely unsuitable for *your* grade of tragacanth. Hence a correction is necessary, which can only be made after experiments to determine *how much* to correct.

In short, if you are completely inexperienced, you can profit greatly by gaining some experience through recourse to experiment. Such products as mouth washes, hair tonics, astringent lotions, need little or no experience, because they are as a rule merely mixtures of simple liquid and solid ingredients, the latter dissolving without difficulty and the whole being a clear solution that is ready for use when mixed. On the other hand, face creams, tooth pastes, lubricating greases, wax polishes, etc., which require relatively elaborate procedure and which depend for their usability on a definite final viscosity, must be made with some skill, and not infrequently some experience is needed.

Figuring

Some prefer proportions expressed by weight, volume or in terms of percentages. In different industries and foreign countries various systems of weights and measures are used. For this reason no one set of units could be satisfactory for everyone. Thus divers formulae appear with different units in accordance with their sources of origin. In some cases, parts instead of percentages or weight or volume are designated. On the pages preceding the index, tables of weights and measures are given. These are of use in changing from one system to another. The following examples illustrate typical units:

Example No. 1

Ink for Marking Glass

Glycerin	40	Ammonium Sulfate	10
Barium Sulfate	15	Oxalic Acid	8
Ammonium Bifluoride	15	Water	12

Here no units are mentioned. When such is the case it is standard practice to use parts by weight, using the same system throughout. Thus here we may use ounces or grams as desired. But if ounces are used for one item then ounces must be the unit for all the other items in the particular formula.

Example No. 2

Flexible Glue

Glue, Powdered	30.9 %	Glycerin	5.15%
Sorbitol (85%)	15.45%	Water	48.5 %

Where no units of weight or volume but percentages are given then forget the percentages and use the same instructions as given under Example No. 1.

Example No. 3

Antiseptic Ointment

Petrolatum	16 parts	Benzoic Acid	1 part
Coconut Oil	12 parts	Chlorthymol	1 part
Salicylic Acid	1 part		

The same instructions as given under Example No. 1 apply to Example No. 3.

It is not wise in many cases to make up too large a quantity of material until one has first made a number of small batches to first master the necessary technique and also to see whether it is suitable for the particular outlet for which it is intended. Since, in many cases, a formula may be given in proportions as made up on a commercial factory scale, it is advisable to reduce the proportions accordingly. Thus, taking the following formula:

Example No. 4

Neutral Cleansing Cream

Mineral Oil	80 lb.	Water	90 lb.
Spermaceti	30 lb.	Glycerin	10 lb.
Glyceryl Monostearate	24 lb.	Perfume	To suit

Here, instead of pounds, grams may be used. Thus this formula would then read:

Mineral Oil	80 g.	Water	90 g.
Spermaceti	30 g.	Glycerin	10 g.
Glyceryl Monostearate	24 g.	Perfume	To suit

INTRODUCTION

Reduction in bulk may also be obtained by taking the same fractional part or portion of each ingredient in a formula. Thus in the following formula:

Example No. 5

Vinegar Face Lotion

Acetic Acid (80%)	20	Alcohol	440
Glycerin	20	Water	500
Perfume	20		

We can divide each amount by ten and the finished bulk is only 1/10th of the original formula. Thus it becomes:

Acetic Acid (80%)	2	Alcohol	44
Glycerin	2	Water	50
Perfume	2		

Apparatus

For most preparations, pots, pans, china and glassware, such as is used in every household, will be satisfactory. For making fine mixtures and emulsions a "malted-milk" mixer or egg-beater is necessary. For weighing, a small, low priced scale should be purchased from a laboratory supply house. For measuring of fluids, glass graduates or measuring glasses may be purchased from your local druggist. Where a thermometer is necessary a chemical thermometer should be obtained from a druggist or chemical supply house.

Methods

To understand better the products which you intend making, it is advisable that you read the complete section covering such products. You may learn different methods that may be used and also avoid errors which many beginners are prone to make.

Containers for Compounding

Where discoloration or contamination is to be avoided (as in light colored, or food and drug products) it is best to use enameled or earthenware vessels. Aluminum, as well, is highly desirable in such cases but it should not be used with alkalies as the latter dissolve and corrode this metal.

Heating

To avoid overheating, it is advisable to use a double boiler when temperatures below 212° F. (temperature of boiling water) will suffice. If a double boiler is not at hand, any pot may be filled with water and the vessel containing the ingredients to be heated is placed therein. The pot may then be heated by any flame without fear of overheating. The water in the pot, however, should be replenished from time to time as necessary; it must not be allowed to "go dry." To get uniform higher temperatures, oil, grease or wax is used in the outer container in place of water. Here of course care must be taken to stop heating when thick fumes are given off as these are inflammable. When higher uniform temperatures are necessary, molten lead may be used as a heating medium. Of course, where materials melt uniformly and stirring is possible, direct heating over an open flame is permissible.

Where instructions indicate working at a certain temperature, it is important that the proper temperature be attained not by guesswork, but by the use of a thermometer. Deviations from indicated temperatures will usually result in spoiled preparations.

Temperature Measurements

In Great Britain and the United States, the Fahrenheit scale of temperature measurement is used. The temperature of boiling water is 212° Fahrenheit (212° F.); the temperature of melting ice is 32° Fahrenheit (32° F.).

In scientific work and in most foreign countries the Centigrade scale is used. On this scale of temperature measurement, the temperature of boiling water is 100° Centigrade (100° C.) and the temperature of melting ice is 0° Centigrade (0° C.).

The temperature of liquids is measured by a glass thermometer. The latter is inserted as deeply as possible in the liquid and is moved about until the temperature remains steady. It takes a little time for the glass of the thermometer to come to the temperatures of the liquid. The thermometer should not be placed against the bottom or side of the container, but near the center of the liquid in the vessel. Since the glass of the bulb of the thermometer is very thin, it can be broken easily by striking it against any hard surface. A cold thermometer should be warmed gradually (by holding over the surface of a hot liquid) before immersion. Similiarly the hot thermometer when taken out should not be put into cold water suddenly. A sharp change in temperature will often crack the glass.

Mixing and Dissolving

Ordinary solution (e.g., sugar in water) is hastened by stirring and warming. Where the ingredients are not corrosive, a clean stick, a fork or spoon is used as a mixing device. These may also be used for mixing thick creams or pastes. In cases where most efficient stirring is necessary (as in making mayonnaise, milky polishes, etc.) an egg-beater or a malted-milk mixer is necessary.

Filtering and Clarification

When dirt or undissolved particles are present in a liquid, they are removed by settling or filtering. In the former procedure the solution is allowed to stand and if the particles are heavier than the liquid they will gradually sink to the bottom. The liquid may be poured or siphoned off carefully and in some cases is then of sufficient clarity to be used. If, however, the particles do not settle out then they must be filtered off. If the particles are coarse they may be filtered or strained through muslin or other cloth. If they are very small particles then filter paper is used. Filter papers may be obtained in various degrees of fineness. Coarse filter paper filters rapidly but will not, of course, take out extremely fine particles. For the latter, it is necessary to use a very fine grade of filter paper. In extreme cases even this paper may not be fine enough. Here it will be necessary to add to the liquid 1–3% of infusorial earth or magnesium carbonate. The latter clog up the pores of the filter paper and thus reduce their size and hold back undissolved material of extreme fineness. In all such filtering, it is advisable to take the first portions of the filtered liquid and pour them through the filter again as they may develop cloudiness on standing.

Decolorizing

The most commonly used decolorizer is decolorizing carbon. The latter is added to the liquid to the extent of 1–5% and heated with stirring for ½ hour to as high a temperature as is feasible. The mixture is then allowed to stand for a while and is then filtered. In some cases bleaching must be resorted to.

Pulverizing and Grinding

Large masses or lumps are first broken up by wrapping in a clean cloth and placing between two boards and pounding with a hammer. The smaller pieces are then pounded again to reduce their size. Finer grinding is done in a mortar with a pestle.

Spoilage and Loss

All containers should be closed when not in use to prevent evaporation or contamination by dust; also because, in some cases, air affects the material adversely. Many materials attack or corrode the metal containers in which they are received. This is particularly true of liquids. The latter, therefore, should be transferred to glass bottles which should be as full as possible. Corks should be covered with aluminum foil (or dipped in melted paraffin wax when alkalies are present).

Materials such as glue, gums, olive oil or other vegetable or animal products may ferment or become rancid. This produces discoloration or unpleasant odors. To avoid this, suitable antiseptics or preservatives must be used. Too great stress cannot be placed on cleanliness. All containers must be cleaned thoroughly before use to avoid various complications.

Weighing and Measuring

Since, in most cases, small quantities are to be weighed, it is necessary to get a light scale. Heavy scales should not be used for weighing small amounts as they are not accurate enough for this type of weighing.

For measuring volumes (liquids) measuring glasses or cylinders (graduates) should be used. Since this glassware cracks when heated or cooled suddenly it should not be subjected to sudden changes of temperature.

Caution

Some chemicals are corrosive and poisonous. In many cases they are labeled as such. As a precautionary measure, it is advised only to sniff a few inches from the cork or stopper. Always work in a well ventilated room when handling poisonous or unknown chemicals. If anything is spilled, it should be wiped off and washed away at once.

Where to Buy Chemicals and Apparatus

Many chemicals and most glassware can be purchased from your druggist. A list of suppliers of all products will be found at the end of this book.

INTRODUCTION

ADVICE
This book is the result of co-operation of many chemists and engineers who have given freely of their time and knowledge. It is their business to act as consultants and, for a fee, to give advice on technical matters. As publishers, we do not maintain a laboratory or consulting service to compete with them.

Please, therefore, do not ask us for advice or opinions, but confer with a chemist in your vicinity.

Extra Reading
Keep up with new developments of new materials and methods by reading technical magazines. Many technical publications are listed under references in the back section of this book.

Calculating Costs
Purchases of raw materials, in small quantities, are naturally higher in price than when bought in large quantities. Commercial prices, as given in the trade papers and catalogs of manufacturers, are for quantities such as barrels, drums or sacks. For example, 1 pound of epsom salts, bought at retail, may cost 10 or 15 cents. In barrel lots its price today is about 2 to 3 cents per pound.

Typical Costing Calculation
Formula for Beer or Milk Pipe Cleaner

Soda Ash	25 lb. @	0.02½ per lb. =	$ 0.63
Sodium Perborate	75 lb. @	0.16 per lb. =	12.00
Total	100 lb.	Total	$12.63

If 100 lb. cost $12.63, 1 lb. will cost $12.63 divided by 100 or about $0.126, assuming no loss.

Always weigh the amount of finished product and use *this* weight in calculating costs. Most compounding results in some loss of material because of spillage, sticking to apparatus, evaporation, etc. Costs of making experimental lots are always high and should not be used for figuring costs. To meet competition, it is necessary to buy in large units and costs should be based on the latter.

Elementary Preparations
The simple recipes that follow have been selected because of their importance and because they can be made readily.

The succeeding chapters go into greater detail and give many different types and modifications of these and other recipes for home and commercial use.

Cleansing Creams
Cleansing creams as the name implies serve as skin cleaners. Their basic ingredients are oils and waxes which are rubbed into the skin. When wiped off they carry off dirt and dead skin. The liquefying type of cleansing cream contains no water and melts or liquefies when rubbed on the skin. To suit different climates and likes and dislikes harder or softer products can be made.

Cleansing Cream (Liquefying)
Liquid Petrolatum (White Mineral Oil)	5½
Paraffin Wax	2½
Petrolatum	2

Melt the ingredients together with stirring in an aluminum or enamelled dish and allow to cool. Then stir in a perfume oil. Allow to stand until a haziness appears and then pour into jars, which should be allowed to stand *undisturbed* over night.

Cold Creams
The most important facial cream is cold cream. This type of cream contains mineral oil and wax which are emulsified in water with a little borax or glycosterin. The function of a cold cream is to furnish a film to take up dirt and waste tissue, which are removed when the skin is wiped thoroughly. Many modifications of this basic cream are encountered in stores. They vary in color, odor, and in claims but, essentially, they are no more useful than this simple cream. The latest type of cold cream is the non-greasy cold cream which is of particular interest because it is non-alkaline and therefore non-irritating to sensitive skins.

Cold Cream
Liquid Petrolatum (White Mineral Oil)	52 g.
White Beeswax	14 g.

Heat the above in an aluminum or enamelled double boiler (the water in the outer pot should be brought to a boil). In a separate aluminum or enamelled pot dissolve

Borax	1 g.
Water	33 cc.

and bring this to a boil. Add this in a thin stream to the melted wax while stirring vigorously in one direction only. When the temperature drops to 140° F. add ½ cc. of perfume oil and continue stirring until the temperature drops to 120° F. At this point pour into jars where the cream will "set" after a while. If a harder cream is desired, reduce the amount of liquid petrolatum. If a softer cream is wanted increase it.

Cold Cream (Non-Greasy)

White Paraffin Wax	1¼
Petrolatum	1½
Glycosterin or Glyceryl Monostearate	2¼
Liquid Petrolatum (White Mineral Oil)	3

Heat the above in an aluminum or enamelled double boiler (the water in the outer pot should be boiling). Stir until clear. To this slowly add, while stirring vigorously:

Water (Boiling)	10

Continue stirring until smooth and then add with stirring, a perfume oil. Pour into jars at 110–130° F. and cover the jars as soon as possible.

Vanishing Creams

Vanishing creams are non-greasy soapy creams which have a cleansing effect. They are also used as a powder base.

Vanishing Cream

Stearic Acid	18 oz.

Melt the above in an aluminum or enamelled double boiler (the water in the outer pot must be boiling). To the above add, in a thin stream, while stirring vigorously, the following boiling solution made in an aluminum or enamelled pot:

Potassium Carbonate	¼ oz.
Glycerin	6½ oz.
Water	5 lb.

Continue stirring until the temperature falls to 135° F., then stir in a perfume oil and stir from time to time until cold. Allow to stand over night and stir again the next day. Pack into jars which should be closed tightly.

Hand Lotions

Hand lotions are usually clear or milky liquids or salves which are useful in protecting the skin from roughness and redness because of exposure to cold, hot water, soap and other materials. "Chapped" hands are a common occurrence. The use of a good hand lotion keeps the skin smooth, soft, and in a normally healthy condition. The lotion is best applied at night, rather freely, and cotton gloves may be worn to prevent soiling. During the day it should be put on sparingly and the excess wiped off.

Hand Lotion (Salve)

Boric Acid	1
Glycerin	6

Warm the above in an aluminum or enamelled dish and stir until dissolved (clear). Then allow to cool and work the above liquid into the following mixture, adding only a little at a time.

Lanolin	6
Petrolatum	8

If it is desired to impart a pleasant odor to this lotion a little perfume may be added and worked in.

Hand Lotion (Milky Liquid)

Lanolin	¼ tsp.	
Glycosterin or Glyceryl Monostearate	1	oz.
Tincture of Benzoin	2	oz.
Witch Hazel	25	oz.

Melt the first two items together in an aluminum or enamelled double boiler. If no double boiler is at hand improvise one by standing the dish in a small pot containing boiling water. When the mixture becomes clear remove from the double boiler and add slowly, while stirring vigorously, the tincture of benzoin and then the witch hazel. Continue stirring until cool and then put into 1 or 2 large bottles and shake vigorously. The finished lotion is a beautiful milky liquid comparable to the best hand lotions on the market sold at high prices.

Brushless Shaving Creams

Brushless or latherless shaving creams are soapy in nature and do not require lathering or water. The formula given below is of the latest type being free from alkali and non-irritating. It should be borne in mind, however, that certain beards are not softened by this type of cream and require the old-fashioned lathering shaving cream.

Brushless Shaving Cream

White Mineral Oil	10
Glycosterin or Glyceryl Monostearate	10
Water	50

Heat the first two ingredients together in a pyrex or enamelled dish to 150° F. and into this run slowly, while stirring, the water which has been heated to boiling. Allow to cool to 105° F. and while stirring add a few drops of perfume oil. Continue stirring until cold.

Mouth Washes

Mouth washes and oral antiseptics are of practically negligible value. However, they are used because of their refreshing taste and deodorizing value.

Mouth Wash

Benzoic Acid	5/8
Tincture of Rhatany	3
Alcohol	20
Peppermint Oil	1/8

Just mix together in a dry bottle until the benzoic acid is dissolved. A teaspoonful is used to a small wine glass full of water.

Tooth Powders

The cleansing action of tooth powders depends on their contents of soap and mild abrasives such as precipitated chalk and magnesium carbonate. The antiseptic present is practically of no value. The flavoring ingredients mask the taste of the soap and give the user's mouth a pleasant after-taste.

Tooth Powder

Magnesium Carbonate	420 g.
Precipitated Chalk	565 g.
Sodium Perborate	55 g.
Sodium Bicarbonate	45 g.
White Soap (Powdered)	50 g.
Sugar (Powdered)	90 g.
Wintergreen Oil	8 cc.
Cinnamon Oil	2 cc.
Menthol	1 g.

Dissolve the last three ingredients together and then rub well into the sugar. Add the soap and perborate, mixing in well. Add the chalk with good mixing and then the sodium bicarbonate and magnesium carbonate. Mix thoroughly and sift through a fine wire screen. Keep dry.

Foot Powders

Foot powders consist of talc or starch with or without an antiseptic or deodorizer. In the following formula the perborates liberate oxygen when in contact with perspiration which tends to destroy unpleasant odors. The talc acts as a lubricant and prevents friction and chafing.

Foot Powder

Sodium Perborate	3
Zinc Peroxide	2
Talc	15

Mix thoroughly in a dry container until uniform. This powder must be kept dry or it will spoil.

Liniments

Liniments usually consist of an oil and an irritant such as methyl salicylate or turpentine. The oil acts as a solvent and tempering agent for the irritant. The irritant produces a rush of blood and warmth which is often slightly helpful.

Sore Muscle Liniment

Olive Oil	6 fl. oz.
Methyl Salicylate	3 fl. oz.

Mix together and keep in a well stoppered bottle. Apply externally but do not apply to chafed or cut skin.

Chest-Rubs

In spite of the fact that chest-rubs are practically useless countless sufferers use them. Their action is similar to that of liniments and they differ only in that they are in the form of a salve.

"Chest-Rub" Salve

Yellow Petrolatum	1	lb.
Paraffin Wax	1	oz.
Eucalyptus Oil	2	fl. oz.
Menthol	1/2	oz.
Cassia Oil	1/8	fl. oz.
Turpentine	1/2	fl. oz.

Melt the petrolatum and paraffin wax together in a double boiler and then add the menthol. Remove from the heat, stir, and cool a little; then stir in the oils, turpentine, and acid. When it begins to thicken pour into tins and cover.

Insect Repellents

Preparations of this type may irritate sensitive skins. Moreover, they will not always work. Psychologically they often are helpful, even though they may not keep insects away, because they give one confidence of protection.

Mosquito Repelling Oil

Cedar Oil	2 fl. oz.
Citronella Oil	4 fl. oz.
Spirits of Camphor	8 fl. oz.

Mix in a dry bottle, and the oil is ready for use. This preparation may be smeared on the skin as often as is necessary to repel mosquitoes and other insects.

Fly Sprays

Fly sprays usually consist of deodorized kerosene, perfuming material, and an active insecticide. In some cases they merely stun the flies who may later recover and begin buzzing again.

Fly Spray

Deodorized Kerosene	89 fl. oz.
Methyl Salicylate	1 fl. oz.
Pyrethrum Powder	10 oz.

Mix thoroughly by stirring from time to time; allow to stand covered over night and then filter through muslin.

Caution: This spray is inflammable and should not be used near open flames.

Deodorant Spray

(For public buildings, sick-rooms, lavatories, etc.)

Pine Needle Oil	2
Formaldehyde	2
*Acetone	6
*Isopropyl Alcohol	20

One ounce of the above is mixed with 1 pt. of water for spraying.

Cresol Disinfectant

†Caustic Soda	25½ g.
Water	140 cc.

Dissolve the above in a pyrex or enamelled dish and warm it. To this add slowly the following warmed mixture:

†Cresylic Acid	500 cc.
Rosin	170 g.

Stir until dissolved and add water to make 1000 cc.

Ant Poison

Sugar	1 lb.
Water	1 qt.
‡Arsenate of Soda	125 g.

Boil and stir until uniform; strain through muslin; add 1 spoonful of honey.

Bedbug Exterminator

*Kerosene	90 fl. oz.
Clove Oil	5 fl. oz.
§Cresol	1 fl. oz.
Pine Oil	4 fl. oz.

Simply mix and bottle.

Mothproofing Fluid (Non-Staining)

Sodium Aluminum Silicofluoride	½
Water	98
Glycerin	½
Sulfatate (Wetting Agent)	¼

Stir until dissolved.

Fly Paper

Rosin	32
Rosin Oil	20
Castor Oil	8

Heat the above in an aluminum or enamelled pot on a gas stove with stirring until all the rosin has melted and dissolved. While hot pour on firm paper sheets of suitable size which have been brushed with soap water just before coating. Smooth out the coating with a long knife or piece of thin flat wood and allow to cool. If a heavier coating is desirable increase the amount of rosin used. Similarly a thinner coating results by reducing the amount of rosin. The finished paper should be laid flat and not exposed to undue heat.

Baking Powder

Bicarbonate of Soda	28
Monocalcium Phosphate	35
Corn Starch	27

Mix the above powders thoroughly in a dry can by shaking and rolling for ½ hour. Pack into dry airtight tins as moisture will cause lumping.

Malted Milk Powder

Malt Extract (Powdered)	5
Skim Milk (Powdered)	2
Sugar (Powdered)	3

Mix thoroughly by shaking and rolling in a dry can. Pack in an air-tight container.

Cocoa Malt Powder

Corn Sugar	55
Malt (Powdered)	19
Skim Milk (Powdered)	12½
Cocoa	13
Vanillin	⅛
Salt (Powdered)	⅜

Mix thoroughly and then run through a fine wire sieve.

Sweet Cocoa Powder

Cocoa	17½ oz.
Sugar (Powdered)	32½ oz.
Vanillin	¾ g.

Mix thoroughly and sift.

Pure Lemon Extract

Lemon Oil U.S.P.	6½ fl. oz.
Alcohol	121½ fl. oz.

Shake together in 1 gal. jug until dissolved.

Artificial Vanilla Flavor

Vanillin	¾ oz.
Coumarin	¼ oz.
Alcohol	2 pt.

---

* Inflammable.
† Do not get this on skin as it is corrosive.
‡ Poison.
§ Corrosive to skin.

INTRODUCTION

Stir the above in a glass or china pitcher until dissolved. Then stir into the following solution:

Sugar	12 oz.
Water	5¼ pt.
Glycerin	1 pt.

Color brown by adding sufficient "burnt" sugar coloring.

Canary Bird Food

Yolk of Eggs (Dried and Chopped)	2
Poppy Heads (Coarse Powder)	1
Cuttlefish Bone (Coarse Powder)	1
Sugar (Granulated)	2
Soda Crackers (Powdered)	8

Mix well together.

Writing Ink (Blue-Black)

Naphthol Blue Black	1 oz.
Gum Arabic (Powdered)	½ oz.
Carbolic Acid	¼ oz.
Water	1 gal.

Stir together in a glass or enamelled vessel until dissolved.

Laundry Marking Ink (Indelible)

A.
Soda Ash	1 oz.
Gum Arabic (Powdered)	1 oz.
Water	10 fl. oz.

Stir the above until dissolved.

B.
Silver Nitrate	4 oz.
Gum Arabic (Powdered)	4 oz.
Lampblack	2 oz.
Water	40 fl. oz.

Stir this in a glass or porcelain dish until dissolved. Do not expose this to strong light or it will spoil. Finally pour into a brown glass bottle. In using these solutions wet the cloth with solution A and allow to dry. Then write on it with solution B using a quill pen.

Marking Crayon (Green)

Ceresin	8
Carnauba Wax	7
Paraffin Wax	4
Beeswax	1
Talc	10
Chrome Green	3

Melt the first four ingredients in any container and then add the last two slowly while stirring. Remove from the heat and continue stirring until thickening begins. Then pour into molds. If other color crayons are desired, other pigments may be used. For example, for black, use carbon or bone-black; for blue, Prussian blue; for red, orange chrome yellow.

Antique Coloring for Copper

Copper Nitrate	4 oz.
Acetic Acid	1 oz.
Water	2 oz.

Dissolve by stirring together in a glass or porcelain vessel. Pack into glass bottles.

To Use: Wet the copper to be colored and apply the above solution hot.

Blue-Black Finish on Steel

A. Place object in molten sodium nitrate (700–800° F.) for 2–3 minutes. Remove and allow to cool somewhat; wash in hot water; dry and oil with mineral or linseed oil.

B. Place in the following solution for 15 minutes:

Copper Sulfate	½ oz.
Iron Chloride	1 lb.
Hydrochloric Acid	4 oz.
Nitric Acid	½ oz.
Water	1 gal.

Then allow to dry for several hours; place in above solution again for 15 minutes; remove and dry for 10 hours. Place in boiling water for ½ hour; dry and scratch brush very lightly. Oil with mineral or linseed oil and wipe dry.

Rust Prevention Compound

Lanolin	1
*Naphtha	2

Mix until dissolved.

The metal to be protected is cleaned with a dry cloth and then coated with the above composition.

Metal Polish

Naphtha	62 oz.
Oleic Acid	⅓ oz.
Abrasive	7 oz.
Triethanolamine Oleate	⅓ oz.
Ammonia (26°)	1 oz.
Water	1 gal.

In one container mix together the naphtha and oleic acid to a clear solution. Dissolve the triethanolamine oleate in water separately, stir in the abrasive, if it is of a clay type, and then add the naphtha solution. Stir the resulting mixture at a high speed until a uniform creamy emulsion results. Then add the ammonia and mix well, but do not agitate as vigorously as before.

Glass Etching Fluid

Hot Water	12
†Ammonium Bifluoride	15
Oxalic Acid	8

---

* Inflammable—keep away from flames.
† Corrosive.

Ammonium Sulfate	10
Glycerin	40
Barium Sulfate	15

Warm the washed glass slightly before writing on it with this fluid. Allow the fluid to act on the glass for about 2 minutes.

Leather Preservative

| Neatsfoot Oil (Cold Pressed) | 10 |
| Castor Oil | 10 |

Mix.

This is an excellent preservative for leather book bindings, luggage and other leather goods.

White Shoe Dressing

Lithopone	19	oz.
Titanium Dioxide	1	oz.
Shellac (Bleached)	3	oz.
Ammonium Hydroxide	¼	fl. oz.
Water	25	fl. oz.
Alcohol	25	fl. oz.
Glycerin	1	oz.

Dissolve the last four ingredients by mixing in a porcelain vessel. When dissolved stir in the first two pigments. Keep in stoppered bottles and shake before using.

Waterproofing for Shoes

Wool Grease	8
Dark Petrolatum	4
Paraffin Wax	4

Melt together in any container.

Polishes

Polishes are usually used to restore the original lustre and finish of a smooth surface. As a secondary purpose they are expected to clean the surface and also to prevent corrosion or deterioration. There is no one polish which will give good results on all surfaces.

Most polishes contain oil or wax for their lustering or polishing properties. Oil polishes are applied easily but the surfaces on which they are used attract dust and show finger marks. Wax polishes are more difficult to apply but are more lasting.

Oil or wax polishes are of two types: waterless and with water. The former are clear or translucent and the latter are milky in appearance.

For use on metals, abrasives of various kinds such as tripoli, silica dust or infusorial earth are incorporated to grind away oxide films or corrosion products.

Shoe Polish (Black)

| Carnauba Wax | 5½ | oz. |
| Crude Montan Wax | 5½ | oz. |

Melt together in a double boiler (the water in the outer container should be boiling) then stir in the following melted and dissolved mixture:

| Stearic Acid | 2 oz. |
| Nigrosine Base | 1 oz. |

Then stir in

| Ceresin | 15 oz. |

Remove all flames and run in slowly, while stirring

| Turpentine | 90 fl. oz. |

Allow the mixture to cool to 105° F. and pour into air-tight tins which should be allowed to stand undisturbed over night.

Auto Polish (Clear Oil Type)

Paraffin (Mineral) Oil	5	pt.
Raw Linseed Oil	2	pt.
China Wood Oil	½	pt.
*Benzol	¼	pt.
Kerosene	¼	pt.
Amyl Acetate	1	tbsp.

Mix together in a glass jug and keep stoppered.

Auto and Floor Wax (Paste Type)

Yellow Beeswax	1	oz.
Ceresin	2½	oz.
Carnauba Wax	4½	oz.
Montan Wax	1¼	oz.
*Naphtha or Mineral Spirits	1	pt.
*Turpentine	2	oz.
Pine Oil	½	oz.

Melt the waxes together in a double boiler. Turn off the heat and run in the last three ingredients in a thin stream with stirring. Pour into cans; cover and allow to stand undisturbed overnight.

Furniture Polish (Oil and Wax Type)

Thin Paraffin (Mineral Oil)	1	pt.
Carnauba Wax (Powdered)	¼	oz.
Ceresin Wax	⅛	oz.

Heat together until all of the wax is melted. Allow to cool and pour into bottles before the mixture turns cloudy.

Polishing Wax (Liquid)

| Beeswax, Yellow | 1 | oz. |
| Ceresin Wax | 4 | oz. |

Melt together and then cool to 130° F.; turn off all flames and stir in slowly:

| *Turpentine | 17 | fl. oz. |
| Pine Oil | ½ | fl. oz. |

Pour into cans or bottles which are closed tightly to prevent evaporation.

Floor Oil

Mineral Oil	46	fl. oz.
Beeswax	½	oz.
Carnauba Wax	1	oz.

* Inflammable—keep away from flames.

Heat together in double boiler until dissolved (clear). Turn off the flame and stir in
*Turpentine   3 fl. oz.

Lubricants

Lubricants in the form of oils or greases are used to prevent friction and wearing of parts which rub together. Lubricants must be chosen to fit specific uses. They consist of oils and fats often compounded with soaps and other unctuous materials. For heavy duty, heavy oils or greases are used and light oils for light duty.

Gun Lubricant

White Petrolatum	15 oz.
Bone Oil (Acid Free)	5 oz.

Warm gently and mix together.

Graphite Grease

Ceresin	7 oz.
Tallow	7 oz.

Warm together and gradually work in, with a stick:

Graphite	3 oz.

Stir until uniform and pack in tins when thickening begins.

Penetrating Oil
(For freeing rusted bolts, screws, etc.)

Kerosene	2 oz.
Thin Mineral Oil	7 oz.
Secondary Butyl Alcohol	1 oz.

Mix and keep in a stoppered bottle.

Molding Material

White Glue	13	lb.
Rosin	13	lb.
Raw Linseed Oil	⅓	qt.
Glycerin	1	qt.
Whiting	19	lb.

This mixture is prepared by cooking the white glue until it is dissolved. Then cook separately the rosin and raw linseed oil until they are dissolved. Add the rosin, oil, and glycerin to the cooked glue, stirring in the whiting until the mass makes up to the consistency of a putty. Keep the mixture hot.

Place this putty mass in the die, pressing it firmly into the same and allowing it to cool slightly before removing. The finished product is ready to use within a few hours after removal. Suitable colors can be added to secure brown, red, black or other color.

In applying ornaments made of this composition to a wood surface, they are first steamed to make them flexible; in this condition they can be glued to the wood surface easily and securely. They

* Inflammable.

can be bent to any shape, and no nails are required for applying them.

Grafting Wax

Wool Grease	11
Rosin	22
Paraffin Wax	6
Beeswax	4
Japan Wax	1
Rosin Oil	9
Pine Oil	1

Melt together until clear and pour into tins. This composition can be made thinner by increasing the amount of rosin oil and thicker by decreasing it.

Candles

Paraffin Wax	30
Stearic Acid	17½
Beeswax	2½

Melt together and stir until clear. If colored candles are desired a pinch of any oil soluble dye is dissolved at this stage. Pour into vertical molds in which wicks are hung.

Adhesives

Adhesives are sticky substances used to unite two surfaces. Adhesives are specifically called glues, pastes, cements, mucilages, lutes, etc. For different uses different types are required.

Wall Patching Plaster

Plaster of Paris	32
Dextrin	4
Pumice Powder	4

Mix thoroughly by shaking and rolling in a dry container. Keep away from moisture.

Cement Floor Hardener

Magnesium Fluosilicate	1 lb.
Water	15 pt.

Mix until dissolved.

In using this compound, the cement should first be washed with clean water and then drenched with the above solution.

Paperhanger's Paste

A. White or Fish Glue	4 oz.
Cold Water	8 oz.
B. Venice Turpentine	2 fl. oz.
C. Rye Flour	1 lb.
Cold Water	16 fl. oz.
D. Boiling Water	64 fl. oz.

Use a cheap grade of rye or wheat flour, mix thoroughly with cold water to about the consistency of dough or a little thinner, being careful to remove all lumps. Stir in a tbsp. of powdered alum to 1 qt. of flour, then pour in boiling water, stirring rapidly until the flour is thoroughly cooked. Let this cool before using and thin with cold water.

Soak the 4 oz. of glue in the cold water for 4 hours. Dissolve on a water bath (glue-pot) and while hot stir in the Venice turpentine. Make up C into a batter free from lumps and pour into D. Stir briskly, and finally add the glue solution. This makes a very strong paste, and it will also adhere to a painted surface, owing to the Venice turpentine in its composition.

Aquarium Cement

Litharge	10
Plaster of Paris	10
Powdered Rosin	1
Dry White Sand	10
Boiled Linseed Oil	Sufficient

Mix all together in the dry state, and make a stiff putty with the oil when wanted for use.

Do not fill the aquarium for 3 days after cementing. This cement hardens under water, and will stick to wood, stone, metal, or glass, and, as it resists the action of sea-water, it is useful for marine aquaria.

Wood Dough Plastic

*Collodion	86
Ester Gum (Powdered)	9
Wood Flour	30

Allow the first two ingredients to stand until dissolved, stirring from time to time. Then while stirring add the wood flour a little at a time until uniform. This product can be made softer by adding more collodion.

Putty

Whiting	80
Raw Linseed Oil	16

Rub together until smooth. Keep in closed container.

Wood Floor Bleach

Sodium Metasilicate	90
Sodium Perborate	10

Mix thoroughly and keep dry in a closed can. Use 1 lb. to 1 gal. of boiling water. Mop or brush on the floor, allow to stand ½ hour, then rub off and rinse well with water.

*Paint Remover

Benzol	5	pt.
Ethyl Acetate	3	pt.
Butyl Acetate	2	pt.
Paraffin Wax	½	lb.

Stir together until dissolved.

* Inflammable.

Soaps and Cleaners

Soaps are made from a fat or fatty acid and an alkali. They lather and produce a foam which entraps dirt and grease which is washed away with water. There are numerous kinds of soaps depending on the uses to which they are to be put.

Cleaners contain a solvent such as naphtha with or without a soap. Abrasive cleaners are soap pastes containing powdered pumice, stone, silica, etc.

Liquid Soap (Concentrated)

Water	11
†Caustic Potash (Solid)	1
Glycerin	4
Red Oil (Oleic Acid)	4

Dissolve the caustic in water, add the glycerin and bring to a boil in an enamelled pot. Remove from heat, add the red oil slowly while stirring. If a more neutral soap is wanted, use more red oil.

Saddle Soap

Beeswax	5
†Caustic Potash	0.8
Water	8

Boil for 5 minutes while stirring. In another vessel heat

Castile Soap	1.6
Water	8

Mix the two ingredients with good stirring; remove from heat and add:

Turpentine	12

while stirring.

Mechanics' Hand Soap Paste

Water	1.8 qt.
White Soap Chips	1.5 lb.
Glycerin	2.4 oz.
Borax	6 oz.
Dry Sodium Carbonate	3 oz.
Pumice Powder (Coarse)	2.2 lb.
Safrol	To suit

Dissolve the soap in ⅔ of the water by heat. Dissolve the last three in the rest of the water. Pour the two solutions together and stir well. When it begins to thicken, sift in the pumice, stirring constantly till thick, then pour into cans. Vary amount of water, for heavier or softer paste (water cannot be added to the finished soap).

Dry Cleaning Fluid

Glycol Oleate	2 fl. oz.
Carbon Tetrachloride	60 fl. oz.
Varnoline (Naphtha)	20 fl. oz.
Benzine	18 fl. oz.

An excellent cleaner that will not injure the finest fabrics.

† Do not get on skin as it is **corrosive**.

Wall Paper Cleaner
Whiting	10 lb.
Calcined Magnesia	2 lb.
Fuller's Earth	2 lb.
Pumice (Powdered)	12 oz.
Lemenone or Citronella Oil	4 oz.

Mix well together.

Household Cleaner
Soap Powder	2
Soda Ash	3
Trisodium Phosphate	40
Finely Ground Silica	55

Mix well and put up in the usual containers.

Window Cleanser
Castile Soap	2
Water	5
Chalk	4
French Chalk	3
Tripoli Powder	2
Petroleum Spirits	5

Mix well and pack in tight containers.

Straw Hat Cleaner
Sponge the hat with a solution of
Sodium Hyposulfite	10 oz.
Glycerin	5 oz.
Alcohol	10 oz.
Water	75 oz.

Lay aside in a damp place for 24 hours and then apply a mixture of:
Citric Acid	2 oz.
Alcohol	10 oz.
Water	90 oz.

Press with a moderately hot iron after stiffening with gum water if necessary.

Grease, Oil, Paint and Lacquer Spot Remover
Alcohol	1
Ethyl Acetate	2
Butyl Acetate	2
Toluol	2
Carbon Tetrachloride	3

Place the garment with spot over a piece of clean paper or cloth and wet with the above fluid; rub with a clean cloth toward the center of the spot. Use a clean section of cloth for rubbing and clean paper or cloth for each application of the fluid. The above product is inflammable and should be kept away from flames. Cleaners of this type should be used out-of-doors or in well-ventilated rooms as the fumes are toxic.

Paint Brush Cleaner
A. Kerosene	2
Oleic Acid	1
B. Strong Liquid Ammonia, 28%	¼
Denatured Alcohol	¼

Slowly stir B into A until a smooth mixture results. To clean brushes, pour into a can and stand the brushes in it overnight. In the morning, wash out with warm water.

Rust and Ink Remover
Immerse the part of the fabric with the rust or ink spot alternately in Solution A and B, rinsing with water after each immersion.

Solution A
Ammonium Sulfide Solution	1
Water	19

Solution B
*Oxalic Acid	1
Water	19

Javelle Water (Laundry Bleach)
Bleaching Powder	2 oz.
Soda Ash	2 oz.
Water	5 gal.

Mix well until the reaction is completed. Allow to settle overnight and siphon off the clear liquid.

Laundry Blue (Liquid)
Prussian Blue	1
Distilled Water	32
*Oxalic Acid	¼

Dissolve by mixing in a crock or wooden tub.

"Glassine" Paper
Paper is coated with or dipped in the following solution and then hung up to dry.
Copal Gum	10 oz.
Alcohol	30 fl. oz.
Castor Oil	1 fl. oz.

Dissolve by letting stand overnight in a covered jar and stirring the next day.

Waterproofing Paper and Fiberboard
The following composition and method of application will render uncalendered paper, fiberboard, and similar porous material waterproof.
Paraffin (Melting Point about 130° F.)	22.5
Trihydroxyethylamine Stearate	3.0
Water	74.5

The paraffin wax is melted and the stearate added to it. The water is then heated to nearly the boiling point and vigorously agitated with a suitable mechanical stirring device while the above mixture of melted wax and the emulsifier is slowly added. This mixture is cooled while it is stirred.

The paper or fiberboard is coated on

---
* Poison.

the side which is to be in contact with water. This method works most effectively on paper pulp moulded containers and possesses the advantages of being much cheaper than dipping in melted paraffin as only about a tenth as much paraffin is needed. In addition, the outside of the container is not greasy, and can be printed upon after treatment which is not the case when treated with melted wax.

Waterproofing Liquid
Paraffin Wax	⅖ oz.
Gum Dammar	1⅕ oz.
Pure Rubber	⅛ oz.
Benzol	13 oz.
Carbon Tetrachloride to make	1 gal.

Dissolve the rubber in benzol; add the other ingredients and allow to dissolve. (Inflammable.)

The above is suitable for wearing apparel and wood. It is applied by brushing on two or more coats, allowing each to dry before applying another coating. Apply outdoors as vapors are inflammable and toxic.

Waterproofing Heavy Canvas
Raw Linseed Oil	1 gal.
Crude Beeswax	13 oz.
White Lead	1 lb.
Rosin	12 oz.

Heat the above, while stirring, until all lumps are gone and apply warm to the upper side of the canvas, wetting the canvas with a sponge on the underside before applying.

Cement Waterproofing
Chinawood Oil Fatty Acids	10 oz.
Paraffin Wax	10 oz.
Kerosene	2½ gal.

Stir until dissolved. This is painted or sprayed on cement walls, which must be dry.

Oil and Greaseproofing Paper and Fiberboard

This solution applied by brush, spray, or dipping will leave a thin film which is impervious to oils and grease. Applied to paper or fiber containers, it will enable them to retain oils and greases.

Starch	6.6
Caustic Soda	0.1
Glycerin	2.0
Sugar	0.6
Water	90.5
Sodium Salicylate	0.2

The caustic soda is dissolved in the water and then the starch is made into a thick paste by adding a portion of this solution. This paste is then added to the water. This mixture is placed on a water bath and heated to about 85° C. until all the starch granules have broken. The temperature is maintained for about ½ hour longer. The other substances are then added and thoroughly mixed. The composition is completed and ready for application. A smaller water content may be used if applied hot and a thicker coating will result.

Fireproof Paper
Ammonium Sulfate	8
Boric Acid	3
Borax	1¾
Water	100

Mix together in a gallon jug, by shaking, until dissolved.

The paper to be treated is dipped into this solution in a pan, until uniformly saturated. It is then taken out and hung up to dry. Wrinkles can be prevented by drying between cloths in a press.

Fireproofing Canvas
Ammonium Phosphate	1 lb.
Ammonium Chloride	2 lb.
Water	½ gal.

Impregnate with the solution; squeeze out the excess and dry. Washing or exposure to rain will remove fireproofing salts.

Fireproofing Light Fabrics
Borax	10 oz.
Boric Acid	8 oz.
Water	1 gal.

Impregnate; squeeze and dry. Fabrics so impregnated must be treated again after washing or exposure to rain as the fireproofing salts wash out easily.

Dry Fire Extinguisher
Ammonium Sulfate	15
Sodium Bicarbonate	9
Ammonium Phosphate	1
Red Ochre	2
Silex	23

Use powdered materials only; mix well and pass through a fine sieve. Pack in tight containers to prevent "lumping."

Fire Extinguishing Liquid
Carbon Tetrachloride	95
Solvent Naphtha	5

The inclusion of the naphtha minimizes the production of toxic fumes when extinguishing fires.

Fire Kindler
Rosin or Pitch 10
Sawdust 10 or more
Melt, mix, and cast in forms.

Solidified Gasoline
*Gasoline ½ gal.
White Soap (Fine Shaved) 12 oz.
Water 1 pt.
Ammonia 5 oz.
Heat the water, add soap, mix and when cool add the ammonia. Then slowly work in the gasoline to form semi-solid mass.

Boiler Compound
Soda Ash 87
Trisodium Phosphate 10
Starch 1
Tannic Acid 2
Use powdered materials, mix well and then pass through a fine sieve.

Anti-Freezes
The materials listed below are the basic ingredients used in all good anti-freeze liquids. Of these, alcohol is the only one that evaporates. Radiators containing alcohol should be tested from time to time to be sure of protection. A hydrometer for testing alcohol solution strength can be bought from sellers of denatured alcohol.

* Inflammable.

Soldering Flux (Non-corrosive)
Rosin (Powdered) 1
Denatured Alcohol 4
Soak overnight and mix well.

Photographic Solutions
Developing Solution
Stock Solution A
Dissolve the following, separately, in glass or enamel dishes.
Pyro 4 oz.
Sodium Bisulfite (Pure) 280 gr.
Potassium Bromide 32 gr.
Distilled Water 64 oz.
Stock Solution B
Sodium Sulfite (Pure) 7 oz.
Sodium Carbonate (Pure) 5 oz.
Distilled Water 64 oz.
Take the following proportions:
Stock Solution A 2
Stock Solution B 2
Distilled Water 16
At a temperature of 65° F. this developer requires about 8 minutes.

Acid Hardening Fixing Bath
A. Sodium Hyposulfite 32
Distilled Water 8
Stir until dissolved and then add the following chemicals in the order given below, stirring each until dissolved:
B. Distilled Water (Warm) 2½
Sodium Sulfite (Pure) ½
Acetic Acid (28%) (Pure) 1½
Potassium Alum Powder ½
Add Solution B to A and store in dark bottles away from light.

Anti-Freeze Liquids
Pints of anti-freeze per gallon of water for protection at:

	+10° F.	0° F.	—10° F.	—20° F.
Denatured Alcohol 180 proof	3.4	4.9	6.5	8.3
Denatured Alcohol 188 proof	3.3	4.7	6.0	7.7
Glycerin 95%	3.3	5.3	7.1	9.0
Radiator Glycerin 60%	10.0	18.7	39.0	106.5
Ethylene Glycol 95%	2.7	4.0	5.1	6.5

Specific gravity for protection at:

	+10° F.	0° F.	—10° F.	—20° F.	—30° F.
Denatured Alcohol	0.968	0.959	0.950	0.942	0.921
Glycerin	1.090	1.112	1.131	1.147	1.158
Ethylene Glycol	1.038	1.048	1.056	1.064	1.069

Chapter II

ADHESIVES

Glue Solutions

In preparing glue for use, temperature rules must be observed to obtain a solution of maximum strength. It is advisable at all times to weigh both glue and water before mixing. Individual glue flakes and grains vary in size, and weighing is necessary to ensure that the proportion of glue to water is constant. Glue should then be poured into cold, pure water. The colder the water, the faster and greater the swelling. Pouring glue into hot water is not recommended as the best practice. Where possible, water used for soaking should be between 30 and 60°F. Moreover, the soaking operation should be carried out in a cool place in the plant.

Ground glue should be soaked for at least one hour, and flake glue for at least six hours. When the glue is ready to be melted, it is heated in a water jacketed kettle. The temperature of the melting glue should not go above 140°F. since overheating breaks down some of the protein and results in high loss of strength. Should the glue solution be heated for 10 hours at 200°F., for example, more than 50% of its adhesive strength is lost. A melting temperature of 140°F. is a normal one. Loss of strength through overheating can be avoided by preparing only enough glue for immediate use under proper temperature control in thoroughly cleaned utensils.

Gum Arabic Mucilage

Gum Arabic (Gum Acacia)	10
Rice Starch	10
Sugar	40
Water	100
Preservative (Moldex)	⅕–⅖

Casein Mucilage

Powdered Casein	85
Bentonite	15

Mix

The addition of 60 parts dried powdered blotting paper reduces the drying time.

Water is added to the above before use.

High-Strength Starch Adhesive

Starch	100.0
Caustic Soda	3.2
Water	300.0
Urea	24.0

The urea must be added along with the caustic soda. A mixture of potato and corn starch is best.

Remoistening Label Adhesive
Formula No. 1

Water	47.4
Tapioca Dextrin	47.4
Glucose	4.8

Scent, e.g., Wintergreen Oil, Methyl Salicylate, Sassafras Oil, or Safrol   0.2
Alcohol   0.2

To cold water, in a cool container, the dextrin is slowly added with stirring to break up lumps. The smooth mixture is heated to 180°F. with stirring and kept at this temperature for at least ½ hour or until a complete solution is obtained. Add the glucose and stir into solution. Cool, while stirring, to 140°F. The scent mixed with the alcohol is then added slowly with stirring until thoroughly mixed. Use at room temperature.

No. 2

Water	53.4
Bone Glue (80 gram)	43.5
Glucose	2.0
Zinc Sulfate	0.5
Scent (as above)	0.3
Alcohol	0.3

To 95% of the (cold) water add the bone glue slowly while stirring. Stir until the glue is thoroughly wet and swells, and until stirring is difficult. Heat to 140 to 160°F. with stirring until a smooth adhesive is obtained. Add zinc sulfate dissolved in 5% of water, glucose, and glycerin to the hot solution and mix in thoroughly. Cool to 140°F. and add the scent dissolved in the alcohol, mix in thoroughly. Use at 140°F. as at slightly lower temperatures thermophyllic bacteria decompose the glue and at much higher temperatures degradation of the glue is rapid.

Remoistening adhesives as gummings on paper are sensitive to moisture changes in the air and tend to cause the paper to curl. In commercial practice the curling is prevented or greatly reduced by breaking the gumming into small diamond shapes by pressing over a sharp angle. The sheets can then be printed for labels with little trouble.

Waxed Paper Adhesive
U. S. Patent 2,373,597

Latex (60%)	100
Corn Syrup	25
Sodium Lauryl Sulfate	2

Apply to each waxed surface; partially dry and then bring them together to dry completely.

Paper to Tin Adhesive

Caustic Potash	5
Water	56
Rosin	50

Heat and stir until uniform. Then add

Balata Resin	5
Water	50

Stir until uniform.

Adhesive for Paper and Metal Foil

Glycol Bori-Borate	25
Invert Sugar Syrup	50
Powdered Calcium Carbonate	25

Paper Pad End Binding Cement
U. S. Patent 2,387,967

Polyvinyl Acetate Emulsion (55% Solids)	1	gal.
Dibutyl Phthalate	½	lb.
Glycerin	2	oz.

Adhesive for Cellophane

Gum Acacia (Gum Arabic)	16.5
Glycerin	29.5
Water	49.5
Formaldehyde	4.5

Thermoplastic Adhesive Paper Coating
U. S. Patent 2,394,254
Stearic Acid	2
Ethyl Cellulose	1
Paraffin Wax	1

Adhesive Tape Coating
U. S. Patent 2,382,731
Triethyleneglycol Ester of Hydrogenated Rosin	160
Rubber	160
Zinc Oxide	125

Pressure Sensitive Transparent Adhesive Tape
U. S. Patent 2,332,265

A transparent pressure sensitive sheet suitable as adhesive tape is obtained by applying the adhesive coating to a transparent backing film of plasticized gamma-polyvinyl chloride containing less than 20% of a plasticizer which may be dibutyl phthalate, tricresyl phosphate, tributyl aconitate, dioctyl phthalate, butyl phthalyl butyl glycolate, etc. The adhesive coating may consist of 50 to 75% rubber, or latex, or a polybutadiene, interpolymer of butadiene with other polymerizable compounds, etc., together with 15 to 35% of a polyisobutylene, either the oily plastic or elastic high molecular weight polymer, and 5 to 15% of a tacky resin, such as ester gum, etc., rosin, melted rubber. For example, a 10% cyclohexanone solution from a mixture consisting of 100 parts of polyvinyl chloride, 5 parts of dioctyl phthalate, and 2 parts of the monoester of gylcerine with cottonseed fatty acid, is deposited on a polished metal drum heated at 150°C. to form a transparent film of the required thickness. This film is dried, stripped from the drum, and coated on one side with a pressure sensitive adhesive containing 20 parts of natural rubber or a synthetic butadiene polymer, 7 parts of an elastic polyisobutylene, and 3 parts of ester gum dissolved in a liquid solvent consisting of 20 parts of benzene and 80 parts of gasoline. This coating is dried at 30° to 50°C. into a glossy, transparent layer which is highly resistant to light and air influences.

Thermoplastic Adhesive Tape
Formula No. 1
Canadian Patent 432,344
Ethyl Cellulose	4
Staybelite	7
Hercolyn	2

The above is melted and mixed until uniform. Then it is coated on paper or cloth tape.

No. 2
U. S. Patent 2,389,469

A backing sheet is coated with the following:
Crepe Rubber	10	lb.
Cumarone Resin	2	lb.
Zinc Oxide	½	lb.

Mix in a rubber mill and add benzol. Then mix in
Diglycol Laurate	1–2%
Water	2–4%

Anti-Stick Coating for Adhesive Rolls
U. S. Patent 2,364,875
Hydroxypolyvinyl Acetate	64
Sodium Oleate	32
Sodium Hydroxide	4

Asphalt Adhesive
U. S. Patent 2,409,258
Asphalt Emulsion (50% Solids)	20
Burgundy Pitch	3–6

Increasing Adhesiveness of Asphalt
British Patent 560,716
Add 0.3% laurylamine and
0.4–1.6% oleic acid

Plywood Adhesive
U. S. Patent 2,385,374

Resorcin	2500
Formaldehyde	1250
Oxalic Acid	15

Reflux together and add
Water   2600
Adjust with caustic soda to pH 7–9.
To make thermosetting at 200–225°F. add

Pyridin	342
Water	158

Up to 1800 g. walnut shell flour may be added as filler.

Rubber Latex Adhesive
U. S. Patent 2,365,878

Bentonite	60
Rubber Latex	25
Magnesium Chloride	10
Water-Soluble Soap	4
Preservative	1

Add water slowly, with stirring, until the desired consistency is reached.

Cellulose Ether Adhesives
U. S. Patent 2,362,761

Formula No.	1	2	3	4
Glycol Borate (Aquaresin)	1	2	1	1
Cellulose Ether of Sodium Glycollate	2	..	..	1
Methyl Cellulose	..	6	4	3
Sodium Caseinate	..	2	1	..
Sodium Tetraborate	..	0.4	..	..
Phenol	..	0.1	0.1	..
Water	97	81.5	85.9	96
Acetone	..	8	8	..

Tire-Leak Seal
Formula No. 1
U. S. Patent 2,355,977

Wood Flour	10–20	lb.
Fuller's Earth	30	lb.
Bentonite	7½	lb.
Asbestos Floats	7½	lb.
Lampblack	4–8	oz.
Glucose Syrup (60% Glucose)	9	pt.

No. 2
U. S. Patent 2,347,925

Water	5
Salt	¼
Alcohol	1
Linseed Meal	4

Gelatin Seal
(For bottles and jars)

Gelatin	6	lb.
Glycerin	275	cc.
Water	5000	cc.
Potassium Dichromate (25%)	15	cc.
Tricresol	50	cc.

All the ingredients are put into a double boiler and heated until the gelatin is dissolved (about 1 hour). After this period the heating is continued for another hour with occasional stirring. The mixture is then poured into a large shallow pan and placed in the ice-box until solid. It may be kept in this state until ready for use.

Automobile Radiator Leak Seal
Formula No. 1
U. S. Patent 2,391,737

Sulfite Liquor	50
Asbestos (Ground)	3
Water	47

No. 2
U. S. Patent 2,315,321

Water	700
Glue	40
Rosin	40
Asbestos Fiber	40
Sodium Silicate	25
Tetrasodiumpyrophosphate	10

Phosphoric Acid	8½
Monoethanolamine	7½

Stopping Boiler Leaks

Sodium Silicate (O Brand)	55 gal.
Cottonseed Meal	50 lb.
Water	12 gal.

Mix well and put into boiler heated to 200°F. Then add a slurry of

Linseed Meal	59 lb.
Water	4 gal.
Molasses	55 gal.

Then close boiler and fire to full pressure. After a half hour shut down fire. Hold under slight pressure for 24 hours. The above quantities are for a 5000-gal. boiler.

Sealing Glass Into Brass Fittings

Whiting	3
Water Glass	7

The water glass should be a grade having a soda to silica ratio of 1 to 3.22 (41° Baumé).

Bonding Glass to Aluminum

A sheet of paper soaked in water glass may be used as a gasket in bonding glass to aluminum. The water glass-impregnated paper forms the bond. A grade of sodium silicate (water glass) having a soda to silica ratio of 1 to 3.90 (33.5° Baumé) is recommended.

Sealer for Lenses

Carnauba Wax	406
Beeswax	1840
Burgundy Pitch	945
Pure Gum Rubber (Unvulcanized)	101
Lamp Black	205

Melt 1, 2 and 3 together. Heat very high and add 4, cut into thin strips. Stir gradually with stirring. Dispense when hot.

Cement for Joining Metal to Glass

Formula No. 1

This preparation is used for tanks with glass sides or bottom to make them water tight by cementing the glass to the iron frame or to repair leaks that may occur.

Litharge	260 g.
Glycerin Solution (Glycerin 2 Parts, Water 1 Part)	100 cc.

Place the litharge in a mortar, add the diluted glycerin slowly while grinding. Mix thoroughly by grinding a short time. Heat will be evolved and the mixture will begin to set. While still soft, pour it into place and by means of a spatula work it in as in the case of a putty. Allow to stand for a day, when it will be thoroughly hard.

If desired, cover with a layer of white lead or aluminum paint.

No. 2

A mixture of sodium silicate with quartz meal makes a good cement for bonding metal and glass. A thin coat of the mixture should be applied to both the glass and the metal. The parts should be pressed firmly together and baked for 14 to 20 hours depending on their thickness.

No. 3

Fresh Milk Curd	1
Chalk (Powdered)	4

This combination makes a bonding material for immediate use.

No. 4

Mastic	3
Shellac	2–4
Alcohol (90%)	To make a liquor of medium consistency

This compound is impervious to moisture.

No. 5

Vinylite VMCH	20
Dibutyl Phthalate	1
Methyl Isobutyl Ketone	1
Ethyl Methyl Ketone	20
1-Nitropropane	20

Furnace Cement

Fireclay	23
Silicon Carbide Firesand	77
Water Glass	9
Water	8

Magnesium oxide or chrome ore can be used in place of the silicon carbide. The water glass used should be a sodium silicate with a soda to silica ratio of 1 to 2.40 (52° Baumé). The exact amount of water required depends somewhat upon the particle size of the ingredients. Slightly more or less water than the amount stated may therefore give the desired consistency to the cement.

Coke Oven Cement

Raw Clay	8
Calcined Clay	12
Silicate of Soda	2
Water	11

The sodium silicate should be a hydrated silicate with soda to silica ratio of 1 to 2. The cement is applied with a cement gun, and is useful for repairing the charging holes in coke ovens while still hot.

Dry Furnace Cement

Fireclay	50
Silicon Carbide Firesand	50
Powdered Silicate of Soda	17½

The powdered sodium silicate should be a grade with a soda to silica ratio 1 to 3.22. This cement is mixed with water when applied, and is suitable for boiler settings and brickwork and other refractory uses. The proportions of fireclay and silicon carbide firesand can be varied over a wide range, provided the total of the two ingredients in the formula is kept at 100 parts.

Acidproof Cement
Formula No. 1

Litharge	8
Water Glass	4
Glycerin	1

The water glass should be a grade having a soda to silica ratio of 1 to 2.40 (30° Baumé). Setting time about 3 minutes.

No. 2
U. S. Patent 2,396,509

Powdered Quartz Sand	200
Sodium Titanium Fluoride	3
Sodium Silicate (40° Bé.)	90

Mix before use. This cement hardens in about 5 minutes.

No. 3

Ground Quartz	100
Water Glass	70–100

About half of the ground quartz should be 20 mesh grade and the balance 100 mesh and fines. The amount of sodium silicate varies with the fineness of the quartz used. Larger proportions of quartz fines require larger quantities of the water glass. The sodium silicate (water glass) should be the grade in which the ratio soda to silica is 1 to 3.90 (33.5° Baumé). Enough water glass is used to give the mixture a thick creamy consistency. Long drying periods, up to 30 days, are recommended for acid proof cement applications.

Acid Proof Cements (Höchst)

Name	Application	Resistant to	Powder	Composition %	Liquid
SW 10 and 20	Acid Proof Brick Linings. Cements are Not Liquid Tight	All Acids Except Sulfuric and Hydrofluoric.	Na_2SiF_6 Clay Quartz Powder	4.0 2.0 94.0	Sodium Silicate Solution Containing 8% Na_2O and 26% SiO_2
SWK	Masonry Work, Linings, Joints.	Sulfuric Acid, All Concentrations; Salts, Etc.	K_2SiF_6 Soluble Silicic Acid Clay Quartz Powder	6.0 3.0 2.0 89.0	Potassium Silicate Solution Containing 10% K_2O and 23.5% SiO_2
SWD	Brick Work, Tile Linings, Etc.	All Acids Except Sulfuric and Hydrofluoric.	Na_2SiF_6 K_2SiF_6 $CaSiF_6$ Clay Soluble Silicic Acid Gypsum Quartz Powder	3.0 3.0 1.0 6.5 2.0 0.3 84.2	Sodium Silicate Solution Containing 11.6% Na_2O and 31.5% SiO_2
SWD-Z	Ceramic Linings for Pressure Vessels; e.g., Cellulose Digesters	Especially Resistant to Sulfurous Acid.	Na_2SiF_6 K_2SiF_6 $CaSiF_6$ Clay Soluble Silicic Acid Gypsum Slag Wool Powder Quartz-Sand and Powder	4.0 2.0 0.8 5.2 3.2 1.5 20.0 63.3	Sodium Silicate Solution Containing 11.6% Na_2O and 31.5% SiO_2
Z-Acid Proof Mortar	Masonry Work, Tile Linings, Jointing of Brick Work.	All Acids Except Sulfuric and Hydrofluoric.	Na_2O SiO_2 H_2O	20.0 53.0 27.0	Water Mixed in Ratio of 79 pt. per 21 pt. of Powder
K-Refractory Cement	Acidproof and Fireproof Masonry Work; e.g., Drying Kilns.	All Acids Except Hydrofluoric.	K_2SiF_6 $BaSO_4$ Soluble Silicic Acid Fire Brick Powder Quartz Powder	3.0 12.0 3.0 50.0 32.0	Potassium Silicate Solution Containing 10.5% K_2O and 23.5% SiO_2

Acidproof Cement for Sulfite Digesters

Portland Cement	1
Ground Quartz (20 mesh)	2
Water Glass	As required

The sodium silicate should be the grade having a 1 to 2.40 soda-silica ratio. It can be diluted with water up to 10 per cent. Enough of the water glass is used to furnish the desired consistency. The cement should be mixed in very small batches, only enough for a brick or two at a time.

Acidproof Digester Lining Cement

Pulverized Silica	20
Ball Clay	4
Sodium Silicate (Powdered)	4
Water	9

The sodium silicate required is the grade having a soda to silica ratio of 1 to 3.22.

Cement for Pipe Joints
Formula No. 1

Paris White (Ground)	4	lb.
Litharge (Ground)	10	lb.
Yellow Ochre (Ground)	½	lb.
Short Cut Hemp	½	oz.
Linseed Oil	As required	

Mix to a stiff putty with the linseed oil and use on pipe joints as needed.

No. 2

Castor Oil	24
Bardol B	12
Clay	32
Whiting	31
Aluminum (Powdered)	1

Make a rough mixture by adding the powdered ingredients to the blend of oil and Bardol B, and pass through a paint mill to obtain desired fluidity.

Threaded Joint Seal
U. S. Patent 2,393,929

Blown Castor Oil	4.5–6
Calcium Stearate	1 –2
Ethyl Ricinoleate	0.3–1.5

This seal is not affected by gasoline.

Cement for Closing Leaks in Iron Pipes

Iron Borings (Coarsely Powdered)	5 lb.
Sal Ammoniac (Powdered)	2 oz.
Sulfur	1 oz.
Water	Enough to make a thick paste

This cement, thoroughly mixed, is tamped tightly into the leak. By cutting down on the amount of sulfur the cement will set more firmly, but will require a longer time to harden.

Caulking Composition for Metal Joints
U. S. Patent 2,396,607

Polyvinyl Butyral	73.0
Dibutyl Sebacate	27.0
Oleic Acid	5.0
Caustic Soda	1.0
Zinc Oxide	1.0
Casein	2.4
Water	160.0

Heat, together to 185°F. and mix until uniformly emulsified. On cooling a putty is formed.

Sealing or Plugging Cement

Toluene	40
Alcohol	10

Synthetic Resin Cements (Asplit)

Name	Application	Resistant to	Powder	%	Liquid	%
Asplit	Jointing of Tile and Brick Work.	All Acids Except Oxidizing Acids; Also Resistant to Sodium Carbonate.	p-Toluene Sulfone Chloride Silica Quartz Powder	10.0 20.0 70.0	Condensation Product of Phenol and Formaldehyde Benzyl Alcohol	90 10
Asplit A	Jointing of Tile and Brick Work.	All Acids Except Oxidizing Acids; Also Resistant to Caustic Soda.	p-Toluene Sulfone Chloride Barium Sulfate Quartz Powder	10.0 70.0 20.0	Condensation Product of Phenol and Formaldehyde Benzyl Alcohol *Triethyl Phosphate	75 5 20
Asplit El	Coating of Larger Tanks Especially Thick Walled Concrete Reservoirs for Gasoline in Submarine Stations.	Gasoline and All Common Solvents.	p-Toluene Sulfone Chloride Polymer of Vinyl Chloride and Maleic Anhydride Kieselguhr Silica Quartz Sand Quartz Powder	6.5 1.5 1.0 10.0 41.0 40.0	Condensation Product of Phenol and Formaldehyde Benzyl Alcohol *Triethyl Phosphate	75 5 20

* Dichloropropanol or chlorobenzaldehyde can replace this compound.

ADHESIVES

Ethyl Cellulose	20
Tricresyl Phosphate	5
Indian Red (Pigment)	17
Ochre	8

Aquarium Cement

Glazier's Putty	10 lb.
Litharge	1 lb.
Red Lead	1 lb.
Asphaltum	4 fl. oz.

Mix with boiled linseed oil to the proper consistency. Lamp black may be added to give a gray color.

Adhesive Cement for Tins
U. S. Patent 2,381,946

	%
Rosin	74– 9
Hydrogenated Oil	10–15
Microcrystalline Wax	5–10
Paraffin Wax	2– 5
β-Naphthylamine	1

Can Seam Sealing Compound
U. S. Patent 2,326,966

Glyceryl Sebacate	100
Magnesium Silicate	50–150
Pyroxylin	10– 15
Acetone	200–300

Cement Adhesive

Vistac	1.5
Staybellite A-1	1.5
Vistanex Polybutene Medium	30.0
Solvesso #4	170.0

Spark Plug Cement

Kaolin	30
Water Glass	14

Water glass having a soda to silica ratio of 1 to 3.22 (41° Bé.) should be used. The product is effective in sealing the metallic electrodes in porcelain spark plug bodies after heating to 1,000°C.

Cements for Thermoplastics

Celluloid	Butyl acetate; acetone; alcohol (equal parts). For film, the proportion of alcohol must be increased, a suitable formula being butyl acetate 3, acetone 6, alcohol 7.
Cellulose Acetate	Ethyl lactate 1, acetone 1, benzene 1, alcohol 1. A simpler cement is acetone 9, benzene 1; but this dries more quickly
Cellulose Triacetate	Methylene chloride 9, alcohol 1
Ethyl Cellulose	Solvent naphtha 4, alcohol 1
PVC Copolymer	Methyl ethyl ketone
Formvar	Trichlorethylene or benzene 7, alcohol 3
Methyl Methacrylate	Glacial acetic acid, especially warm Monomer proprietary cements
Polystyrene	Benzene

All these cements may with advantage contain up to 10% of the plastic material. This gives body to the cement, but delays drying somewhat.

Mechanical Packing Installing Paste

Potassium Soap	16
Glycerin	4
Water	1
Powdered Mica	To suit
Asbestos Fibers	

Plastic Putty
U. S. Patent 2,346,408

Reclaimed Rubber	10
Whiting	75
Linseed Oil	5
Gasoline	10

Emulsion Adhesive for Vinylite Sheet

Chlorinated Rubber (125 C.p.)	25.4
Chlorinated Paraffin (60% Chlorine)	12.7
Toluene	41.3
Sodium Oleate	1.0
Water	19.6

Add water and soap to the chlorinated rubber and chlorinated paraffin solution with rapid agitation.

Vinylite Sheeting Adhesives

	Formula							
	No. 1	No. 2	No. 3	No. 4	No. 5	No. 6	No. 7	No. 8
Vinylite Resin VYNS	10		12		20			2
Vinylite Resin VYNW						10		
Poly-n-Butyl Methacrylate	10	32	12	20				
Rosin			6					
Chlorinated Rubber (1000 Cp)							25.0	
Plasticizer E-60							12.5	
Methyl Ethyl Ketone	80		70	70	32	36	62.5	96
Cyclohexanone					32	36		
Propylene Oxide					16	18		
Trolucil		34						
Solvesso No. 2		34						
Petrex No. 21					10			
Acetic Acid								2

	Formula		
	No. 9	No. 10	No. 11
Resin VMCH	10.0		10.0
Hycar OR-15	10.0	13.0	10.0
Cumar P-25		4.3	
Dioctyl Phthalate		4.3	
Nitroethane		12.9	7.5
Methyl Ethyl Ketone	80.0	65.0	16.1
Ethyl Acetate			16.0
Methyl Isobutyl Ketone			20.0
Toluene			20.0
Sulfur		0.3	0.2
Captax		0.2	0.2

Air drying for 24 hours is desirable, but a short force dry at 200 to 225°F. is equally satisfactory. A VYHH, or VYNS resin primer can be used, but these solutions require baking at 350 to 375°F., and are somewhat limited in their usefulness since they may lose their adhesion to the metal in the presence of an excess of active Vinylite resin solvent.

ADHESIVES

Adhesives for Vinylite Resin Coated Cloth

	Formula		
	No. 1	No. 2	No. 3
Vinylite VYNS		7.14	
Vinylite XYHL (Dry)			9
Bakelite Resin XR-9396		7.14	
Bakelite Resin XV-16530			5)
Hycar OR	15.2		
Stabelite Hydrogenated Rosin			18
Cumar P-25	4.9		
Whiting	15.15		
Iron Oxide Black	3.6		
Iron Oxide Yellow	3.6		
Zinc Oxide	3.6		
Dioctyl Phthalate	4.9		18
Nitroethane	18.9		
Methyl Ethyl Ketone	14.9		
Ethyl Acetate (95%)	14.9		
Isophorone		14.29	
Propylene Oxide		71.43	
Acetone			5
Sulfur	0.35		

Bonding Vinylite to Cloth

The choice of an adhesive for bonding plasticized sheeting to cloth depends largely upon the fabrication method preferred. One of the most satisfactory methods involves the priming of the cloth with a VYNW, or a VYNS solution, drying, and then bonding the cloth and sheeting under heat and pressure. Typical cloth primers are:

	Formula	
	No. 1	No. 2
Vinylite Resin VYNS	18	
Vinylite Resin VMCH	2	
Vinylite VYNW		14
Flexol DOP	18	5.5
Methyl Ethyl Ketone	25	40.5
Cyclohexanone		40
Hexone	23	
Solvesso No. 1	14	

Penetration of the coating into the cloth will improve adhesion, but it will also lower the flexibility of the combination. Therefore, some variation in solvent and plasticizer contents of these coatings will have to be made to fit individual requirements and coating conditions. The VYNW coating is preferred where the coating must withstand extremes of temperature.

Cellulose Acetate Adhesive
Formula No. 1

Cellulose Acetate	0.2
Dry Zinc Chloride	2.4
Water	2.4

Mix until uniform, then add the following previously made solution:

Cellulose Acetate	0.52
Triacetin	9.48

A thinner for the above consists of:

Triacetin	2.50
Dry Zinc Chloride	1.25
Water	1.25

No. 2

Cellulose Acetate	0.5
Acetone	8.5
Triacetin	32.5
Diacetone Alcohol	8.5

Cements for Cellulose Acetate
Formula No. 1
Fast Drying type

Acetone	200
Methyl Acetate	200
Methyl Cellosolve Acetate	30

No. 2
For a slower drying cement

Acetone	100
Methyl Acetate	100
Methyl Cellosolve Acetate	100
Diacetone Alcohol	50

No. 3

Acetone	3
Methyl Cellosolve	2
Methyl Cellosolve Acetate	4
Diacetin	1

Adhesive for Lucite and Plexiglas

Lucite or Plexiglas (Clean, clear, colorless scrap)	3
Ethylene Dichloride	37
Methylene Dichloride	60

The scrap is reduced to small pieces, placed in the solvent mixture, and periodically stirred at room temperature until the scrap is dissolved.

Cement for Thermosetting Plastics

Resorcinol	1
Paraformaldehyde	1
Denatured Alcohol	2

Mix and dissolve before using. Apply and heat to 100°C. to make it set.

Laminating Adhesive for Glassine or Cellophane
U. S. Patent 2,325,584
Formula No. 1

Paraffin Wax (M.P. 155°F.)	61
Rosin	27
Rubber	5
Petrolatum (M.P. 125°F.)	7

No. 2

Chlorinated Rubber	40–50
Resin	25–38
Paraffin Wax	2– 7
Plasticizer (e.g., Dibutyl Phthalate)	15–21

The resin which has been found particularly adaptable for such use comprises gylcerol abietate 20–25% and para-coumarone 5–13%. The paraffin wax preferably has a melting temperature of 143–145°F.

No. 3

Chlorinated Rubber (65–68% Chlorine)	46
Glycerol Abietate	19
Paraffin Wax	6.1
Dibutyl Phthalate	16.9
Para-Coumarone	12

After these ingredients are combined in these approximate proportions the composition is dissolved in any suitable hydrocarbon solvent, such as benzol, toluol, xylol, etc., to give a solution of the desired viscosity.

Paper Laminating and Waterproofing Composition
U. S. Patent 2,408,297

	%
Asphalt (Softening Point 155°F.)	73–95
Vistac	3–12
Acrawax C	2–15

Leather Adhesive
Canadian Patent 431,616

Butadiene-Acrylonitrile Copolymer	100
Vinyl Chloride-Vinyl-Acetate Copolymer	100
Stearic Acid	1.5
Zinc Oxide	5
Sulfur	2
Calcium Silicate	20
Naphtha or Other Solvents	600–800

Sealing Corks into Glass Bottles

Clear Rosin	2
Ether	4
Collodion	3

Dissolve ½ lb. clear rosin in 1 lb. technical grade ether, add to ¾ lb. collodion and mix thoroughly. If color is required use oil-soluble dyes.

Peelable (Removable) Adhesive Label
U. S. Patent 2,376,777

Labels that become pressure sensitive on heating to 200–400°F. are made by coating paper or textile tape with:

Paraffin Wax	80
Crepe Rubber	40
Cyclo Rubber	40
Benzol	200
Naphtha	360
Alcohol	24
Antioxidant	1.5

Heat or Pressure Sensitive Adhesive
Formula No. 1
U. S. Patent 2,375,163

Cumarone Resin	200
Zinc Stearate	10

Kneed the above into

Milled Reclaim Rubber	100

This adhesive is resistant to cold-flow.

No. 2
U. S. Patent 2,381,946

	%
Resin	74–79
Hydrogenated Oil	10–15
Amorphous Wax	5–10
Paraffin Wax	2–5
Antioxidant	1

No. 3

Paraffin Wax	8
Vistanex U	12
Balata Gum	14
Piccolite Resin	16
Acrawax C	1

Melt together and apply at about 325°F. This is not tacky at ordinary temperatures.

No. 4

Piccolastic A-5	15
Ethocel (Standard ethoxy, 13 cp)	10

Heat Piccolastic A-5 resin to about 300°F. and then stir in the ethyl cellulose. Care must be observed that the ethyl cellulose is not heated too high as to cause darkening. Use at about 300°F.

Laboratory Adhesives
Formula No. 1

Nitrated Cotton (5–6 sec.)	30 g.

Make up a solution of:

Acetone	100 cc.
Amyl Acetate	45.0 cc.
Butyl Acetate	15.0 cc.
Ethyl Acetate	15.0 cc.
Ethyl Abietate	1.5 cc.

Using the latter solution as solvent, add the nitrated cotton until the solution is of the consistency of a syrup. Dissolving the cotton takes about 2 hours. If the mixture becomes too thick, a little more of the above solvent, which should be kept on hand, is added.

In case a more flexible film is desired the amount of the plasticizer may be doubled. In case of blushing, increase the amount of amyl acetate.

No. 2
Dry Yellow or Orange

Shellac	3
Pine Tar	1

Place the shellac in a double boiler using water in the outer member. Add the pine tar and permit to digest with occasional stirring until the mass is homogeneous; this will take about 5 hours. Pull out like taffy and form into sticks. The cement can be made harder or softer by varying the amount of tar.

No. 3
De Khotinsky Type
(Benzene Resistant)

Note: Most recipes for de Khotinsky cement call for 40 to 50% pine tar according to the material with which shellac is plasticized for application in question. Recent investigation indicates that pine tar is inferior to the creosote plasticizer recommended below.

Prepare the plasticizer by mixing one volume of terpineol with three volumes of beechwood creosote (alkali-soluble). Coaltar creosote, which is not completely alkali-soluble, will not do.

Heat 12 to 25 g. of the plasticizer to about 130°C. With constant stirring add 85 g. of shellac as fast as it dissolves smoothly. When the mixture is homogeneous, allow to cool until it will barely flow from the vessel, and pour into molds which have been lightly but completely covered with petrolatum. The use of only 12 g. of plasticizer gives a very hard cement; 25 g. give a very soft product.

No. 4
(Benzene Soluble)

Rosin	35 g.
Shellac	20 g.
Beeswax	20 g.
Fibrous Talc	
(Asbestine pulp)	0 to 30 g.

Melt the rosin in a large (6 or 8 in.) hemispherical iron pan, add the shellac and beeswax with stirring. Heat with a large Bunsen flame so that the temperature reaches 360°C. in 6 minutes; then extinguish the burner. When the temperature has reached about 275°C. add the talc, if any is to be used. Finally pour into metal molds which have previously been very thoroughly scoured with washing powder and thickly coated with aqueous dextrin paste which is still wet. With talc, a more viscous cement is obtained.

Self-Curing Rubber Cement
(Overnight at Room Temperature)

	A	B
Smoked Rubber Sheets	68.25 lb.	64.00 lb.
Roll Brown	19.00 lb.	19.25 lb.
Para Flux	3.25 lb.	3.25 lb.
Zinc Oxide	7.00 lb.	7.00 lb.
Stabilite	1.00 lb.	1.00 lb.
Ulto	1.50 lb.	—
Phenex	—	0.50 lb.
Sulfur (Spider Brand)	—	5.00 lb.

	A—	B—
Cement	89 lb.	89 lb.
Smoked Sheets	54 lb.	54 lb.
Gasoline	225 gal.	225 gal.

Blend A and B 50 to 50 as required. Do not store the blend as it is self-curing.

Alloy for Joining Glass, Ceramics and Metals
Norwegian Patent 63,110
Formula No. 1
Zinc	95
Magnesium	5

No. 2
Zinc	86
Aluminum	9
Magnesium	5

Borating Solution
(To make wire adhere to glass)
Water	2400 cc.
Caustic Potash	300 g.
Boric Acid	660 g.

Glass Seal for High Temperature Vacuum Work

A very efficient glass to glass seal for high temperature vacuum experimentation can be made using crystalline silver chloride. The crystals of silver chloride are fused on the glass and the sections joined.

The crystals of silver chloride are formed in the following manner. Add ammonium hydroxide to ordinary silver chloride to dissolve it; use a slight excess of the ammonium hydroxide to insure complete solution of the silver chloride. Allow the solution to stand at room temperature and as the ammonium hydroxide volatilizes the silver chloride crystals will form.

Chapter III

COSMETICS AND DRUGS

Cosmetic Formulation

Although new raw materials have made it possible to produce new cosmetic products, as well as to improve old ones, the fundamental principles of careful manufacture and cleanliness have not changed.

It is the purpose here to give information concerning the best utilization of certain types of raw materials. While only general formulae will be given, these will be of such a nature as to be easily modified according to individual preference. Any desired modification should be made and tested as would be a completely new formula, that is, the product should be tried on the laboratory scale first, under completely controlled conditions and then submitted to adequate shelf testing. Only when the preparation has successfully passed laboratory tests should it be made on a commercial scale.

In many instances the quality of the finished product will vary, sometimes by a little, occasionally by a great deal, depending upon variations in raw material quality. As a single instance, the impure borax used in cold creams may have a critical effect upon stability. Where a minimal quantity has been used, the presence of more than a very small amount of calcium salts may well lead to inversion or destruction of the emulsion; only a highly purified borax should be used in cosmetics. Beeswax, lanolin, emulsifying agents, etc., from different sources of supply may lead to considerable variation in identically formulated and processed preparations. Thus it is imperative to start and continue with standardized and controlled materials from trustworthy sources.

Operating Procedure. Every step in the manufacture of cosmetics must be done carefully under trained supervision. Errors frequently occur in the fundamental operations of weighing and measuring. Balances and scales must be accurate, must be kept clean, and must be frequently checked. Where possible, each weighing operation should be checked by a second person, both for weights and for the ingredients. Measuring vessels must be checked to determine exact capacities and must be discarded, when they become battered. Errors often occur in the use of thermometers; make sure that the thermometer is immersed to the proper extent while it is being read and that the liquid is stirred so that a uniform temperature is attained.

Unless water is quite impure, it ordinarily has little effect upon the quality of the finished product. This must be ascertained, however, and not merely assumed. Organic matter in water may promote putrefaction in the cream or lotion; excessive proportions of calcium and magnesium salts (hard water) may affect emulsion stability. Iron and copper, even in small amounts, may lead to discoloration and the promotion of rancidity. In extreme cases it may be necessary to use softened, or even distilled water.

Equipment. For creams, steam or hot-water jacketed kettles are most popular. Parts coming in contact with the product should be enamelled, or made of stainless steel or aluminum. Tinned equipment is common, but the coating must be kept in good condition by retinning when necessary.

Agitation is of paramount importance. Small commercial batches of creams are sometimes made by hand stirring with a paddle but this is a hazardous procedure. High or low-speed propeller agitators are often satisfactory, but only if the propeller blades are large enough to assure complete turnover of the mass. For batches up to one hundred pounds, eccentric paddle mixers used in baking and food processing are very satisfactory. They usually can operate at several speeds, although the slowest speed is the most useful and provides thorough mixing and emulsification with minimum incorporation of air. Some varieties of these mixers have gas-heated hot water jackets which are extremely convenient. A colloid mill or homogenizer is quite useful, since it tends to give uniformly emulsified products which are therefore smoother and more stable. Properly used, a colloid mill can enormously increase the capacity of ordinary mixing equipment.

Filling. Creams are most conveniently filled into jars by pouring while warm. The temperature of pouring is important in determining the final finish of the cream in the jar. Certain creams, notably those containing substantial amounts of pigment, and those vanishing creams based wholly upon stearic acid must be filled cold, however, and smoothed by hand.

Perfume and Color. An indispensible element in the saleability of cosmetics is the perfume they contain. This must be selected carefully not only for odor, but also for chemical and physiological activity. Poorly selected perfumes may well discolor creams, especially those that are alkaline. Furthermore, certain oils and aromatics are distinctly irritating to the skin even when present in small amounts, and their use must be avoided in creams and face lotions.

Dyes are to be selected from the Food and Drug Administration list of certified colors. Water-soluble dyes are used with water based creams and lotions and are mixed into the finished cream; the same procedure is used in adding solutions of oil-soluble colors to oil based creams and lotions. In the case of water-soluble dyes, care should be taken to use only such dyes that are stable in the pH range of the liquid or cream to which they are to be added. Thus an alkali stable dye should be used for preparations on the alkaline side, and an acid stable dye should be used for products on the acid side.

Cosmetic Emulsions

Cosmetic creams and milk lotions may be classified as emulsions, namely, dispersions of small droplets of one liquid in another liquid, the liquids, one of which is generally water, being insoluble in each other. Liquid is used loosely in this connection, and embraces various mixtures of fats and waxes as well as oils and solvents as one phase of the emulsion and water as the other phase. The emulsion is described as being of the oil-in-water type when droplets of the oily components are distributed throughout the water, while in water-in-oil emulsions the reverse holds true. Oils will not form a stable emulsion with water alone, the small droplets combining to form larger drops and finally two distinct layers will separate. A stabilizing agent, known as an *emulsifying agent*, is required for stabilization. The emulsifying agent forms a film around each drop, preventing it from combining with others to form larger drops. In some cases one emulsifying agent alone may not be sufficient to form a stable emulsion and separation may occur. An additional emulsifying agent or a stabilizer may be necessary to insure a stable emulsion.

The type of emulsion can be tested in two convenient ways. First, if the emulsion quickly and easily mixes with water, but not with oil, it is of the oil-in-water type. The water-in-oil emulsions of course show the opposite action. Second, the electrical conductivity of the mixture can be determined. If the emulsion exhibits very low conductivity, it is of the water-in-oil type, in which the oil forms the continuous phase and will not conduct the electric current. In oil-in-water emulsions, on the other hand, water, usually containing some electrolytes, is the continuous phase and is a moderately good conductor of the current. This testing is of more than academic interest, since the type of emulsion formed has very great effects upon the properties of the preparation. For example, a cream with certain proportions of oils, waxes, and water may be a smooth, white, fairly soft, stable cream when the emulsion is of the oil-in-water type, but when the emulsion is inverted or changed over to the water-in-oil type it will be much duller in appearance, its color may change, it may be definitely softer or harder, and may eventually separate. In fact, some creams may at first be of one emulsion type and then spontaneously invert on standing to the other type. Beeswax-borax creams containing too little borax may thus change over from a normal oil-in-water emulsion to a much less attractive and relatively unstable water-in-oil emulsion. It must be remembered that a choice of the correct emulsifying agents for a particular purpose is very important. Prevention of discoloration of the emulsion on standing over periods of time also depends to a large extent on the correct emulsifying agent. Certain emulsifying agents, while perfectly satisfactory as regards stability, tend to oxidize and turn yellow, after some time, thus rendering the product unsaleable.

A good emulsion should not separate into layers, it should not discolor on aging, and it should not change in consistency. A cream or base emulsion should not become thin or semi-fluid, a liquid emulsion should not thicken and become unpourable. Temperature changes may be of great importance. A cream or liquid emulsion which is perfectly satisfactory in cold climates might be far too thin or liquid in warmer climates.

Soaps of one kind or another, or alkali metal salts of higher fatty acids, are common emulsifying agents, and are satisfactory to a certain extent, particularly for pastes or heavy creams. However, they suffer from a number of disadvantages, notably, where fairly long shelf life is an important factor. Ordinary soaps do not, in themselves, give too high a degree of stability, particularly in liquid emulsions. Furthermore, their effectiveness can be lowered by a variety of conditions. Since the soap acts as the emulsifying agent maintaining the immiscible components in a homogeneous state in the emulsion, any factor tending to inactivate the emulsifying action of the soap will adversely affect the stability of the emulsion. Fatty acid esters of certain polyhydric alcohols, while in themselves poor emulsifying agents, become excellent emulsifying agents when modified by the addition of small quantities of emulsion stabilizers.

Modern Emulsifying Agents

Monostearin (Glyceryl Monostearate). Monostearin is the commercial grade of glyceryl monostearate to which no stabilizer has been added. It consists of a mixture of mono-, di- and tri-stearates, and is a light cream-colored wax-like solid with a melting point of 55°–57°C. It has a specific gravity of 0.970 (25°/25°C.). Monostearin is insoluble in water but readily forms a milky emulsion in water containing a small percentage of soap or wetting agent, such as sodium lauryl sulfate. With the addition of a wetting agent Monostearin gives an emulsifying agent particularly recommended for paste creams stable in the presence of dilute acids, salts and other electrolytes. Deodorant creams containing benzoic acid, aluminum sulfate, etc., can very readily be made with this type of emulsifying base (see Deodorant Creams). In conjunction with petrolatum, Monostearin is an effective aid to absorption base creams of the water-in-oil type. It is also of interest by itself as an edible synthetic wax.

Glyceryl Monostearate S. This is essentially glyceryl monostearate modified with a small percentage of soap. It acts as an effective emulsifying agent by itself in the manufacture of creams of the greaseless and vanishing cream type where straight emulsification of oils and waxes is desired in the absence of electrolytes. In conjunction with cetyl alcohol or a colloidal clay such as bentonite, Glyceryl Monostearate S makes an excellent greaseless ointment base into which boric acid, tannic acid, calamine, etc., can be incorporated (see Medicated Creams). Glyceryl Monostearate S is a white wax-like solid with a melting point of 56°–57°C. A 3% dispersion in water has a pH of 9.3 to 9.7.

Diglycol Stearate S. This is a mixed mono- and di-stearate modified with soap and containing an excess of free fatty acid. It is a white colored wax-like solid with a melting point of 51°–54°C. (capillary tube). It has a specific gravity of 0.96 (25°/25°C.). It is readily dispersible in hot water forming on cooling with stirring, a stable fluid or paste emulsion. depending on the amount used. A 3% dispersion in water has a pH of 6.8–7.1. Diglycol Stearate S acts as an excellent emulsifying agent for oils, solvents and waxes where a stable, neutral viscous cream-like product is desired. It is specially suitable for the manufacture of greaseless creams, brushless shaving creams and similar products. It is also of interest as a suspending medium for titanium dioxide for skin whiteners and night creams.

Glaurin (Diethylene glycol monolaurate). This is a light yellow liquid with a faint pleasant odor. It has a specific gravity of 0.960 (25°/25°C.), and is insoluble in water but miscible with alcohol, hydrocarbons and oils in certain proportions. It is useful in many instances for its properties, as an oil, as a solvent and as a lubricant. In conjunction with small amounts of soap, it acts as an excellent emulsifying agent for the manufacture of hand lotions and similar liquid emulsions. Used in conjunction with an acid stable wetting agent such as Wetanol, Duponol, etc., it can be employed as an emulsifying agent stable in the presence of dilute acids, salts and other electrolytes.

Diglycol Laurate S. This is a partial lauric acid ester of diethylene glycol, modified with a small amount of soap, and can be used directly as an emulsifying agent for liquid emulsions in the absence of electrolytes. It is a light straw-colored liquid, dispersible in water, and miscible with alcohol, hydrocarbons and oils. A 3% dispersion in water has a pH of 9.0–9.2. Added to gum karaya in the making of finger wave concentrates with alcohol, Diglycol Laurate S prevents the gum from settling to the bottom and caking.

Modern Cold Creams

The term Cold Cream is properly limited to traditional beeswax-borax, oil and water creams, but it can also be applied to any white cream that contains sufficient water to exert a cooling effect on the skin because of water evaporation. The regular type of cold cream generally contains from 25% to 35% water. Small amounts of water tend to favor the production of water-in-oil emulsions. In the case of the modern type of cold cream the water content may range from 40% to over 60% depending on the other ingredients present. Both Glyceryl Monostearate S and Diglycol Stearate S, which are somewhat similar chemically to naturally occurring fats, are used as emulsifying agents and require no additional emulsifying aids. They are both excellent emulsifying agents and are also of real utility as skin softeners. Because of their fat-like qualities they may be subject to attack by certain molds. Preservatives should always be included in formulations using fatty acid esters, particularly where shelf life is an important factor. In this connection we recommend the preservative Moldex.

Formula No. 1

Glyceryl Monostearate S	12.0
Beeswax	3.0
Spermaceti	3.0
Mineral Oil	30.0
Glycerin	8.0
Water	43.5
Moldex (Preservative)	0.1
Perfume	0.4

All glyceryl monostearate creams may be made by either of two alternative procedures. According to the simpler, but not necessarily better method, all of the ingredients, with the exception of perfume, are heated together to about 85°C. until the waxes, fats and oils have been completely melted and float on the water as a liquid mixture. Stirring is now started and continued until the mixture becomes a smooth cream. Unless particular care is taken in this procedure, some wax lumps may remain undispersed. A more usual method involves melting the waxes and oils in one pot, heating the water with the water-soluble components in another pot, and mixing both at the same temperature, 70°C. Stirring is continued until a smooth cream results and the perfume is added at 45°C. with stirring.

For creams with different consistency, texture, oiliness, etc., proportions may be varied within relatively wide limits. The oil generally should not exceed thirty per cent of the total; Glyceryl Monostearate may in special cases go as high as 25%. Where the oil content is relatively high, passage of the cream, while it is still warm, through either a colloid mill or homogenizer almost always improves the product.

No. 2

Glyceryl Monostearate S	14.0
Petrolatum	6.0
Ozokerite	2.0
Mineral Oil	25.0
Glycerin	5.0
Water	47.5
Moldex	0.1
Perfume	0.4

The type of petrolatum, whether short or long fiber, has a definite effect upon the qualities of the cream: long fiber petrolatum gives an oilier cream, while the preparation containing the short-fibered grade will show greater drag.

No. 3

Glyceryl Monostearate S	12.0
Petrolatum	9.0
Paraffin Wax	6.0
Mineral Oil	14.0
Glycerin	3.0
Water	55.5
Moldex	0.1
Perfume	0.4

Paraffin wax usually used in cosmetic formulation is the 125°/127° F. grade; higher melting varieties give a somewhat harder cream. Paraffin wax has the ability to harden the cream yet at the same time permits it to melt quickly on the skin.

No. 4

Glyceryl Monostearate S	15.0
Petrolatum	4.0
Lanolin	10.0
Mineral Oil	5.0
Water	65.0
Moldex	0.1
Perfume	0.9

The high lanolin content of this cream, along with Glyceryl Monostearate S, suggests its use as a night cream. Sufficient perfume must be present to cover the lanolin odor.

No. 5

Diglycol Stearate S	13.0
Spermaceti	8.0
Paraffin Wax	5.0
Mineral Oil	27.0
Glycerin	5.0
Water	41.5
Moldex	0.1
Perfume	0.4

This cream uses Diglycol Stearate S instead of Glyceryl Monostearate S, but the manufacturing procedure is unchanged.

This cream containing more than fifteen per cent of mineral oil is useful and effective as cleansing cream.

Where sufficient emollient, in addition to the Glyceryl Monostearate, is present, especially with reduced oil content, the creams may qualify as dry skin cleansers or all-purpose creams.

These creams are oil-in-water emulsions and may be washed off the skin with water.

No. 6
(For dry skin)

A. Lanolin	16 g.
Mineral Oil, U.S.P. Light	120 cc.
Glyceryl Monostearate S	80 g.
B. Glycerin	20 cc.
Distilled Water	400 cc.
Moldex	0.8 g.

Heat 1 and 2 in separate containers to 150°F. Add B to A with good mixing, while cooling slowly. Add perfume to suit at 105°F. and stir until uniform.

No. 7

Absorption Base	73.5
Liquid Paraffin	7.5
Soft Paraffin Wax	49.0
Water	30.0
Ozokerite	10.0

No. 8
For Tropical Climates

Beeswax	23.75
White Oil	25.00
Borax	1.25
Water	50.00

No. 9

Lanolin (Anhydrous)*	20.0
White Beeswax	16.0
White Mineral Oil	33.0
Stearic Acid	30.0
Carbitol	8.0
Propylene Glycol	8.0
Triethanolamine	4.0
Terpineol	0.2
Water	95.0

Melt the stearic acid, lanolin, and beeswax in the mineral oil, heat to 70°C., and then add the terpineol.

Heat the water to 70°C. in a separate kettle, add the triethanolamine, and then add this solution to the hot mixture of wax and oil. Stir vigorously until a creamy emulsion is obtained.

Add the perfume to the Carbitol

---

*The best grade of light-colored material should be used.

and propylene glycol and add this solution to the emulsion.

Continue stirring until the emulsion is smooth and quite viscous, and then stir occasionally until room temperature is reached.

It is possible to pour this cream into jars while still warm and thin enough to pour, but the resulting cream may not have the smooth texture of a cream that is packaged when cold. A pressure filler is usually necessary to fill the containers with the emulsion at room temperature.

This formula should serve as a starting point for making a cold cream to suit the individual preference, and should not be considered as necessarily the best product obtainable. Great variation in the wax and oil constituents is possible with little change in the basic ingredients. For example, vegetable oils, such as sweet almond and olive oils, may be substituted for all or a part of the mineral oil to produce an excellent product.

Absorption Base Creams

Creams formulated on a water-in-oil basis seem to assure more effective skin absorption of the active product. Water-in-oil creams when rubbed into the skin remain smooth and oily throughout the rubbing process. The reverse type, oil-in-water creams, are first smooth; then as they are further rubbed out and the water evaporates, they undergo phase inversion and exhibit a peculiar type of streakiness.

Water-in-oil creams formulated with beeswax, borax, oil, and water (low water and borax content) are not notably stable and are inclined to leak oil on standing. More satisfactory preparations may usually be made by the use of absorption bases. These latter ordinarily contain substantial proportions of lanolin extractives which act as water-in-oil emulsifiers. Monostearin (glyceryl monostearate, free from soap or other oil-in-water emulsifying agent), is quite effective in combination with petrolatum.

The following will make a good cream which may be modified by changes in proportions.

Formula No. 1

Lanolin Absorption Base	25.0
Petrolatum	10.0
Mineral Oil	10.0
Beeswax	5.0
Monostearin	10.0
Moldex (Preservative)	0.1
Perfume	0.4
Water	39.5

All of the ingredients except water are melted together and cooled to 45°C. Water at 40°C. is slowly added with vigorous stirring which is continued until the mixture has reached 30°C. If the cream is to be remelted for filling, heating must be very slow and the temperature must not be allowed to rise to the point where the cream becomes completely liquid and is likely to separate. A colloid mill is advisable in making absorption base creams.

No.2
Night Cream

Absorption Base	60.0
Water	40.0
Glycerin	10.0
Hard Paraffin Wax	5.0

Liquid Creams

Liquid creams, with and without oils, can be made with the assistance of emulsifiers such as Glyceryl Monostearate S, Diglycol Stearate S and Diglycol Laurate S.

Hand creams should contain little oil, since an excessive proportion of oil would tend to leave the hands greasy. A small amount of alcohol can be included in the preparation to accelerate drying. The use of emollients, such as lanolin, in addition to fatty acid esters is advantageous. In any event, lotions based upon Glyceryl Monostearate and similar compounds will be far more beneficial to the skin than customary soap-stearic acid products. Some glycerin should be present for its skin softening effect; excessive amounts, however, will retard drying on the hands. Finally, small proportions of gums in the finished product give smoothness of application.

The general directions for the manufacture of greaseless lotions are simple. The emulsifying agent (Glyceryl Monostearate S or Diglycol Stearate S), oleic acid, mineral oil, and whatever waxes or wax-like materials are present in the formula, are all melted together and stirred. The water soluble materials with the exception of the alcohol and gum solution are mixed with the water and heated to approximately the same temperature as the wax-oil mixture and added with high speed stirring to the melted wax oil mixture. Stirring is continued. When the temperature is about 50°C. the alcohol, gum solution and perfume are then added and stirring continued until the emulsion is cool. A colloid mill is preferable to ordinary high speed stirring and should be used wherever possible. In this case the emulsion should be passed through the mill before the alcohol, gum solution and perfume are added. These materials should be added as before at about 50°C. A colloid mill gives a much finer emulsion and a smoother and more stable product. A high speed mixer also gives excellent results, but no reliance should be placed on ordinary hand mixing. The results obtained are erratic and the shelf life of the finished product is always much shorter.

Formula No. 1

Glyceryl Monostearate S	3.5
White Oleic Acid	2.0
Glycerin	5.0
Triethanolamine	0.8
Moldex	0.1
Water	81.2
Alcohol	7.0
Perfume	0.4

No. 2

Diglycol Stearate S	5.0
Cetyl Alcohol	1.0
White Oleic Acid	2.0
Mineral Oil	1.0
Moldex	0.1
Water	89.5
Perfume	0.4

No. 3

Diglycol Stearate S	1.7
White Oleic Acid	0.7
Triethanolamine	0.3
Mineral Oil	2.0
Glycerin	3.0
Spermaceti	0.5
Moldex	0.1
Water	74.2
Karaya Gum	0.1
Water	10.0

Perfume	0.4
Alcohol	7.0

Liquid cleansing creams containing substantial amounts of oil can be made with the help of Diglycol Laurate S, a liquid, emulsifying agent. The procedure is again extremely simple. The self-emulsifying Diglycol Laurate S is mixed with the non-aqueous ingredients including perfume, applying gentle heat if necessary to obtain a clear solution. Add the water slowly with high speed agitation and pass through a colloid mill.

No. 4

Diglycol Laurate S	3.0
Cetyl Alcohol	1.0
Mineral Oil	10.0
Moldex	0.1
Perfume	0.4
Water	85.5

No. 5

Diglycol Laurate S	10.0
Mineral Oil	30.0
Moldex	0.1
Perfume	0.6
Water	59.3

No. 6

Beeswax	4.0
Paraffin Wax	10.0
White Petrolatum	11.0
Mineral Oil	55.0
Water	18.0
Borax	1.2
Glycerin	1.0

Greaseless Creams

A greaseless cream is an emulsion cream which seems to disappear when rubbed into the skin or leaves a non-greasy film on the skin. The most popular and oldest type of such creams is the vanishing cream; foundation creams, hand creams, protective creams, brushless shaving creams, medicated creams, deodorant creams, etc., all fall into the same class.

Vanishing Creams. The conventional type of vanishing cream consists essentially of a combination of free stearic acid and one or more stearate soaps. Potassium and ammonium soaps were first used but, more recently, have been substituted by Trigamine and triethanolamine soaps. The standard vanishing cream sometimes proves to be too drying for sensitive skins. Creams made with Glyceryl Monostearate S are definitely emollient in action and are slow to dry out or shrink in the jar.

Vanishing Cream
Formula No. 1

Glyceryl Monostearate S	12.0
Spermaceti	5.0
Glycerin	5.0
Titanium Dioxide	2.0
Water	75.5
Moldex	0.1
Perfume	0.4

All of the components, except the titanium dioxide and perfume are heated together to 90°C. and stirred until a homogeneous mixture results. The titanium dioxide is ground thoroughly with a small portion of cream taken from the batch and then stirred into the rest. Continue stirring until the mixture passes through its pasty stage and becomes a smooth cream. Stir in the perfume at about 50°C. Titanium dioxide makes a white rather than translucent cream and has a slight whitening effect upon the skin.

A similar cream can be made with Diglycol Stearate S.

No. 2

Diglycol Stearate S	11.0
Lanolin	3.0
Sesame Oil	3.5
Glycerin	5.0
Water	76.9
Moldex	0.1
Perfume	0.5

Again, all of the ingredients except the perfume are heated and stirred together and the perfume finally added at 50°C.

Glyceryl Monosterate creams are sometimes modified by the addition of a small amount of soap made during the process by combination of potassium hydroxide with stearic acid. In this case it is advisable to use the two pot manufacturing procedure.

No. 3

Glyceryl Monostearate S	10.0
Beeswax	2.0
Peanut Oil	2.0
Stearic Acid	2.0
Potassium Hydroxide	0.1
Glycerin	3.9
Water	79.5
Moldex	0.1
Perfume	0.4

The oil soluble ingredients are melted together to about 75°C. while the water soluble components are dissolved in water at the same temperature. The two solutions are mixed together and stirring continued until the mixture is homogeneous; perfume is added, as usual, at 50°C. when the cream is still soft, but the temperature low enough to prevent excessive volatilization.

No. 4

Glyceryl Monostearate	13
Beeswax	2
Sesame Oil	3
Lecithin	2
Glycerin	10
Water	70

No. 5

Diglycol Stearate	11
Lanolin	3
Peanut Oil	3
Glycerin	3
Water	80

No. 6

Stearic Acid	20.00
Caustic Potash	0.75
Water	60.75
Cholesterol	1.30
Lecithin	0.50
Cetyl Alcohol	1.70
Peanut Oil	5.00
Glycerin	10.00

No. 7

Stearic Acid	24
Zinc Stearate	1
Glycerin (C.P.)	44
Aqua Ammonia	3
Borax	1
Agar Agar	1
Perfume	1
Water to make	200

No. 8

A.	Pure Stearic Acid	234
	Anhydrous Lanolin	12
B.	Triethanolamine	13
	Glycerin (C.P.)	102
	Crystallized Borax	5
	Distilled Water	634

Melt A. together; at 65°C. add a boiling solution of B. Stir until cool. A cream with a very lustrous, smooth appearance will result.

Formula No.	Wax or Grease	Stearic Acid	Tri-ethanol-amine	Carbi-tol	Propyl-ene Glycol	Water
9.	*Lanolin (Anhydrous)	9.0	50	2.7	9	9	120
10.	*Lanolin (Anhydrous)	20.0	40	2.7	9	9	120
11.	Cetyl Alcohol / *Lanolin (Anhydrous)	5.3 / 4.5	21	1.9	8	7	120
12.	Carbowax Compound 4000 / *Lanolin (Anhydrous)	5.3 / 4.5	40	2.7	8	7	120
13.	Carbowax Compound 4000	10.0	40	2.7	9	9	120
14.	Carbowax Compound 4000 / Borax	3.6 / 1.0	32.6	1.3	3	3	120
15.	White Mineral Oil / Borax	3.6 / 1.0	32.6	1.3	3	3	120

* Add 0.1 to 0.2 lb. terpineol to these lanolin formulas.

In a hot water bath or steam-jacketed kettle heat to 70°C. the stearic acid and the wax-like or water insoluble ingredients such as the lanolin, mineral oil, cetyl alcohol, or Carbowax compound 4000. When lanolin is used in the cream, add 0.1 pound of terpineol to the hot mixture of stearic acid and lanolin.

Heat the water to 70°C. in a separate container and add the triethanolamine and also the borax when it is included in the formula.

While constantly stirring, add the melted fatty acid solution to the amine solution. A paddle-type stirring device is suggested to prevent aeration of the cream.

When a smooth mixture is obtained, stir in the Carbitol and propylene glycol to which the perfume has been added.

Continue to stir constantly while the emulsion is cooling until a heavy, smooth cream is obtained and then stir occasionally until cooled to room temperature. The cream will be quite stiff at first, becoming thinner as the stearic acid crystallizes on cooling.

The speed of stirring has an important effect upon the body of the cream. During the cooling, and as soon as a stiff, smooth emulsion is obtained, stirring speed should be reduced until just sufficient to prevent crusting on top of the emulsion. Rapid stirring, after this stage has been reached, will usually cause aeration and yield a thin cream.

The grade of stearic acid used has some effect upon the consistency of a vanishing cream and, if the cream is hard and waxy, more water can be used to produce a desirable consistency. If the creams produced are thinner than desired, 0.5 to 2 lb. of borax can be added with the amine to increase the body. When borax is added, the stearic acid content is usually decreased by 6 to 10 lb.

Cetyl alcohol can replace part of the stearic acid in any of the above formulas. Sweet almond oil can be used in place of part or all of either the lanolin or mineral oil.

Finishing Cream
Triple Pressed

Stearic Acid	500
Potassium Hydroxide (C.P.)	30
Glycerol (C.P.)	200
White Oil (Medium Viscosity)	200
Water	2800
Rose, Lily or Other Perfume Oil	40

Melt stearic acid in an enamelled double boiler. Dissolve the potassium hydroxide in about 50 g. of water and add this to the stearic acid with stirring. Then add the glycerol dissolved in the balance of the water. Add the mineral oil. Turn off the heat and allow to cool. Stir in the perfume. Allow to stand 18 hours before placing in jars.

Camphor Peppermint Cream

Finishing Cream (Without Perfume)	3730
Camphor Peppermint Oil	35

Camphor peppermint oil is prepared by saturating oil of peppermint with camphor.

All Purpose Cream
Formula No. 1

(A)	Paraffin Wax	13.6
	White Petrolatum	3.5
	Mineral Oil	13.6
	Diglycol Stearate S	13.6
(B)	Water	53.6
	Triethanolamine	1.6
	Moldex	0.1
(C)	Perfume	0.4

Heat (A) to 70°C. and stir until complete solution is obtained. Heat (B) to the same temperature and add (A) to (B) stirring continuously while cooling. Add (C) at about 55°C. Stir and pour at 45° to 46°C.

	No. 2	No. 3
Absorption Base	4.0	15
Soft Paraffin Wax	7.7	40
Lanolin	4.0	7
Liquid Paraffin	1.5	3
Glycerin	—	8
Water	20.0	70

Sports Cream

Formula	No. 1	No. 2
Absorption Base	24.0	24.0
Cetyl Alcohol	1.0	1.2
Liquid Paraffin	5.0	10.5
Lanolin	3.0	—
Glycerin	3.0	3.0
Water	64.0	56.3
Hard Paraffin Wax	—	5.0

"Nourishing" Cream

Beeswax	6.0
Paraffin Wax	12.5
Petrolatum	8.0
Mineral Oil	58.0
Water	15.0
Potassium Carbonate	0.3
Borax	0.2

The fats and oils are melted at 65 to 70°C. The liquid solution of potassium carbonate and borax, previously brought to the same temperature, is then added under constant stirring. If lecithin, vitamins, etc., are also used, they and the perfume oil are dissolved in a small amount of mineral oil and added to the cream after the temperature has dropped to about 35°C.

Cleansing Creams

Cleansing or massage creams contain a fairly high content of mineral oil which should be of a highly refined grade. This mineral oil dissolves grease and suspends dirt particles so that they may be removed readily from the skin by a cloth or absorbent paper. Carbitol and propylene glycol assist in coupling the oil, lanolin and soap into the water, and thus improve the cleansing action of the cream. Carbitol and propylene glycol also maintain the physical properties of the cream during storage and enhance the emollient action of the preparation upon the skin. The high percentage of triethanolamine in this type of cream serves, with the fatty acid, to emulsify completely the mineral oil, aids in its penetration into the pores to remove dirt, and forms a cream that is readily removed with water. It is important that the proportion of triethanolamine be correct. A deficiency of the amine is indicated by a thin emulsion that is usually less stable and may not be readily removed from the skin by washing with water. An excess of the amine may result in a granular cream which tends to separate on cooling.

Melt the stearic acid in the mineral oil; add the lanolin, where required, and about 0.15 parts of terpineol; and bring the temperature of the mixture to 70°C.

In a separate container, add the triethanolamine to the water and bring this solution to 70°C.

Add the hot oil mixture to the heated amine solution and stir vigorously to obtain a uniform emulsion.

Mix the perfume with the Carbitol or propylene glycol, and add this solution to the emulsion. However, for the preparation of formula No. 3 using ethanol, continue with even, but less rapid, stirring until the temperature has reached 48°C. before adding the ethanol-perfume solution. The viscosity of formula No. 3 depends largely upon the temperature of the emulsion when the ethanol is added. If the temperature is higher than 52°C. or lower than 45°C., the emulsion will be almost fluid instead of a desirable cream.

Continue with even stirring until a viscous, smooth cream is obtained, and then stir occasionally

Ingredients	Formula No. 1	Formula No. 2	Formula No. 3
White Mineral Oil	50	100	70
Lanolin (Anhydrous)*	8*	14*	..
Stearic Acid	29	20	20
Triethanolamine	3.8	2.0	8.0
Carbitol or Propylene Glycol	10	10	..
Ethanol (Ethyl Alcohol)	..	..	8.0
Water	100	54	100

* Add 0.1 to 0.2 parts of terpineol to formulas containing lanolin.

until the cream has cooled to room temperature. Too rapid stirring causes undesirable aeration of the cream.

Various vegetable oils, such as sweet almond and olive oils, can be used in this type of cream in place of the lanolin or in place of part or all of the mineral oil. When making such substitutions, it may be necessary to lower the proportion of stearic acid to produce a cream of desired consistency.

The water content of the cleansing creams can be increased or decreased slightly to change the consistency of the cream as desired.

No. 4

Lanolin (Anhydrous)	34.00
White Mineral Oil	57.00
Stearic Acid	25.00
Quince Seed Mucilage	19.00
Carbitol	25.00
Propylene Glycol	50.00
Triethanolamine	9.5
Terpineol	0.35
Water	315.00

Melt the stearic acid in the mineral oil, add the lanolin and terpineol, and bring the temperature of the solution to 70°C.

In a separate container, bring the solution of the triethanolamine and water to 70°C.

Add the hot oil mixture to the heated amine solution, stir vigorously until a good emulsion is formed.

Add the Carbitol and the quince seed mucilage. The quince seed mucilage is made by adding 9.5 oz. of quince seed to 20 lb. of water at 80°C., soaking overnight, and then straining through a cloth. A suitable preservative should be added to the quince seed mucilage to prevent its molding over a period of time. The mucilage can then be stored for use as needed.

Mix the perfume in the propylene glycol and stir this solution into the cream when it has cooled to about 50°C. The stirring should be fast enough to keep the cream well mixed but not to aerate it. The stirring should be continued at low speed until the emulsion has cooled to room temperature. If the cream is allowed to cool without stirring, it will thicken upon standing a few days.

The mineral oil can be replaced in its entirety or in part with a vegetable oil, such as olive or sweet almond oil. The lanolin content can be decreased slightly where these oils replace some of the mineral oil.

Cellosize hydroxyethyl cellulose WS has been found to be an excellent thickening and stabilizing agent for liquid creams. It requires no preservative or special preparation. A dispersion of karaya gum or sodium alginate may also be used in place of the quince seed mucilage. A dispersion of desirable consistency, and with the slippery feel of the quince seed mucilage, can be prepared by stirring 0.5 lb. of sodium alginate into 50 lb. of hot water containing 1 lb. of triethanolamine. The alginate is added slowly, with rapid stirring, until a smooth dispersion is obtained. A preservative should be added to the dispersion. Karaya gum can be dispersed in a similar manner, but the dispersion is thinner, so that less water should be used with this gum.

No. 5
U. S. Patent 2,361,756

White Wax	5
Spermaceti	5
Stearyl Alcohol	5
White Mineral Oil	40
Sulfated Hydrogenated	
Castor Oil	10
Triethanolamine	1
Water	40

No. 6

Absorption Base	15.0
Lanolin	5.0
Liquid Paraffin	45.0
Water	45.0

Foundation Creams
Powder Creams; Liquid Make-Up

Glyceryl Monostearate greaseless creams are especially well suited for use as tinted foundation and make-up bases, since they do not have the high degree of oiliness associated with cold creams or petrolatum and are free from the tendency to roll on the skin sometimes shown by stearate creams. Since relatively large amounts of pigment must be included for covering and tinting purposes, the cream itself should be relatively soft. Red, yellow, and brown oxides are most satisfactory for tinting, being free from any tendency to bleed. Titanium dioxide provides covering or hiding power; zinc oxide, because of its ability to react with soaps and destroy the emulsion, should not be used. The amount of titanium dioxide used will depend upon the hiding power desired, more rather than less being desirable; the proportion of colors necessary to attain a desired shade is entirely dependent upon the amount of titanium dioxide used, more colors being needed to give a certain depth of tone as the amount of titanium dioxide is increased.

The foundation cream can be manufactured by first preparing the cream base, adding the white and colored pigments while the cream is still soft and then passing the mixture through a colloid mill. Better color dispersion (freedom from color streaks) is generally obtained, however, by passing the solidified cream, containing pigment, through a roller mill.

Formula No. 1

Glyceryl Monostearate S	20.0
Spermaceti	5.0
Glycerin	5.0
Moldex	0.1
Water	66.9
Pigments	3.0

Melt the Glyceryl Monostearate S and spermaceti at 160°F. Add the glycerin, Moldex and water previously heated to 160°F. and stir until cool. Add the pigment and pass through a roller mill.

No. 2

A. Lanolin Alcohols	3.5
Cocoa Butter	5.2
Lanolin (U.S.P.)	5.5
Cetyl Alcohol	12.0
Beeswax	2.0
Spermaceti	10.0
Glyceryl Monostearate S	32.0
B. Water	225.0
Moldex (Preservative)	0.75
Perfume	To suit

Melt A at 70°C.
Heat B to 70°C.
Add A to B and stir until the mixture is uniform.

Powder Cream

Absorption Base	73.5
Soft Paraffin Wax	49.0
Ozokerite	10.0
Liquid Paraffin	7.5
Water	30.0
Powder	80.0

The powder is ground into the emulsion after the latter has been prepared.

Tropical Powder Cream
Mineral Oil	50
Kaolin	30
Zinc Oxide	20

Liquid Foundation Make-Up
Sesame Oil	64.0
Zinc Oxide	11.0
Titanium Oxide	16.0
Oxidized Cholesterol	2.0
Glyceryl Monostearate	1.0
Perfume and Color	6.0
Moldex (Preservative)	0.1

Theatrical Grease Paint
Stearic Acid	12.25
Diglycol Stearate	6.12
Caustic Potash	0.50
Glycerin	2.10
Distilled Water	13.34
Erythrosin or Tartrazine	0.12
Zinc Oxide	6.12
Lake Color	3.10
Perfume	0.75

Hand Creams

These are usually greaseless creams containing substantial amounts of skin softeners, one of the most effective of which is glycerin. The usual fatty emollients, which are not necessary in the presence of Glyceryl Monostearate, tend to make the cream too greasy on the hands. The use of a little titanium dioxide will leave a whiter deposit on the skin and give the illusion of bleaching.

Formula No. 1
Glyceryl Monostearate S	12.0
Cetyl Alcohol	1.0
Stearic Acid	5.0
Glycerin	12.5
Titanium Dioxide	1.0
Water	68.0
Moldex	0.1
Perfume	0.4

This cream is made in the usual way, either with one or two vessels. It should be mentioned at this point that when only small amounts of titanium dioxide are used to whiten the cream, grinding is not necessary to obtain color dispersion; on the other hand, passage through a colloid mill is almost invariably beneficial as regards stability and texture of the finished product.

No. 2
Cetyl Alcohol	0.25
Avocado Oil	2.50
Oxycholesterol	0.30
Stearic Acid	1.50
Triethanolamine	0.50
Water	78.00
Glycerin	6.00
Alcohol	6.00

	No. 3	No. 4
Cholesterol	1.3	5.0
Soft Paraffin Wax	—	10.0
Stearic Acid	15.0	—
Liquid Petrolatum	—	60.0
Cetyl Alcohol	1.7	—
Peanut Oil	5.0	—
Caustic Potash	0.8	—
Glycerin	16.0	4.0
Water	60.2	80.0
Hard Paraffin Wax	—	24.0

No. 5
Absorption Base	22.5
Soft Paraffin Wax	15.0
Ozokerite	3.5
Liquid Paraffin	5.0
Lanolin	4.0
Water	50.0
Zinc Stearate	0.5
Zinc Oxide	5.0
Color	To suit

No. 6
Stearic Acid (Triple Pressed)	22.00

Methyl p-Hydroxy Benzoate	0.05
Propyl p-Hydroxy Benzoate	0.05
Triethanolamine	1.80
Glycerin	7.50
Propylene Glycol	7.50
Distilled Water	40.60
Veegum	20.00
Perfume	0.50

Weigh the stearic acid into a suitable container. Heat to 75°–80°C., add the weighed preservatives and stir until dissolved.

In a separate vessel measure out the water, Veegum, and triethanolamine, and heat to 75°–80° C.

Add the Veegum mixture to the stearic acid mixture with slow speed stirring and continue stirring until the temperature falls to 50°–55°C. and add the perfume. Stir very slowly till cold and either fill into jars cold or rewarm, stir and fill at about 35°C.

No. 7

A. Stearic Acid (Triple Pressed)	14.00
Light Mineral Oil	1.70
Oleic Acid	1.80
Propyl p-Hydroxy Benzoate	0.05
Veegum	25.00
Water (Distilled)	44.45
B. Potassium Hydroxide Solution (35%)*	1.80
Glycerin	10.00
Perfume	0.50

Heat the stearic acid, light mineral oil, and oleic acid to 75°–80°C. in a suitable container. Add the preservative and stir until dissolved.

35% Potassium Hydroxide Solution
Potassium Hydroxide (86%)	100 g.
Distilled Water to make	250 g.

Measure the glycerin, water, and Veegum into a separate container, heat to 75°–80°C., add the potassium hydroxide solution and immediately add to A with slow speed stirring.

Continue with slow speed until the temperature drops to 50°–55°C., add the perfume and stir until cold. The cream may be reheated to 60°C. and poured hot, or warmed to 35°C. and filled while warm.

Face Lotion
Formula No. 1

Isopropyl Rubbing Alcohol (70%)	3
Synthetic Camphor (U.S.P.)	1/6
Water	3

The camphor is first crumbled or crushed in a bowl, and dissolved in the 70% isopropyl alcohol which is then diluted by adding the prescribed amount of water.

No. 2

Cholesterol	1.3
Stearyl Alcohol	2.0
Potassium Stearate	3.0
Water	193.7

Smooth Cosmetic Mucilage

Finely Milled Tragacanth Powder	6
Alcohol (95%)	6
Glycerin	4
Water	984

The tragacanth powder is triturated with the alcohol and glycerin and the water quickly added. By the addition of a few drops of 33% potash solution the acetic acid inherent in the tragacanth is neutralized whereupon the mildly alkaline mucilage is heated gently and constantly stirred until absolutely smooth and clear. To prevent spoilage a suitable preservative should be added.

Beauty Milk
Cholesterol	1.3
Cetyl Alcohol	0.5
Liquid Petrolatum	10.0
Stearic Acid	5.0
Triethanolamine	2.5
Water	100.0

Skin Cleaner and Softener
A.	Water	80
	Glyceryl Monostearate S	3
	Spermaceti	15
B.	Beeswax	3

Mix A into another vessel. Melt B. Pour the fatty mixture (melted) into the water mixture heated near to its boiling point and stir until a paste has formed.

Hand Lotions
Formula No. 1
White Mineral Oil	100
Glycerin	60
Span #85	3
Tween #85	3
Karaya Gum Solution (1%)	160

No. 2
Cetyl Alcohol	0.50
Stearic Acid (Triple Pressed)	2.30
Mineral Oil	0.90
Methyl p-Hydroxy Benzoate	0.015
Propyl p-Hydroxy Benzoate	0.015
Veegum	4.60
Triethanolamine	0.23
Water	91.44
Perfume	To suit

Mix the first five ingredients and heat to 70°–80°C. Separately mix the Veegum, triethanolamine, and water and heat to 70°–80°C. Slowly pour the fatty phase into the water phase with stirring. Continue stirring until cool. The perfume can be added at about 50°C.

No. 3
Lanolin	0.50
Cetyl Alcohol	0.50
Stearic Acid (Triple Pressed)	2.30
Mineral Oil	0.50
Methyl p-Hydroxy Benzoate	0.01
Propyl p-Hydroxy Benzoate	0.01
Veegum	4.70
Triethanolamine	0.23
Water	91.25
Perfume	To suit

No. 4
Lanolin	6.0
Carbowax Compound 1500	30.0
Stearic Acid	18.0
Quince Seed Mucilage	20.0
Carbitol	30.0
Propylene Glycol	20.0
Triethanolamine	1.0
Potassium Hydroxide	0.5
Water	480.0
Terpineol	0.1

Melt the stearic acid, lanolin, and the Carbowax compound 1500 and heat to 70°C.

In a separate container, heat the solution of the triethanolamine and the water to 70°C.

Add the hot wax mixture to the heated triethanolamine solution and stir vigorously until a smooth emulsion is formed.

Add the Carbitol and the quince

seed mucilage and stir slowly and continously.

When the temperature reaches 50°C., add the propylene glycol to which the perfume has been added. *Stirring should be continued at slow speed until the lotion has completely cooled,* if the desired stability and consistency are to be obtained. Avoid too rapid stirring, which will cause aeration and foaming.

Numerous changes can be made in the ratio of the components of this lotion to vary its fluidity and absorbability. In general, its stability is not seriously affected by moderate alterations in the suggested formula.

Glycerin Hand Jelly

A.	Sodium Pectate	1.00
	Boiling Water	35.00
	Glycerin	50.00
	Sodium Benzoate	0.10
B.	Calcium Citrate	0.15
	Water	10.00
C.	Alcohol	5.00
	Perfume	Sufficient

Add B to A, stirring continuously add C. This sets to a firm, semi-transparent jelly.

Sun Protection Cosmetics
Formula No. 1

A.	Sunscreen No. 52	
	(Menthyl Salicylate)	7.0
	Alcohol	60.0
	Glycerin	10.0
	Perfume	0.5
B.	Water	22.5

Weigh the ingredients of A into a container and stir until a clear solution results. Upon obtaining a clear solution, add the required amount of water, stirring until the addition of the water is complete.

No. 2
Emulsified Lotion Type

A.	Diethylene Glycol Monostearate	2.0
	Stearic Acid	1.5
	Cetyl Alcohol Pure	0.5
	Sunscreen No. 2	
	(Menthyl Salicylate)	4.0
	Benzyl Alcohol	0.6
	Triethanolamine	1.0
B.	Water	89.9
C.	Perfume	0.5

Weigh the ingredients of A into container No. 1 and heat to about 80°C. Weigh the water of B into container No. 2 and heat to 80°C. When both temperatures are alike, add B to A, stirring at a medium rate of speed. Continue to stir the lotion until cold. Add C when the temperature is reduced to about 45°C.

No. 3

Butyl Benzal Acetone Oxalate	2.0
Sesame Oil	10.0
Tannic Acid	1.0
Alcohol	86.0
Perfume	1.0
Hydroquinone	0.2

The physical light screens block the passage of all light rays and, therefore, prevent both sunburn and tanning. A cream or makeup containing 15–20% of zinc oxide, or titanium oxide, or calamine, will protect the skin from all sunlight. Those who suffer from constitutional or skin disorders in the treatment of which sunlight is contraindicated, should use physical light screens on the affected and exposed parts of the body.

Sun Tan Lotion

Carboxymethyl Cellulose	10
Isopropanol	20
Tween No. 20	½
Carbowax 1500	5
Phenyl Salicylate	2
Menthyl Salicylate	1
Titanium Dioxide	13
Mineral Black	½
Magnesium Oxide	1
Bentonite	3
Petrolatum	2
Water	42
Perfume	To suit

A water solution of the carboxymethyl cellulose is made with a high speed stirrer and to this is added Tween, Carbowax, bentonite, magnesium oxide, mineral black, titanium dioxide, petrolatum and isopropanol in the order named. Perfume is then added to suit.

Sun Tan Lotion Base

Mineral Oil	35
Peanut Oil	10
Diglycol Laurate S	5
Cetyl Alcohol	½
Water	48

Warm and mix until uniform. Cool while stirring and add perfume and color.

Sun Tan Oils
Formula No. 1

Sun tan oils are made by dissolving menthyl salicylate in a concentration of 4%, either in mineral oil or in a mixture of mineral oil and vegetable oil. This product, of course, then has to be perfumed.

No. 2

Menthyl Salicylate	10.0
Sesame Oil	45.0
White Mineral Oil	44.0
Hydroquinone	0.2
Perfume	1.0

Sunscreen Cream
Formula No. 1

A. Stearic Acid	20.0
Cetyl Alcohol Pure	0.5
Sunscreen No. 2	4.0
B. Ammonium Hydroxide (26° Bé.)	1.0
Sodium Hydroxide	0.4
Glycerin	10.0
Water	63.8
C. Perfume	0.3

Weigh the ingredients of A into container No. 1 and heat to about 80°C. Weigh the ingredients of B into container No. 2 and heat to about 80°C. When both temperatures are alike, add B to A stirring at a medium rate of speed. Continue stirring the cream until cold. Add C when the temperature is reduced to about 45°C. The concentration of perfume can vary.

Sun Tan Cream

Menthyl Anthranilate	5.0
Sesame Oil	15.0
Cholesterol	2.0
Cold Cream	39.0
Vanishing Cream	39.0

Sun Protective Cream
Formula No. 1

Phenyl Salicylate	5.0
Ethyl Aminobenzoate	2.0
Titanium Dioxide	1.0
Neocalamine	1.0
Yellow Ferric Oxide	0.1
Coumarin	0.1
White Wax	2.0
Triethanolamine	0.5
Stearyl Alcohol	8.0
Stearic Acid	2.0
Glycerin	10.0
Distilled Water to make	100.0

Place the triethanolamine and the stearic acid in a 250-cc. beaker and heat together on a water bath for 10 minutes. Add the white wax and the stearyl alcohol and melt completely. Add the phenyl salicylate, ethyl aminobenzoate and the coumarin. After all the ingredients have completely melted, add the titanium dioxide, neocalamine and yellow ferric oxide in a fine state of division and mix well. Heat the distilled water and glycerin together to about 70°C and then add to the other ingredients with constant stirring. Continue the stirring until the emulsion forms and has an ointmentlike consistency.

No. 2
U. S. Patent 2,376,884

A. Stearic Acid 20.00
   Cetyl Alcohol 2.00
   Hydroquinone 5.00
   Laevo-Ascorbic Acid ... 0.25
B. Borax 1.0
   Sodium Carbonate 2.0
   Glycerin 6.0
   Water .... To make 110.0

Mixture A is melted, and then emulsified with a heated solution of mixture B. By stirring until cold, a fine vanishing cream is obtained.

Sunburn Preventive Ointment

	Formula		
	No. 1	No. 2	No. 3
Homomenthyl Salicylate	10	20	20
Ethyl p-Aminobenzoate	..	..	..
Titanium Dioxide	10	10	10
Magnesium Stearate	10	10	5
Butyl Stearate	..	..	5
Mineral Oil	10	5	..
Diethylphthalate	10	5	..
Sodium Lauryl Sulfate	0.25	0.3	..
Bentonite Ointment[a]	47.75	47.7	..
Pigment Base[b]	2.0	2.0	2
Vanishing Cream[c]	..	..	..
Methyl Cellulose[d]	..	..	58

[a] Bentonite 15%, distilled water 85%.
[b] Made up of burnt umber 50%, red ferric oxide 25%, yellow ferric oxide 25%.
[c] Made up of stearic acid 15%, cetyl alcohol 1%, triethanolamine, 1.5%, Carbitol 10%, distilled water 72.5%.
[d] Methyl cellulose in water 5%.

These ointments can be prepared by simple admixture, and no special sequence of procedures is necessary. For small quantities it is most convenient to use a spatula and glass slab.

The bentonite ointment base is prepared by mixing the bentonite with the water in a mortar. When all the water has been taken up the lumpy mass is transferred to a glass slab and rubbed smooth with a spatula. It is almost impossible to obtain a smooth mixture in the mortar alone.

The magnesium stearate-butyl stearate mixture is prepared on a slab. Any amount of this mixture can be prepared at one time and used as a stock mixture. It saves considerable time in compounding.

The vanishing cream base is prepared in the usual manner. The stearic acid and cetyl alcohol are mixed in one dish and melted on a water bath; to this is added the triethanolamine, Carbitol, and water mixture, heated to the same temperature, and the mixture is stirred until emulsified.

In each and all of the cases the sunscreens can be added by simple admixture at any time. Homomenthyl salicylate is a viscous fluid and can be added to base or pigment. Ethyl p-aminobenzoate may be powdered finely before incorporating with pigments or base.

No. 4

An excellent ointment for sunburn may be prepared by gently heating and mixing:

Peanut Oil	60 g
Spermaceti	8 g
White Beeswax	7 g

Then add:

Calcium Acetate	5 g
Distilled Water	20 g

Cheap Sunburn Ointment

Red petrolatum (veterinary's grade) works excellently, as tested by the U. S. Army Air forces. It does not wash off readily.

Sun Tan Compact Powder

Kaolin	20.0
Zinc Oxide	20.0
Talc	49.0
Magnesium Stearate	7.0
Cetyl Alcohol	1.0
Magnesium Carbonate	2.0
Perfume Compound	0.2
Brown 19289 ⎫ equal	
Yellow 103794 ⎬ parts	0.8
Orange 101943 ⎭	

Leg Make-Up
Liquid Stocking Formula No. 1

Mineral Oil	1.0
Glyceryl Monostearate S	6.0
Water	48.0
Glycerin	15.0
Titanium Dioxide and Earth Pigments	21.0
Talc	9.0

Heat the mineral oil and Glyceryl Monostearate S to 160°F., add the water and glycerin previously heated to 160°F., then stir until cool and mill in the titanium dioxide and earth pigments, and the talc.

No. 2

Glyceryl Monostearate S	0.4
Water	78.1
Glycerin	0.5
Titanium Dioxide and Earth Pigments	20.0
Methyl Cellulose	1.0

Dissolve the glycerin in the water and make a 5% dispersion of the Glyceryl Monostearate S and glycerin in hot water, stirring until cool. Soak the methyl cellulose in the balance of water cold until dissolved. Add this to the Glyceryl Monostearate dispersion and mill in the pigments.

No. 3

Propylene Glycol	1.03
Iron Oxide (Cosmetic Grade)	5.12
Umber	.51
Pure Titanium Dioxide	3.59
Zinc Stearate	1.58
Water	61.51
*Veegum	20.51
Isopropyl Alcohol	6.15
Perfume and Preservative	To suit

The isopropyl alcohol and an equal weight of water are added to the propylene glycol. This mixture is slowly stirred into the dry blend of zinc stearate, titanium dioxide, umber, and iron oxides, until a smooth paste is obtained. Now add the water and then the Veegum. Perfume as desired and use a water soluble preservative. The product is then passed through a homogenizer or colloid mill to effect complete dispersion of the pigments, being careful to avoid entrainment of air in the product.

The viscosity of these prepara-

---

* Magnesium aluminum silicate gel.

tions may be altered by increasing or decreasing the amount of Veegum at the expense of the water. If pigments other than the above are used, the consistency should be adjusted accordingly.

No. 4
Paste Leg Make-Up

Propylene Glycol	1.03
Iron Oxide (Cosmetic Grade)	5.12
Umber	.51
Pure Titanium Dioxide	3.59
Zinc Stearate	1.58
Veegum	82.02
Isopropyl Alcohol	6.15
Perfume and Preservative	To suit

How to Apply Leg Make-Up

The paste leg make-up applies without streaking if the leg is first moistened with water. When the make-up has dried, buffing with a soft cloth will bring out a silky sheen and remove any excess of pigment.

Lip Preparations
Lipsticks

One of the principal requirements of a good lipstick is that the ingredients used, including the dyes or pigments, be absolutely harmless and must not irritate the skin. The lipstick must apply easily and should produce neither a dull and lifeless nor a greasy effect. The lipstick itself should not be so brittle as to break off on application nor should it have the tendency to sweat, shrink, or show other external effects upon storage or use. The color used must be thoroughly distributed and should not be affected by climatic changes such as humidity and temperature.

The leading lipsticks sold show a melting point of between 53 and 63°C. Naturally any lipsticks destined for export to tropical countries must not have too low a melting point. The melting point is of course adjusted by suitable additions of carnauba wax or ozokerite. The lipstick base is made up of the usual oils, fats, and waxes customarily used in cosmetics, cetyl alcohol, butyl stearate, diglycol stearate, triethanolamine stearate and diethylene glycol. The exclusive use of hydrocarbons such as paraffin, mineral oils and ceresin as originally used is avoided in modern formulations. Among the most important ingredients should be mentioned beeswax which is particularly valuable in making so-called indelible lipstick. Paraffin and a small quantity of mineral oil produce a highlight effect, but should be used in moderation in order to avoid any possible greasy appearance. Castor oil may well be replaced by butyl stearate or diethylene glycol, while the presence of cetyl alcohol will generally improve the satiny texture of the lipstick itself. Cocoa butter is a very valuable ingredient but also should not be present in too large a quantity lest it produces a cracking effect upon application.

Formula No. 1

White Beeswax	20.0
Glyco Wax A	15.0
Lanolin	14.0
Ozokerite	8.0
Cetyl Alcohol	3.0
White Petrolatum	8.0

Heavy Paraffin Oil	10.0
Castor Oil	4.0
Butyl Stearate	4.0
Preservative (Moldex)	0.2

No. 2

Lanolin	17.0
White Beeswax	25.0
Spermaceti	8.0
Carnauba Wax	2.5
Cocoa Butter	9.0
White Petrolatum	10.0
Heavy Paraffin Oil	9.0
Castor Oil	6.0
Preservative (Moldex)	0.2

Years ago alkannin and carmine were the principal, if not the only colors, that were used in lipsticks. They were later substituted by water-soluble and still later by oil-soluble colors, followed by indelible pigments and finally by so-called changeable and indelible special lipstick colors. The stability of the color effect on the lips is largely dependent on the colors used. Insoluble pigments are comparatively easily removed without much rubbing, especially in a soft lipstick. Oil-soluble colors alone have on the other hand insufficient coloring and covering properties. Water-soluble colors are of course not soluble in the base. It is however possible to triturate them in the oil and to produce a stable suspension in the molten mass immediately prior to pouring into the mold. To enliven or subdue certain shades of color, zinc oxide or preferably titanium dioxide is used. All other insoluble fillers such as talc, kaolin, chalk, etc., have the disadvantage of rendering the lipstick less indelible.

Greaseless lipsticks are usually made from glycerin boric acid with a suitable addition of water-soluble aniline dye. These types of lipsticks are hygroscopic for which reason they should be carefully packed. They are characterized by their transparent appearance. When they are used, the lips are first moistened and then the stick applied, with as many coats as desired until the desired color effect is achieved.

In selecting the perfume attention must be paid to its covering power with particular regard to lanolin, cocoa butter and castor oil. Among the basic aromatics usually present in such compositions might be mentioned heliotropin, methyl ionone, geraniol, phenylethyl alcohol and rhodinol, as well as linalool, the musks, neroli, ethyl vanillin and coumarin. Sometimes certain flavor effects such as strawberry, peach and pineapple are successfully incorporated.

Liquid Lip Stain

Triethanolamine Oleate	15.0
Glycerin	10.0
Water	10.0
Beeswax	10.0
White Mineral Oil	40.0
Oil-Soluble Scarlet	2.5
Perfume	To suit

Anti-Chap Lipsticks

Anti-chap lipsticks are formulated from a suitable base with the addition of a substance or substances that are soothing and healing to the skin. Usually it is preferable to have a base containing lanolin and cacao butter. Such a formula would be as follows:

Formula No. 1

Beeswax	40

Castor Oil	30
Cacao Butter	20
Anhydrous Lanolin	5
Mineral Oil (210/220)	5

No. 2

Beeswax	32
Castor Oil	33
Sesame Oil	22
Anhydrous Lanolin	13

No. 3

Beeswax	10
Anhydrous Lanolin	10
Mineral Oil (210/220)	43
Ceresin	37

No. 4

Mineral Oil (210/220)	55
Ceresin	45

The anti-chap lipstick base should contain some soothing substance. Eucalyptol, menthol, camphor or lavender used alone or in combination may help to heal a chapped condition. Benzocaine (U.S.P. XII) has an analgesic effect when applied locally and may be used in ointments in small amounts. It is believed that the addition of vitamin A and D to local applications may help to heal traumatic tissue.

Hair Preparations

Hair Tonics or Lotions

There are almost as many types of hair tonics as there are hair and and scalp conditions requiring treatment. For falling hair there is one group of preparations, for oily hair there is another, and still another type of hair tonic is designed to promote the growth and give lustre to the hair.

The latter type is probably the most popular and may contain such ingredients as glycerin, alcohol, castor oil, sulfonated castor oil, deodorized mineral oil, diethyl phthalate, potassium sulfate, resorcinol, formaldehyde, formic acid, sodium lauryl sulfonate, saponine, etc. As solvents, water, alcohol, glycerin, mineral oil and chloroform may be used. Excessive irritation caused by some solvents can be counteracted or reduced by the addition of diethyl phthalate. An emollient effect is achieved by sulfonated castor oil while the formation of dandruff is checked by potassium sulfate. Quinine is often present in French formulations especially designed for the prevention of dandruff. A germicidal effect is imparted by benzoic acid, esters of benzoic acid, carbolic acid, resorcinol, etc., while saponine and soap provide foaming qualities. Sodium lauryl sulfonate and sodium cetyl sulfonate, soap, sulfonated oils, etc., act as detergents. Following are a few suggested formulations:

Formula No. 1

Chloral Hydrate	2.0
Deodorized Castor Oil	18.0
Alcohol	79.5
Perfume Compound	0.5

A preparation of this type would be especially suitable for the treatment of dry hair.

No. 2

Diethyl Phthalate	4.0
Alcohol	60.0
Formaldehyde	0.2
Water	35.0
Perfumed Compound	0.8

A mixture of this type would be especially suitable for the treatment of oily hair.

As a suitable base often incorporated in more recent formulations, cholesterin, lecithin, vitamins (Pro-vitamin A, Vitamin B_1), hormones, and possibly even sulfur come into question. Difficulty is sometimes experienced in obtaining a clear solution when using cholesterin of which an addition of 0.3 to 0.5% may often be made. Such difficulties may be prevented by an addition of glycerin in combination with certain essential oils and fatty oils while the solution will be hastened by warming the alcohol slightly. A formulation for cholesterin-containing hair tonic would be as follows:

No. 3

Cholesterin	3
Glycerin	27
Isopropyl Alcohol	200
Alcohol	710
Perfume Compound	15
Water	45

No. 4

Cholesterin	5
Tincture of Capsicum	10
Lecithin	10
Alcohol	900
Perfume Compound	10
Water	65

Sometimes emulsions, possibly also containing cholesterin, are preferred and the following suggestions might provide some ideas for further experimentation:

No. 5

Cholesterin	1.0
Lanolin	1.0
Peanut Oil	1.0
Cocoa Butter	1.0
Glycerin	5.0
Sodium Choleate	0.5
Borax (0.5% solution)	100.0

The cholesterin is dissolved in a mixture of the lanolin, peanut oil, and cocoa butter which has been kept at a fairly high temperature. The glycerin and sodium choleate are heated and the two solutions combined while hot, and during constant stirring the borax solution previously brought to the same temperature is then added. The mixture must be well agitated until cool.

Further formulations for hair tonics containing skin-regenerating media such as cholesterin and lecithin are as follows:

No. 6

Cholesterin	1.0
Lecithin	0.3
Alcohol	200.0
Castor Oil	3.0

The ingredients are dissolved while heating gently, then allowed to cool and the following mixture added:

Alcohol	60.0
Distilled Water	35.0
Quinosol	0.5
Salicylic Acid	0.2

Isopropyl alcohol is a better solvent for cholesterin than ethyl alcohol for which reason the former is often used preferably but it is advisable to mask its odor by the addition of a suitable masking agent. The following formulation is suggested:

No. 7

Alcohol (95%)	150.0
Isopropyl Alcohol	10.0
Carbon Tetrachloride	6.0
Glycerin	5.0
Castor Oil	4.0
Cholesterin	0.5
Choline or Egg Lecithin	0.3
Distilled Water	22.0
Perfume Compound	2.0

Among others, vitamins A, B_1, and D, as well as certain hormones, gynodermin, androdermin and hormodermin, have been mentioned but inasmuch as these hormones and vitamins are usually available only in an oil base they would be more suitable for the manufacture of emulsions because if they were used in the manufacture of regular tonics the mixtures would have to be shaken thoroughly before use. If used, the quantity of hormones would be about 6%.

Experimentation has also been carried out concerning the use and effectiveness of radium-containing substances such as radium bromide which would have a stimulating effect on the hair follicles but their use naturally would result in a more or less clinical type of hair tonic and further experimentation and tests would be necessary before general use in hair tonics.

A tonic for oily hair should have an emulsifying and fat-dissolving quality possibly combined with a lathering effect. The foaming quality is usually achieved by the addition of saponine or alkalies.

Sulfur-containing hair tonics should be made with especially dispersed sulfur in the preparation.

No. 8

Powdered sulfur 25 is heated with butyl glycol 100 until the sulfur is almost completely dissolved. This solution is allowed to cool, then added to the alcohol which will cause a partial precipitation of the sulfur while the balance remains in solution in the alcohol. In this method preparations are made which will contain sulfur either in complete solution or in semi-colloidal suspension. The sulfur content should be about 0.05%.

A hair tonic containing sulfur but very low in alcohol content can be made as follows:

No. 9

Colloidal Sulfur (24% Solution in Glycerin)	10
Resorcinol	4
Neutral Turkey Red Oil	2
Alcohol	5
Water	79

The resorcinol is first dissolved in the water to which the mixture of Turkey red oil and alcohol is added, finally incorporating the sulfur-glycerin solution.

Hair preparations with low alcohol content have the disadvantage of not being readily absorbed and showing the tendency to run off too easily. This can be remedied by suitable additions of fatty materials such as 0.1% sulfonated lauryl alcohol or from 0.5 to 0.8% cocoa butter soap which has the added advantage of rendering the hair pliable and glossy.

Small additions of menthol will add a cooling effect to any hair tonic but should not exceed approximately 0.2%. The use of lactic acid (0.2%) will tend to inhibit the secretions of the scalp and reduce the itching it causes.

The selection of the perfume compound should be studied carefully, not only for the odor but also for the suitability of the various essential oils used. Certain oils provide not only an agreeable fragrance but also possess a germicidal effect so that they offer a double contribution to the finished preparation. Oil of thyme, lavender, cloves, bay, cinnamon, estragon and basilicum fall into this classification. The terpeneless oils are frequently used in view of their greater solubility.

The importance of the perfume compound used cannot be overemphasized as many an otherwise excellent hair preparation goes unused and un-sold due to an unpleasant basic odor which could be overcome by expert perfuming.

No. 10

Cholesterol	1.0
Lecithin	0.5
Vitamin F (So-called)	3.0
Benzyl Alcohol	0.4
Amyl Salicylate	0.8
Jasmin Compound	0.1
Carnation Compound	0.2
Ethyl Acetate	0.5
Beta-Naphthol	0.1
Alcohol	To make 100.0

No. 11

Sulfur-Cholesterol	0.8
Lecithin	0.5
Oxyquinoline Sulfate	0.2
Peru Balsam	0.8
Salicylic Acid	0.2
Castor Oil	1.0
Alcohol	96.5

No. 12

Salicylic Acid	4.0
Spirits of Camphor	10.0
Castor Oil	2.5
Alcohol (95%)	To make 200.0

No. 13

Pilocarpine Nitrate	0.02
Quinine Salicylate	0.46
Tannic Acid	0.22
Castor Oil	0.20
Water	24.10
Rosewater	25.00
Alcohol (95%)	50.00

No. 14

Cinchona Tincture	7
Cantharides Tincture	1
Eau de Cologne	10
Glycerin	4
Clary Sage Water	2
Rose Water	3
Orange Flower Water	6
Alcohol (95%)	67

Practically all hair lotions require to be mixed, matured for a while, chilled and filtered.

	No. 15	No. 16
Liquid Petrolatum	4.5	—
Cholesterol	0.5	0.5
Sodium Choleate	0.5	—
Glycerin	5.0	4.0
Borax	0.5	—
Water	89.0	4.5
Alcohol (95%)	—	91.0

No. 17

Glyceryl Monostearate	12.0
Cholesterol	1.5
Lanolin	1.5
Lecithin	1.0
Triethanolamine Stearate	8.0
Castor Oil	3.0
Water	73.0

No. 18

Sulfanilamide	60 gr.
Fluid Extract of Thyme	6 dr.
Glycerin	6 dr.
Bay Rum	8 oz.
Water	To make 16 oz.

Dissolve the sulfanilamide in boiling water and then add the fluidextract, glycerin and bay rum. The lotion should be applied once or twice daily, rubbing well into the scalp. The preparation is particularly effective with young persons.

Anti-Dandruff Lotion
Formula No. 1

Euresol	0.5
Refined Sulfur Oil	1.5
Peru Balsam	0.5
Specially Denatured Alcohol	70.0
Distilled Water	27.5

No. 2

Resorcinol or Euresol	2.0
Acetic Acid	1.0
Cologne Compound	1.0
Tincture of Capsicum	5.0
Castor Oil	0.1
Specially Denatured Alcohol	To make 100.0

No. 3

Menthol	0.1
Tea-Tree Oil	0.2
Lavender Oil	0.5
Castor Oil	1.0
Salicylic Acid	3.0
Oxyquinoline Sulfate	0.2
Rosemary Oil	0.3
Specially Denatured Alcohol	94.7

Balsamic Hair Lotion

Resorcinol or Euresol	2.0
Benzoin Resinoid	0.4
Peru Balsam Oil	1.0
Tolu Balsam Oil	1.5
Distilled Water	5.0
Glycerin	2.1
Alcohol	88.0

Anti-Dandruff Ointment
Formula No. 1

Precipitated Sulphur	3.3
Salicylic Acid	1.0
White Wax	6.7
Hydrous Wool Fat	33.3
Petroleum Jelly	To make 100.0

No. 2

Resorcinol	4.0
Petroleum Jelly	To make 100.0

Dandruff-Removing Lotion
Formula No. 1

Resorcinol	6.0
Glycerin	10.0
Alcohol	9.6
Water	To make 100.0

It should be noted that although resorcinol has no appreciable effect on dark hair, this drug will stain blond, grey or white hair. Euresol (acetyl-resorcinol), is less apt to discolor the hair, and may be used in the place of resorcinol in equal amounts. Euresol is soluble in alcohol and in acetone, but not in water. It is readily miscible with the various ointment vehicles. A much-used Euresol combination mentioned contains:

No. 2

Euresol	3.3
Alcohol (95%)	56.0
Distilled Water	To make 100.0

No. 3

Tincture of Cantharides	1
Dilute Acetic Acid	1
Spirit of Rosemary	2
Glycerin	1
Rosewater	To make 16

No. 4

Cade Oil	2.0
Thymol	1.0
Green Soap	33.0
Denatured Alcohol	To make 100.0

This shampoo should be rubbed well into the scalp without any previous mixing with water or wetting of the hair, left for 5 minutes, and then rinsed off.

Cold Permanent Wave Lotion
Formula No. 1
For strong hair

Thioglycollic Aid	8%
or Ammonium Thioglycollate calculated as the thioglycollic acid	8%
Ammonia	To bring the pH to 9.5

Make up a 20% solution of thioglycollic acid or ammonium thioglycollate, add concentrated ammonium hydroxide to pH 9.5, then adjust the final concentration with water to bring the thioglycollate strength to 8% as thioglycollic acid adding small amounts of ammonia to keep the pH up to 9.5. It is essential that the water used be free from iron and copper, distilled or deionized water is recommended.

No. 2
For normal hair

Ammonium Thioglycollate calculated as the thioglycollic acid	6.8%
Ammonia	To bring the pH to 9.2

Process same as above.

No. 3
For dyed or bleached hair

Ammonium Thioglycollate calculated as the thioglycollic acid	4.5%
Ammonia	To bring the pH to 9.0

Process same as above.

Neutralizer for Cold Wave Lotions

Potassium Bromate	5 g.
Citric Acid	5 g.
Distilled Water	1 qt.

No. 4

The following solution is applied to the hair after it has been wound upon rods:

Thioglycollic Acid	10 g.
Sodium Hydroxide	7 g.
Hydroxylamine Sulfate	10 g.
Water	100 cc.

At the end of fifteen minutes, the hair is treated with the following solution:

Water	500 cc.
Hydrogen Peroxide (20 volume)	500 cc.
Tartaric Acid	30 g.

The hair is left for five minutes and grows warmer than body temperature and is then washed after it has cooled. The resulting curls are shiny, soft and permanent.

If standard extract of witch hazel is substituted for the water, the slight odor of thioglycollic acid almost disappears and the pH is lowered, making it useful for home use.

No. 5
U.S. Patent 2,389,755

Water	500
Ammonia	35
Ammonium Sulfocyanide	45
Ammonium Sulfite	70
Thioglycollic Acid	20
Carbitol	4

Allow to stand overnight and then add water 4700–6000.

No. 6
U.S. Patent 2,389,755

Water	75.5
Ammonium Hydroxide	5.0
Ammonium Thiocyanate	7.0
Ammonium Sulfite	11.0
Monochloracetic Acid	1.0
Diethyleneglycol Monoethyl Ether	0.5

Permanent Wave Cream

A. Diglycol Stearate S	9.4
Lanolin	1.7
Wetanol (Wetting Agent)	0.1
Water	37.3
B. Ammonia (28° Bé)	7.0
Ammonium Carbonate	4.7
Sodium Sulfide	9.5
Water	30.3

Melt the Diglycol Stearate S and lanolin together to 160°F. Dissolve the Wetanol in the water, heat to 160°F. and add to the oil phase and stir until cool.

Dissolve the ammonium carbonate in the water allowing to stand overnight if necessary for complete solution. Add the ammonia and then add the complete mixture cold with agitation to A, and stir until homogeneous.

Hair Setting (Waving) Mix
U.S. Patent 2,383,990

Zein	20
Sugar	3
Aerosol (Wetting Agent)	2
Water	20
Alcohol	55
Perfume	To suit

Chemical Heating Composition for Hair Waving
U.S. Patent 2,350,926

Mannitol (Crystallized)	15
Potassium Permanganate	45
Talc	40

In certain cases sucrose or dextrose or both can be added to such compositions, mainly for the purpose of extending the duration of the reaction.

At the time of use a suitable absorbent sheet may be moistened with about 4 cc. of water and applied to the previous envelope; the assembly then being draped and clamped around a pre-formed tress.

Curly Hair Straightener
Formula No. 1

White Wax	300 g.
Heavy Liquid Petrolatum	700 g.
Perfume	To suit

No. 2

Wool Fat	250 g.
Petrolatum	750 g.
Perfume and Color	To suit

If these preparations are not sufficiently heavy, it is suggested that a small amount of wax or rosin be added.

No. 3
U.S. Patent 2,390,073

The natural curl or lack of curl is changed by chemically altering the keratin present so that wetting with water will not restore the natural condition. Kinky hair is straightened by application of a solution containing cresolsulfonic acid 7.5, isopropylnaphthalenesulfonic acid 7.5, formaldehyde 10, and water 75%. After standing for 30 min. the hair is heated and combed with a pressing comb, while the triethanolamine oleate is applied as lubricant and heat conductor. Straight hair is curled when an alkaline reacting mixture, e.g., triethanolamine 4% to 10% and formaldehyde 4% to 10%, are used.

Hair Cream
Hair Conditioning Cream
Formula No. 1

Borax	4
Sulfonated Oil	1
Precipitated Sulfur	4
Camphor	½
Beeswax	6
Absorption Base	54
Distilled Water	30
Perfume *	To suit

Perfume

Bay Oil	10
Pimento Oil	5
Bergamot Oil	8
Ti-Tree Oil	½

No. 2

Borax	1
Monostearin	5
Glyceryl Monostearate	10
Beeswax	3
Liquid Paraffin Oil	60
Glycerin	5
Distilled Water	30
Cantharidin	0.001%
Rosemary Oil	½
Eau de Cologne Essence	To suit

No. 3

Stearamide	4.0
Soap Chips	3.5
Mineral Oil	24.5
Beeswax	1.5
Water	66.5

The constituents are heated together, stirred until cool, and then preferably homogenized. Apart from the additional stability that stearamide, judiciously incorporated, imparts, it also serves to facilitate subsequent shampooing, owing to its ability to re-emulsify mineral oil residues if still present in sufficiently reasonable amounts.

No. 4

A. Sodium Pectate		1.25
Sodium Benzoate		0.10
Boiling Water		50.00
B. Calcium Citrate		0.10
Water		45.00
C. Alcohol		5.00
Perfume		Sufficient

This hair cream is nearly solid when cold, but on shaking becomes readily pourable.

	No. 5	No. 6
Absorption Base	4	12.0
Stearyl Alcohol	1	—
Beeswax	—	6.0
Lanolin	—	0.5
Liquid Paraffin	50	130.0
Water	90	60.0

Glycerin	—	3.0
Triethanolamine	—	0.5
Isopropanolamine	0.5	—

No. 7

Glyceryl Monostearate	12.0
Wool Wax Cholesterol	1.5
Lanolin	1.5
Lecithin	1.0
Triethanolamine Stearate	8.0
Castor Oil	3.0
Water	73.0

No. 8

Wool Wax Cholesterol	0.4
Linseed Oil Fatty Acid	1.0
Lanolin	2.0
Lanette Wax SX	9.0
Lecithin	0.2
Avocado Oil	3.5
Cetyl Alcohol	3.0
Lactic Acid	1.0
Water	78.5

Hair Pomade

Carnauba Wax	4
Paraffin Wax	15
Yellow Petrolatum	45
White Petrolatum	35
Perfume	1

Liquid Hair Dressing

Castor Oil	1.5 g.
Potassium Soap of Castor Oil	2.0 g.
Isopropyl Alcohol	60 cc.
Water	40 cc.
Perfume	To suit

Brilliantine
Formula No. 1
Cream

Absorption Base	530
Water	300
Triethanolamine	8

The absorption base is melted and to it is added the triethanolamine diluted by water; the mixture is then stirred. This yields a semi-fluid cream, in spite of the small amount of water used. This emulsion is formed by small globules of oils crowded together without intervening spaces, and surrounded by a film of water which prevents them from coalescing. Once applied, the water evaporates, the emulsion breaks down and the hair is both brilliantined and fixed.

Liquid
No. 2

Viscous Mineral Oil	400.0
Regular Mineral Oil	700.0
Perfume	3.0
Ultramarine Blue Liquid V (Mineral Oil-Soluble)	0.5

No. 3

Viscous Mineral Oil	400
Light Mineral Oil	500
Soluble Castor Oil	100
Perfume and Color	To suit

Solid
No. 4

White, Long-Fibered Petroleum Jelly	950
White Ceresin	50
Color and Perfume	To suit

Hair Bleaching Paste

Hydrogen Peroxide (20 vol.)	3
Diglycol Laurate S	1

Shampoo

The following shampoo concentrates are liquid soaps of a clear, reddish color. They contain a slight excess of amine, which improves both the lathering and the rinsing properties of the soaps. They should be diluted with water to the desired consistency or concentration. A solution of 1 part of either formula with 3 parts by weight of water makes an excellent shampoo.

Shampoo formula No. 2 will produce a better lather and will rinse more readily than formula No. 1 because of the combination of amine soaps as well as the presence of the Tergitol wetting agent. If this wetting agent is added to formula No. 1, it will produce the same effect in this shampoo also.

	Formula No. 1	No. 2
Coconut Oil Fatty Acids	42	42
Oleic Acid	56	56
Carbitol	40	40
Propylene Glycol	15	15
Triethanolamine	58	28.5
Monoethanolamine	—	12.6
Tergitol Wetting Agent 7	—	10

Formulas No. 1 and 2 are concentrates, and should be diluted, as directed, with water before use.

The proportions of fatty acids in these formulas are based upon an equivalent weight of 210 for coconut oil fatty acids and 282 for oleic acid. A good grade of oleic acid should be used in the shampoos and it will usually have an equivalent weight of about 282. However, the equivalent weight of coconut oil fatty acids may vary considerably.

To obtain a completely neutral soap, the fatty acids and the amines should be analyzed and the above proportions altered as indicated by the analyses. If terpineol is used to mask the odor of the soap in the shampoo, less perfume will be required to produce the same strength of fragrance.

Mix the fatty acids and Carbitol, and add the amines.

Stir until a clear solution is obtained, and add the Tergitol wetting agent 7. No heating is required.

Dilute with water to any desired consistency. When the water is first added, the soap becomes of a petrolatum-like consistency, which gradually dissolves to a clear, water-thin solution of a pale amber color. If the water solution is cloudy, stir in more amine, a little at a time, until it becomes clear.

Mixed isopropanolamine may be used to replace triethanolamine in these formulas to produce greater color stability.

Coconut Oil Shampoo
Formula No. 1

Coconut Potash Soap	35.0
Olive Oil Soft Soap	10.0
Glycerin	5.0
Alcohol	10.0
Perfume	0.2
Distilled Water	40.0

No. 2

Coconut Oil	14.0
Olive Oil	3.0
Castor Oil	3.0
Caustic Potash	4.7
Glycerin	2.0
Industrial Alcohol	4.0
Calgon	1.0
Perfume	0.4
Distilled Water	68.0

In formula No. 1, the soap is dissolved in half the given quantity of hot water, the rest of the water being added cold. The glycerin and alcohol, to which the perfume has been added, are then stirred in. The resultant liquid soap is stored for 3 weeks, and the clear liquid drawn off from any sediment. It is filtered bright through diatomaceous earth.

In formula No. 2, in which the soap itself is prepared, the potash is dissolved in 9 parts of the water and allowed to stand for a few hours. The clear lye is then run slowly into a steam-jacketed pan containing the previously melted oils. The lye must be added carefully and the rate of addition controlled to prevent excessive foaming. If the batch threatens to overflow, cold soft water is sprayed over the mass. After stirring until reaction appears to be complete, the pan is covered and allowed to stand for 1 hour.

The soap should now be tested for incomplete saponification or excess alkalinity. Dissolve about 2 g. of soap sample in 6 g. of distilled water with warming. Turbidity shows the presence of unsaponified fat, in which case further caustic potash should be added. To test for free alkali, add 2 drops of a 1% phenolphthalein solution to a sample solution. If a decided red color results, add coconut or castor-oil fatty acids.

More accurately, the test sample is dissolved in neutral alcohol and titrated to neutrality with 0.1 normal potassium hydroxide or standard acid as required. If off appreciably from neutrality, calculate the amount of lye or oils required and add these to the kettle. Continue the boiling and repeat the analysis.

If nearly neutral on the alkaline side, and a clear appearance indicates little or no free oil, boiling may be continued without any adjustment. If a sample drawn later checks the previous one, the reaction has gone to completion.

After sampling and correction, the remaining portion of the charge is added slowly and gradually. If large quantities of water are added, lumps will form and float about in the thin soap solution. These lumps will disappear only after prolonged boiling and stirring. If possible, the speed of stirring should be gradually reduced as the soap gets thinner, down to about 20 rotations per minute.

The finished soap is allowed to cool in the kettle, when it should be drawn off and stored in vats. The longer it is kept, the clearer it becomes. Filtration subsequently is desirable.

No. 3
Coconut Oil Potash Soap

(45% Fatty Acids)	45
Water	52
Potassium Carbonate	1
Cologne	2

The cologne is made by mixing 1 part each of bergamot oil and lavender flower oil (40% esters) in 20 parts of alcohol.

No. 4

Coconut Oil	130
Potash Lye (28° Bé.)	135
Borax	2
Water	220
Glycerin, Alcohol (S.D. 3A) or Propylene Glycol.	50
Perfume	3

Adding 6 drams of oil of pine tar and 1 dram of oil of tar to each gallon of this preparation yields a tar shampoo.

Where a high proportion of alcohol is desired, the following rather simple formula is suggested:

No. 5

Coconut Oil	45	lb.
Potash Lye (34° Bé.)	30¾	lb.
Alcohol (S.D. 3A)	44	lb.
Perfume	10	oz.

A heavy shampoo, made without coconut oil, consists of:

No. 6

Olive Oil	88	lb.
Corn Oil	22	lb.
Alcohol	92	lb.
Glycerin	18	lb.
Potash Lye (39° Bé.)	33	lb.

To make these shampoos, first saponify the oils with the potash lye and add the borax dissolved in the water. Continue mixing, while hot, until the soap is in solution. If alcohol or sugar are added do not make the addition until the temperature drops to around 115°F. The perfume should also be added at this point as well as any color. When the solution is complete and the charge is cool, it is pumped to storage tanks for clarification.

Transparent Shampoo Jelly

Coconut Oil (Low Free Fatty Acid Content)	256
Potash Lye (28° Bé.)	274
Glycerin	25
Alcohol (S.D. 3A)	15
Perfume	3

Golden Shampoo

Toilet Soap Powder	25	lb.
Borax	22	lb.
Henna Powdered	3	lb.
Aubepine Powder	1	oz.

Silver Shampoo

Toilet Soap Powder	28	lb.
Borax	22	lb.
Aubepine Powder	1	oz.

Cream Shampoo
Formula No. 1

Glyceryl Monostearate S	10
Polyglycol 400 Monostearate	20

Lanolin	10
Duponol ME	40
Sodium Sulfate	5
Water	106
Perfume	2

No. 2

Mineral Oil	24
Liquid Coconut Oil Soap (40%)	68
Diglycol Laurate Synthetic	8
Perfume	To suit

No. 3
Shampoo Cream

White Mineral Oil	38
Diglycol Laurate S	10
Liquid Soap (33–35%)	52
Perfume	To suit

No. 4

Veegum	60
Foaming Agent (Duponol)	30
Soap	8
Sodium Chloride	2

Dissolve the soap in the Veegum, then add the remaining ingredients and stir until uniform. Allow the paste to stand overnight, then homogenize to make smooth. No heat is required for this formulation.

Soapless Shampoo

Sulfonated Castor Oil (75%)	60
Sulfonated Olive Oil (75%)	20
Mineral Oil (Light)	3
Distilled Water	14
Glycerin	3½
Perfume	½

Mix the perfume with the glycerin and add the mixture to the rest of the ingredients which have already been blended together.

Dandruff-Removing Shampoo

Olive Oil Soap	3.25
Eucalyptus Oil	1.00
Water	35.75

Bath Preparations

Bubble Bath Powder
Formula No. 1

Saponin	15.0
Sodium Lauryl Sulfate	82.0
Perfume	3.0

No. 2
U. S. Patent 2,382,732

Crystallized Aluminum Sulfate	1,718 g.
Sodium Bicarbonate	1,300 g.
Aluminum Oxide (Finely Powdered)	195 g.
Saponin	30 g.

The mixture is used in the proportion of 373 g. to 30 gal. of warm water.

Depilatories

Depilatory Cream
Formula No. 1

Distilled Water	51.8
Duponol Paste (28%)	1.8
Stearyl Alcohol (Stenol)	8.0
Precipitated Chalk	12.6
Magnesium Hydroxide	7.4
Calcium Hydroxide	5.1
Methocel (300 Cp.) (5% Solution)	10.2
Thioglycollic Acid (100%)	2.6
Perfume	0.5

Add the thioglycollic acid to 18 parts of the water, heat to 65°C. Add the calcium and magnesium hydroxide. Stir for 15 minutes at 70°C. Cool to 30°C and add the Methocel and about one half of the chalk. Heat 33.8 parts of the water and the Duponol and Stenol to 70°C. and agitate. Cool to 60°C. and add the balance of the chalk. Mix the two mixtures with constant stirring while cooling to 40°C. Add the perfume at 40°C. Cool to 30°C. and pass through a colloid mill. Avoid introduction of air. Tube immediately.

No. 2

Wheat Starch	15
Strontium Sulfate	60
Pure Zinc Oxide	8
Lithium Carbonate	8
Menthol	2

Leave a layer of paste 2 mm. thick on the skin for 20 minutes, remove with a solution consisting of perfume, alcohol and borax.

No. 3

Strontium Sulfate	60.0
Wheat Starch	15.0
Pure Zinc Oxide	15.0
Lithium Carbonate	8.0
Menthol	2.0

A layer of the paste 1 to 2 mm. thick is left on the skin for from 10 to 20 minutes and is then removed with an aqueous solution containing borax, alcohol and perfume.

No. 4
U. S. Patent 2,352,524

Strontium Hydrate	50.0 g.
Calcium Oxide	12.0 g.
Colloidal Clay	102.0 g.
Methyl Cellulose	11.0 g.
Mercapto-Acetic Acid	12.0 cc.
Water	300.0 cc.
Perfume	0.8 cc.

Shaving Preparations

Brushless Shaving Cream

These are greaseless creams which keep the beard moist and support the individual hairs to permit easy shaving. In addition, the cream should be of minimum alkalinity to decrease sensitive skin irritation and be easily washed off the razor. A small amount of mineral oil should be present to improve lubrication. A wetting agent such as Wetanol added in small percentages increases spread and wetting characteristics.

Brushless Shaving Cream
Formula No. 1

Diglycol Stearate S	15.0
Cocoa Butter	4.0
Stearic Acid	5.0
Mineral Oil	1.0
Glycerin	5.0
Water	69.5
Moldex	0.1
Perfume	0.4

No. 2

Glyceryl Monostearate S	12.0
Spermaceti	4.5
Mineral Oil	2.5
Glycerin	7.0
Water	73.5
Moldex	0.1
Perfume	0.4

No. 3

Diglycol Stearate S	14.0
Stearic Acid	6.0
Mineral Oil	1.0
Glycerin	5.0
Water	73.5
Moldex	0.1
Perfume	0.4

The Diglycol Stearate S or Glyceryl Monostearate S is melted with the stearic acid and other waxes and oils at 75°C. while the glycerin, water and preservative are heated together to about the same temperature. The two solutions are mixed together and stirring continued until homogeneous. The perfume is added at about 50°C. when the cream is still soft and the temperature is low enough to prevent excessive volatilization.

No. 4

Oleic Acid	2
Triethanolamine	1¼
Stearic Acid	26½
White Mineral Oil	3
Water	66½
Perfume	½
Neutral Sodium Silicate	¼

Add the triethanolamine to the water and saponify the 2 parts of oleic acid at 95°C. Add the mineral oil and melted stearic acid at 85°C. and mix. Next add the silicate, followed by the addition of the perfume. Let stand to set-up.

No. 5

Stearic Acid	15.70
Cetyl Alcohol	0.30
Lanolin	0.40
Potassium Hydroxide	0.22
Sodium Hydroxide	0.02
Glycerin	10.00
Veegum	25.00
Propyl p-Hydroxy Benzoate	0.05
Distilled Water	48.31
Perfume	To suit

Heat the stearic acid, cetyl alcohol, lanolin, and propyl p-hydroxy benzoate to 75–80°C. Separately heat the potassium hydroxide, sodium hydroxide, glycerin, Veegum, and distilled water to 75–80°C. Add the melted oil phase to the aqueous phase with slow stirring, so that a minimum of air is entrapped. Continue slow-speed stirring until the temperature falls to 50°C., and add the perfume. Continue stirring to about 40°C. The cream may either be milled when cold or reheated to 60°C., and poured.

No. 6

Anhydrous Lanolin	6.5
White Mineral Oil	7.5
Stearic Acid (Triple-Pressed)	35.0
Carbitol	2.0
Propylene Glycol	2.0
Triethanolamine	2.2
Borax	2.2
Water	145.0
Terpineol	0.1

No. 7

Carbowax Compound 1500	45.0
Sodium Alginate	3.5
Stearic Acid (Triple-Pressed)	37.5
Carbitol	15.0
Propylene Glycol	12.0
Triethanolamine	3.0
Potassium Hydroxide	1.6
Water	180.0
Perfume	0.8

Method for No. 6

Melt the stearic acid with the lanolin and mineral oil, bring the temperature to 70°C., and add the terpineol.

In a separate container, heat the water containing the triethanolamine and borax to 70°C.

The hot stearic acid-mineral oil solution is then added to the hot amine solution with vigorous stirring. A paddle-type stirring device is suggested to prevent aeration and foaming of the emulsion.

Continue with stirring until a smooth emulsion is obtained and add the Carbitol.

When the emulsion has cooled to about 50°C., add the propylene glycol to which the perfume has been added.

The cream should not be permitted to cool without occasional stirring, as it becomes quite stiff and when re-stirred to package may become thin and grainy. The cream should be covered when not being stirred to prevent the formation of a hard top layer that will produce a grainy texture when it is stirred into the cream.

It is best to allow the cooled cream to stand overnight and to re-stir for about 1 minute before packaging into tubes or jars. If the cream is packaged while still warm, it may become thin after standing several days.

Method for No. 7

Melt the stearic acid and Carbowax compound 1500 together and bring the temperature to 70°C.

Dissolve the potassium hydroxide in an equal weight of water and add this solution and the triethanolamine to the hot mixture of stearic acid and Carbowax compound 1500. Stir until a clear solution is obtained.

Add the sodium alginate and stir well.

Add the water (70°C.) and stir until thoroughly incorporated and then add the propylene glycol and two-thirds of the Carbitol. When a smooth mixture is obtained, stir slowly but continuously until the cream has cooled to about 50°C.

Dissolve the perfume in one-third of the Carbitol and stir into the cream.

Pour into jars or tubes when the cream has cooled to about 45°C.; the cream reaches a desirable consistency after standing about 24 hours and a slight pearliness develops after several days. This cream has less body if it is stirred at intervals until cooled to room temperature before packaging.

A white, pearly product, somewhat like a vanishing cream, can be made by omitting part or all of the mineral oil from formula No. 6 and using about 1 lb. less of both triethanolamine and borax. This type of brushless shaving cream is sometimes preferred for use on oily skins. A cream with heavier body, for an extremely dry skin, can be made by increasing the amount of mineral oil to about 18 lb. and adding about 1 lb. more of both the amine and borax. The consistency of the creams can also be varied by altering the proportion of water.

Menthol can be added if a cooling effect is desired in the shaving cream by dissolving the desired amount in the propylene glycol.

Brushless Shaving Stick U. S. Patent 2,366,759	
Sesame Oil	38.75
Spermaceti	45.00
Stearin	7.50
Tallow Soap	1.50
Polyglycerol Monostearate	4.00
Titanium Dioxide	2.00
Perfume	1.25

Lather Shaving Cream	
Stearic Acid	15.70
Cetyl Alcohol	0.30
Lanolin	0.40
Potassium Hydroxide	0.22
Sodium Hydroxide	0.02
Glycerin	10.00
Veegum	25.00
Propyl p-Hydroxy Benzoate	0.05
Water	48.31
Perfume	To suit

Heat the stearic acid, cetyl alcohol, lanolin and the preservative to 75–80°C. Separately heat the potassium hydroxide, sodium hydroxide, glycerin, Veegum and water to 75–80°C. Add the melted oil phase to the aqueous phase with slow stirring, so that a minimum of air is entrapped. Continue slow speed stirring until the temperature falls to 50°C. when the cream may be either milled cold or reheated to 60°C. and poured.

Lather Shaving Cream Improver

Lathering creams of poor lathering power or lather stability can be improved by adding a foam or emulsion stabilizer.

Lather Type Shaving Cream	75–85
Foam Stabilizer*	25–15

Pre-Shave Cream

Absorption Base	30
Liquid Paraffin	5
Glycerin	2
Water	60

Nail Preparations

Nail Lacquer Remover
Clear

Gamma Valero Lactone	50
Water	50

Cream
Formula No. 1
Canadian Patent 426,997

Ethyl Acetate	40.0
N-butyl Acetate	40.0
Castor Oil	4.0
Perfume	0.1
Ethyl Cellulose	2.8
Stearic Acid	11.0
Ammonium Hydroxide	3.8

*Starch, gum arabic or glue with a suitable preservative, e.g., Moldex.

No. 2
U.S. Patent 2,351,195

	%
Acetone	75
Stearic Acid	10
Ethyl Cellulose	2– 4
Castor Oil	2–10
Ammonia (26° Bé.)	2– 4.6

Nail Polish Drier
U. S. Patent 2,366,260

Olive Oil	24
Castor Oil	1
Denatured Alcohol (70%)	2
Coloring	To suit

Cream for Brittle Nails

Triethanolamine	2.0
White Petrolatum	1.5
Beeswax	0.5
Anhydrous Lanolin	0.5
Water	15.0

This cream is to be applied at night, and during the day if possible.

Cosmetics for the Eyes
Mascara
Formula No. 1

Triethanolamine Stearate	30
Paraffin Wax (High Melting Point)	40
Beeswax	12
Lanolin	8
Lampblack	10

The method is to melt, mix and mill the ingredients and afterwards cast or extrude them into stick or tablet form.

This gives the usual black type of mascara. Where a dark brown mascara is desired, the following formula will serve.

No. 2

White Beeswax	300

Montan Wax	100
Stearic Acid	300
Triethanolamine	130
Lampblack	20
Burnt Umber	150

Melt the waxes and grind in the color in a warm mill. Stir in the ethanolamine and pour into molds.

Cake cosmetics are very easy to apply, so long as they are made with the proper oil-in-water emulsifying agents possessing appropriate wetting properties. They are therefore still the most popular representatives of their class. Even so, formulation is not without its problems. The product has to adhere to the eyelashes when the brush puts it there, it must dry so as not to stain the lower eyelid and, when dry, must not peel off and fall as dust. Soaps possess all these qualities, but to avoid irritation milder triethanolamine stearate is used.

Consistency, gloss and reduced solubility are imparted to the final product by means of substances with a higher melting point: wax, ceresin, etc. Resins and gums may also be added without danger of incompatibility.

No. 3

Stearic Acid	28
Triethanolamine	14
Water	90
Powdered Coloring Matter	25
Beeswax Or Ceresin	25–30

Mix thoroughly, run the mass through a 3-roller mill, then pass it through a small plodder which will produce sticks of the required thickness. Cut lengthwise and allow to dry before stamping out, which can be done by using a bath cube press or an automatic soap stamping machine, duly adjusted and fitted with appropriate dies. The coloring matter may be lampblack or bone black, umber, burnt umber, raw sienna, etc., modified if desired with suitable cosmetic lakes.

More recently there have appeared a number of cream mascaras on the market. These have certain advantages over the cake but also certain disadvantages. It is true that the paste sold is ready for instant use, without the user having to add the necessary water but, at the same time, some people have ideas of their own as to what the consistency of the paste should be. In constitution these newer types are very much the same as the cake type, but with the addition of the requisite amount of water necessary for a paste of the proper consistency before application. A binder is unnecessary in the cream because there is enough soap present to form a strong, stable emulsion.

A reliable cream mascara which does not dry up in the tube and which stays on the eyelashes very well can be made according to the following formula:

No. 4

Mucilage of Quince Seed	350
Sugar Syrup	350
Gum Arabic	75
Ivory Black	225

The gum and the ivory black are ground up with the syrup (3 parts of sugar to 2 parts of water), then the mucilage, containing a preservative, like methyl parahydroxybenzoate, is added.

Oil base mascaras are not so

satisfactory. Black mascara preparations are made from 140 parts of good black, such as lampblack, which is finely ground in 210 parts of simple tincture of benzoin (20% strength). Then there is added 12.5 parts of gum lac dissolved in 630 parts of alcohol, 5 parts of castor oil, and 2.5 parts of rose perfume or like. Ordinary mascara in aqueous mixture can be made from 10 parts of lampblack, ivory black or drop black, mixed with 10 parts of powdered gum acacia and rubbed out thoroughly with 80 parts of rose water or orange flower water. Gum acacia may be replaced by 2 parts gum tragacanth rubbed down with 10 parts perfumed alcohol. Chinese or India ink are also used for this purpose.

Spirituous mascaras may contain either rosin or benzoin (approximately 2% rosin or 4% benzoin, on the total weight), together with 12 to 15% lampblack, 1 to 2% gum lac, 2% castor oil; and the balance industrial spirit or isopropyl alcohol.

No. 5

Trihydroxyethylamine Stearate	390
Beeswax	300
Carnauba Wax	50
Bone Black	250

Melt together, mix well and pass through a heated ointment mill and run into forms.

No. 6

Stearic Acid	270
Triethanolamine	120
Beeswax	280
Diethyleneglycol Distearate	20
Carnauba Wax	50
Bone Black	250

Warm and mix until uniform. Then pass through a heated ointment mill and fill into forms.

Cake mascara should easily come off onto a moistened brush, be applied easily to the lashes, and be water-resistant when on the lashes. Glyceryl Monostearate S is highly satisfactory as a base for products of this type. Triethanolamine stearate is added to increase water solubility (discharge to the brush) and waxes to provide water resistance. About 10% of pigment is generally required, and aniline colors must not be used.

No. 7

Glyceryl Monostearate S	50.0
Stearic Acid	20.0
Triethanolamine	10.0
Beeswax	10.0
Pigment	10.0

These ingredients are simply melted and stirred together. Grinding is usually not needed when finely ground cosmetic pigments are utilized.

For a black mascara use ivory black cosmetic grade. Burnt umber can be used for dark brown, and cosmetic ultramarine for blue. Green mascara can be made by using a cosmetic grade of hydrated chrome oxide, and purple is obtained by mixing cosmetic ultramarine with cosmetic carmine.

Eye Shadow

The ideal eye shadow is a product which is fairly hard at room and body temperature, but with enough lubricating material in it to allow for a thin film. Stickiness and drag must be avoided. The following are representative formulas for eye shadow:

Formula No. 1

Hydrogenated Oil	45.0
Mineral Oil	10.0
Beeswax	20.0
Paraffin Wax	5.0
Ozokerite	5.0
Color	15.0

No. 2

Short Fiber Petroleum Jelly (43–48°C.)	23.0
Ceresin (64°C.)	21.0
Spermaceti	3.0
Mineral Oil (65–75)	53.0

No. 3

White Beeswax	4.5
Spermaceti	9.5
Lanolin Absorption Base	13.0
White Soft Paraffin Wax	73.0

From 78 to 85% of such a base is used with 12 to 15% of zinc oxide (to bring out the color) and 2–5% of a suitable mineral or earth pigment or cosmetic lake, e.g., 3% middle green lake, plus 2% gold-bronze. The metallic powders (gold bronze and aluminum) impart a pleasantly scintillating effect, but the most important material of this type to include is the background brightening agent, zinc oxide or titanium dioxide.

Eye Antiseptic Ointment

Sulfanilamide	10.000
Sulfathiazole	10.000
Proflavine	0.025
Distilled Water	5.000
Anhydrous Lanolin	To make 100.000

Eye Drops

Sodium Sulfacetamide	10–60
Chlorocresol	0.025
Distilled Water	To make 100

Eyelid Scale or Crust Remover

Sodium Bicarbonate	5 gr.
Boric Acid	10 gr.
Solution of Epinephrine	1 min.
Distilled Water	To make 1 fl. oz.

Cosmetic Anti-Perspirants and Deodorants

Formula No. 1

Monostearin	170
Spermaceti	90
Glycerin	39
Distilled Water	380
Preservative (Moldex)	1
Aluminum Sulfate	150
Distilled Water	150
Titanium Dioxide	20
Duponol ME	5

No. 2

Glyceryl Monostearate	15
Spermaceti	5
Methenamin	3
Glycerin	8
Water	69

No. 3

Veegum	50.0
Glyceryl Monostearate S	14.0
Sodium Lauryl Sulfate	2.0
Titanium Dioxide (Pure)	2.0
Aluminum Sulfocarbolate	8.0
Aluminum Sulfate (Cryst.)	14.0
Buffer	10.0

Dissolve the buffer in Veegum, then make a dispersion of the first three ingredients by heating and stirring slowly until homogeneous. Blend the titanium dioxide, aluminum sulfocarbolate, and aluminum sulfate; stir this dry mixture into the warm dispersion above.

No. 4

Aluminum Sulfate	25
Orange Flower Water	75

The solution is used by dabbing on the skin or, diluted to 5% with water, for soaking the feet.

No. 5

Hexamine	3
Denatured Alcohol	40
Orange Flower Water	To make 100

Hexamine, in contact with acid secretions, is decomposed, liberating formaldehyde, which is antiseptic and astringent. The acid becomes neutralized, and in the neutral medium the hexamine is no longer decomposed; there is thus no danger of irritation from the use of formalin itself.

No. 6

Zinc Peroxide	10
Zinc Oxide	20
Soft Paraffin Wax	70

A non-aqueous base is necessary with peroxides, to prevent decomposition.

No. 9

Hexamine	5
Diglycol Stearate	20
Orange Flower Water	75

Melt the stearate and add to the water, both being at about 78°C. When emulsified, add the hexamine dissolved in about 15 parts of warm water. Stir till cold.

No. 8

Chloramine-T	3
Boric Acid	20
Talc	50
Prepared Chalk	To make 100

Powders are easy to make but do not effectively stop the production of perspiration. It is sometimes objected that these preparations, which diminish or stop secretion of sweat, are harmful, but the stoppage over such a small area has no effect on the general health.

No. 9
U. S. Patent 2,350,047

Aluminum Chloride	15
Glyceryl Monostearate (Acid Stable Grade)	15
Spermaceti	3
Beeswax	2
Magnesium Stearate	2½
Water	62½

No. 10
U. S. Patent 2,350,047

Aluminum Sulfate	18
Zinc Oxide Of Magnesium Oxide	3
Alcohol	10
Water	69

Deodorant Creams

Deodorant creams may be divided into two categories: those that are merely deodorant, and those that have anti-perspirant action in addition. The first type is customarily formulated around a mixture of benzoic acid and zinc oxide in which deodorant action is due to the antiseptic and preservative properties of the chemicals. They are presumed to prevent bacterial decomposition of perspiration and thus retard development of odor. The oxide and acid are usually used in a petrolatum or lard base, but a better and greaseless cream can be made with glyceryl monostearate.

Formula No. 1

Monostearin	13.5
Duponol C	1.5
Benzoic Acid	4.0
Zinc Oxide	10.0
Glycerin	5.0
Water	61.5
Perfume	0.5

All of the components, with the exception of the zinc oxide, benzoic acid, and perfume, are heated together to 85–90°C. until the glyceryl monostearate melts and stirring gives a homogeneous liquid emulsion. The mixture is allowed to cool to 50°C. at which point the perfume is stirred in. The cream is allowed to cool to room temperature, preferably by standing overnight and the benzoic acid and zinc oxide are ground into the cream by means of a roller mill.

A truly anti-perspirant cream must include an astringent aluminum salt, the sulfate being generally selected. Although this salt is less strongly acid than aluminum chloride, it is still substantially acid and will cause deterioration of fabrics. For this reason, a buffer to reduce acidity, such as urea or similar compounds, may well be included. Urea content should be about 50% of the aluminum sulfate. Manufacture of an anti-perspirant cream is somewhat difficult and directions must be rigorously followed.

No. 2

Monostearin	20.0
Duponol C	2.0
Ceresin	5.0
Titanium Dioxide	1.0
Moldex	0.1
Glycerin	8.0
Water	39.5
Perfume	0.4
Aluminum Sulfate	12.0
Water	12.0

Heat together to about 90°C. all of the ingredients with the exception of the perfume, the aluminum sulfate, and the water to dissolve the aluminum sulfate. Stir until homogeneous, allow to cool to 50°C., and add the perfume. The perfume should be specially selected so that it is stable in the presence of aluminum sulfate. Meanwhile heat the aluminum sulfate with an equal weight of water until it dissolves and filter through cloth. An iron-free grade of aluminum sulfate should be used to minimize discoloration. The cream and aluminum sulfate solution must be cooled to 40°C. (or preferably allowed to stand overnight at room temperature) and thoroughly mixed. The cream will not be stable if this temperature is exceeded.

No. 3

A.	Monostearin	5.0
	Stearic Acid	15.0
	Moldex	0.1
B.	Glycerin	8.0
	Duponol C	0.5
	Water	30.0
C.	Perfume	0.4
D.	Aluminum Sulfate	15.0
	Urea	8.0
	Water	18.0

Melt the ingredients of group A together at 80°C. Add the glycerin and Duponol to water at 85°C. Thoroughly mix A and B and continue stirring until the solids are dissolved and strain through cloth. Mix the cream with solution D at 40°C., not higher. The perfume is added at about 50°C. with stirring.

Liquid Body Deodorant
Formula No. 1

Zinc Sulfocarbolate	2.0
Urea	1.0
Nipagin M. Or Moldex	0.1
Distilled Water	91.9

No. 2

Hexamethylenetetramine	4 oz.
Isopropyl Alcohol	3 pt.
Water	To make 1 gal.
Perfume	⅓ fl. oz.

Face Powder and Compacts

Rouge Compact Powder
Formula No. 1

Kaolin	40.0
Titanium Dioxide	5.0
Talc	35.0
Magnesium Stearate	10.0
Magnesium Carbonate	5.0
Rice Starch	5.0
Orange Lake	4.0
Scarlet Lake	2.5

No. 2

Talc	36
Zinc Oxide	20
Kaolin	20
Zinc Stearate	5
Bromo Acid Lake	11
Petrolatum (Heavy)	3
Gum Tragacanth (1% Solution)	3
Perfume Oil	2

Rouge, Mandarin

Colloidal Kaolin	30
Magnesium Oxide (Heavy)	10
Precipitated Chalk	15
Zinc Oxide	20
Magnesium Stearate	10
Italian Talc	10

This formula is colored with eosine lake 3 parts, mimosa yellow lake 1, natural sienna 1; and perfumed with rhodinol, backed up with benzophenone, diphenyl oxide and musk ketone.

Tropical Face Powder

Starch	300–400
Talc	300
Kaolin	0–100
Zinc Oxide	200
Magnesium Stearate	50
Zinc Stearate	50
Color and perfume	To suit

Foot Preparations

Foot Powder
Formula No. 1

Fuller's Earth	2
Boric Acid (Powdered)	2

The formula is useful in combating athlete's foot infections and aids in removing or reducing perspiration of the feet.

No. 2

Paraformaldehyde	6
Boric Acid	10
Kaolin	25
Kieselguhr	To make 100

Foot Balm
Formula No. 1

Glyceryl Monostearate	20 g.
Glycerin	5 g.
Paraffin Oil	5 g.
Formaldehyde	15 cc.
Water	55 cc.

Melt the ingredients together and stir until cold. The solution of formaldehyde in the above may be replaced by a solution of any other *compatible* active ingredient.

A standard foot balm which is not too greasy may be made along the following lines:

No. 2

Zinc Oxide	5.0
Cetyl Alcohol	5.0
Lanolin Absorption Base	30.0
Petroleum Jelly	27.0
Water	20.0
Camphor	2.5
Thymol	0.3
Menthol	0.2
Alcohol	10.0

Melt together the absorption base, cetyl alcohol and petroleum jelly. Dissolve the camphor, thymol and menthol in the alcohol. Stir the water gradually into the base, then add alcohol, and finally grind in the zinc oxide through an ointment mill.

Dental Preparations

Dental Cleanser and Gum Hardener

Sodium Chloride	55
Sodium Bicarbonate	20
Calcium Carbonate	20
Corn Starch	5
Flavoring	To suit

Mix the flavoring with the corn starch, add to the other ingredients and pass through a sieve until uniform. Use in place of tooth powder.

Denture Cleaner
Formula No. 1

Trisodium Phosphate	120.0 g.
Cinnamon Oil	0.3 cc.
Liquor Amaranth	2.0 cc.

Dissolve a scant ¼ tsp. in ½ glass of water and use with a brush. Note: Cellulose acetate type dentures are decomposed by alkaline substances. Formaldehyde and phenol formaldehyde types withstand most chemical agents.

No. 2

Sodium Perborate	240 gr.
Sodium Chloride	480 gr.
Exsiccated Magnesium sulfate	30 gr.
Calcium Chloride	30 gr.
Anhydrous Sodium Carbonate	30 gr.
Methyl Salicylate	1 min.

Menthol	2 gr.
Peppermint Oil	12 min.

Add a small amount of kaolin or magnesia to the powder to prevent it from caking. To use, dissolve a small portion in water, and allow the dental plate to remain in the solution overnight.

Denture Adhesive

Boric Acid	5
Gum Tragacanth, Powdered	70
Gum Acacia, Powdered	25
Vanillin	¼

Tooth Cavity Filling
U.S. Patent 2,361,161

Zinc Oxide	4 dr.
Thymol Iodide	5 gr.
Creosote	10 drops
Clove Oil	12 drops
Lanolin	Enough to make a solid paste

Toothache Oil

Clove Oil	12
Camphor	6
Chloroform	6
Phenol	2
Menthol	2
Cinnamon Oil	3

Tooth Paste
Formula No. 1

In a tinned-copper, jacketed kettle heat a mixture of 20 lb. of propylene glycol and 51½ lb. of glucose (Sweetose type) to about 220°F. Dissolve 4 oz. of salt in 8¼ lb. of distilled water and then stir in 9 lb. redried starch. Stir well to wet the starch and add this mixture gradually to the hot glycol-glucose mix, continuing the agitation with a wooden paddle if it is so desired. Continue to heat at boiling temperature until the mass is transparent and smooth.

Make an Irish moss mucilage by adding 6 lb. of Irish moss, picked free from extraneous material, in 384 lb. of boiling hot distilled water. Stir well to extract the moss and permit the temperature to drop to about 100°F. Strain the mucilage through a muslin cloth and dissolve 12 oz. of benzoic acid therein. Bring the yield up to exactly 360 lb. with cold distilled water. This batch represents sufficient Irish moss mucilage for 5 batches of tooth paste.

Transfer 72 lb. of Irish moss mucilage into a suitable mixer. Add the glycol-glucose solution while it is still warm and sift in while agitating:

Neutral White Soap	13¾ lb.
Benzoic Acid	3¼ lb.
Saccharin	½ lb.

Stir until smooth and add the flavor. If menthol is used be sure to dissolve it in the flavoring oils before adding. About 4 lb. of flavor are used for a batch this size. Continue the mixing, then add gradually in 25-lb. lots, stirring between additions

Light Precipitated Chalk	115 lb.
Calcium Sulphate	8½ lb.

When this has all been mixed in, finally add 45 lb. magnesium hydroxide paste (hydro magma paste); mix until uniform after all the ingredients have been added. This takes about ½ hour. Let stand over night and then cool and fill.

It will be noted that the above formula contains no glycerin. If it is desired to use this, glycerin may be used in amount equal to the sum of the weight of the propylene glycol and glucose indicated. The methods of making are exactly the same.

A tooth paste with a high soap content and containing no water has the following composition:

No. 2

Neutral White Soap	96½ lb.
Light Precipitated Chalk	82 lb.
Alcohol Specially Denatured 31-A	88 lb.
Glycerin USP	80½ lb.
Benzoic Acid	35 oz.

First mix the alcohol, glycerin, benzoic acid and flavor. Charge into a mixer, preferably covered, the soap and the chalk and blend them by mixing. Add the alcohol solution gradually to this soap-chalk mixture and continue stirring until the mass is smooth. Cool the paste and fill as soon as possible.

No. 3

Veegum	26.42
Tragacanth, Powdered	1.00
Saccharin, Soluble	0.18
Sodium Benzoate	1.00
Glycerin	19.80
Superlight Chalk	43.85
Neutral White Soap	5.00
Light Mineral Oil	0.20
*Flavor	1.80
Calcium Sulfate Dihydrate	0.75

*Flavor	Parts by Volume
Peppermint Oil (Double Distilled)	70
Eucalyptol	17
Anethol	8
Methyl Salicylate	3
Cassia Oil (or Cinnamic Aldehyde)	2

Add the gum tragacanth, soluble saccharin and sodium benzoate to the glycerin. Mix well in a pony mixer and while mixing add the Veegum and one-half of the chalk. When homogeneous, add the soap and the flavor oil dissolved in the mineral oil. Continue mixing on the pony mixer until uniform and then add the remainder of the chalk and finally the calcium sulfate dihydrate. Mix well and mill finely with an ointment, roller, or colloid mill.

The amount of calcium sulfate dihydrate is very critical. If the calcium sulfate content in the above formula is increased from 0.75% to 1%, the resulting paste is much stiffer in consistency. This ingredient, therefore, provides a most satisfactory method of controlling the stiffness or viscosity of a paste of this type. In pastes containing no calcium sulfate the viscosity is most easily controlled by adjusting the chalk to the proper value.

No. 4
Calcium Phosphate	27
Calcium Carbonate (Precipitated)	27
Powdered Soap	3
Gum Tragacanth	½
Glycerin, C.P.	42
Peppermint Oil	½

No. 5
Precipitated calcium carbonate	42.02	g.
Soap (Powdered)	6.5	g.
Heavy White Mineral Oil	1.18	g.
Potato Starch	3.5	g.
Tragacanth (Powdered)	0.25	g.
Glycerin	23.08	g.
Saccharin	0.25	g.
Methyl Salicylate	1.5	g.
Eucalyptol	0.344	g.
Peppermint Oil	0.25	g.
Carmine Color	0.25	g.
Thymol	0.028	g.
Water	20.86	g.

Tooth Paste Flavors
Formula No. 1
Peppermint Oil	67
Thyme Oil	12
Methyl Salicylate	12
Anise Oil	9

No. 2
Peppermint Oil	60
Anise Oil	30
Clove Oil	10

No. 3
Peppermint Oil	67
Clove Oil	7
Cinnamon Bark Oil	7
Cassia Oil	5
Lemon Oil	10
Benzaldehyde	4

Tooth Powder
Formula No. 1
Dense Precipitated Chalk	160 lb.
Calcium Sulphate	20 lb.
Netural White Soap Powder	30 lb.
Saccharin	5 oz.
Flavor	3 lb.

No. 2
Dense Precipitated Chalk	44 lb.
Light Precipitated Chalk	20 lb.
Silica Air-Floated (000 grade)	25 lb.
Zinc Chloride	1 lb.
Castile Soap Powder	5 lb.
Borax	5 lb.
Saccharin (Soluble)	4 oz.
Flavor	1 lb.

No. 3

Willow Charcoal	100 lb.
Dense Precipitated Chalk	90 lb.
Neutral White Soap Powder	30 lb.
Saccharin	6 oz.
Flavor	4 lb.

No. 4

Light Precipitated Chalk	100
Dense Precipitated Chalk	100
Powdered Sugar	10
Sodium Perborate	25
Flavor	4

No. 5

Dense Precipitated Chalk	150 lb.
Magnesium Carbonate	25 lb.
Saccharin	5 oz.
Flavor	4 lb.
Calcium Peroxide	25 lb.

The procedure is first to pre-mix the flavor, color, sweetener and other ingredients present in small quantities into a portion of the chalk or other abrasive by rubbing the liquids in by hand and then through a sieve or preferably through a small pre-mixer equipped with a brush sieve. This mixture is then screened and blended in a larger mixer into which the rest of the powdered ingredients have been added. The agitator is run until the mass is uniform, when it is run into storage and filled by the common types of dry filling equipment.

Anti-Pyorrhea Tooth Powder
British Patent 572,352

Calcium Glycerophosphate	2
Salt	2
Animal Charcoal	1

Liquid Dentifrice
Formula No. 1

Glycerin	180
Crystal Sugar	160
Saccharin	1
Wetting Agent (Lauryl Sulfate)	50
Salt	40
Amaranth Red Color in Water (1%)	40
Distilled Water	466
Quince Seed Infusion in Water (5%)	500
Alcohol S.D. 38-B	540
Flavor	To suit

No. 2

Irish Moss Extract	1½
Alkyl Aryl Sulfate (Nacconol LAL)	8
Alcohol S.D. 38-B	75
Distilled Water	300
Salt	½
Amaranth Red Color in Water (1%)	5
Flavor	2

To make either of these preparations the Irish moss or quince seed is first dispersed in hot water and filtered into the mixing tank, then the wetting agent is dissolved in this dispersion followed by the sugar, salt and color. Lastly the alcohol, in which the flavor has been dissolved, and the glycerin are mixed in. This solution is permitted to stand for from 5 to 10 days for thorough clarification, after which it is filtered and fillled.

No. 3

Sodium Alginate	4.0 g.
Distilled Water	200.0 cc.

Allow this to stand overnight and add:

Saccharin, Soluble	0.25 g.
Distilled Water	20.00 cc.

To the above add:

Flavor	8.0 cc.
Alcohol	100.0 cc.
Glycerin	40.0 cc.

Then mix and add to:

Sodium Lauryl Sulfate	20.0 g.
Distilled Water	150.0 cc.

Finally add:

Amaranth Solution (5%)	1.0 cc.
Distilled Water	To make 500.0 cc.

Coloring and Perfuming
Use of Certified Colors

Liquids and gels are customarily colored with 0.0005% to 0.001% of dye.

Similar low concentrations of dyes will color greases and waxes.

Stains are made at various concentrations, about 1% to 2% solutions being used.

With four exceptions, all the water soluble certified dyes are in the application group known as acid dyes. They are sulfonates or salts of color acids. As a group they offer a full range of bright hues of good tinctorial power. They are reasonably resistant to light, heat and acid reagents. Individually, however, they may vary widely in one or more respects, as to fastness, solubility and degree of affinity. They possess direct affinity for amphoteric materials, proteins and fibres of animal origin, particularly if applied in warm, weakly acidic dye baths. Strangely enough, although having little direct affinity for cellulose, many of them are excellent stains. Although water soluble, many are greatly affected by temperature changes. Many of them are also soluble to a useful degree in alcohols, glycols, glycerin and glycol ethers. Generally they will be found to be insoluble in hydrocarbons, particularly in alipatic straight-chain derivatives. All of them tend to form complexes with the basic dyes, which are poorly soluble in water. Accordingly they are considered incompatible with basic dyes. It is often advantageous, however, to mix basic and acid dyes in other liquids, if the complex formed is soluble in the liquid used.

Although most of the certified acid dyes have soluble color acids (sulfonates) there are sixteen dyes that are not sulfonates and whose color acids tend to be insoluble or are only sparingly soluble in water. None of the dyes in this group should be used in strongly acid solutions. Rather it is the practice to use some sodium bicarbonate with them to guard against the effect of tramp acids. The following dyes are in this group:

Soluble Derivative	Corresponding Color Acid
D. & C. Red No. 22	D. & C. Red No. 21
D. & C. Red No. 23	D. & C. Red No. 21
D. & C. Red No. 25	D. & C. Red No. 24
D. & C. Red No. 26	D. & C. Red No. 24
D. & C. Red No. 28	D. & C. Red No. 27
D. & C. Red No. 29	
D. & C. Orange No. 6	D. & C. Orange No. 5
D. & C. Orange No. 7	D. & C. Orange No. 5
D. & C. Orange No. 9	D. & C. Orange No. 8
D. & C. Orange No. 11	D. & C. Orange No. 10
D. & C. Orange No. 12	D. & C. Orange No. 10
D. & C. Orange No. 13	D. & C. Orange No. 10
D. & C. Orange No. 14	
D. & C. Yellow No. 8	D. & C. Yellow No. 7
D. & C. Yellow No. 9	D. & C. Yellow No. 7
D. & C. Green No. 8	

It is interesting to note that when D. & C. Red No. 21 is dispersed in a lipstick and applied to the lips, the alkali in the blood tends to convert this yellowish red color acid into D. & C. Red No. 22. The eosine thus formed imparts its characteristic bluish pink hue to the mucous membrane.

The four water soluble, certified dyes that are not acid dyes but basic dyes are:

D. & C. Red No. 19
Ext. D. & C. Blue No. 1

D. & C. Red No. 20
Ext. D. & C. Blue No. 2

In contrast to the acid dyes, of which the color nucleus has a negative charge in water and is a color acid, these four dyes are the chlorides or acetate salts of color bases. Here the color nucleus ionizes with a positive charge, hence their name basic dyes. Because of their basic nature they tend to unite with acid mordants, notably with tannins. An outstanding feature is their greater resistance to reducing conditions than all the other listed dyes, with the possible exception of D. & C. Yellow No. 10. As the color nucleus is an anion they do not combine with positive iron, so are useful for coloring compounds containing iron salts. Actually for internal use, D. & C. Red No. 19 and D. & C. Red No. 20 are the only certified dyes at once permissible and suitable in drug preparations containing reduced iron, except that occasionally D. & C. Yellow No. 10 can be used. All four dyes are almost universal stains having affinity for a wide variety of synthetic and natural products. Many clays will absorb them to approximately 2%, which percentage can be increased by using tannins with them, as their tannates are insoluble. The water soluble derivatives of these dyes are ordinarily considered to be relatively fugitive to light, yet in some applications they have shown considerable resistance. They are very soluble in water, glycerin, glycol ethers and alcohols, although in most cases Ext. D. & C. Blue No. 2 is much less soluble than the others. They are insoluble in hydrocarbons. The addition of alkalies to their solutions in water tends to precipitate their sparingly soluble bases or basic complexes. The rhodamine base precipitated by alkalies from water solutions of D. & C. Red No. 19 and D. & C. Red No. 20, if filtered off and dried, can then be dissolved in melted stearic acid to form rhodamine stearate. Rhodamine stearate when certified is called D. & C. Red No. 37. Rhodamine base is soluble in glycerin, glycols, glycol ethers, alcohols and ketones, and imparts its characteristic color to these liquids. In ethyl ether rhodamine base dissolves to give a colorless solution. Although the ether solution is colorless, almost any absorbent material wet with it will appear colored after the ether evaporates. All four of these basic dyes are tinctorially among the most powerful colorants and are outstanding in the brilliance and beauty of their hues.

D. & C. Red No. 37, Rhodamine B stearate, is in one of the five subgroups of the oil-soluble dyes, and is a stearate of a xanthine dye base. It is practically insoluble in water, is soluble in alcohols, ketones, aromatic hydrocarbons and vegetable oils. It is a convenient means of introducing rhodamine into creams and emulsions to obtain attractive, delicate pink tints. Usually 0.001% will produce a strong color. This dye is sometimes used in nail polish lacquers, but tends to penetrate into the body of the nail.

The following dyes are also known as oil soluble dyes of the water insoluble azo, quinoline and anthraquinone groups:

F. D. & C. Red No. 2   D. & C. Yellow No. 10
F. D. & C. Orange No. 2   D. & C. Green No. 6
F. D. & C. Yellow No. 3   Ext. D. & C. Blue No. 5
F. D. & C. Yellow No. 4   D. & C. Violet No. 2

These dyes are insoluble in water, largely because they lack salt forming radicals. They are soluble in aromatic hydrocarbons, and to a much lesser degree in aliphatic hydrocarbons. They are soluble in chloroform and most other halogenated solvents. Organic esters, vegetable and mineral oils and fatty acids are strongly, and in most cases, readily colored by these dyes. They also color waxes. However, care should be taken to run storage tests of 8 to 12 weeks' duration whenever D. & C. Green No. 6, Ext. D. & C. Blue No. 5 and D. & C. Violet No. 2 (anthraquinone derivatives) are used to color waxes, for the reason that some waxes slowly decolorize or turn gray when colored with these products. These dyes all show anti-oxidant effects in drying and unsaturated oils. They are soluble in most lacquers. When used in lacquers as finger and toe nail polishes they show little tendency to migrate into and permanently stain the body of the nail. (Mention is made of the fact that some unusually hypersensitive persons develop dermatitis when they use lacquer nail polishes, but this may be due not to the dye, which is present in small amount, but to other ingredients of the lacquer.) These dyes are tinctorially strong and moderately bright in hue. Their fastness to light varies with the medium in which they are used.

When finely dispersed in suitable media the insoluble coloring agents have a moderately high degree of translucency or transparency. About 0.1% of these colorants will impart an appreciable color to most powders. For this reason it is customary first to make mastermixes or base-blends by milling the toners with a portion of the product to be colored, and then to add the requisite amount of these master-mixes to the bulk of the product.

The toners also can be ground into greases, oils, waxes, plasticizers, and mixed products on ink rolls, in ball mills or in mortars. Here again it is advantageous first to make a concentrated dispersion of the colorant and then add as much as is needed of this paste to the main portion of the product. This is the preferable procedure for coloring nail polish lacquers.

Sometimes it is advantageous to use both soluble and insoluble coloring agents in the same product. For example, a water-oil emulsion containing an abrasive might be formulated with water soluble, oil soluble and insoluble colorants present.

Blend Formulae for Some Popular Colors (Water Soluble)

Dyes	Dark Bluish Red, Bordeaux	Raspberry	Cherry	Strawberry	American Beauty	Pink	Red Orange	Ade Orange
F. D. & C. Red No. 1	—	15.5	47	—	—	—	—	—
F. D. & C. Red No. 2	90	77.0	53	83.5	100	—	20	—
F. D. & C. Red No. 3	—	—	—	—	—	95	—	—
F. D. & C. Orange No. 1	—	7.5	—	16.5	—	—	80	50
F. D. & C. Yellow No. 5	—	—	—	—	—	5	—	50
F. D. & C. Blue No. 1	10	—	—	—	—	—	—	—

Dyes	Egg Yellow	Amber	Lemon and Canary	Lime (Sol'n)	Leaf Green	Mint Green	Pistachio Green	Deep Blue
F. D. & C. Red No. 2	—	10	—	—	—	—	—	40
F. D. & C. Orange No. 1	10	—	—	—	9	—	—	—
F. D. & C. Yellow No. 5	90	90	100	99	69	53	20	—
F. D. & C. Green No. 2	—	—	—	—	22	47	—	—
F. D. & C. Blue No. 1	—	—	—	1	—	—	80	60

Dyes	Grape	Wine	Caramel	Chocolate	Black	Fluorescent Soap Green
F. D. & C. Red No. 2	72.5	60.0	25	25	20	—
F. D. & C. Orange No. 1	—	27.5	20	37	30	—
F. D. & C. Yellow No. 5	—	—	50	30	—	—
F. D. & C. Blue No. 1	27.5	12.5	5	8	50	—
D. & C. Yellow No. 8	—	—	—	—	—	100

Popular Color Blends of Oil and Wax Soluble Dyes

Dyes	Rose	Red	Red Orange	Butter	Lemon	Green	Blue	Red Violet
D. & C. Red No. 18	100	50	—	—	—	—	—	20
F. D. & C. Red No. 32	—	50	25	—	—	—	—	—
F. D. & C. Orange No. 2	—	—	75	—	—	—	—	—
F. D. & C. Yellow No. 3	—	—	—	40	—	—	—	—
F. D. & C. Yellow No. 4	—	—	—	60	—	—	—	—
D. & C. Yellow No. 11	—	—	—	—	75	100	—	—
D. & C. Green No. 6	—	—	—	—	25	—	85	—
D. & C. Violet No. 2	—	—	—	—	—	—	15	80

Dyes	Chocolate (Solution)	Chocolate (Coating)	Black	Jet Black
D. & C. Red No. 18	30	25	13	13
F. D. & C. Orange No. 2	—	—	15	10
F. D. & C. Yellow No. 3	40	60	—	7
D. & C. Green No. 6	30	15	72	70

The Relative Stability and Fixing Power of Aromatics and Essential Oils

The importance of a thorough study of the tenacity of the basic materials used in perfume composition has often been emphasized and many a comparative study of the subject has been made. The importance attached to this subject can be easily understood for the stability of the perfume used determines to a great extent the value of the finished preparation.

Experiments concerning the relative stability of a number of essential oils and aromatics which were carried out by two independent investigators are of great interest, the more so as these altogether independently conducted studies resulted in practically identical results.

Several years ago A. Ellmer determined the stability of various essential oils and aromatics by immersing a strip of white blotting paper 2 cm. wide to the depth of 2 cm. in a 5% solution in absolute alcohol of the essential oil or aromatic under observation. The strips of blotting paper were then withdrawn and left to evaporation in air kept at constant room temperature.

Subsequently A. B. Luque carried on similar tests proceeding somewhat differently: 1 gr. of a 1% solution in alcohol of each oil and aromatic in question was placed in a glass tube which was suspended in a water bath constantly kept at 40°C. with stirring equipment attached to insure uniformity of temperature, and the solution left to evaporate.

One interesting observation brought out by both sets of experiments was that substances such as benzyl cinnamate, methyl anthranilate, cinnamic alcohol, coumarin, and oil sage clary, which are all good fixatives for other aromatics, show only limited or mediocre odor stability by themselves. It is also interesting to note that apparently no recognizable connection is shown between boiling point and odor tenacity, which confirms the conviction gained by practical experience that the fixative as well as odoriferous effect of aromatics is also dependent on other conditions and is not solely determined by their stability. On the other hand, other experiments and actual production have confirmed the theory that the increased boiling point of an aromatic enhances its odor tenacity.

The following table giving in detail the experiments referred to above will be of interest as it gives pertinent facts as to the relative tenacity of certain aromatic and essential oils which should prove very useful under actual manufacturing conditions.

Tenacity of Aromatics

	According to Ellmer	According to Luque
Cineol	2	360
Methyl Salicylate	6	1 130
Amyl Acetate	24	4 680
Heptyl Alcohol	24	4 690
Anethol	24	4 730
Benzyl Cinnamate	24	4 790
p-Methyl Acetophenone	24	4 820
n-Octyl Alcohol	24	4 820
Borneol	48	8 460
Terpineol	48	9 260
Linalyl Acetate	48	9 340
Bornyl Acetate	48	9 380
Bromstyrol	48	9 440
Anisic Aldehyde	48	9 540
Menthol	48	9 580
Phenylethyl Alcohol	72	12 800
Nerolin	72	1 320
Citronellol	120	31 600
Nerol	120	31 800
Methyl Anthranilate	120	32 400
Nonyl Aldehyde	120	62 600
Heptyl Aldehyde	144	37 400
Methyl Heptine Carbonate	144	38 200
Eugenol	192	37 200
Geraniol	192	38 200
Amyl Salicylate	168	39 000
Geranyl Formate	216	41 400
Cinnamic Aldehyde	246	46 800
Hydroxy Citronellal	336	65 900
Cinnamic Alcohol	360	69 800
Coumarin	384	75 900
Phenylacetaldehyde	504	9 600
Isopulegol	600	10 800
Santalol	1 368	265 000
Vanillin	1 440	279 000
Cuminic Aldehyde	2 472	408 000

Tenacity of Essential Oils

	According to Ellmer	According to Luque
Oil Dill	24	4 680
Oil Anise	24	4 760
Oil Lemon	72	13 600
Oil Sage	72	13 800
Oil Chamomile, Roman	72	..

	According to Ellmer	According to Luque
Oil Petitgrain, Paraguay	96	17 400
Oil Geranium, Bourbon	96	18 500
Oil Marjoram	120	20 700
Oil Cinnamon	120	21 500
Oil Angelica	168	30 200
Oil Citronella, Java (85%)	192	38 400
Oil Hyssop	216	
Oil Sage Clary	240	46 500
Oil Patchouly from leaves	324	63 200
Oil Sandalwood, E.I.	672	129 000
Oil Chamomile, German	840	
Oil Rosemary	1 032	224 000
Oil Arnica Flowers	5 640	

Perfume Fixing Agents

The prerequisite of the modern successful perfume creation is successful fixation achieved by so-called fixing agents which tend to equalize the different evaporation curves of its various ingredients. The art of successful fixation is based on the careful selection of suitable materials of fixative value, of which modern science has developed a great many in addition to those which have been known and used for centuries. The fixative must be adapted to the particular type of odor of the entire combination and must under no circumstances change that basic conception.

Fixatives may be divided into natural, which would include essential oils, resins, oleoresins, resinoids, balsams, etc., and synthetic fixatives, largely represented by certain chemical compounds comprising alcohols, aldehydes, esters, acetals, ketones, lactones, nitrogen-containing materials, etc. Animal fixatives, such as musk, amber, castoreum, and civet, belong of course to the classification of natural fixatives but are at present largely substituted by synthetic developments many of which offer distinct advantages.

Below is a tabulation of fixatives of all types classified by odor groups.

	Synthetic Fixative	Natural Fixative
Acacia	Coumarin Methyl Naphthyl Ketone Vanillin Heliotropin Musk Tonkin Substitute	Resinoid Benzoin (Siam) Resinoid Storax Resinoid Myrrh Oil Guaiac Wood
Carnation	Benzyl Isoeugenol Eugenol Isoeugenol Vanillin Cinnamic Alcohol Musk Tonkin Substitute	Resinoid Storax Resinoid Balsam Tolu Resinoid Peru Balsam Resinoid Benzoin (Siam)
Fougère	Coumarin Vanillin Musk Tonkin Substitute Heliotropin	Resinoid Benzoin (Siam) Resinoid Balsam Tolu Resinoid Peru Balsam Resinoid Tonka Bean Resinoid Oakmoss
Chypre	Musk Tonkin Substitute Coumarin Vanillin Heliotropin Eugenol Isoeugenol Cinnamic Alcohol	Oil Sandalwood Oil Patchouly Oil Vetiver Resinoid Labdanum Resinoid Balsam Tolu Resinoid Peru Balsam Resinoid Oakmoss
Eau de Cologne	Methyl Naphthyl Ketone Anthranilic Acid Methyl Ester Cinnamic Acid Methyl Ester Musk Tonkin Substitute Coumarin	Resinoid Benzoin (Siam) Resinoid Storax Oil Sage Clary

COSMETICS AND DRUGS

	Synthetic Fixative	Natural Fixative
Gardenia	Musk Tonkin Substitute	Resinoid Storax
	Isoeugenol	Resinoid Benzoin (Siam)
	Cinnamic Alcohol	Resinoid Peru Balsam
	Hydroxy Citronellol	Resinoid Balsam Tolu
Hyacinth	Phenyl Acetic Acid-Paracresyl Ester	Resinoid Storax
	Phenyl Acetaldehyde	Resinoid Peru Balsam
	Vanillin	Resinoid Balsam Tolu
	Musk Tonkin Substitute	Resinoid Benzoin (Siam)
	Cinnamic Alcohol	
	Eugenol	
Lavender	Vanillin	Lavander (Concrete)
	Coumarin	Oil Guaiac Wood
	Musk Tonkin Substitute	Resinoid Benzoin (Siam)
		Resinoid Storax
		Resinoid Oakmoss
		Resinoid Labdanum
Lilac	Vanillin	Civet
	Hydro Quinone Dimethyl Ether	Oil Sandalwood
	Coumarin	Oil Linaloe
	Musk Tonkin Substitute	Resinoid Benzoin (Siam)
	Hydroxy Citronellol	Resinoid Tonka Bean
	Hydroxy Cinnamic Alcohol	
Jasmine	Hydroxy Citronellol	Oil Sandalwood
	Anthranilic Acid Methyl Ester	Civet
	Indol	Resinoid Storax
	Skatol	Resinoid Balsam Tolu
		Resinoid Peru Balsam
Mimosa	Coumarin	Resinoid Benzoin (Siam)
	Methyl Naphthyl Ketone	Resinoid Myrrh
	Vanillin	Resinoid Balsam Tolu
	Heliotropin	Resinoid Peru Balsam
		Oil Vetiver
		Oil Sage Clary
Narcissus	Eugenol	Resinoid Benzoin (Siam)
	Isoeugenol	Resinoid Storax
	Benzyl-Isoeugenol	Resinoid Peru Balsam
	Cinnamic Alcohol	Resinoid Balsam Tolu
	Musk Tonkin Substitute	Civet
	Phenyl Acetic Acid-Paracresyl Ester	
Oriental Perfume	Coumarin	Resinoid Myrrh
	Vanillin	Resinoid Labdanum
	Musk Tonkin Substitute	Resinoid Olibanum
	Heliotropin	Resinoid Oakmoss
	Benzal Acetone	Oil Sandalwood
	Hydro Quinone Dimethyl Ether	Oil Patchouly
		Oil Vetiver
Violet	Heliotropin	Oil Costus Root
	Vanillin	Oil Sandalwood
	Musk Tonkin Substitute	Oil Orris (Concrete)
	Methyl Ionone	Resinoid Orris
	Ionone	Resinoid Cedar
Rose	Benzophenone	Oil Sandalwood
	Musk Tonkin Substitute	Oil Guaiac Wood
	Cinnamic Alcohol	Oil Patchouly
	Phenyl Ethyl Alcohol	Oil Linaloe
	Hydroxy Citronellol	Resinoid Storax
		Resinoid Balsam Tolu
		Resinoid Peru Balsam

	Synthetic Fixative	Natural Fixative
Wallflower	Eugenol	Resinoid Storax
	Isoeugenol	Resinoid Balsam Tolu
	Musk Tonkin Substitute	Resinoid Peru Balsam
	Vanillin	Resinoid Benzoin (Siam)
	Heliotropin	
	Hydro Quinone Dimethyl Ether	Oil Orris (Concrete)

It is very seldom that any one of these fixatives is used alone, as the experienced perfume chemist usually combines a number of those best suited to his end product.

It can readily be understood that perhaps more than any other phase of perfume compounding the successful mastery of competent fixation of the odor complex is dependent upon a thorough study of this difficult subject. Many an otherwise successful creation fails because it lacks the necessary stability which is one of the basic demands in modern perfumery. It can therefore be readily understood that the demand is frequently made that a perfume compound be reformulated in order to increase its stability. It is at this point that the experienced essential oil and perfume chemist well versed in the intricate behavior characteristic of these complicated structures is often called upon to help, giving the perfume its final Midas touch.

Rose Concentrate for Face Powder
Geraniol Extra	40
Geranyl Acetate	30
Citronellol	20
Geranium Bourbon Oil	30
Benzyl Acetate	10
Phenyl Acetaldehyde (50%)	10
Tincture of Artificial Musk (1 oz. per gallon)	20

Cheap Jasmine Perfume Base
Benzyl Acetate	90
Linalyl Acetate	50
Benzyl Alcohol	20
Methyl Anthranilate	3

Lilac Perfume Base
Terpineol	100
Benzyl Acetate	15
Tincture of Artificial Musk (1 oz. per gallon)	15
Artificial Jasmine	5
Phenyl Acetaldehyde (50%)	5
Cananga Oil	2

Violet Perfume Base
Violet Concrete, Soluble	10
Orris Concrete	5
Rose Concrete	2
Cassie Concrete	2
Mimosa Concrete	2
Jasmine Concrete	2
Heliotropin	5
Alpha Ionone	10
Benzyl Acetate	2
Methyl Ionone	2
Manilla Ylang Ylang	2
Neroli (Petales)	2
Iraldeine	2

To change the character of the violet and give it a definite note, one may add 1 g. of aldehyde C.12 * or ¼ g. of aldehyde C.14. Others may prefer 1 g. of aldehyde C.16 or 18, while a fresher note can be obtained by the use of 10 g. ethyl acetate.

* C stands for carbon atom.

The wood violet note is imparted to a violet perfume by the addition of oil of violet leaves. To give such a perfume a much more pronounced effect, the quantity of orris concrete and of cassie concrete can be doubled.

Bay Rum

Bay Oil	2.0
Clove Oil	0.3
Pimento Oil	0.3
Petitgrain Oil	0.3
Ethyl Acetate	0.7
Quassia Extract	0.5
Specially Denatured Alcohol	60.0
Glycerin	4.0
Menthol	0.2
Distilled Water	To make 100.0

Tint amber or yellow-brown, either with caramel or a soluble dyestuff.

CHAPTER IV

DRUG PRODUCTS

Aluminum Hydroxide Gel

Seventy ounces of sodium carbonate is dissolved in 2 gal. of hot water and strained through fine linen into a cask of approximately 40 gal. capacity. Fifity-six ounces of ammonia alum is dissolved in 1 gal. of hot water and slowly strained into this solution with constant stirring. It is essential to run the alum into the sodium carbonate to ensure keeping the carbonate in excess throughout. A further 2 gal. of hot water are added and the whole stirred briskly to promote complete evolution of gas. Some ammonia is evolved due to excess carbonate. To reduce possible adsorption to a minimum the cask is immediately filled with cold water. The precipitate is allowed to settle for about 24 hours and the supernatant liquid siphoned off as completely as possible. The precipitate (volume about 4 gal.) is collected in a fine linen bag of about 4 gal. capacity. It is washed in the bag until the washings are free from sulfate (6-7 washings). The volume is allowed to concentrate to slightly less than 1 gal., 0.1% sodium benzoate and 0.1% oil of peppermint is added and the volume made up to 1 gallon.

Amino Acid Elixir

Aminoacetic Acid	131.5	g.
Raspberry Syrup	75.0	cc.
Syrup	60.0	cc.
Alcohol	53.0	cc.
Benzoic Acid	2.0	g.
Compound Orange Spirit	1.5	cc.
Vanillin	0.15	g.
Distilled Water To make	1000.0	cc.

Dissolve the aminoacetic acid in 700 cc. of distilled water; add the syrup and the raspberry syrup and mix well. Dissolve the benzoic acid in the alcohol and the compound orange spirit; add to the previously prepared mixture. Filter, if necessary, and add sufficient distilled water to make 1000 cc.

Anti-Acid for Stomach and Digestion Disorders

Calcium Lactate	5 g.
Sodium Bicarbonate	10 g.
Peppermint Oil	10 drops
Sugar	1 tbp.
Water	To make 1 pt.
Dose:	1 or 2 teaspoonful

Antiseptic Baby Oil

Olive Oil	4 oz.
Chlorothymol	30 gr.
Rose Oil	30 min.
White Mineral Oil	To make 20 oz.

Antiseptic Ear Powder
(For treating discharging ears)
Iodine	¾
Ether	10
Boric Acid (Powdered)	100

Mix in a closed cooled container.

Ear Oil
(For drying up discharging ears)
Guaiacol	1.0 cc.
Creosote	2.0 cc.
Iodoform	7.0 g.
Ether	50.0 cc.
Olive Oil (Sterilized)	40.0 cc.

Asthma Inhalant
N-(γ,γ-Diphenylpropyl) Piperidine Hydrochloride	50.0
Racemic Dihydroxyephedrine Hydrochloride	5.0
4-Hydroxyephedrine Hydrochloride	10.0
Acetone Bisulfite	2.0
Vanillin	0.5
Glycerin	200.0
Water	To make 1000.0

The spasmolytic action of diphenylpropylpiperidine is said to be slower than that of ephedrine, but more prolonged. Acetone bisulfite is used as a stabilizer.

Atabrine Tablets

Atabrine (100%) 57 kg., maize starch 22 kg., amylose 1 kg. and talc 29 kg. The ingredients are mixed, granulated with cold water, forced through a 1.2 mm. sieve, and dried. A separate granulation is then prepared by adding 3 kg. of Atabrine to a mixture of 11 kg. of talc and 0.72 kg. of melted cocoa butter and forcing the mixture through a 0.4 mm. sieve. The two granulations are then mixed and passed through a 1.2 mm. sieve; after determination of the water content, the mixture is made up to 132 kg. The finished tablets, each weighing 0.22 g., contain 0.100 g. of Atabrine.

Athlete's Foot Preparations
Ointments
Formula No. 1
Zinc Undecylenate	20.0
Talc (U. S. P.)	76.0
Undecylenic Acid (Grade AA)	2.0
Pigment	2.0

No. 2
Undecylenic Acid (Grade AA)	5.0
Triethanolamine	3.0
Zinc Undecylenate	18.0
Propylene Glycol	10.0
"Carbowax 1500"	19.0
"Carbowax 4000"	29.6
Distilled Water	15.0
Pigment	0.4

No. 3
Salicylic Acid	6
Light Liquid Petrolatum	5
Wool Fat	12
White Petrolatum	To make 100

Liquids
Formula No. 1
Salicylic Acid	10 g.
Acetone	33 cc.
Ethyl Alcohol (85%)	33 cc.
Glycerin	33 cc.

The ingredients are mixed in the order given. The alcohol and the acetone act as solvents for the salicylic acid, and the glycerin acts to prevent the dehydration and other undesirable effects of the alcohol on the raw blistered skin surfaces. It is best dispensed in screw capped bottles.

The solution is applied by freely mopping the infected areas, and then allowing them to dry for about 5 minutes.

No. 2

Phenylmercuric Chloride	½ %
Calamine	15 gr.
Zinc Oxide	30 gr.
Glycerin	30 min.
Water To make	1 fl. oz.

No. 3

Formaldehyde Solution (40%)	10 min.
Salicylic Acid	10 gr.
Alcohol ⎫ Equal	
Water ⎭ parts,	
To make	1 fl. oz.

No. 4

The feet may be painted once or twice a week with a sodium propionate lotion, containing 8.2% sodium propionate, 1.2% propionic acid and 10.0% N-propyl alcohol. The feet should be kept clean, clean socks worn every day, and if excessive perspiration is serious, a drying agent, like aluminum acetate, should be used occasionally.

Powder

Formula No. 1

Salicylic Acid	5
Menthol (Pulverized)	2
Camphor (Pulverized)	8
Boric Acid	50
Corn-Starch	35

Grind to a fine powder and dust on the feet and in shoes.

No. 2

Phenyl Salicylate	1
Chloral Hydrate	1
Sterilized, Purified Siliceous Earth	
To make	100

No. 3

Sodium Propionate	20.0
Talc (U. S. P.)	79.5
Pigment	0.5

No. 4

Undecylenic Acid	2
Zinc Undecylenate	20
Talc	78

No. 5

A dusting powder, such as the following, should be used in the shoes and socks and dusted on the feet:

Salicylic Acid	2.0
Zinc Stearate	3.0
Boric Acid	6.0
Starch	10.0
Powdered Talc	79.0

Treatments

Flash Burn Preventive Cream (U. S. Navy formula)

Bleached Dewaxed Shellac	13.70
Isopropyl Alcohol (99%)	28.48
Bodied Linseed Oil (Z–3)	3.50
Stearic Acid	0.15
Triethylene Glycol Di-2-Ethylhexoate	0.80
Carbitol	1.10
Titanium Dioxide	37.00
Sodium Bicarbonate	2.25
Magnesium Stearate	8.00
Menthyl Salicylate	2.50
Wetting Agent (Wetanol)	0.30
Yellow Iron Oxide	1.60
Black Iron Oxide	0.62

The cream is smeared over hands, face and neck before exposure to flame.

Burn Film With Sulfagel

Sodium Sulfadiazine	25
Pharmagel B (Gelatin)	50

Sulfadiazine	45
A-3 Water	100

This preparation is autoclaved for 30 minutes and placed in sterile containers.

The A-3 water, is distilled water containing 0.52 g. methyl-p-hydroxybenzoate and 0.28 g. propyl-p-hydroxybenzoate per 1000 cc. used to prevent mold growth, etc., in the gelatin.

Burn Ointment

Sulfanilamide	5
Urea	20
Water-Soluble Base *	To make 100

The advantages claimed are theoretically that: (1) the sulfonamide action is enhanced by the destructive action of urea on p-aminobenzoic acid, normally present in serum; (2) there is a combined local analgesic and epithelial stimulant effect of the urea; (3) there is an anti-adhesive factor in the base.

Practical experience has shown this ointment to be inexpensive and not to deteriorate on storage, and it can be adequately sterilized. The ointment is suitable for dispensing in collapsible tubes for use as an emergency first-aid dressing for first degree burns and for superficial grazes.

Burn Sprays
Formula No. 1

Tannic Acid	2
Alcohol	20
Water	78

* Water-Soluble Base

Sodium Alginate	10
Liquid Paraffin	10
Water	To make 100
Phenyl Mercuric Nitrate for Preservation	1:600

No. 2

A spray film for coating and protecting burned and otherwise denuded areas of the body.

Paraffin Wax	265
White Wax (Beeswax)	215
Sulfonated Liquid Petrolatum	330
Sodium Lauryl Sulfate	10
Water	30
Sulfonamide	50
Triethanolamine	100

Burn Treating Emulsions
Formula No. 1

Acriflavine	0.1 g.
Cod Liver Oil	32.0 cc.
Petrolatum	32.0 g
Wetting Agent (Wetanol)	3.0 g.
Water	To make 100 cc.

No. 2

Acriflavine	0.1 g.
Lanette Wax SX	5.0 g.
Wetting Agent (Wetanol)	2.0 g.
Cod Liver Oil	40.0 cc.
Water	53.0 cc.

No. 3

Spermaceti	15.0 g.
Soft Paraffin Wax	12.0 g
Liquid Petrolatum	12.0 cc.
Sodium Lauryl Sulfate	1.0 g
Water	60.0 cc.

Triple Dye—Soap Mixture for Burns

A. Soft Soap	50
Water	50
B. Gentian Violet	1
Brilliant Green	1
Acriflavine	1

Sterilize A by boiling, then add B.

Calamine Lotion

Calamine	80 g.
Zinc Oxide	80 g.
Glycerol	20 cc.
Calcium Hydroxide (saturated solution)	1000 cc.
Phenol	10 g.

Mix the calamine and zinc oxide together until the mass is homogeneous. Add the glycerol and thoroughly mix in. Dissolve the phenol in the saturated calcium hydroxide solution and add the calcium solution to the mixed dry ingredients. This lotion is recommended for skin rashes.

Camphor Solution

Camphor	10
Isopropanol	50
Triethanolamine Oleate	40

Stir until dissolved. This mixture will form a clear solution when diluted 1:9 with water.

Crab Lice Treatment

Benzyl Benzoate	68
Polyethylene Glycol Laurate	14
Benzocaine	12
DDT	6

One part of above is emulsified with 5 parts of water and applied to all hairy parts of the body.

Enteric Coating For Pills, etc.
U. S. Patent 2,373,763

Myristic Acid	68
Hydrogenated Castor Oil	25
Castor Oil	2
Cholesterol	1
Sodium Taurocholate	4

Impetigo Lotion

Sulfathiazole	15
Tragacanth	4
Physiological Salt Solution	To make 100

Apply twice daily.

Itch Remedy

This is a doughy non-greasy cakelike material which is applied by hand in a thick layer over the itching regions, but not rubbed in.

N-Butyl-p-aminobenzoate	100.0 g.
Benzyl Alcohol	170.0 cc.
Anhydrous Lanolin	20.0 cc.
Cornstarch	640.0 g.
Sodium Lauryl Sulfate	64.0 g.

More benzyl alcohol is added in mixing the other ingredients, according to the following directions:

Warm the benzyl alcohol, and dissolve in it the n-butyl-p-aminobenzoate making an approximately saturated solution.

Add melted lanolin, keeping the mixture warm and stirred until as much of the lanolin as will dissolve is in solution.

Mix well the cornstarch and sodium lauryl sulfonate, and add slowly to this powder, a little at a time, the warm lanolin mixture. Knead this mixture to distribute the liquid evenly through the powder.

Add benzyl alcohol about a tenth of that already used to produce a doughy, non-greasy, cakelike ointment that can be packed in ointment jars or other suitable containers. It is better not to use containers made of metal.

Liniments
Formula No. 1

Camphor	2

Isopropyl Alcohol	4
Lecithin	4
Distilled Water	278
Glycerin	10
Tincture of Arnica	2

The camphor is worked into the lecithin, dissolved in the isopropyl alcohol, and then 38 parts of water are added, stirring constantly until a gel is formed which is then thinned with the rest of the water, the glycerin and arnica tincture being added last.

No. 2

Olive Oil	1000 cc.
Oleic Acid	6 cc.
Lime Water	1000 cc.

No. 3
Stoke's Liniment

Turpentine	1500 cc.
Water	2250 cc.
Acetic Acid	375 cc.
Eggs	12
Starch	45 g.

No. 4
Turpentine Liniment (Emulsified)

Turpentine	25 cc.
Oleic Acid	1 cc.
Dilute Ammonia	3 cc.
Ammonium Chloride	1 g.
Water	To make 100 cc.

Nasal Preparations with Ephedrine
Formula No. 1

Ephedrine Sulfate	2.0 g.
Chlorobutanol	0.5 g.
Menthol	0.1 g.
Alcohol	2.0 cc.
Dextrose	8.0 g.
Glycerin	3.0 cc.
Distilled Water	To make 100.0 cc.

No. 2

Ephedrine Hydrochloride	7.00 g.
Sodium Chloride	22.00 g.
Methyl p-Hydroxybenzoate	0.52 g.
Sodium Bicarbonate	17.00 g.
Glycerin	22.00 cc.
Water	1000.00 cc.

No. 3

Ephedrine Sulfate	0.3 g.
Tragacanth Mucilage (3%)	20.0 g.
Rose Water	10.0 cc.

Plaster Dressing Deodorizer

Iodoform	3
Compound Tincture of Benzoin	To make 240

Applied daily to the odorous portion of the plaster cast. The solution has been tested on more than 100 cases and has proven effective in counteracting the foul smell encountered with such plaster dressings.

Pill Excipient

Glucose	60
Dextrin	20
Starch	20

Penicillin Pastilles
Formula No. 1

Sodium Citrate	2.0 g.
Starch	20.0 g.
Gelatin	40.0 g.
Sucrose	60.0 g.
Distilled Water	140.0 cc.
Peppermint Oil	6.0 min.
Calcium Penicillin	100,000 units

The solids and distilled water are boiled together for 3 minutes with constant stirring. The mixture is then cooled to about 80°F., and the penicillin added with constant stirring. The mixture is then poured into a small waxed container, 8 by 16 by 2 cm., covered and stored in the refrigerator. The jell may be divided into 40 cubes or lozenges.

No. 2

Tragacanth (Powdered)	42.6 mg.
Gum Acacia (Powdered)	42.6 mg.
Sodium Carboxymethylcellulose	19.0 mg.
Sugar (Powdered)	745.0 mg.
Calcium Penicillin	1,000 units
Flavor (Creme De Menthe)	0.66 mg.
Magnesium Stearate	2.56 mg.

Prickly Heat Preparations
Formula No. 1

Menthol	2 gr.
Camphor	10 gr.
Eucalyptus Oil	3 min.
Paraffin Wax	To make 1 oz.

Warm together and mix until uniform.

No. 2

Sublimed Sulfur	10
Zinc Oxide	10
Boric Acid	10
Starch	10

No. 3

Menthol	1
Glycerin	1
Salicylic Acid	4
Alcohol	94

No. 4

Menthol	2
Camphor	2
Bismuth Subnitrate	10
Zinc Oxide	10
Alcohol	120
Lime Water	120

Rash Lotion

Thymol	5 g.
Cinnamon Oil	2 g.
Alcohol	100 cc.
Water	20 cc.

The ingredients are all mixed together. This solution is used in common skin rashes.

Scabies Lotion
Formula No. 1

Soft Soap	4 oz.
Water	40 fl. oz.
Benzyl Benzoate	20 fl. oz.
Alcohol	10 fl. oz.

Dissolve the soft soap in the water by means of heat. Adjust the volume of this solution to 2 pt. and place it in an 80-oz. bottle. Add the benzyl benzoate and shake. Finally add the alcohol, which will remove any froth. Shake well before using.

No. 2
("Favorin")

Benzyl Benzoate	35
Triethanolamine	50
Benzyl Alcohol	17

Sulfadiazine Tablets, 7.7 gr. (0.5 g.)

Quantity:	105,000
Die-size:	7/16 in.
Sulfadiazine	115 lb. 8 oz.
Sugar (Powdered)	3 lb. 7 oz.
Sugar (Granular)	2 lb. 15 oz.
Starch	9 lb.
Moistener:	Water
Starch	12 lb.
Talc	10 oz. 4 dr.
Magnesium Stearate	4 oz. 1 dr.

Sulfathiazole Dusting Powder

Sulfathiazole	5.0

Zinc Peroxide 20.0
Boric Acid   To make 100.0

Sulfonamide Chewing Wafer
(For Mild Tonsillitis)

Sulfanilamide   2 g.
Sodium Bicarbonate   2 g.
Sulfathiazole   4 g.
Paraffin Wax   16 g.
Peppermint Oil   10 min.

The solids are ground into a fine powder, and under constant stirring are added to the melted wax and peppermint oil, poured into a mold and divided into 20 tablets.

The patient chews the substance as he would chew gum. One tablet is administered every 2 hours. Subjectively there is an almost immediate soothing effect on the inflamed mucous membranes. Resolution of the inflammatory process usually occurs in 2 or 3 days.

Styptic Cotton

Solution of Ferric Chloride
  U.S.P.   80
Glycerin   16
Water   225
Purified Absorbent Cotton   100

Mix the solution of ferric chloride, glycerin and water, immerse the purified cotton in this solution, and allow it to remain 1 hour. Then remove it, press it and spread it out in thin layers, in a warm place protected from dust and light. When dry transfer it to well-closed glass containers.

Wart Removal
No. 1

Using a 25% suspension of podophyllin in liquid petrolatum, the mixture is thoroughly shaken and applied liberally to the affected area, care being taken that the numerous crevices of the warts are penetrated. This treatment is simple and requires no special skill or apparatus, but the application should be thorough and prolonged. It is suitable not only for single soft warts but also for more difficult cases, such as warts of the urinary meatus and multiple warts of the penis, vulva, and anus. In addition multiple warts of the face disappear under this treatment.

No. 2

Zinc Choride   30 gr.
Salicylic Collo-
  dion   To make 1 fl. oz.

Vaginal Douche

Chlorthymol   1.3 g.
Menthol   2.0 g.
Peppermint Oil   2.0 cc.
Lactic Acid To make 180.0 cc.

Use 1 teaspoonful to 2 quarts of warm water; morning and evening.

Vaginal Suppositories

Penicillin Vaginal Suppository
Gelatin   40.000
Butyl p-Hydroxy-
  benzoate   0.025
Distilled
  Water   To make 100.000
Penicillin (Sodium
  or Calcium)   the prescribed
                      quantity

Dissolve the penicillin in about 3 cc. of the distilled water. Add the butyl parahydroxybenzoate to the remainder of the distilled water, heat the mixture to boiling, pour it upon the gelatin, and stir until the mixture is of uniform consistency. When the mixture has cooled to a temperature of 50°C., add the solution of the penicillin, mix well, and pour the product into molds which have not been chilled. Allow to stand at room temperature until the suppositories have hardened. Remove the suppositories from the molds, dust them with talc, and place them in individual glass containers. *Store in a refrigerator* at a temperature not above 10°C.

Vaginal Sulfa Suppository

Sulfanilamide	2
Sulfathiazole	2
Zinc Peroxide	1
Sodium Tetradecyl Sulfate	4
Cocoa Butter	8

Venereal Prophylactic

Mercury Oxycyanide	0.2
Sulfathiazole Powder	10.0
Tragacanth	10.0
Water	225.0

The amount of water required to obtain a heavy, creamy product may vary somewhat, but the consistency should be such that the resulting mixture can be easily aspirated and injected through a penile syringe. In using this preparation, 5 cc. are injected into the urethra by means of a penile syringe. After the syringe is removed, the mixture is retained for exactly five minutes; more than this will cause a painful urethritis. The mixture is then removed mechanically, by means of a syringeful of liquid petrolatum and by urination. As a further protective measure, additional portions of the prophylactic mixture are rubbed on the skin of the penis, scrotum and thighs.

Medicated Creams and Ointments

These are creams selected according to texture and consistency required along with medicaments. In some cases care must be exercised in avoiding incompatible combinations among the active ingredients themselves as well as with the emulsifying agent. When the added medicaments are electrolytic in nature, Monostearin should be used, together with an electrolyte-stable wetting agent, such as Wetanol. Glyceryl Monostearate S can also be used for this purpose together with a colloidal clay such as bentonite.

Medicated Cream

Glyceryl Monostearate S	18.0
Spermaceti	5.0
Carbitol	5.0
Water	70.8
Menthol	0.2
Camphor	0.5
Phenol	0.1
Clove Oil	0.2
Eucalyptus Oil	0.2

Base for Medicated Creams

Glyceryl Monostearate S	10.0
Glycerin	25.0
Bentonite	2.0
Water	63.0

The bentonite is mixed with 50 parts of the water. The Glyceryl Monostearate S is melted in the glycerin on a water bath. The bentonite mixture and the balance of the water are added at the same temperature and the mixture stirred until cool. This base can be used with various medicaments such as tannic acid, iodine, phenol, benzoic acid, calamine and potassium iodide.

Ointment Base
Formula No. 1

Diglycol Monostearate	10
Sulfated Hydrogenated Castor Oil	20
White Petrolatum	30

Distilled Water	40

No. 2

Beeswax U.S.P.	10
Anhydrous Lanolin	5
Glyceryl Monostearate	12½
Stearic Acid	2
White Petrolatum	75

No. 3

Diglycol Stearate	14
White Petrolatum	6
Paraffin Wax	2
Liquid Petrolatum	30
Distilled Water	48

The solids and oils are melted together and to them the water (heated to 50–60°C.) is added slowly with good mixing until completely emulsified. Stir occasionally while cooling.

No. 4

Stearyl Alcohol	250
White Petrolatum	250
Glycerin	120
Sodium Lauryl Sulfate	10
Methyl Parahydroxy-benzoate	0.25
Propyl Parahydroxy-benzoate	0.15
Distilled Water	369

Melt the stearyl alcohol and white petrolatum on a water bath and warm to 75°C. Dissolve the sodium lauryl sulfate and methyl parahydroxybenzoate in the glycerin and water, warm to 75°, and add this solution, with stirring, to the warm stearyl alcohol and white petrolatum. Stir the ointment until congealed.

No. 5

Spermaceti	15	g.
Paraffin Wax, Soft	12	g.
Mineral Oil	12	cc.
Wetting Agent (Wetanol)	1	g.
Water	60	cc.

The first four ingredients are melted together. The water may be added either while the mixture is warm, in which case the product is stirred until cool; or the oily mixture may be allowed to cool and the water incorporated at room temperature when required.

No. 6

Wool Wax (Lanolin) Alcohols	60
Paraffin Wax	240
Petrolatum	100
Mineral Oil (Refined)	570

No. 7

Triethanolamine Stearate	5
Diglycol Stearate	95
Stearic Acid	100
Triethanolamine	25

Up to 775 parts of water can be added to Formula No. 7.

Vanishing Ointment Base
Formula No. 1

Diglycol Stearate	12
Glycerin	3
Spermaceti	3
Distilled Water	42

No. 2

Stearic Acid	13
Diglycol Stearate	20
Water	67

Proceed as for making the ointment base.

White Ointment Base

Anhydrous Lanolin	5
White Beeswax	5
White Petrolatum	90

Water-in-Oil Ointment Base

Anhydrous Lanolin	5
Sorbitan Monooleate	5
White Petrolatum	32½
Mineral Oil, U.S.P.	7½
Glycerin	5
Water	45

Oil-in-Water Ointment Base

Formula	No. 1	No. 2	No. 3	No. 4	No. 5
Cetyl Alcohol	15	6	—	—	2
Propylene Glycol	10	—	5	5	5
White Beeswax	1	—	—	—	—
Sodium Lauryl Sulfate	2	2	—	—	—
Water	72	50	50	75	44
Lanolin	—	6	4	—	10
White Petrolatum	—	14	—	—	—
Mineral Oil, U.S.P.	—	22	25	—	25
Stearic Acid	—	—	15	—	—
Triethanolamine	—	—	2	—	—
Glyceryl Monostearate	—	—	—	18	12
Carbowax 1500	—	—	—	2	2

Absorption Base
Formula No. 1
Anhydrous Lanolin	50
Cholesterol	2¼
Petrolatum, U.S.P.	246¾
Diglycol Oleate	2

Warm together and mix until uniform.

No. 2
British Patent 572,318

Hydrogenated Cottonseed Oil	86
Wool Wax Alcohols	7
Lanolin	7

No. 3
Cholesterol	10
Stearyl Alcohol	30
White Wax	80
Wool Fat (Anhydrous)	150
White Petrolatum	730

Melt the ingredients together on a water bath, mix thoroughly, remove from the water bath, and stir until the mixture congeals.

Washable Ointment Base
Diethylene Glycol Monostearate	10 g.
Sulfated Hydrogenated Castor Oil	20 g.
Petrolatum	30 g.
Water	40 g.

Antiseptic Cream
Euflavin	0.5
Tragacanth Powder	3.0
Glycerin	16.5
Water	80.0

Anti-Vesicant (War Gas) Ointment
Triacetin	73
Dichloramine T	15
Cellulose Acetate Butyrate	12

Coal Tar Ointment
Crude Coal Tar	2.0
Zinc Oxide	2.0
Starch	15.0
Special Base *	15.0

The coal tar ointment prepared in this manner is easily removed from body surfaces with dry tissues or fabrics or by the use of soap and water.

*The special base contains 5% of diglycol stearate, and 95% petrolatum and is prepared by fusing these two substances together and stirring the product until it congeals.

Penicillin Cream

This product should be kept at 0°C. to maintain its potency.

Glyceryl Monostearate	12.0
Absorption Base	10.0
Spermaceti	5.0
Liquid Petrolatum	2.0
Glycerin	3.0
Chlorocresol	0.1
Distilled Water	To make 100.0

To prepare 10 g. of penicillin cream (500 units per g.), 9 g. of the base may be placed in a ½-oz. jar, and, with the cap (preferably non-metallic) placed lightly on top, autoclaved at 20 lb. pressure for 20 minutes. When cool, but not yet set, add a solution of penicillin, prepared by adding 1 cc. of sterile water to a 500-unit ampoule of penicillin; screw cap down firmly, shake vigorously until the solution is incorporated, then immediately place in a refrigerator.

Penicillin Glycerin Jelly

Glycerin	110 cc.
Tragacanth	8 g.
Methyl p-Hydroxy-benzoate (0.25%)	332 cc.
Penicillin	250,000 units
Sterile Distilled Water	To make 500 g.

Special advantages and superior action are obtained when penicillin is applied on a nongreasy glycerin jelly. The jelly is made by gradually mixing 110 cc. of glycerin with 8 g. of powdered tragacanth in a dry mortar and pouring the mixture slowly into a jar containing 332 cc. of an 0.25% solution of methyl-para-hydroxy-benzoate. It is placed then in an autoclave and allowed to steam for 10 minutes, after which it is sterilized at 10 lb. pressure for 30 minutes. When cool, 250,000 units of penicillin, dissolved in enough sterile distilled water (about 25 cc.), is added to make a total weight of 500 g. of jelly.

Sulfa Drug Ointments

Sulfathiazole, sodium sulfathiazole, sulfanilamide and sulfadiazine have good bacteriostatic activity upon the *S. aureus* organism, with sodium sulfathiazole showing greatest activity. The optimum concentration of the drug appears to be between 5.0% and 7.5%. Optimum pH values for ointments of sulfathiazole, sulfanilamide and sulfadiazine were shown to be around pH 9.0 For sodium sulfathiazole, the optimum pH value in an ointment was observed to be around 10.0. The presence of water in a sulfa drug ointment seems desirable for greater bacteriostatic activity. A water content of at least 50% is recommended for optimum activity. The addition of surface tension reductants (e.g., Duponal WA and Aerosol OT) to sulfa ointments improves the penetration power as shown by the increase in the zone of inhibition, particularly in the case of sodium sulfathiazole. A concentration of the surface tension reductant to 0.5% was observed to have beneficial effects.

Sulfathiazole Creams
Formula No. 1

Glyceryl Monostearate S	12.0
Yellow Wax, U.S.P.	2.0
Glycerin	6.0
Corn Oil	4.0

Water	76.0
Sulfathiazole	5.0

No. 2

Glyceryl Monostearate S	12.0
Mineral Oil	20.0
Petrolatum	20.0
Water	48.0
Sulfathiazole	5.0

No. 3

Glyceryl Monostearate S	12.0
Mineral Oil	15.0
Petrolatum	20.0
Stearic Acid	5.0
Lanolin	5.0
Water	43.0
Sulfathiazole	5.0

Melt all the components except the glycerin, water and sulfathiazole, together at about 160°F. Heat the water containing glycerin (where present in the formula) to 160°F. and add to the melted oils, etc. Stir until cool, and add the sulfathiazole powder with stirring. Age at least overnight, and stir well before packaging.

No. 4

Sulfathiazole	5
Sodium Lauryl Sulfate	1
Stearyl Alcohol	10
Cetyl Alcohol	3
Spermaceti	10
Glycerin	10
Water	61

Sulfonamide Ointments
Formula No. 1

All-Grease Base:

Petrolatum	95
Sulfonamide	5

No. 2

Washable Jelly Base:

Water	70.6
Glycerin	18.0
Pectin	6.0
Sulfonamide	5.0
Methyl Parahydroxybenzoate	0.2
Sodium Bisulfite	0.2

No. 3

Wax-Oil Base: White Ointment (U.S.P.)

White Petrolatum	85
Sulfonamide	5
White Wax	5
Wool Fat	5

No. 4

Washable Oil-in-Water Base:

Mineral Oil	64.00
Water	23.45
Beeswax	5.00
Sulfonamide	5.00
Triethanolamine	2.00
Methocel	0.30
Sodium Bisulfite	0.20
Aerosol OT	0.05

No. 5

Washable Oil-in-Water Base:

Water	71
Cetosten	15
Glycerin	5
Sulfonamide	5
Duponol C	2
White Wax	2

No. 6

Washable Oil-in-Water Base:

Water	74
Glycerin	10
Cetyl Alcohol	5
Stearic Acid	5
Sulfonamide	5
Amino Glycol (2-Amino-Methyl-1, 3-propanediol)	1

No. 7

Bentonite Base:

Water	85

Bentonite (BC Volclay)	10
Sulfonamide	5

No. 8

Sulfonated Hydrogenated Castor-Oil Base:

Water	40
Petrolatum	25
Diglycol Stearate	15
Sulfonated Hydrogenated Castor Oil	15
Sulfonamide	5

No. 9

Vanishing-Cream Base:

Water	65.00
Stearic Acid	15.00
Glycerin	10.00
Sulfonamide	5.00
Tegin-P (Propyleneglycol Monostearate)	4.15
Triethanolamine	0.85

No. 10

U.S.P. Cold Cream (Modified Base):

Peanut Oil	54.0
Water	19.0
White Wax	11.0
Spermaceti	10.5
Sulfonamide	5.0
Sodium Borate	0.5

Heavy grease and cold cream bases yield little sulfonamide to an aqueous medium, whereas oil-in-water emulsions, stearate vanishing cream, bentonite and pectin jelly bases yield high levels of sulfonamide to such a medium.

Tropical Ulcer and Wound Ointment

Formula No. 1

Cod Liver Oil	40
Zinc Oxide Ointment	60
Copper Sulfate (Powdered)	1

Change dressings daily; then leave on for 4–5 days.

No. 2

Clean the ulcer with a weak phenylmercuric nitrate solution and apply a water-in-oil emulsion consisting of:

Sulfathiazole	12
Calcium Oleate	2
Beeswax	3
Cod Liver Oil	60
Water	40

The ingredients are sterilized separately and worked into a paste under aseptic conditions. Sulfanilamide-impregnated dressings are then applied on top. In some instances, 10% sulfanilamide paste in vitaminized oil is used with satisfactory results.

No. 3

Zinc Oxide	1
Iodoform	2
Mineral Oil	2–3

Vitamin A and D Ointment

A vitamin A and D ointment which does not have the disagreeable odor of cod liver oil may be readily made by incorporating an A and D concentrate with a mixture of wool fat and petrolatum.

Vitamin A	100,000 U.S.P. units
Vitamin D	20,000 U.S.P. units
Wool Fat	20 g.
White Petrolatum To make	100 g.

Wound Healing Ointment

Absorption Base	2.850
Light Mineral Jelly	45.000
Hydrous Lanolin	3.045
Amber Petrolatum	45.000
Thyme Oil	0.100
Basic Phenyl Mercuric Nitrate	0.005

Oil Soluble Fish Liver
  Extract   3.000
Water Soluble Yeast
  Extract   1.000

Medicinal Lubricating Jelly

Tragacanth (Selected ribbons)	25.0 g.
Boric Acid	12.5 g.
Alcohol	40.0 g.
Glycerin	80.0 g.
Formaldehyde	4.0 cc.
Methyl Salicylate (1% Alcoholic solution)	40.0 cc.
Water	To make 1000.0 g.

Dissolve the boric acid in the water and macerate the tragacanth in this solution. Strain through a fine cheese cloth and add the remaining ingredients and, finally, sufficient water to bring up to the required weight. Mix thoroughly.

Industrial Skin Protective Preparations; Barrier Creams; Dermatitis Lotions

Industrial Protective Creams

A special type of hand cream is the industrial protective cream. As is now well known this cream is rubbed into the skin before work that may irritate or make the hands hard to clean. After the work is completed the hands may be easily cleaned by simply washing in soap and water, the dirt having been prevented from penetrating to any extent. The secret of preparations of this type is the inclusion of a *barrier* substance in the cream to prevent the penetration of paints, inks, oils, etc.

Formula No. 1

Glyceryl Monostearate S	12.0
Petrolatum	3.0
Beeswax	5.0
Talc	10.0
Glycerin	5.0
Water	64.5
Moldex	0.1
Perfume	0.4

No. 2

Glyceryl Monostearate S	12.0
Beeswax	12.0
Hydrous Wool Fat	6.0
Cholesterol	1.0
Sodium Silicate (commercial solution)	5.0
Ammonium Hydroxide (10% solution)	0.5
Water	63.5

Melt together beeswax, Glyceryl Monostearate S, hydrous wool fat, and cholesterol. Heat the water to the same temperature as the wax mixture and add the sodium silicate and ammonium hydroxide solutions. Stir the aqueous solution into the wax mixture and continue stirring until it congeals. This formula gives a non-greasy preparation which dries on the skin and does not rub off. Its use is indicated for dry work as a protection against dust-borne irritants, or where the material or object must be guarded against being soiled.

No. 3

Stearic acid	20.0
Mineral Oil (Low Viscosity)	10.0
Triethanolamine	1.0
Water	To make 100.0

Melt the solid, add to mixture of warmed fluids and mix intimately.

No. 4

Stearic Acid	10.0
Mineral Oil (Low Viscosity)	5.0
Triethanolamine	0.5
Water	To make 50.0

Melt the solid, add to mixture of warm fluids; mix intimately with the following mixture:

Bentonite	10.0
Water	To make 50.0

Heat and run through homogenizer or ointment mill.

No. 5

Octyl-Stearyl Alcohol	16
White Petrolatum	8
Hydrogenated Cottonseed Oil	8
Stearic Acid	8
Water	To make 100

Heat with constant stirring.

No. 6

Mineral Oil (Low Viscosity)	20
Glyceryl Monostearate	12
White Petrolatum	20
Water	To make 100

Heat with constant stirring.

No. 7

Anhydrous Lanolin	400 g.
Castor Oil	400 g.
Bentonite	232 g.
Corn Starch	240 g.
Aerosol—OT (100%)	30 g.
Colloresin	10 g.
Grape (Technical) OW Perfume Oil	¾ cc.
Trichlorobutyl Alcohol	8 g.
Water	2200 cc.

No. 8

Stearic Acid	10
Glyceryl Monostearate	3
Glycerin	12
Starch	8
Preservative (Moldex)	⅓
Perfume	½
Water	To make 100

No. 9

Stearic Acid	10.0 g.
Beeswax	2.0 g.
White Petrolatum	4.5 g.
Triethanolamine	1.5 cc.
Glycerin	8.0 cc.
Distilled Water	54.0 cc.
Talc	20.0 g.
Perfume	To suit

This cream has been tested in industries where workers are exposed to mechanical irritants and various types of grime, and it has been found to reduce the incidence of dermatitis due to such causes.

Dermatitis Treatments

The chief measures in preventing dermatitis are the careful selection of workers and emphasis on personal cleanliness. Also of great importance is the use of protective lotions which should be made available to all workers. If working with soluble oils, employees should use a wash of the type of Formula No. 1; if working with paraffin oil, they should use a wash of the type of Formula No. 2.

Dermatitis Preventives
Formula No. 1

Chlorinated Lime	175 gr.
Sodium Bicarbonate	350 gr.
Boric Acid	35 gr.
Water	To make 30 oz.

No. 2

White Snow Flakes	7.48
Glycerin	26.40
Sodium Silicate	24.20
Tragacanth	0.21
Lemon Oil	0.16
Water	41.60

The paste thus formed is dissolved in water in proportion of 1 lb. of paste to 2 gal. of water.

Workers are advised to use these washes after getting home from work and before starting out in the morning, as follows:

1. Dilute the wash 1 part to 10

parts of water. 2. Apply the wash to the hands, arms, etc. 3. Wash these parts with hot water and soap (not medicated). Dry thoroughly. Such treatment will remove all traces of oil from the skin and will prevent the onset of dermatitis.

A corrosive and irritant alkali can be neutralized by the use of a weak acid, such as acetic acid or citric acid, or even boric acid compresses. The following preparation may be used as a protection against alkaline solutions:

No. 3

Lanolin	9.0
Stearic Acid	50.0
Triethanolamine	2.7
Carbitol	18.0
Saturated Solution of Boric Acid	120.0

Conversely, the use of weak alkalies, like sodium bicarbonate or sodium borate, is indicated in the neutralization of irritant acids.

A second valuable procedure is the use of detoxifying chemicals which act on the irritating materials to form either insoluble or non-irritating chemical compounds. Thus the dichromates can be detoxified by sodium or potassium bisulfite. Dermatitis from tincture of iodine can be neutralized by wet dressings of 5 to 10% sodium thiosulfate. Dermatitis due to fluorides should be helped by repeated applications of a paste made of glycerin or liquid petrolatum and magnesium oxide, followed by wet dressings of 10% calcium gluconate.

Mercurial preparations and chlorinated hydrocarbons belong to a group of preparations which may be absorbed by the external use of McNally's general antidote, made as follows:

No. 4

Magnesium Oxide	2
Activated Charcoal	1
Tannic Acid	1
Purified Fullers' Earth	1

No. 5

Precipitated Sulfur	10.0 g.
Spirit of Camphor	10.0 cc.
Alcohol	80.0 cc.
Solution of Methyl Cellulose (2%)	30.0 cc.
Rose Water	To make 240.0 cc.

In contrast to the curdy mixtures ordinarily seen, this product has a smooth, creamy appearance. The medicaments are finely divided and hence act more efficiently.

Dermatitis Ointment

Sulfur-salicylic acid combinations, among the most widely used in dermatology, are often dispensed in petrolatum, such ointments being exceedingly greasy and with a pronounced odor of sulfur. A preparation with none of these disadvantages is given as follows:

No. 6

Precipitated Sulfur	2.0 g.
Salicylic Acid	2.0 g.
Oil of Rose	0.25 cc.
Non-greasy Base *	To make 30.0 g.

The non-greasy character of this ointment makes it comparable to a cosmetic and permits its use on areas not suitable for a greasy preparation. Since it is of the so-

---

* *The non-greasy base is prepared from:*

Sodium Lauryl Sulfate	0.8 g.
Cetyl Alcohol	15.0 g.
Glycerin	5.0 cc.
White Petrolatum	14.0 g.
Water	35.0 cc.

called washable type, it is easily removed from body surfaces, even hairy parts. The addition of a perfume oil, like oil of rose, helps to disguise the sulfur odor.

No. 7
(For Tetryl Dermatitis)

Shellac	13
Isopropyl Alcohol	31
Linseed Oil	4
Titanium Oxide	12
Sodium Perborate	13
Talcum	20
Carbitol	3

Antiseptics and Germicides
Antiseptic Skin Lotion

Lauryl Pyridinum Chloride	0.1
Propylene Glycol	6.0
Nonaethylene Glycol Monostearate	5.0
Water	84.0

Surgical Skin Antiseptic

Prontosil	3
Acetone	34
Alcohol	62
Glycerin	1

This mixture is more bacteriostatic and less dangerous to tissues than iodine.

Antiseptic Sticks
Formula No. 1

White Gelatin	70.00
Sodium Chloride	1.00
Phenyl Mercuric Acetate	5.38
Glycerin	200.00
Red Bole	20.00
White Bole	120.00
Water	Sufficient

No. 2

Gelatin	70
Euflavin	1
Glycerin	200
White Bole	140
Water	Sufficient

Surgical Instrument Germicide
U.S. Patent 2,347,012

Formaldehyde	4.0
Ethyl Alcohol	70.0
Water	10.0
Thymol	0.063–0.5
Hexamethylenamine	0.063–0.5
Solvent	15.0

The solvent is a mixture of methyl alcohol and acetone, the proportions being 1 to 0.5–2.

Starch Surgical Sponge

A 5% suspension of purified corn starch is pasted by heating and then sterilized by heating in an autoclave for 15–20 minutes at 15 lb. pressure per square inch. The autoclave paste is then placed in shallow pans or other suitable containers and slowly frozen, preferably at a temperature just below 0°C. When freezing is complete, the paste is removed from the freezer and allowed to thaw. The resulting spongy mass may be cut to the requisite sizes and shapes. Sponges with different textures may be prepared by varying the pasting and freezing conditions and the kind and concentration of the starch used. Sterile sponges are prepared by autoclaving and maintaining aseptic technic or the final produce may be immersed in 70% alcohol.

Starch sponges are highly absorbent and there are several methods by which medicinals, such as penicillin, sulfa-drugs or the like may be introduced into a starch sponge. For example, the sponges may be squeezed nearly dry and dipped into the solution. If desired, they may be filled with medicament, dried, and remoistened just prior to use.

Hydrogen Peroxide Stabilizer

35% hydrogen peroxide is stabilized with 0.1 g. 8-oxyquinoline and 0.1–0.3 g. sodium pyrophosphate per lb.

Fumigating Sickrooms

Hang up all bedding, rugs, draperies and clothing; open closets and bureaus; close up all cracks and keyholes where gas can escape from the room (this may conveniently be done by pasting strips of paper over the crevices). To fumigate 800 cu. ft., place 14 oz. of permanganate of potash evenly over the bottom of a wide bucket or basin (an ordinary dishpan will do) which has been raised a short distance off the floor by means of a box or bricks, to prevent injury to the floor from the heat which will be evolved. Pour 1 lb. of formaldehyde into the basin and leave the room immediately. Keep the room tightly closed for 24 hours and then ventilate thoroughly. The temperatrue during fumigation should be above 60°F. In dry weather, steam should be released or wet sheets hung in the room prior to fumigation to keep relative humidity at 65% or above.

Fumigating Cone (Incense)

Powdered Wood Charcoal	960 gr.
Compound Tincture of Benzoin	960 min.
Powdered Tragacanth	60 gr.
Saturated Solution of Potassium Nitrate	To make 2 fl. oz.

Mix the powders with the tincture and make into a soft mass with the solution of potassium nitrate. Form into small cones, allowing them to dry on a radiator, or in cool oven, guarding against overheating. This quantity is sufficient for about 36 cones. When in use, the cones should be placed in a saucer or tin lid.

Alternatively, large filter-papers may be damped with a 1 in 20 solution of potassium nitrate and compound tincture of benzoin, allowed to dry, and afterwards cut into strips.

Chapter V

EMULSIONS, DISPERSIONS AND FOAMS

Hydrocarbon Solvent Emulsions

Hydrocarbon Solvent		Amine		Oleic Acid	Water
Kerosene	89	Triethanolamine	3.2	10	100
Kerosene	89	Monoethanolamine	1.4	10	100
Kerosene	89	Morpholine	2.0	10	100
Naphtha	82	Triethanolamine	4.3	14	100
Naphtha	82	Monoethanolamine	2.0	10	100
Naphtha	82	Morpholine	2.7	10	100
Creosote Oil	88	Triethanolamine	6.0	12	100

Two solutions are made up, to be called the *oil solution* and the *water solution*, respectively.

Mix the solvent with the oleic acid; as they are mutually soluble, a clear *oil solution* forms.

In a separate container, the water and the amine are similarly mixed at ordinary temperature to form a clear *water solution*.

Add the *oil solution* in its entirety to the *water solution*, agitating vigorously. A white emulsion results instantly.

It is important that the stirring be as rapid as possible at the start and be continued for only a short time after the emulsion is formed. A mechanical mixer or homogenizer is essential to produce an emulsion of good stability.

The water-in-oil type of emulsion can usually be produced with triethanolamine oleate if the amount of oleic acid is increased and the proportion of water to solvent is decreased. Any further dilution, then, is more easily accomplished by the addition of solvent rather than water.

Chlorinated Solvent Emulsions

Chlorinated Solvent		Amine		Oleic Acid	Water
Ethylene Dichloride	86	Triethanolamine	4.3	10	100
Ethylene Dichloride	86	Monoethanolamine	1.6	8	100
Ethylene Dichloride	86	Morpholine	2.7	10	100
Dichlorethyl Ether	83	Triethanolamine	7.0	14	100
Propylene Dichloride	86	Triethanolamine	6.0	14	100

According to the following method, which has proved to be the most practical one for making chlorinated solvent emulsions of good stability, the solvent and water are stirred alternately into the soap made from the amine and oleic acid. A rapid mechanical stirrer is desirable to make stable emulsions.

Adequate ventilation should be provided, and special care should be taken to avoid inhaling vapors and repeated contact with the skin wherever chlorinated solvents are used.

Mix the oleic acid, amine, and one-third of the chlorinated solvent and stir until the mixture is homogeneous.

Stirring vigorously, slowly add a volume of water equal to the chlorinated solvent present until a thick, creamy emulsion results.

Add one-half of the remaining chlorinated solvent in small portions, then one-half of the remaining water while stirring continuously.

Then add the rest of the chlorinated solvent and the rest of the water in a similar manner. Emulsification is complete when the chlorinated solvent and water are evenly distributed. Prolonged stirring is usually not needed.

Successful water emulsions of dichlorethyl ether, ethylene dichloride, and other chlorhydrocarbon solvents can be conveniently prepared with Tergitol wetting agent 7, which is an effective emulsifying agent even at concentrations as low as 1 to 2% based on the weight of the chlorhydrocarbon. For example, an emulsion may be prepared by adding a chlorinated solvent to a 5 to 10% aqueous solution of Tergitol wetting agent 7 and then diluting the emulsion with water until the desired concentration of chlorinated solvent is obtained. In general, these emulsions have been found to be less stable, but if separation occurs on storage, the dispersion can be readily restored to complete uniformity by slight agitation. Tergitol wetting agent 7 is particularly adapted for such conditions because it is unaffected by hard water or by dilute solutions of acids, alkalies, and most salts. Hence these conditions could be present at the time of use without affecting the stability of the emulsion.

Oil Emulsions

Oil		Amine		Oleic Acid	Water
Castor Oil	82	Triethanolamine	2.1	16	80
Cottonseed Oil	88	Triethanolamine	2.1	10	80
Lard Oil	87	Triethanolamine	3.2	10	80
Linseed Oil	88	Triethanolamine	2.1	10	80
Olive Oil	88	Triethanolamine	2.0	8	80
Neatsfoot Oil	88	Triethanolamine	2.1	10	80
Neatsfoot Oil	88	Monoethanolamine	0.9	8	80
Neatsfoot Oil	88	Morpholine	1.3	8	80
Pine Oil	91	Triethanolamine	3.2	6	100
Pine Oil	91	Monoethanolamine	1.5	6	100
Pine Oil	91	Morpholine	2.0	6	100
Lubricating Oil	89	Triethanolamine	2.1	7	100
Lubricating Oil	89	Monoethanolamine	0.8	7	100
Lubricating Oil	89	Morpholine	1.2	6	100
White Mineral Oil	82	Triethanolamine	4.3	12	100
White Mineral Oil	82	Monoethanolamine	1.6	10	100
White Mineral Oil	82	Morpholine	2.3	10	100
White Mineral Oil	82	Mixed Isopropanolamine	3.8	10	100

The procedure consists of stirring the oil and water alternately into the soap made from the amine and oleic acid.

At ordinary temperatures, add the oleic acid and amine to one-third of the oil and stir until the mixture is fairly homogeneous.

Stirring vigorously, slowly add a volume of water equal to the oil present in the mixture, or sufficient water to produce a thick, creamy emulsion.

Add one-half of the remaining oil in small portions with constant stirring, and then one-half of the remaining water, a little at a time.

The rest of the oil and then the remaining water are added in the same manner. Emulsification is complete when uniformity is attained.

Stearic acid can be used in place of oleic acid, but the mixture of stearic acid and one-third of the oil must be heated sufficiently to melt the acid before the amine is added. A volume of water, equal to the volume of oil that is added to the soap-and-oil mixture, should be heated to the same temperature as the oil and stearic acid. The rest of the oil and the rest of the water can be added at room temperature.

When the emulsion is to be used shortly after preparation, the proportions of amine and oleic acid can be considerably reduced. This is done by stirring additional oil and water alternately into the original emulsion by the procedure given above.

Cocoa Butter Emulsion

Cocoa Butter	1
Glycerin or Water	1

Warm the cocoa butter and then add glycerin or water slowly while whipping. This gives a foamy mass.

Benzyl Benzoate Emulsion

Benzyl Benzoate	25
Sulfonated Castor Oil	10
Water	To make 100

Methyl Salicylate Emulsion

Methyl Salicylate	25
Sulfonated Castor Oil	10
Water	65

Sulfathiazole Emulsion

Sulfathiazole	5.0
Triethanolamine	2.0
Distilled Water	24.0
Beeswax	5.0
Liquid Paraffin	64.0

Glycol Ester Resin Emulsions

Formula No.	1	2	3	4
Resin: Solvent[1] 80:20	95		100	
Resin (Without Solvent[2])		84		80
Potassium Hydroxide	0.6	0.6		
Oleic Acid	2.3	2.5		
Sulfated Castor Oil	1.7	2	0.8	2
Duponol ME			0.8	2
Water	100.44	110.9	138.4	136

[1] For Flexalyn or Flexalyn C, Staybelite Esters No. 1 or 2, or Polypale Esters No. 1, 2, or 3. Solvent used for the 80:20 resin-solvent solution is usually a hydrocarbon such as Varsol, toluene, or Tollac.

[2] Staybelite Ester No. 3; this resin, being a liquid, requires no solvent.

DDT Emulsion
Formula No. 1
(Non-Solvent Type)

It is often desired to apply DDT in its pure state without dissolving it in any of the various solvents. The following method and formula produces a very finely dispersed product.

DDT	25
Gum Arabic	10
Water	65

Dissolve the gum in water and heat to 195°F. Heat the DDT to 195°F. Pour both into a colloid mill or homogenizer simultaneously.

* Other formulas for DDT emulsions will be found in the section on insecticides.

No. 2

DDT	25
Xylol	90

Mix until dissolved, then add Nonaethylene Glycol Oleate S 10.

The above is a concentrate which is mixed with three or more times its volume of water.

Viscous Wax and Grease Emulsions

Wax		Amine		Stearic Acid	Water
Beeswax	88	Triethanolamine	3.2	9	300
Carnauba Wax	87	Triethanolamine	4.9	9	400
Japan Wax	85	Triethanolamine	3.2	12	400
Paraffin Wax	88	Triethanolamine	3.5	9	300
Lanolin	80	Triethanolamine	5.4	15	200
Petrolatum	88	Triethanolamine	3.5	9	100
Petrolatum	88	Monoethanolamine	1.8	9	100
Petrolatum	88	Mixed Isopropanolamine	3.7	8	100
Petrolatum	88	Morpholine	2.2	8	100

Any one of the wax- or grease-emulsion formulas suggested with triethanolamine can be used with monoethanolamine, mixed isopropanolamine, or morpholine substituted on an *equivalent weight* basis. When one of these amines is used, the amount of fatty acid can often be decreased without affecting the stability of the emulsion.

The wax or grease is emulsified by means of a water solution of the soap made from the amine and stearic acid:

Measure the water, amine, and stearic acid into a container or kettle and heat to about 100°C. When heated, the stearic acid gradually melts and can be stirred into the water to give a smooth soap solution.

After the stearic acid has melted completely, allow the mixture to boil gently. Stir carefully so that a smooth soap solution is obtained with a minimum amount of foam.

In a separate steam-heated container, melt the wax or grease and bring the temperature between 85° and 90°C.

Add the hot wax or grease to the boiling soap solution and stir vigorously until an even dispersion is obtained.

Stir gently, but continuously, until the emulsion has cooled to room temperature in order to prevent the emulsion from becoming too viscous or the wax from graining out.

The substitution of oleic acid for stearic acid in these formulas produces less viscous emulsions. It therefore permits a considerably higher concentration of wax or grease to be used. When other ingredients are to be incorporated in the emulsion it is best to include them before emulsification. Ingredients soluble in water are added to the soap-and-water solution, the others to the melted wax or grease, or with a solvent if necessary.

Low Viscosity Wax Emulsions

This method is especially adapted to making emulsions of low viscosity and good stability without the difficulties of wax separation encountered in diluting a viscous wax emulsion to a very low concentration. It is applicable to waxes such as carnauba or beeswax, and to oils and greases or low melting waxes that contain saponifiable oils or free fatty acids. A small proportion of carnauba wax, beeswax, or casein is used with paraffin wax to make emulsions that can be diluted with water to low wax content and still maintain good stability. The film of the morpholine emulsions will be water-resistant, as morpholine leaves the film with the water, producing a dried film not re-emulsifiable upon the addition of water.

The following seventeen typical formulas (reading across) can be used to make low viscosity wax emulsions.

Waxes		Amine		Fatty Acid	Water
				STEARIC ACID	
Paraffin Wax.. 45	Carnauba Wax. 3	Triethanolamine ..	6.0	13.5	225
Paraffin Wax.. 45	Carnauba Wax. 3	Monoethanolamine	2.6	9.0	225
Paraffin Wax.. 45	Carnauba Wax. 3	Morpholine	3.7	9.0	225
Paraffin Wax.. 45	Beeswax 3	Triethanolamine ..	6.0	13.5	225
Paraffin Wax.. 45	Beeswax 3	Monoethanolamine	2.6	13.5	225
Paraffin Wax.. 45	Beeswax 3	Morpholine	3.7	13.5	225
Paraffin Wax.. 40	Casein 5	Monoethanolamine	2.6	13.5	225
Paraffin Wax.. 40	Casein 5	Morpholine	3.7	13.5	225
				OLEIC ACID	
Paraffin Wax.. 40	Carnauba Wax. 4	Triethanolamine ..	6.0	11.0	225
Paraffin Wax.. 40	Carnauba Wax. 4	Monoethanolamine	2.6	9.0	225
Paraffin Wax.. 40	Carnauba Wax. 4	Morpholine	3.7	9.0	225
Carnauba Wax 15	———	Triethanolamine ..	1.5	2.3	90
Carnauba Wax 15	———	Monoethanolamine	1.0	1.7	90
Carnauba Wax 15	———	Morpholine	1.3	1.7	90
Beeswax 15	———	Triethanolamine ..	1.5	2.6	100
Japan Wax.... 15	———	Triethanolamine ..	1.5	3.0	100
Ester Gum.... 23	———	Triethanolamine ..	0.8	4.2	100

The emulsions of paraffin wax with carnauba wax and stearic acid are the most stable of all those suggested and show good stability even in extremely high dilutions. Casein can be used in place of carnauba or beeswax, although the resulting emulsions are less stable than those containing carnauba wax. For best results, oleic acid is not recommended with paraffin wax when either casein or beeswax is used. A mold inhibitor should be added to the hot melted wax mixture just before the water is added when casein is used in the formulation. About 5% of phenol, based on the casein, may be used.

Melt the waxes in a hot water or steam-jacketed kettle, add the oleic or stearic acid, and bring the temperature to 95°C. When casein is used, the paraffin wax-stearic acid mixture is heated to only 75°C. before the dry casein is stirred in.

Add the amine slowly, stirring constantly until the solution clears.

Heat the water to boiling and add it slowly, a little at a time, thoroughly incorporating each small portion before another addition is made. When the water is first added, the mixture becomes increasingly viscous. When from one-third to two-thirds of the water has been added, it begins to thin out again. The emulsions containing casein, however, do not become as viscous during the addition of water as those made with the waxes. The water can be added more rapidly to emulsions without casein, but the rate of addition should be adjusted by experiment.

After the mixture has become definitely thinned and is of *smooth, even* consistency, the rest of the water can be added slowly, but continuously, with constant stirring.

The emulsion should be stirred slowly, but continuously, until it has cooled to room temperature.

Stearic acid can be substituted for oleic acid in all the formulas. While the resulting emulsions may be slightly more viscous, the stability should be equally as good or better than when oleic acid is used.

If a more dilute emulsion is desired, more water can be added either at the time of making the emulsion or when it is ready for use.

Irish Peat Wax Emulsion

Irish Peat Wax	0.2
Mineral Oil	20.0
Water (Hot)	<79.0

Warm until wax is dissolved and then add the water in small amounts with good mixing.

Acrawax Emulsion

Acrawax	28
Stearic Acid	6

Melt together and add slowly, with good stirring a mixture of:

Water (Boiling)	62
Monoethanolamine	4

Acrawax C Emulsion

Acrawax C	50
Triethanolamine	5
Oleic Acid	5
Water	150

The Acrawax C and oleic acid are heated together and when melted, the triethanolamine is added. When the reaction is complete, the temperature is brought down as close to the setting point as possible. The stirrer is introduced, and while stirring, the water is added. There is a certain amount of steaming, but this subsides rapidly. The resulting grainy mixture is then passed through the colloid mill and a smooth homogeneous product results.

Fatty Acid-Amide Emulsion

Commercial Octadecane Amide	
Stearic Acid	10.0
Ammonia Water (28%)	10.0
Water	940.0

Mix the amide and the stearic acid together and melt at 110°C.

Heat the water to boiling in a separate container and shut off the heat. Add the 28% ammonia to the water. Stir vigorously while the fatty acid-amide mixture is poured slowly into the water and ammonia. Stir occasionally until cool. To make a better emulsion put through a homogenizer.

Amides may be easily emulsified by this formulation or by suitable modifications.

The emulsifying agent may be varied at will to give different properties to the emulsion. Oleic acid, lauric acid, linoleic acid or any other fatty acids or mixtures of fatty acids may be substituted for the stearic acid shown in the formula. The neutralizing alkali may be changed from ammonia to triethanolamine, sodium hydroxide or potassium hydroxide in appropriate amounts, to neutralize the fatty acid used.

Other emulsifying agents such as casein or vegetable protein may be used.

Soluble Mineral Oil Emulsions

Emulsions of the *soluble* mineral oils are the oil-in-water type. Properly made, these emulsions are stable, neutral, and non-corrosive while providing good lubrication. produce a slightly superior *soluble* oil, that is, one that will remain clear on long standing and that will produce a stable emulsion. The amounts of oleic acid are to-

Oil		Amine		Oleic Acid
Light Mineral Oil	88	Triethanolamine	3.8 to 4.0	8.0 to 10.0
Light Mineral Oil	88	Monoethanolamine	1.6 to 1.8	7.6 to 9.7
Light Mineral Oil	88	Morpholine	2.1 to 2.3	7.3 to 7.5
White Paraffin Oil	85	Triethanolamine	4.2 to 4.5	10.0 to 12.0
White Paraffin Oil	85	Monoethanolamine	1.7 to 1.9	8.0 to 8.5
White Paraffin Oil	85	Morpholine	2.2 to 2.5	7.0 to 7.5
White Paraffin Oil	85	Mixed Isopropanolamine	3.7 to 3.9	8.0 to 8.5

It is always necessary to derive an optimum formula for a specific oil to be emulsified because commercial petroleum products vary greatly, and because formulation by this method requires great exactness. The procedure, however, is comparatively simple and easy to follow.

Derivation of the Optimum Formula

For triethanolamine, for instance, weigh 88 g. of the oil to be *solubilized* in a glass container, add 8.0 g. of oleic acid, and stir to a clear solution. Weigh 4.0 g. of triethanolamine into this solution and stir thoroughly. If the container is held to the light, the mixture will usually appear cloudy or show minute suspended droplets. Now add oleic acid, drop by drop, stirring thoroughly after each addition, until the mixture becomes clear. It should now emulsify in water, but a few more drops of oleic acid may

taled and the formula can be reduced to the basis of 100 lb. for large batch production. This procedure is also followed for the other amines.

If the emulsion obtained from the formulated *soluble* oil does not have the necessary stability, increase slightly the amine content of the oil and then make clear with oleic acid, adding a little at a time, as described above. When less stability is required of the emulsion, the amine and oleic acid contents can be reduced. This is accomplished by merely adding more mineral oil to the *soluble* oil until an emulsion of the desired stability is obtained.

Weigh out the oleic acid as determined above and 10 lb. of the mineral oil. Stir to obtain a uniform solution.

Add the exact amount of amine as determined above and stir until the solution is *clear*. Some warming of the solution will occur because of the reaction of the oleic acid and the amine.

This mixture, which is the *soluble* oil base, can be diluted immediately by stirring in the rest of the mineral oil, or it can be stored and diluted when desired. Both the *soluble* oil base and the resulting *soluble* oil are stable indefinitely when made in the proper proportions.

The *soluble* oil emulsifies spontaneously when poured into water. However, the best method of emulsifying is accomplished by stirring the *soluble* oil with an equal volume of water until a creamy mass is obtained, and then diluting further with water as desired.

Soluble Vegetable and Animal Oil Emulsions

Vegetable or animal oils are not made *soluble* with the same success as mineral oils. However, good *soluble* oils of this type can be made with morpholine. These emulsions are sufficiently stable, if made as directed, for most purposes. They should not separate on long standing, if stored in closed containers. When considering the use of a *soluble* vegetable or animal oil made with morpholine, it must be borne in mind that morpholine will evaporate from a thin film of the oil on exposure to the air, leaving the film water-resistant so that it cannot be re-emulsified later by water.

The following three formulas are given as suggestions for some of the oils that can be made soluble with morpholine.

Oil Type	Oleic Acid	Morpholine
Linseed Oil	88	8	2.8
Neatsfoot Oil	88	8	2.8
Olive Oil	88	8	2.8

Different shipments of the same oil will vary in free fatty acid content so that a definite formula cannot be given for a specific oil, such as olive oil. The same method of procedure is used as described for formulating a *soluble* mineral oil. Mix the oil and oleic acid, add the morpholine, and stir until thoroughly mixed. If the mixture is not clear, add oleic acid, drop by drop, stirring well after each addition, until the mixture becomes clear. The exact amount of oleic acid can then be calculated.

The best emulsions of the *soluble* oils are made by adding water, a small amount at a time, to the *soluble* oil, each addition being thoroughly incorporated before another addition is made.

When the mixture begins to thin, the remainder of the water can be added slowly but continuously, with constant stirring, until the desired dilution is obtained.

Soluble Solvent Emulsions

Kerosene can be made *soluble* with monoethanolamine or morpholine and oleic acid. This is an economical method for producing emulsions where the product is to be shipped, but if the kerosene is to be emulsified directly, use the procedure given under Hydrocarbon Solvent Emulsions, which is more efficient since less soap is needed. The emulsions are the water-in-oil type, which may change to an oil-in-water upon standing several weeks.

In chlorinated solvents, triethanolamine hydrochloride slowly precipitates; hence, the *soluble* solvent should be emulsified promptly. *Soluble* ethylene dichloride is stable for a longer time when made with either monoethanolamine or morpholine and oleic acid. These emulsions are the oil-in-water type.

The proportions of ingredients are given by weight in the five suggested formulas:

Adequate ventilation should be provided, and special care should be taken to avoid inhaling vapors and repeated contacts with the skin whenever chlorinated solvents are used.

Add the oleic acid to the solvent. Stir in the amine. If the mixture is not clear, add oleic acid, a little at a time, until the mixture becomes clear.

Emulsions of good stability are obtained by stirring 1 part by volume of this *soluble* solvent into 1

Solvent		Amine		Oleic Acid
Kerosene	85.5	Monoethanolamine	1.7	12.8
Kerosene	86.0	Morpholine	2.6	10.4
Ethylene Dichloride	92.0	Triethanolamine	3.0	8.5
Ethylene Dichloride	92.0	Monoethanolamine	1.4	7.2
Ethylene Dichloride	92.0	Morpholine	2.1	7.2

part of water, using a mechanical stirrer to produce the most stable emulsions.

When a creamy emulsion is obtained, further dilution can be made as desired by stirring in the required amount of water.

Industrial Mineral Oil Emulsion
U.S. Patent 2,359,503

A mixture of 2.4 kg. mineral oil and 1.6 kg. talloil acids are heated to a temperature of about 100°C. and 0.3 kg. of a 47% aqueous solution of potassium hydroxide heated to approximately the same temperature is stirred into the heated oil mixture in small doses and with vigorous stirring. Thereafter 5.6 kg. warm water is similarly vigorously stirred in, and during or after the addition of the water 0.1 kg. sodium diphosphate is added. After the temperature has fallen to about 50°C., 0.17 kg. of a 47% aqueous solution of potassium hydroxide is carefully but vigorously stirred in.

Cutting Oil Emulsion
U.S. Patent 2,174,907

Mineral Oil	54.08
Palm Oil Fatty Acids	7.28
Triethanolamine	3.64
Water	35.00

Soluble Oils

Formula	No. 1	No. 2	No. 3	No. 4	No. 5	No. 6
Mahogany Soap	...	...	...	...	7.0	5.0
Sulfur Oil	...	...	4.0	2.0	...	...
Lard Oil Or Tallow	4.0	...	...	...	...	...
Tall Oil	6.0	12.0	12.5	12.0	7.0	7.0
Sulfuric Acid (66° Bé)	1.5	1.0	...	...	...	...
Caustic Potash (45%)	5.3	5.3	4.3	3.4	1.9	2.35
Diethylene Glycol	3.0	3.0	2.0	2.0	1.5	1.1
Texas Oil (100–300)	80.1	78.6	77.1	80.5	82.5	84.45
Cresylic Acid	0.1	0.1	0.1	0.1	0.1	0.1

Soluble Neatsfoot Mineral Oil		Polyglycol (400) Mono-	
Mineral Oil	42	laurate	5
Neatsfoot Oil	42	Water	1
Diglycol Laurate S	10		

Soluble Pine Oils

Formula	No. 1	No. 2	No. 3	No. 4
Wood Rosin	20.8	...	18.8	...
Tall Oil	...	16.0	...	12.0
Caustic Soda (50%)	4.9	4.2	4.4	...
Caustic Potash (45%)	...	...	...	3.7
Pine Oil (Phenol Coefficient 7)	...	72.0	...	5.0
Pine Oil (Phenol Coefficient 4.5)	67.7	...	70.0	...
Low-Viscosity Mineral Oil	...	...	...	77.3
Diethylene Glycol	...	...	...	2.0
Water	6.6	7.8	6.8	...
Phenol Coefficient Of Disinfectant	3	5	3	...

Emulsions with Pectin

One gram of pectin can be substituted for 12.5 grams of gum arabic in the preparation of emulsions and the resulting emulsions can be diluted with either oil or water. However, each oil requires its own formula and the primary emulsion cannot contain more than 50% oil. For a stable emulsion, the pectin must be given at least 10 minutes' time to swell.

To make a mineral oil emulsion, 1 gram of pectin is triturated with 25 cc. of mineral oil, then 25 cc. of water is added and trituration continued until emulsification starts. Then the mixture is allowed to stand for 10 minutes or more and mixed until a thick creamy emulsion forms. After this 25 cc. of mineral oil are added slowly, followed by 10 cc. of fruit syrup, then by 9 cc. of water and 0.004 g. vanilla in 6 cc. of ethyl alcohol. For a cod liver oil emulsion, the alcohol and vanilla are omitted and 0.4 cc. of methyl salicylate and a total of 40 cc. of water are used; following the same procedure.

Zein Dispersion

Formula No. 1
U.S. Patent 2,360,081

Zein	10
Alcohol (90% Vol.)	20
Sulfated Stearyl Alcohol	10
Water	100

No. 2
U.S. Patent 2,377,237

Zein	15
Sulfonated Castor Oil	8–15
Urea	5–15
Glycerin	2–6
Butyl Alcohol	1–5
Water	To make fluid

Chapter VI

FARM AND GARDEN PREPARATIONS

Seedless Tomato Culture

To produce seedless, and better and larger tomatoes, the blossoms should be sprayed just as soon as they open (by using an atomizer and taking care not to spray the foliage of the plant), with the following solution:

2,4-Dichlorophenoxy-acetic Acid	1 g.
Water	1 gal.

A second spraying is usually helpful.

Improving Tomato Yield (Plant Hormone)

Naphthylacetic Acid	3
Cyclohexanone	27
Dimethyl Ether	270

Spray on tomato blossoms.

Preventing Sprouting of Stored Potatoes

Best compound for the purpose is one of the growth-controlling hormones, the methyl ester of alpha-naphthylacetic acid. This can be applied as a spray, dust, or an emulsion. Nine-tenths of 1 g. per bushel is sufficient. This works out as about three ounces of the chemical for 100 bushels of potatoes.

Plant Growth Regulator Formula No. 1

Indole Butyric Acid	0.2 g.
Lanolin, Anhydrous	20.0 g.
Pharmagel B	2.0 g.
Sorbitol	0.5 g.
Sodium Bicarbonate	0.5 g.
Water	80.0 ml.

No. 2

Indole Butyric Acid	0.2 g.
Wax Blend No. 1 *	20.0 g.
Pharmagel B	2.0 g.
Sorbitol	0.5 g.
Sodium Bicarbonate	0.5 g.
Water	80.0 ml.

No. 3

Indole Butyric Acid	0.2 g.
Wax Blend No. 1	20.0 g.
Polyvinyl Alcohol (RH–403)	2.0 g.
Sorbitol	0.5 g.
Sodium Bicarbonate	0.2 g.
Water	80.0 ml.

Hydroponic Plant Food (U.S. Army)

Potassium Nitrate	9 lb. 3 oz.
Calcium Sulfate	6 lb. 6 oz.
Magnesium Sulfate	4 lb. 6 oz.
Monocalcium Phosphate	2 lb. 10 oz.
Ammonium Sulfate	1 lb. 3 oz.
Water	1000 gal.

Use cheapest commercial grades.

Cleaning and Disinfecting Fruits
U.S. Patent 2,374,209

Products are prepared for shipment to market by scrubbing and

---

* Wax Blend No. 1.

Carnauba Wax	45
Anhydrous Lanolin	50
Cetyl Alcohol	5

soaking them for four minutes in an aqueous solution containing 0.2% soap and 0.5% sodium carbonate as detergents, 0.15% sodium orthophenylphenolate as a disinfectant, and 1% of sodium pyrophosphate as a buffer to maintain the pH above 10. The surface solution is then rinsed off. Blemishes and parts with decay are penetrated permanently.

Tree Wound Dressing
Formula No. 1

Rosin	7
Fish Oil	3
Copper Naphthenate	3

Warm and mix until uniform.

No. 2

Rosin	8
Sardine Oil	3

No. 3

Rosin	7
Sardine Oil	3
Copper Oleate	3

Stimulating Gum Rosin Yields

Applying 40% sulfuric acid to very shallow streaks which remove only the bark but do not penetrate the wood. Experimentally this method increases gum yields 60 to 75%. Another method, applying acid to streaks made every third week, shows promise of about normal production with a third of the labor.

Cut Flower Preservative
Formula No. 1
U.S. Patent 2,317,631

Hydrazine Sulfate	23–43
Manganese Sulfate	42–82
Calcium Hypochlorite	3–5
Sugar	3125–5125

Use 1¼ oz. of the mixture per gallon of water. A little alum may be added to harden the stems.

No. 2
German Patent No. 554,512

1.5 parts of methylcellulose are dissolved in 100 parts of water. Cut flowers dipped in this solution retain their original appearance for a considerable length of time. The thin film of methylcellulose imparts some mechanical strength to tender flower leaves.

Preservative for Cut Orchids
U.S. Patent 2,367,795

Anhydrous Sodium Sulfate	12–18
Aluminum Sulfate	38–40
Hydrazine Sulfate	0.015
Activated Carbon	0.015
Water	42–54

Preventing Seed Damping-Off

1. Standard Drench Method

Ten days or 2 weeks before planting drench the soil with a solution of 1 part formaldehyde in 50 parts of water by volume at the rate of ½ gal. to each square foot of surface. The soil may be covered with paper for 12 to 24 hours to partially delay escape of the formaldehyde. Work the soil well several times, allowing the excess of formaldehyde to escape, and do not plant until 10 days have elapsed.

2. Improved Method
(Developed by New York State Agricultural Experiment Station)

Essentially, this method consists of mixing known amounts of rather concentrated formaldehyde with known amounts of soil, substituting thorough mechanical mixing for the excess water employed in the drench method. This treatment is applicable particularly to greenhouse work where seedlings are grown from seed in flats for transplanting or resale.

Mix 2½ level tbsp. of formaldehyde with 6 times this amount of water (1 part 37% formaldehyde to 6 parts water). Sprinkle this solution over one bushel of soil consisting of one-half sand and one-half soil. Mix thoroughly and place in flats or other containers and allow to stand for 12 to 24 hours before seeds are sown. Soon after sowing the seeds, water the flats thoroughly. This serves to release the formaldehyde fumes and avoids injury to the seeds. The above amount of treated soil is sufficient for a flat of 700 sq. in., 2¾ in. deep. A pound of formaldehyde is sufficient for 62 sq. ft., 2¾ in. deep.

Precautions: Before applying the formaldehyde solution, the moisture content and friability of the soil should be in proper condition for sowing the seeds. Thoroughly mix the solution with the oil. Wait 12 to 24 hours before sowing the seeds, and at least twice as long before pricking off or transplanting seedlings. If the soil is kept in a pile more than 3 in. deep it must be allowed to air proportionately longer. *Water the flats thoroughly soon after the seeds are sown.* With soil mixtures in which the ratio of sand to soil is greater than one to one, *reduce* slightly the amount of formaldehyde and increase it slightly if high in leaf mold, peat, muck and similar materials.

The advantages of this method are that it is relatively inexpensive, the labor is not excessive, no unusual equipment is required, it assures the application of a known amount of formaldehyde to a known amount of soil, its effectiveness in the control of damping-off has been demonstrated for many vegetables and finally it enables one to obtain the disinfection of the soil, the seeds and the flats or containers in a single operation.

3. Post-Seedling Method

This method is perhaps the easiest of all ways to improve germination and stands of seedlings by protecting them against soil fungi which cause damping-off. It is simple and is applicable to flats, hot beds, cold frames, greenhouses and plant houses. It has the further advantage of being a safe and effective chemical treatment which can be applied to the soil shortly before germination, i.e., immediately after the seeds are sown and covered with soil.

Immediately after the seeds are planted and covered, wet soil with 1 or 2 tbsp. (not over 2 tbsp.) formaldehyde solution in 3 gal. of water at the rate of 1 gal. to 5 sq. ft. of seed bed.

A less discriminating method but one effective with most vegetables except crucifers consists in applying a solution of formaldehyde, 1 tsp. in 1 gal. of water, immediately after seeding without determining the exact rate of application of the solution.

Treatment of Vegetable Seeds

Many diseases of vegetable crops are caused by bacteria or fungi that are carried in or on the seed. Seed disinfection is one of the important ways of fighting these diseases.

Beets—Leaf Spot
Use 15 parts 37% (U.S.P.) formaldehyde to 1,000 parts of water (1 lb. to 8 gal. or 4 tbsp. to 1 gal.). Dip the seed for 7 minutes, rinse in water and plant at once or dry.

Celery and Celeriac—Bacterial Blight
Pre-soak the seed in lukewarm water 15 to 30 minutes; then soak in a solution (1 tbsp. 37% formaldehyde to 1 gal. water) for 15 minutes. Rinse 15 minutes. This treatment usually causes some retardation.

Onion—Smut
For the prevention of onion smut in dry soil, apply a solution of 1 lb. of 37% formaldehyde to 16 gal. of water in the furrow with the seed; use about 16 gal. of this solution per 3,000 ft. of row (about 200 gal. per acre). In wet soil, apply a solution of 1 lb. to 10 gal. of water, and use 10 gal. per 3,000 feet of row (about 125 gal. per acre). This solution can be conveniently applied by means of an attachment on the seeder.

Sweet Potatoes
Although formaldehyde is not generally employed for the treatment of the seed stock, it is recommended for sterilizing hot bed frames, containers, utensils and storage houses as a part of the program for control of black rot, foot rot, Java black rot, dry rot, rhizoctonia, etc.

Sand or soil that is not contaminated with sweet potato disease organisms should be used in the hotbed. Previously used hotbed frames should be cleaned and then sterilized by wetting them thoroughly with a solution of 1 lb. of 37% formaldehyde solution to 15 gal. of water. The solution may be sprayed on the frames, applied with a broom or any other convenient method may be used whereby the frames are thoroughly wetted with the formaldehyde solution.

Rhubarb (Roots)—Foot Rot and Crown Rot
Soak rhubarb roots for ½ hour in a solution of 1 lb. of formaldehyde to 30 gal. of water.

Mushroom Disease Control— Bubbles Disease
Two lines of attack—sanitation and fumigation—are absolutely essential in order to control the bubbles disease. Sanitation is of primary importance and without it fumigation is only half effective. All diseased material must be removed, disinfected or destroyed as soon as the disease appears in order to prevent the reproduction and spread of the disease.

Places where the fungus appears, the houses, ground about the composting yard, the beds—wherever the disease might be carried—should be thoroughly sprayed or otherwise treated with 1 gal. of formaldehyde to 45 gal. of water. Fumigation with formaldehyde has been found to be the most satisfactory means of control of the bubbles disease.

Tobacco

Prevention rather than cure is the keynote in the control of most tobacco diseases. Several of the more important diseases, such as mosaic and bacterial leaf spots, commonly originate in the seedbed. Other diseases wholly or partially harbored in the soil are damping-off or bed rot, black root rot, black shank, sore shin, and possibly brown root rot. No old tobacco material of any sort should be allowed to reach the seedbed and the seedbed should not be located near curing barns, tobacco fields or weedy areas. Unless woodland is used for the seedbed the soil should be sterilized preferably by thorough steaming or by treatment with formaldehyde. Unless new, the frames and covers of the seedbed should be disinfected by spraying or washing with a solution of 1 part 37% formaldehyde solution in 25 parts of water. Cover with paper, canvas or bags for 24 hours to hold in the formaldehyde gas. To sterilize the soil, drench with 1 part of 37% formaldehyde to 50 parts water.

Ornamentals

Many flowers and ornamentals are subject to damping-off and seed-borne diseases which can be controlled by the use of formaldehyde.

A few of the ornamentals which have responded well to methods of damping-off control are: Calendula, China aster, clarkia, gypsophils, kochia, larkspur, lunaria, marigold, scabiosa, stock, straw flower, sweet pea and zinnia.

Anchusa, campanula and snapdragon are susceptible to formaldehyde injury, and the improved method should be employed, with a full 24-hour period allowed before planting seeds of these ornamentals.

Narcissus Bulbs

For control of basal rot fungus and to facilitate the killing of nematodes, soak the bulbs when in the maximum state of dormancy for 4 hours in a solution of 1 lb. (1 pt.) 37% formaldehyde solution in 25 gal. of water at 110°–111.5°F.

Seed Potato Treatment

Many seed potatoes planted each spring are infected with diseases which cut down the stand and reduce the yield. In many instances, such potatoes can be made into good seed by treating them with formaldehyde. This treatment is recommended for black scurf or rhizoctonia, and under certain conditions, common scab, black leg and bacterial ring rot. The treatment will not make good seeds out of culls; little and badly diseased potatoes should be discarded.

For Seed Surface Borne Diseases
When to Treat:
1. Just before planting.
2. In advance of planting provided potatoes are properly dried and stored in disinfected containers or bins. Treatment some time in advance of planting is advantageous, so that if the sprouts are injured, new ones can form before planting time.

Presprinkling:
The black scurf or sclerotia of rhizoctonia on tubers is composed

of hard, compact fungous tissue. Sprinkle the tubers with water and keep them covered about 48 hours before treatment. This procedure starts the growth of the fungus and formaldehyde treatment is much more effective.

Hot Treatment:

The hot-formaldehyde method is especially satisfactory when large quantities of seed potatoes are to be treated, for it saves a considerable amount of time. The solution may be used in wood or metal tanks. It does not weaken with use and can be used indefinitely if more solution is added at intervals to replace that carried off on the tubers.

Soak uncut seed potatoes no longer than 3 minutes in a solution of 1 pound formaldehyde to 15 gal. of water heated to 124—126°F. It is essential to keep the temperature within these limits during the treatment; use an accurate thermometer. Remove potatoes from the solution, drain and cover them with canvas for 1 hour, then dry.

Preparation of the Solution

Add water to the tank until it is two-thirds full. Measure or estimate in gallons amount of water used (1 gal. = 231 cu. in.). Mark water level on the tank. Add 1 pt. of formaldehyde to every 15 gal. of water. Prepare a reserve supply of formaldehyde solution in prescribed strength in a barrel. Keep this close at hand for use in replenishing losses of solution in tank due to adherence to potatoes during treatment and also to lower quickly the temperature in the tank. Condensation from live steam dilutes the solution and to maintain the proper solution strength in the tank, full strength formaldehyde (1 lb. for every 55 bushels of potatoes treated) should be added at intervals. Keep solution to proper level by adding water up to the water line previously marked on tank. Dirt washed off potatoes accumulates and interferes with the heating of the solution. For best results, treating equipment should be cleaned out at the end of each day and a fresh solution started the following day.

Tank and Equipment

Small lots of potatoes can be readily treated in a common washboiler on a cookstove. For larger lots, a 60–80 gal. tank of wood, metal or concrete may be used. Place slat floor a few inches from bottom of the tank to prevent potatoes from scorching and provide space for settlement of mud washed off the tubers.

A large tank makes it easy to maintain the solution at a uniform temperature. Convenient size for large capacity treatment is a tank 12–14 ft. long, 3½–4 ft. wide, 2½–3 ft. deep, which permits 8 crates to be in solution all the time. Every half-minute a crate at one end is removed and another is added at the other end. This gives the desired 3-minute treatment. A carload is treated in 3–5 hours.

Potatoes may be treated in crates, wire baskets or in bulk. If containers are used, the work is less laborious if there is an overhead scaffolding and pulleys for removing the potatoes from solution. Very large quantities of bulk or sack potatoes can be handled efficiently by a slow-moving belt conveyor, time-regulated to keep the potatoes in the hot formaldehyde solution for the entire treatment time.

Special Precautions
1. Control the temperature accurately by means of a tested thermometer. A floating dairy thermometer is most satisfactory.
2. Time the treatment exactly—do not guess. By using care, any injury or delay in germination can be minimized.

Heating Solution
Solution can be heated by steam, gas or oil burners and wood or coal fires, the last being least desirable. Either live high pressure steam applied direct or through a steam coil in the tank is suitable. Use an oversize coil, so that the temperature can be brought up quickly. Steam is available in any community from portable engine boilers, or the boilers at creameries, mills, power plants, etc. Place treating tank close to source of steam, with cut-off valve on steam line near tank.

Cold Method Treatment
For many years formaldehyde has been used in cold water solution for the elimination of potato scab on the seed tubers. This form of treatment is efficient for this disease but it is not recommended for black scurf or rhizoctonia.

Soak the uncut tubers 1½ hours in a solution of 1 lb. formaldehyde and 30 gal. of water. This soaking period may be decreased to ½ hour and effectiveness increased if the tubers are first dipped in or sprinkled with water and covered with burlap sacks to keep them moist for a day or two before treatment.

Ring Rot
Most ring rot comes from use of seed infected with the disease. Ring rot is so infectious, however, that it may be spread also by means of barrels, racks, planters, or any other equipment such as a cutting knife that is used with infected potatoes and later used in handling ring-rot-free tubers that are to be planted.

Ring rot does not live over the winter in the soil in rotted material, but lightly infected potatoes may live over and come up as volunteers. It is not wise to replant on land where ring rot was found in potatoes the previous year. Ring rot apparently is not spread from plant to plant by insects, nor is it carried by the wind.

With clean, uninfected seed, ring rot can be eliminated in one season provided that: all infected potatoes are sold or removed from the farm; every piece of equipment should be carefully disinfected before the seed is brought on the farm; planting of new seed is done on sod land to avoid infection from volunteer plants.

Growers who cannot make a complete change of seed can rid their farm of ring rot in two years. This method requires purchase of enough uninfected seed to plant a seed plot large enough to supply all seed needed the following year. The following precautions must be observed: plant the seed plot before any other potatoes are planted; use disinfected barrels, knives and equipment; plant the seed plot by hand or with a disinfected planter; spray, hoe and cultivate the seed plot first, and disinfect all equipment before working in the seed plot; dig the seed plot first, using disinfected barrels and storing the crop in a disinfected bin; be sure that the clean seed does not come in contact with anything that may carry disease; when planting the clean seed the following year, don't take a chance and fill out the field with any other seed; the seed plot should be one-tenth of the acreage to be planted the following year.

To Disinfect Equipment With Formaldehyde

Mix 1 pint of 37% formaldehyde in 15 gal. of water. To kill the ring rot organism, disinfectants must come in contact with it, so make sure that all dirt is removed by scraping or washing before barrels are dipped in the disinfecting solution, and before bins or machinery are sprayed. An oil drum split lengthwise makes a good container for the disinfecting solution. A barrel can be placed in this tank and rolled.

Prevention of Storage Rots and Infections in Storage

Disease and decay producing organisms which affect sweet potatoes, white potatoes, etc., may remain alive for many months in the storage house, in soil and refuse from previously stored crops. These organisms should be eliminated before the new crop is stored by removal of all soil and refuse from the storehouse and also by thorough cleaning and sterilization of the surfaces including bins, crates, baskets, machinery and tools. Sterilization may be accomplished by the application of appropriate fungicides, as a spray, wash or gas. Formaldehyde solution is one of the most commonly used fungicides, as it can be employed as a spray, dip, wash or as a source of formaldehyde gas. Since formaldehyde is a volatile gas it can be removed from the warehouse or utensils by airing after sterilization.

Wet Method

Mix 1 lb. of 37% formaldehyde solution with 15 gal. of water. Apply with an ordinary sprayer or apply with a broom, uniformly wetting the surfaces to be sterilized. It is desirable to close the house for several hours to permit the fumes from the formaldehyde to permeate corners and crevices not reached by the spray or wetting process. Air the enclosure until the fumes have been dissipated.

Dip Method

Small articles, such as crates, tools, etc., may be sterilized by dipping in a solution of 1 lb. 37% formaldehyde in 15 gal. of water.

Fumigation Method

After thoroughly cleaning the soil and other refuse from the storeroom, moisten the surface with water by spraying, sprinkling or otherwise and close for several hours to allow the relative humidity to reach 60% or above. Fumigate, using 3 lb. formaldehyde per

1,000 cu. ft. of space. After thorough ventilation the house is ready for storage.

Note: Fumigation of storage rooms in basements of dwellings is not recommended unless the house can be vacated during fumigation.

Grain Smut Control

All seed should be thoroughly cleaned and smut balls removed from it before treating. Light immature seeds, sick kernels, weed seeds and trash can be removed by a fanning mill.

Sow the treated seed immediately. If the seed is to be stored, dry it thoroughly by spreading out, aeration, or passage through a fanning mill. Never let the treated seed freeze while damp or wet. Handle the seed carefully, avoid injuring it.

Grain Seed to be Treated

Oats—Loose and covered smuts, use dip, sprinkle or spray method.

Barley—Covered and black loose smut, use dip or sprinkle method.

Wheat (Spring and Durum)—Stinking smut, sometimes called bunt of wheat, use dip method.

Spray Method (Dry Method)

(Applicable only to oats for smut control.)

Make a solution of 1 part of 37% formaldehyde and 1 part of water, and apply uniformly by ordinary hand sprayer at the rate of 1 qt. of the mixture to 50 bushels of seed as the seed flows through the grain sprout or is being shoveled from one pile to another on a clean floor, canvas or in a tight wagon box. Bin or pile the sprayed seed and cover with canvas, blankets, or disinfected sacks for 4 to 8 hours. Disinfect sacks by spraying with the formaldehyde solution. Sow immediately or run through a fanning mill before storing.

This is a convenient method for treating large quantities of seed oats rapidly.

Sprinkle Method

(Applicable to oats and barley for the control of certain smuts. Does not control barley stripe.)

Make a solution of 1 lb. (1 pt.) of 37% formaldehyde with 10, 20, 30 or 40 gal. of water at a temperature of 60° to 70°F. and apply with a sprinkling can. Sprinkle the solution uniformly on 50 bushels of seed grain as it is being shoveled from one pile to another on a clean floor, canvas or in a tight wagon box. Shovel until all the seed is uniformly moist. Pile and cover with canvas, blankets, or disinfected sacks for at least 4 hours, or overnight. Sow immediately or dry for early use. Germination may be injured if the seed is held for some time after treatment or when it is sown in dry soil. Treated seed should not be allowed to freeze before it is dry. If the grain is moist, increase the seeding rate about one-fourth.

Dip Method

(Applicable to spring wheat, durum wheat, oats, and barley for the control of certain smuts. Does not control barley stripe.)

Mix 1 lb. (1 pt.) of 37% formaldehyde with 40 gal. of water in a barrel or tank at 60° to 70°F. Dip loosely filled burlap sacks in this solution until the grain is thoroughly wet. Drain and dry 2 hours, or overnight. Sow immediately or dry and sow as soon as possible. Formaldehyde-treated oats may be held longer than similarly treated wheat without injuring the germination. This treatment sometimes injures germination to some extent, particularly in the case of wheat when held for some time after treatment or when sown in dry soil. Treated seed should not be allowed to freeze while it is damp or wet. If the grain is moist, increase the seeding rate about one-fourth.

Testing Seeds for Germination

Triphenyltetrazolium Chloride	1–2
Water	99–98

Cereal seeds are soaked in the solution for 6–8 hours. Oats require 24 hours. Embryos of seeds that will germinate turn red. This solution is non-toxic.

Dairy Cattle Feeds

For Use with Low-Protein Roughages Or for Cows on Poor Pasture

Containing 18 to 20% of crude protein, the following mixtures should be used with low-protein roughages, such as sorghum or corn silage, grain sorghum fodder, sweet sorghum fodder or hay, corn stover, cottonseed hulls, Sudan, Johnson, Bermuda, prairie and other grass hays, or poor, dry pasture of any type.

To Jerseys or Guernseys, feed 1 lb. of any of the mixtures for each 2½ lb. of milk, or 3½ lb. for each gal. of milk, produced daily. To Holsteins, Ayrshires and Milking Shorthorns, feed 1 lb. of the mixture for each 3 lb. of milk, or 2¾ lb. for each gal. of milk, produced daily.

A simple, practical mixture, under average farm conditions, is: 100 lb. of ground grain with 40 lb. of 41% protein cottonseed meal, or 50 lb. of 36% meal. Keep a mixture of calcium and salt available.

Formula No. 1

Corn Meal, Ground Barley or Wheat, or Sorghum Grain Chops	500
Cottonseed Meal	300
Wheat Bran	200
Ground Oats	100
Ground Limestone or Oyster Shell	22
Salt	11

No. 2

Hominy Feed, Corn Meal, Ground Barley or Wheat	100
Rice Bran or Wheat Bran	100
Ground Oats or Barley	100
Cottonseed Meal	100
Ground Limestone or Oyster Shell	8
Salt	4

No. 3

Corn Meal, Sorghum Grain Chops, or Ground Wheat	200
Wheat Bran or Ground Oats	100
Cottonseed Meal	100
Ground Limestone or Oyster Shell	8
Salt	4

No. 4

Corn Meal, Ground Barley or Wheat, or Sorghum Grain Chops	600
Cottonseed Meal	300
Cottonseed Hulls	100

Ground Limestone or Oyster Shell	20
Salt	10

No. 5

Ear Corn Chops (Crushed snapped corn)	300
Cottonseed Meal	200
Ground Oats or Wheat Bran	200
Ground Limestone or Oyster Shell	14
Salt	7

No. 6

Rolled Barley, Sorghum Grain Chops or Corn Meal	300
Wheat Bran	200
Dried Citrus Peel and Pulp or Dried Beet Pulp	200
Cottonseed Meal	300
Ground Limestone or Oyster Shell	20
Salt	10

No. 7

Ear Corn Chops or Grain Sorghum Head Chops	700
Cottonseed Meal	300
Ground Limestone or Oyster Shell	20
Salt	10

No. 8

Sweet Potato Meal (Dehydrated)	300
Ground Oats, Wheat Bran or Rice Bran	200
Cottonseed Meal	300
Ground Limestone or Oyster Shell	16
Salt	8

No. 9

Whole-Pressed Cottonseed	300
Wheat Bran or Ground Oats	100
Corn Meal, Ground Barley or Sorghum Grain Chops	200
Ground Limestone or Oyster Shell	12
Salt	6

No. 10

Whole-Pressed Cottonseed	400
Ground Barley or Wheat	300
Rice Bran or Wheat Bran	100
Ground Limestone or Oyster Shell	16
Salt	8

For Use with High-Protein Roughages Or for Cows on Good Green Pasture

The following mixtures contain approximately 12% of crude protein and are for use with high-protein roughages, such as alfalfa, lespedeza, peavine, soybean, kudzu, peanut, mungbean and other legume hay, or for cows on good pasture, providing ample grazing. Cows that give 2 gal. of milk daily or less usually require no grain mixture if on good pasture or receiving ample legume hay, but higher-producing cows should receive one of these mixtures.

To Jerseys and Guernseys, feed 1 lb. of the mixture for each 3½ lb. of milk, or 2½ lb. for each gal. of milk, produced daily. To Holsteins, Ayrshires and Milking Shorthorns, feed 1 lb. of the mixture for each 4 lb. of milk, or 2 lb. for each gal. of milk, produced daily.

A simple, practical mixture is: 100 lb. of ground grain and 10 lb. of cottonseed meal.

No. 1

Corn Meal or Sorghum Grain Chops	600
Wheat Bran or Ground Oats	200

Cottonseed Meal	25
Salt	8

No. 2
Ear Corn Chops or Grain Sorghum Head Chops	500
Ground Wheat or Barley	200
Cottonseed Meal	50
Salt	7½

No. 3
Sweet Potato Meal (Dehydrated)	300
Ground Oats, Wheat Bran or Rice Bran	100
Cottonseed Meal	100
Salt	5

No. 4
Sorghum Grain Chops or Corn Meal	300
Ground Wheat or Barley	300
Cottonseed Meal	50
Cottonseed Hulls	100
Salt	7½

No. 5
Ear Corn Chops or Grain Sorghum Head Chops	900
Cottonseed Meal	100
Salt	10

No. 6
Rolled Barley or Wheat	300
Corn Meal or Sorghum Grain Chops	300
Dried Citrus Peel and Pulp or Dried Beet Pulp	300
Cottonseed Meal	100
Salt	10

No. 7
Whole-Pressed Cottonseed	100
Corn Meal or Sorghum Grain Chops	600
Ground Oats or Barley	300
Salt	10

No. 8
Whole-Pressed Cottonseed	100
Dried Citrus Peel and Pulp or Dried Beet Pulp	200
Ground Barley or Wheat	200
Sorghum Grain Chops or Corn Meal	200
Salt	7

For Use with Medium-Protein Roughages Or for Cows on Fair Pasture

The following mixtures contain approximately 15% of crude protein, and are satisfactory to use with a combination of low-protein and high-protein roughages, such as corn or sorghum silage and fodder with alfalfa or other legume hay, Johnson grass or other grass hays with legume hay, equal amounts of cottonseed hulls and legume hay, or fair pasture, such as Bermuda and other native grasses.

To Jerseys or Guernseys, feed 1 lb. of the mixture for each 3 lb. of milk, or 2¾ lb. for each gal. of milk, produced daily. To Holsteins, Milking Shorthorns or Ayrshires, feed 1 lb. of the mixture for each 3½ lb. of milk, or 2½ lb. for each gal. of milk, produced daily.

A simple, practical mixture is: 100 lb. of ground grain with 25 lb. of cottonseed meal.

No. 1
Corn Meal, Ground Wheat or Sorghum Grain Chops	400
Ground Oats or Barley	200
Cottonseed Meal	100
Ground Limestone or Oyster Shell	7
Salt	7

No. 2
Sweet Potato Meal (Dehydrated)	400
Ground Oats, Wheat Bran or Rice Bran	200
Cottonseed Meal	200

FARM AND GARDEN PREPARATIONS

Ground Limestone or
  Oyster Shell   8
Salt   8

No. 3
Corn Meal, Ground Wheat,
  or Sorghum Grain Chops   700
Cottonseed Meal   200
Cottonseed Hulls   100
Ground Limestone or
  Oyster Shell   10
Salt   10

No. 4
Ear Corn Chops or Grain
  Sorghum Head Chops   400
Cottonseed Meal   100
Ground Limestone or
  Oyster Shell   5
Salt   5

No. 5
Corn Meal, Ground Wheat
  or Sorghum Grain Chops 400
Ground Oats or Barley 200
Wheat Bran or Rice Bran 100
Cottonseed Meal   100
Ground Limestone or
  Oyster Shell   8
Salt   8

No. 6
Ground Wheat, Corn
  Meal, or Sorghum
  Grain Chops   200
Ground Oats or Barley 200
Dried Citrus Peel and
  Pulp or Dried Beet
  Pulp   150
Cottonseed Meal   100
Ground Limestone or
  Oyster Shell   6½
Salt   6½

No. 7
Ear Corn Chops (Crushed
  Snapped Corn)   300
Ground Oats or Wheat
  Bran   200
Cottonseed Meal   100

Ground Limestone or
  Oyster Shell   6
Salt   6

No. 8
Rolled Barley or Wheat 300
Ground Oats   100
Dried Citrus Peel and
  Pulp or Dried Beet Pulp 100
Wheat Bran   100
Cottonseed Meal   100
Ground Limestone or
  Oyster Shell   7
Salt   7

No. 9
Whole-Pressed Cottonseed 200
Ground Oats or Wheat
  Bran   200
Corn Meal, Ground Barley
  or Sorghum Grain Chops 400
Ground Limestone or
  Oyster Shell   8
Salt   8

No. 10
Whole-Pressed Cottonseed 200
Ground Wheat, Corn Meal
  or Ground Barley   300
Dried Citrus Peel and
  Pulp or Dried Beet Pulp 200
Ground Limestone or
  Oyster Shell   7
Salt   7

Molasses Cattle Feed
U.S. Patent 2,307,062

No. 1
Corn Sugar Molasses   37
Cane Sugar Molasses   63

No. 2
Cane Sugar Molasses   33
Beet Sugar Molasses   30
Corn Sugar Molasses   37

No. 3
Beet Sugar Molasses   50
Cane Sugar Molasses   50

Improved Sweet Sorghum Silage

Freshly cut sorghum is sprinkled with 10 lb. urea per ton when it enters the silo.

Horse and Mule Feed

Maintaining Idle Stock If Good Pasture Is Not Available

Formula	No.1	No.2	No.3	No.4
Oats, Corn, Sorghum Grain Chops or Coarsely Ground Barley	2	0	0	0
Ear Corn Chops or Grain Sorghum Head Chops	0	2	2½	0
Cottonseed Meal or Cake	1	1	1½	1
Grass Hay, or Sweet Sorghum Fodder or Hay	0	12	6	0
Grain Sorghum Fodder (With Heads)	0	0	0	14
Legume Hay	3	0	0	0
Cottonseed Hulls, Oat Straw or Corn Stover	9	0	6	0

Rations for Light Work

Formula	No.1	No.2	No.3	No.4
Oats, Corn, Sorghum Grain Chops or Coarsely Ground Barley	5	0	0	4
Ear Corn Chops or Grain Sorghum Head Chops	0	6	6	0
Cottonseed Meal or Cake	1	1	1½	1
Grass Hay, or Sweet Sorghum Fodder or Hay	0	11	5	0
Grain Sorghum Fodder (With Heads)	0	0	0	12
Legume Hay	3	0	0	0
Cottonseed Hulls	9	0	6	0

Rations for Heavy Work

Formula	No.1	No.2	No.3	No.4
Oats, Corn, Sorghum Grain Chops or Coarsely Ground Barley	10	0	0	10
Ear Corn Chops or Grain Sorghum Head Chops	0	11½	11	0
Cottonseed Meal or Cake	1½	1½	2	1
Grass Hay, or Sweet Sorghum Fodder or Hay	0	10	5	0
Grain Sorghum Fodder (With Heads)	0	0	0	12
Legume Hay	3	0	0	0
Cottonseed Hulls	8	0	5	0

Hog Feed

Mixtures for Sows

Formula	No.1	No.2	No.3	No.4
Corn Meal or Sorghum Grain Chops	65	50	60	0
Coarsely Ground Wheat or Finely Ground Barley	0	40	0	74
Ground Oats	25	0	15	20
Wheat Gray Shorts, Rice Bran or Polishings	0	0	15	0
Cottonseed Meal, Peanut Meal or Soybean Meal	7	7	6	4
Meat Scraps, Fish Meal or Tankage	3	3	4	2
Limestone, Oyster Shell Flour or Wood Ashes	1½	1½	1½	1½
Salt	½	½	½	½

Mixtures for Growing Pigs

Pigs start eating when about 3 weeks old. Creep-feeding gives good results with the mixtures below or with grain and protein supplement in separate compartments of a self-feeder. If good pasture is not available, replace 10 lb. of grain with 10 lb. of ground legume hay.

	Formula No.1	No.2	No.3	No.4
Corn Meal or Sorghum Grain Chops	55	40	0	70
Coarsely Ground Wheat or Finely Ground Barley	0	45	68	15
Ground Oats	20	0	20	0
Wheat Gray Shorts or Rice Polishings	10	0	0	0
Cottonseed Meal, Peanut Meal or Soybean Meal	9	10	8	10
Meat Scraps, Fish Meal or Tankage	6	5	4	5
Limestone, Oyster Shell Flour or Wood Ashes	1½	1½	1½	1½
Salt	½	½	½	½

Lamb Feed

Average Daily Rations for Finishing 50- to 60-Pound Lambs

Formula No. 1

Ear Corn Chops or Milo Head Chops	1⅓ lb.
Cottonseed Meal or Cake	½ lb.
Sweet Sorghum Fodder or Hay	½ lb.
Cottonseed Hulls	1 lb.
Ground Limestone or Oyster Shell Flour	4/10 oz.

No. 2

Corn, Barley or Milo	1 lb.
Cottonseed Cake	⅓ lb.
Prairie Hay	1½ lb.
Ground Limestone or Oyster Shell Flour	4/10 oz.

No. 3

Rolled Barley or Wheat	½
Sorghum Grain or Dried Beet Pulp	½
Legume Hay	2 to 3

No. 4

Sorghum Grain, Barley or Wheat	1 lb.
Cottonseed Meal	⅓ lb.
Ground Grain Sorghum Fodder	1½ lb.
Ground Limestone or Oyster Shell Flour	4/10 oz.

No. 5

Shelled Corn, Barley or Wheat	1
Cottonseed Meal	⅙
Corn or Sorghum Silage	1½
Legume Hay	1

No. 6

Corn or Sorghum Grain	½
Wheat or Barley	½
Wet Beet Pulp	2½ to 3
Alfalfa Hay (Chopped)	1½ to 2

When wheat or other winter pasture is good and lambs graze long enough, desirable market finish can be produced without other feeds. Salt and a calcium supplement should be available.

Poultry Feed

Chicken Feed

Efficient chick-starter and growing mashes, given below, aid proper development of chicks and pullets. Protein, largely of vegetable origin, in the growing mash encourages normal development; pullets are not likely to lay until fully developed and they lay longer.

Feed the all-mash starter when chicks are 24 hours old; continue until 6 to 10 weeks old. A mixture of equal parts of milo or kafir chops, cracked wheat and yellow corn chops is a good scratch grain, kept in feeders after chicks are a month old. The No. 1 all-mash chick starter may be used as a broiler mash until broilers are 10 or 12 weeks old, with small amounts of grain added 2 to 4 weeks before they are marketed. If chicks are raised for layers, pullets and cockerels should be separated early and

Formula	All-Mash Chick Starters		Growing Mashes	
	No. 1	No. 2	No. 1	No. 2
Yellow Corn Meal or Sorghum Grain Chops	44	38	46	35
Finely Ground Oats or Barley	10	4	12½	15
Wheat Gray Shorts or Rice Polishings	20	8	15	20
Wheat Bran or Rice Bran	0	20	0	0
Cottonseed Meal	6	8	12	10
Peanut Meal or Soybean Meal	0	4	0	10
Alfalfa Leaf Meal	5	5	7	6
Meat Scraps or Fish Meal	6	6	4	0
Dried Milk (Skim or Buttermilk)	6	3	0	0
Ground Limestone or Oyster Shell Flour	2	2½	2	1
Bone Meal or Defluorinated Phosphate	½	0	1	2
Salt	½	½	½	1
Cod-Liver Oil	⅛	1	0	0

pullets kept on green range. Feed the growing mash until pullets are 5 months old. Scratch grain, oyster shell and grit and clean water should always be available.

Cottonseed meal is recommended for young chickens and turkeys. In laying rations, it should be limited to mixtures for layers whose eggs will be consumed fresh, and should not exceed 6% of the total mash. This meal may cause dark yolks in eggs in storage. A practical laying mash, when eggs are not sold for cold storage is: 70 lb. of a combination of any 3 available ground grains, mixed with the following 30-lb. protein concentrate mixture: 6 lb., each, of alfalfa leaf meal and cottonseed meal, 10 lb. of peanut or soybean meal, 3 lb. of fish meal, meat scraps or dried milk, 2 lb., each, of bone meal and ground limestone, and 1 lb. of salt.

If enough liquid skim or buttermilk is fed, in clean troughs, dried milk may be omitted from chicken and turkey rations.

Turkey Food

For rapid gains, high finish and best prices, turkey production requires full feeding from hatching time until sale. Turkeys should always have access to mash, turkey-size grit, fresh water and plenty of green feed, with grain in feeders or fed heavily each night. Raise turkeys on clean range where chickens have not been kept; change feeding grounds weekly; and move roosts frequently.

A good starter mash for poults is: 18 lb. of corn meal, 12 lb. of ground wheat or sorghum grain, 15 lb. of finely ground oats or barley, 8 lb. of wheat bran or rice bran, 8 lb. of alfalfa leaf meal, 20 lb. of cottonseed meal, soybean or peanut meal, 12½ lb. of meat and bone scraps, 5 lb. of dried milk or whey, or dried brewers' yeast, 1 lb. of ground limestone or oyster shell flour, ¼ lb. of salt, and ½ lb. of fortified cod-liver oil. A grain mixture may be kept before poults, in a separate feeder, after the third week. The starter mash should be fed until poults are 8 weeks old, when a gradual change should be made to the growing mash.

A growing mash, to feed with grain, is: 100 lb. of a combination of 3 ground grains, mixed with 8 lb. of alfalfa leaf meal, 8 lb. of meat and bone scraps, 15 lb. of cottonseed meal, peanut or soybean meal, 2½ lb. of calcium supplement and ½ lb. of salt.

Increasing Weight of Poultry

To every 100 lb. of feed add ½ lb. Glyceryl Monostearate (S-928). Mix well and then add to the mixture 250–300 lb. water. Use hot water if available as it will shorten the time to get a uniform mix. Use this mix as soon as possible as it tends to ferment. This is fed for 4 or 5 days before killing.

Tests have shown that this mix causes greater weight gain at a faster rate and that the flesh is firmer than usual. This is due to the easier assimilation of the feed because of the better dispersion and because of a greater amount of water absorbed.

Worm Control in Range Sheep

A mixture of one part of phenothiazine and nine parts of salt can be safely offered sheep as a lick for at least six months consecutively. The most important feature of this method of administering phenothiazine is the sharp reduction or the complete prevention of hatching of round-worm (stomach, nodular and hair-worm) parasite eggs, thus decreasing the larvae on range eventually to a low and even vanishing level. On light to medium infested range and on heavily infested pasture during dry summers, this mixture, licked at free choice, offers a labor-saving, inexpensive means of control.

Cattle Wound Antiseptic Powder

Urea	83
Calcium Phosphate	2
Sulfanilamide	13
Sulfathiazole	2

Cattle Wound Dressing

Pine Oil	3
Bone Oil	25
Pine Tar	35
Sulfonated Bitumen	10
Triethanolamine	2
Zinc Oxide	25

Bull Semen Preserving Medium

Glucose	0.60
Galactose	0.20
Potassium Dihydrogen Phosphate	0.20
Disodium Hydrogen Phosphate	2.00
Lipositol	1–2.00
Gum Acacia	3.00
Sulfathalidine	0.03
Distilled Water	To make 100.00

This should be made when needed.

Anthelmintic for Turkeys

Phenothiazine	50
Nicotine	5
Bentonite	95

Add 18 g. of this mixture to 4 lb. of mash feed.

Chick Coccidiosis Control

To prevent coccidiosis, grow chicks in confinement under strict sanitary conditions for at least the first 4 weeks. Two to 4 days before the chicks are placed on the ground feed a ration containing 5 lb. flowers of sulfur, or 325-mesh unconditioned dusting sulfur, and 2½ lb. No. 10 hardwood charcoal per 100 pounds mash, and continue this ration until chicks have been on the ground 5 to 7 days. Then change to 2½ lb. flowers of sulfur, or 325-mesh dusting sulfur, and 2½ lb. No. 10 hardwood charcoal per 100 pounds mash. When grain is added to the ration, feed 5 lb. each of sulfur and charcoal per 100 pounds of mash. Continue this program until the chicks are 12 to 14 weeks old. If an outbreak occurs when sulfur is not being fed, practice strict sanitation and keep chicks quiet and warm. Feed a mash containing 5 lb. above mentioned sulfur, 2½ lb. No. 10 hardwood charcoal per 100 pounds of feed, exclusively, for 7 to 14 days; then discontinue the sulfur and the charcoal or reduce the amount to 2½ lb. dosage previously given. Chicks fed sulfur should have direct sunlight, or the vitamin D content of the feed should be doubled if cloudy weather prevails.

Poultry Inhalant

This mixture is sprayed in the poultry room with windows closed and left that way for 15 minutes. It clears the respiratory tracts.

Liquid Soap	70 cc.
Eucalyptus Oil	5 cc.
Camphor Oil	5 cc.
Guaicol	10 cc.
Beachwood Creosote	5 cc.
Pine Oil	5 cc.

Mix a tablespoonful with half a pint of water.

Defeathering (Poultry) Compound
U.S. Patent 2,353,869

	Formula No. 1	No. 2
Rosin	38	35–45
Gum Dammar	2	—
Paraffin Wax (M.P. 125–127°F.)	58	50–65
Aluminum Stearate	2	1–3
Candelilla Wax	3	1–5
Carnauba Wax	2	—

Egg Production Increaser

Add 10 g. thyreoprotein per 100 lb. of feed.

Egg Preserving Coating

1	Paraffin Wax	40 g.
2	Aerosol OT	4 g.
3	Diglycol Laurate S	100 cc.
4	Water	200 cc.
5	Dowicide A (10% Solution)	40 cc.

Warm 1, 2 and 3 until dissolved. Bring 4 and 5 to boiling and pour the former mixture into it, slowly with stirring. Stir until cool and package.

For use disperse three pints of above mixture in 5 gal. of hot water with stirring. Cleaned freshly gathered eggs are put in a perforated bucket and the latter is dipped in the diluted coating fluid. The excess of the fluid is allowed to drain off. The eggs are then packed in crates while still wet.

Rabbit Deterrent

Put a quantity of hydrated lime in burlap and then dust on plants, preferably when they are dew-covered. It will keep wild rabbits from damaging gardens. Several applications may be necessary, while plants are in the tender stage and most tasty.

Pigeon Repellent

Oleoresin Capsicum	1
Alcohol	99

Spray over ledges and other places where pigeons congregate. They get this irritant on their feet and body and do not return. The spray is non-poisonous, but very irritating.

Dog Chaser for Shrubs
Formula No. 1
(Fluid)

Isopropyl Alcohol	89 cc.
Anhydrous Lanolin	5 g.
Amyl Mercaptan	10 cc.
Creosote	2 cc.

No. 2
(Paste)

Petrolatum	30
Lanolin	10
Amyl Mercaptan	5

Shampoo for Puppies

The following mixture will produce a dry shampoo especially suitable for puppies too young to be bathed:

Starch	75
Silica Gel	10
Borax or Sodium Bicarbonate	10
Pyrethrum Powder or Paradichlorbenzol	5

Add 1% perfume compound.

The powder is dusted over the fur of the animal after which a thorough brushing and combing is necessary to remove the powder. The resulting cleansing and disinfecting of the fur is reasonably thorough, but obviously does not extend to the skin itself.

Dog Shampoo

Soft Soap	8
Glycerin	2½
Alcohol	2
Phenol	⅜
Eucalyptus Oil	¼
Water	To make 35

Small Scale Casein Manufacture *

It is best to utilize the skim milk as soon as is practical after separating the cream, before souring takes place. It will be convenient to start the preparation of casein in the morning. Skim milk obtained the previous evening and that obtained from the morning milking may then be combined. While it is possible to prepare casein from naturally soured skim milk (or from buttermilk), the following method involving the use of sulfuric acid is more advantageous. It is necessary to remove the cream effectively by means of a good cream separator before making casein, not only because cream is the most saleable portion of milk but also because the presence of butterfat in casein is undesirable.

Equipment and Supplies

The method of making casein described here is based on a unit handling about 45 gal. of skim milk a day, but can easily be adapted for multiples of this unit to handle larger volumes. It is equally suitable for making casein in small dairies.

The equipment required for making casein is largely wooden and can be assembled or constructed with a small amount of labor. The following equipment is recommended, though other available equipment may be substituted.
One 50-gal. wooden barrel
Five small wooden spigots
One large wooden spigot (1-in. internal diameter)
or

---

* U. S. Dept. of Agriculture.

One piece of ⅝-in. garden hose 8 to 10 ft. long
One small faucet for end of hose
One C-clamp
One bucket for dipping curd out of barrel
One wooden paddle
One 1-qt. glass measuring pitcher for measuring acid
One thermometer reading up to 120°F.
One 6-gal. earthenware crock for holding acid
Eight ½-bushel unbleached muslin bags for pressing
Two presses
Wooden trough for catching drippings from bags
One drier
Twenty-one yd. of 32-in. cotton bagging for covering trays
One electric fan
Sacks for storing dried casein
Six 2-qt. bottles of 33.5% sulfuric acid with a specific gravity of 1.247

Precipitating barrel: A 50-gal. wooden barrel is drilled at 5 places on the side and fitted with wooden spigots as shown in Figure 1. A large spigot * (at least 1 in. in internal diameter) is installed flush with the bottom of the barrel. The small spigots are used for draining off the whey and wash water after the curd has settled; the large spigot is used for removing the washed curd. The barrel should stand on a small platform about 2 feet above the ground.

---

* The large spigot may be difficult to purchase or impractical to construct with tools at hand. If so, an extended outlet from the barrel may be made from wood and closed at the outer end with a stopper. The internal diameter of this outlet should be not less than 1 inch.

A barrel with no spigots may also be used. In this case the whey and wash water are removed by siphoning, and the curd is dipped out with a bucket. A simple siphon constructed from a piece of garden hose with a faucet or cork at the outer end is attached to the barrel as shown in Figure 1. To start the siphon, a stream of water is run through it, and the outer end is closed. To siphon out the whey, the outer end of the hose, which must be lower than the level of liquid in the barrel, is opened. As the level of whey in the barrel becomes lower, the position of the siphon is shifted until the end in the barrel is just above the settled curd. The siphon is stopped just before it begins to suck air so that it will not have to be refilled with water before being used again.

Press: The construction and use of a suitable press are shown in Figure 2. Two units of the kind shown are necessary to handle 45 gal. of milk.

Drier: The drier is illustrated in Figure 3. The trays are constructed of cloth stretched over the rectangular frame shown in the figure. Lightweight cotton bagging is the best material to use, but other types of cotton cloth are satisfactory. The drier holds two of these trays at each level. The vanes for distributing the air to the different levels can be made of heavy cardboard tacked to light wooden frames if plywood is not available. A large fan, about 16 in. in diameter, is required for circulating the air. The circular opening at the end of the drier and the shelf on

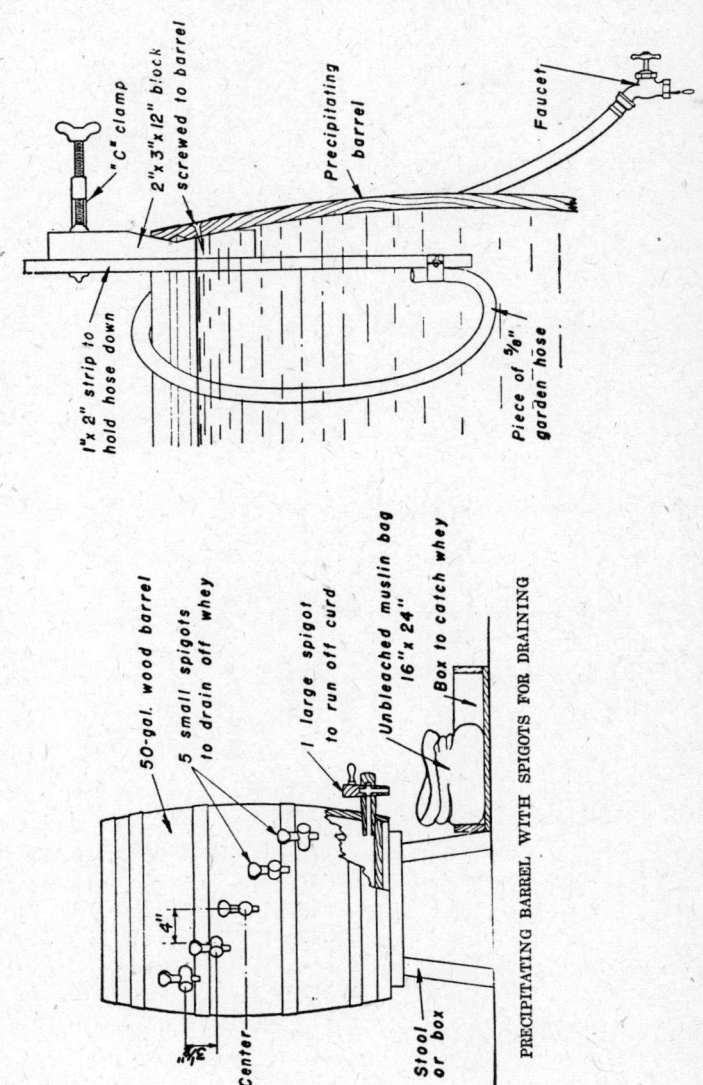

Figure 1. Alternate Methods of Separating Curd and Whey

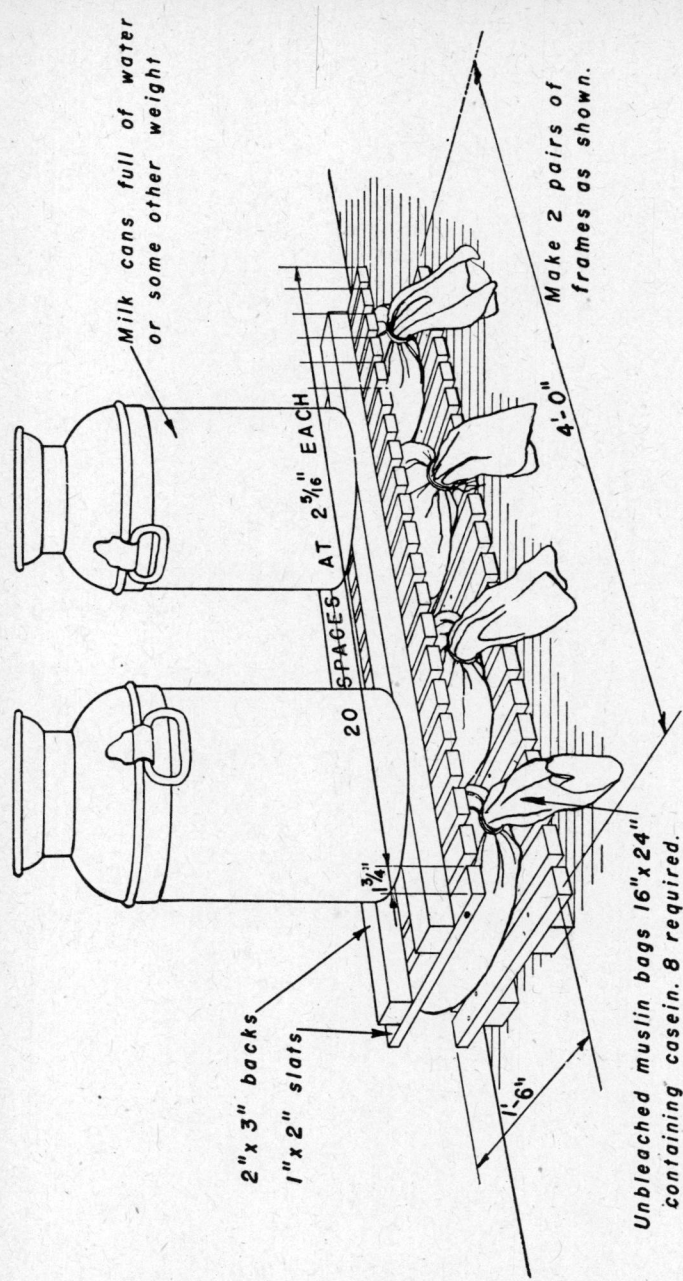

FIGURE 2. PRESSING OUT EXCESS WHEY

FARM AND GARDEN PREPARATIONS

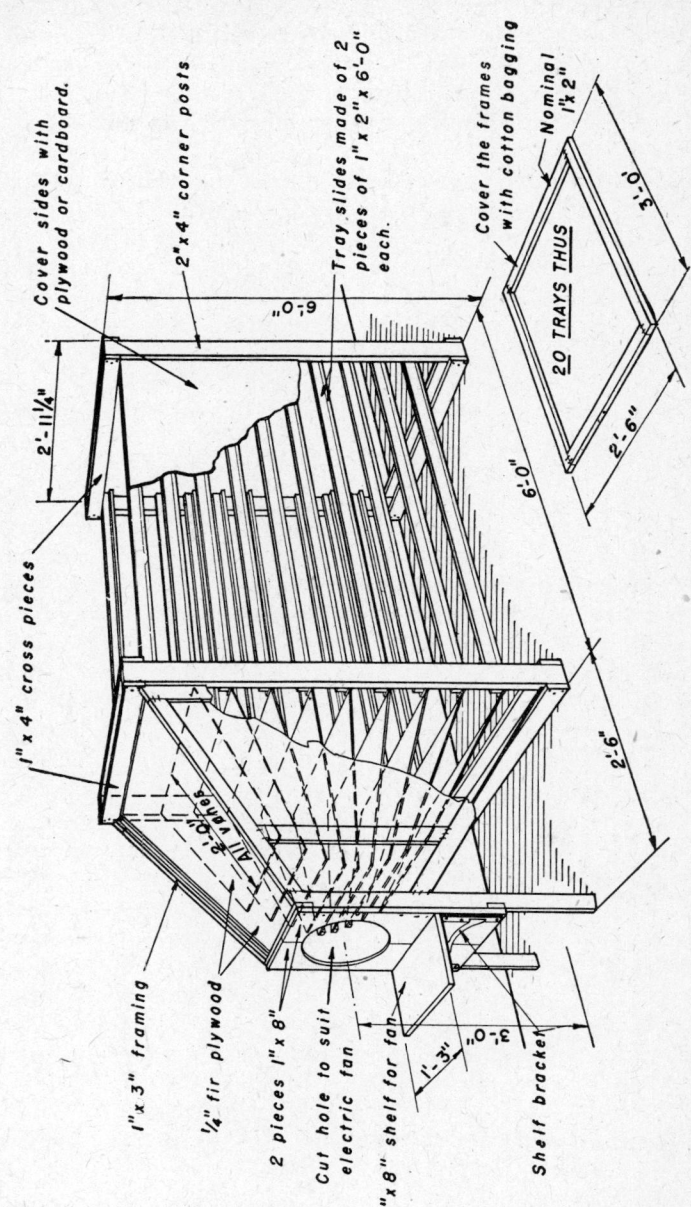

FIGURE 3. DRYING RACK AND TRAYS

brackets are made to fit the particular fan to be used.

One side of the drier is closed by tacking heavy cardboard over it. When the drier is in use, the other side is closed by a wooden frame covered with cardboard. This frame must be easily removable so that the trays can be put in the drier or taken out.

Dilution of the sulfuric acid: The sulfuric acid in a concentration of 33.5% can be purchased in 2-qt. bottles. This must be diluted with water to ten times its volume. Make a mark on a 6-gal. earthenware crock at the level to which it will be filled by 5 gal. of liquid. Fill the crock about halfway to this mark with water. Pour one bottle (2 qt.) of the acid into the water. Fill the crock to the 5-gal. level with water and mix with a wooden paddle. Make all measurements as carefully as possible and avoid spilling any of the acid. Special care should be taken to keep the acid from the skin and clothing, as burns may result from contact with the acid. If acid is spilled, rinse or flush with a solution of bicarbonate of soda.

Coagulation of Casein

The precipitating barrel (Figure 1) is nearly filled with 45 gal. of fresh skim milk which should be at a temperature between 68° and 86°F. Coagulation of the casein should not be attempted with milk at a lower temperature than 68°F. As the diluted acid is used in the proportion of one cup ($\frac{1}{2}$ pt.) to 1 gal. of skim milk, this amount of milk (45 gal.) will require $11\frac{1}{4}$ qt. Nearly two-thirds of this quantity, $7\frac{1}{4}$ qt., is measured out and poured slowly into the milk which is stirred well with a wooden paddle. The remainder of the acid is added 1 qt. at a time, with just enough gentle stirring to mix the acid and the milk. If the stirring is too vigorous, the curd will be finely divided and will not settle well. The milk is allowed to stand for 2 or 3 minutes between each addition of acid. The curd should separate in a granular form just before the last quart of acid is added. After all the acid has been added, the milk is stirred gently for a minute or two.

The amount of acid mentioned above should give the correct acidity for 45 gal. of almost any milk, but milk from Jersey cows may require extra acid. There should be no separation of the curd after the first $7\frac{1}{4}$ qt. of acid have been added. A few small clumps of casein may be noted, but if a clear greenish-yellow whey separates at this point, it is an indication that the milk had started to sour and consequently was already somewhat acid. If the milk has partly soured, only sufficient acid should be added to give a separation of the clear whey, and then one pint more of acid should be added.

If an amount of milk other than 45 gal. is used, the amount of acid required can be calculated on the basis of 1 qt. for 4 gal. of milk. This is equivalent to 1 measuring cup ($\frac{1}{2}$ pt.) of acid for each gal. of milk.

Removing the Whey

In about $\frac{1}{2}$ hour the curd will have settled to the lower half of the barrel. The whey is then removed by opening one of the spigots in the side of the barrel, or by use of the siphon, and collected in a barrel for use as an animal feed. As the whey drains off, it should be allowed to pass through one of the cloth filtration bags to catch any small particles of curd. The whey is drained off as close as possible to the mass of curd.

The curd is then washed by filling the barrel with clear water and stirring gently to mix the curd and the remaining whey with the water. The curd is allowed to settle again, and the wash water is removed in the same manner as the whey was removed. At each settling of the curd, at least half the total volume of whey or wash water should be removed. This washing operation is repeated twice (three washings in all). The number of washings cannot be reduced without producing an inferior grade of casein.

Slightly more than half the total whey is obtained upon the first settling of the casein. The first wash water contains a quarter more and, if desired, may be collected and added to the undiluted whey. The second and third wash waters are discarded, since their nutritive value is low.

Filtering and Pressing

After the last washing is completed, the curd and the remaining wash water are removed through the large spigot at the bottom of the barrel. If a barrel without a spigot is used, the curd is dipped out with a bucket. In either case the last part of the curd is removed by tipping the barrel and washing it out with water.

The curd is run or poured into eight cloth bags (½-bushel size). Each bag in turn is placed in a bucket or wooden trough, and the curd is poured into the bag until it is two-thirds full. It is then tied at the top with a short piece of rope and hung on a peg on the wall or on any suitable rack to drain. Provision should be made for a trough to catch the drippings. Most of the excess water will have drained off in about an hour. The bags are then tied with heavy cord as close to the mass of curd as possible and put in the press. Four bags can be put in a press of the size described. Two 10-gallon milk cans filled with water are then placed on top of the press. The pressing should continue for about 3 hours. The bags are then opened, and the curd, which will be in the form of a firm cake, can easily be removed.

After the cloth bags have been used, they should be washed and then boiled in water to which a little washing soda has been added. This is necessary to prevent the formation of mold, which might contaminate the casein.

Drying

After the cake of curd is removed from the pressing bag, it is crumbled by hand and spread out on the trays of the drier. One tray is used for each bag of curd. The trays are placed in the drier, and the fan is started. The casein should be dry in 24 to 48 hours depending on the temperature and humidity. After the drier has been in operation from 12 to 16 hours, the trays should be removed and the casein mixed with a large spoon or a dull knife. Since the drier described has 20 trays, it will take care of the casein from 90 gal. of milk, so that 45 gal. a day can be processed even if 48 hours are required for drying. The casein is dry when the particles are hard and slightly yellow. Any large pieces should be broken open to make sure there is no undried casein in the center. The casein must be completely dry when it is removed from the trays.

It is possible to dry the casein satisfactorily in the sun in good drying weather. The curd is spread on the trays, and these are supported on any suitable frame or stand in the direct sunlight in such a way that air circulates freely above and below each tray. Casein will dry under these conditions in about 10 hours. This method of drying is very satisfactory, but it is dependent on good drying weather.

The wet pressed curd can be kept for about 24 hours before drying if conditions are such that it will dry rapidly when placed on the trays. It is better, however, if the drying can be started as soon as the pressing is completed.

Storing and Marketing

The dry casein should be sacked and stored in a dry place off the ground or concrete floor and away from rats and other vermin. It is not advisable to store the casein longer than about a month. When 100 lb. have been accumulated it should be marketed. Before the preparation of casein is undertaken, the prospective buyer should, of course, be contacted.

Forty-five gallons of milk will yield 10 lb. of casein. The manufacture of this amount of casein a day is an operation that can be fitted into other work on the farm. It will, however, require about 4 hours a day for one person. Precipitating and washing the curd and placing it in bags for draining is a continuous operation requiring about 2 hours. The other work, such as placing the bags in the press and spreading the pressed curd on the drier trays, can be done whenever convenient and should take about 2 hours altogether.

It would be highly advantageous for several farmers to pool their excess skim milk and make casein on a cooperative basis. By this means the time or cost of making casein can be materially reduced.

Use of Whey as Animal Food

The whey left after the removal of casein is a valuable animal food that contains all the sugar and most of the salts and vitamins of skim milk and approximately one-half the solid material of whole milk. Before it is fed to animals, the acidity should be neutralized by the addition of one cup ($\frac{1}{2}$ pt.) of bicarbonate of soda for each 10 gal. of whey. The bicarbonate of soda should be added cautiously, with stirring, to prevent loss of whey because of foaming. Whey can be best utilized when fed with solid feed, such as ground grain.

Chapter VII

FOOD PRODUCTS

Baking Powder
Formula No. 1
Bicarbonate of Soda	35
Corn Starch	24
Sodium Aluminum Sulfate	29
Acid Calcium Phosphate	12
Albumen	2

No. 2
Bicarbonate of Soda	102
Corn Starch	86
Acid Sodium Phosphate	40
Acid Calcium Phosphate	45
Sodium Aluminum Sulfate	72
Albumen	½

No. 3
Sodium Bicarbonate	28
Corn Starch	41
Sodium Aluminum Sulfate	19
Acid Calcium Phosphate	12

No. 4
Granulated Sodium Bicarbonate	28
Redried Corn Starch	37
Acid Calcium Phosphate	35

No. 5
Bicarbonate of Soda	27
Cream of Tartar	54
Corn Starch	14

No. 6
Acid Sodium Phosphate	40½
Sodium Aluminum Sulfate	45
Corn Starch	64½

No. 7
Acid Sodium Phosphate	27
Sodium Aluminum Sulfate	30
Corn Starch	43

No. 8
Bicarbonate of Soda	28
Sodium Aluminum Sulfate	19
Acid Sodium Phosphate	20
Corn Starch	33

No. 9
Bicarbonate of Soda	35
Sodium Aluminum Sulfate	25
Calcium Acid Phosphate	13
Corn Starch	27

No. 10
Soda Alum	28
Bicarbonate of Soda	29
Corn Starch	43

No. 11
Granular Calcium Acid Phosphate	56
Granular Bicarbonate of Soda	25½
Dried Corn Starch	18½

No. 12
Cream of Tartar	50
Tartaric Acid	3
Bicarbonate of Soda	26½
Corn Starch	20½

The cream of tartar and the soda should be in the granular form. The starch super-dried (otherwise use more starch).

No. 13
Soda Alum	22
Calcium Acid Phosphate	11
Bicarbonate of Soda	27
Corn Starch	40

No. 14
German Patent 599,493
Cream of Tartar	44.0
Tartaric Acid	6.0

Sodium Bicarbonate	27.0
Wheat Flour	20.0
Carbamide	1.5
Magnesium Peroxide	1.5

No. 15

Calcium Biphosphate	34.0
Sodium Bicarbonate	23.0
Wheat Starch Powder	40.0
Carbamide	1.5
Magnesium Peroxide	1.5

No. 16

Sodium Acid Pyrophosphate	44
Sodium Bicarbonate	32
Maize Starch Powder	22
Carbamide	1
Magnesium Peroxide	1

Pie Fillings*

Factors Governing the Amount of Starch Used

The amount of starch used in any kind of pie must be governed by consumer preference, by merchandising conditions, and by established price and quality standards. Pies prepared and served in the home, hotel or restaurant, or those delivered with special handling, may be of more fragile structure than similar products intended for transportation to distant outlets. Pie fillings prepared at a low material cost contain a minimum of expensive fruits, dairy products, or eggs, and a maximum of jelly structure made up of starch and water.

Amount of Starch in Milk Fillings

When milk is thickened with starch, as in preparing a corn starch pudding, 2 oz. of starch is sufficient for 1 qt. (34 oz.) of milk.

* National Starch Products Inc.

If one large egg (2 oz.) is used in cooking the mixture, about ½ oz. of starch may be omitted. However, when such a pudding or custard cream is to be stirred or beaten after cooling, or when it is made richer by adding butter, cream, or high amounts of sugar, about 2¼ oz. of starch will be needed per quart of milk. Fillings for layer cakes and French pastries of average grade require this increased amount of starch.

Amount of Starch in Water Fillings

When starch is cooked in water, 2 oz. in 1 qt. (32 oz.) of water will form a tender, creamy jelly, while 4 oz. of starch will form a very firm jelly. Many operators have adopted, as an average figure, 3 oz. of starch per quart of fruit juice or water. Even in this ratio (about 1:11 by weight) the jelly formed would be unappetizingly firm, were it not for the fact that the fruit acids, added sugar, and fruit pulp and pieces all tend to soften, and to shorten or tenderize the final product.

When a fruit filling is made by thickening all available liquid, such as the juice from a can of fruit, plus some water, the final incorporation of the drained fruit will cause some dilution of the starch jelly. This effect will be greater with a tender, juicy berry than with a firm or mealy-textured apple.

A formula for a fresh filling, such as strawberry, may call for very little water, as such. In a good grade of filling, the formula may show as little as 3 qt. of water per pound of starch. However, when the fruit is stewed, even to partial tenderness, juice accumulates and increases the ratio of liquid to starch.

In fruit fillings of excellent quality, containing 50 to 60% solid fruit by weight, the starch content will range from 2 to 4% of the batch weight, depending on the degree of jell desired. As the fruit content falls, the starch content must rise to maintain body, handling quality, shelf life, and symmetry in the finished pie.

Lemon pie filling, which might be classed as a custard cream made with water and acid instead of milk, will require less starch as its egg content is increased. About 3 oz. of starch per quart (32 oz.) of water will be required in fillings containing little or no egg.

Effect of Other Ingredients on Starch Swelling

While discussing consistency of starch jellies, it must be pointed out that certain other ingredients interfere with the proper clearing and swelling of starch grains, if present at the time the starch is first heated. Sugar is one such ingredient. If 1 lb. of starch is heated in an average ratio of juice or water in the presence of 7 lb. of sugar, cooking is incomplete, regardless of temperature or time, and the cold jelly will be dull, mushy and thin. But if the same starch is cooked in the presence of 3 or 4 times its weight of sugar, with the balance of the required sugar added after swelling is complete, the resulting jelly will be clear, glossy, and creamy, with a satisfactory heavy body.

Similarly, large amounts of egg, fat, or dry milk solids, present at the time the starch is first heated, may inhibit (hinder or prevent altogether) the swelling of the granules. In extreme cases, salt or other chemicals present may act as inhibiting agents.

If starch is heated for a long period in the presence of the stronger fruit acids, a measurable weakening of its jelling power occurs. This is due to actual conversion of some of the starch into nonjelling substances. It is for this reason, chiefly, that lemon juice is added after the starch is swollen in making lemon fillings.

All these factors which affect the cooking and subsequent jelling of starch have to be taken into consideration in the development of dependable formulas and methods for production of pie fillings. Having in mind convenience and time-saving as prime requisites, the recommended methods have been restricted to such steps as are known to be necessary for quality and stability in the fillings.

Handling and Storage of Cooked Starches

Due to adoption of the best of manufacturing methods and quality control, many food starches on the market today show a very low bacteria count. Such starches, in themselves, are usually not sources of spoilage organisms. Nevertheless, the purity of the starch does not make the final starchy preparation invulnerable to spoilage. Contamination by means of faulty handling, air-borne organisms, or the introduction of impure ingredients may result in rapid and costly destruction by fermentation, liquefaction, or molding.

Practically all food products containing cooked starches are moist, if not actually semi-fluid. Since bacteria, yeasts, and molds thrive in a moist environment, starch creams, fillings and jellies, once contaminated, offer ideal conditions for their growth. Fortunately, acids and sugar, found in many pie fillings, hinder the growth of some organisms.

Cleanliness is probably the best and cheapest preventive of spoilage, assuming that all ingredients are above suspicion. Cooking vessels, ladles, paddles and storage containers should be washed and scalded. Dust-laden air or condensed moisture dripping from ice box ceilings may also contaminate a filling left uncovered, by introducing air-borne or water-borne organisms.

Unfortunately, low-temperature storage cannot be recommended for starch fillings, except for short periods of time. Custard creams, which should be consumed the same day they are prepared, are refrigerated both for safety and for greater appetite appeal. But fruit fillings generally show a loss of clarity and brilliance and a decrease in stability (evidenced by cracking and watering) when chilled. Some bakeshops successfully hold pre-cooked fruit fillings at 50–60°F. for three or four days before using. Bakers' supply houses usually add 0.1% sodium benzoate to prepared fillings, and advocate holding the materials at about 70°F. to maintain their stability.

Adjustments in Formulas

The most frequent reason for changing the composition of formulas is raw material cost. If such a change involves any radical alteration of quality, the most economical procedure in way of time and results obtained may be the development of an entirely new formula, balanced for the materials selected. A technical food laboratory is able to render this service to its customers.

Some adjustments, however, can be made without seriously disturbing the balance of a formula. Conversion of the figure for liquid milk to the equivalent in powdered milk and water, replacement of frozen fruit of one type of pack with the same variety packed with more or less sugar—these and other changes can be made if all figures are available. In any such instance, a confirmatory test is advisable.

Handling Fruit for Pie Fillings

The first consideration in making a fine fruit pie is the selection of a high quality fruit—canned, fresh, or frozen. The second consideration is the proper handling of the fruit, before and during cooking.

Crates of fresh fruit should be kept in a cool, dry place until needed. Moldy or musty pieces should be discarded. Fruit which requires washing should be washed and drained shortly before it is used, to avoid possible fermentation or growth of mold.

Canned fruits should be opened only as needed, if their flavor is to be preserved. If left-overs are unavoidable, transfer them from the can to a jar or other sanitary container which can be covered.

Frozen fruits are commonly handled in one of two ways. Many bakers completely defrost such fruits in their original containers, then drain them on sieves or screens. The juice is heated and thickened, and the fruit added at the end of the operation. Some packers and users of frozen fruits advocate another method for maintenance of fruit quality. This consists of slightly thawing the fruit, so that it may be loosened from the can. This fruit is heated rapidly with a little water to about 185–195°F. From this point on, it is thickened and sweetened as usual. Contrary to the belief of many operators, the second method preserves the shape and flavor of most fruits far better than the former, even when as much as 100 lb. of filling is cooked at one time.

In cases where only the juice is heated and thickened, the drained pieces of fruit may be spread in cooling pans and covered with the hot starch jelly. For prevention of *bleeding* and possible fermentation, add the fruit to the kettle mix in the last stage of cooking and allow it to become heated through (to about 180°F.).

The third consideration in the making of fine fruit pies is the proper starch for thickening and stabilizing the filling. Having selected a high grade starch, the baker must handle it as directed to obtain all possible advantages from its use.

Canned Apple Pie Filling
(using heavy-pack, processed apples)

A   Processed Apples
    (No. 10 Can)   7 lb.
    Granulated Sugar 1 lb. 4 oz.
    Salt   5/8 oz.
    Cinnamon   3/16 oz.
    Nutmeg   1/16 oz.
    Water   2 lb.
B   Purity C Bakers'
    Starch   4 3/8 oz.
    Water   8 oz.
C   Granulated Sugar 1 lb. 4 oz.
D   Fresh Lemon Juice   2 oz.
E   Butter   3/4 oz.

Bring the fruit and the other ingredients grouped under A to 185–195°F., preferably in a steam-jacketed kettle. Some bakers prefer to bring fruit to the boiling point, but care must be taken not to break and soften it excessively. Add the cold water and starch suspension, B, in a thin stream, with constant agitation. Cook until clear and thick (to about 185–190°F.). Add balance of sugar, lemon juice or citric acid, and butter, if used.

Fresh Apple Pie Filling

A   Sliced Ripe
    Greenings   5 lb. 7 oz.
    Granulated Sugar   10 oz.
    Salt   1 oz.
    Cinnamon   to flavor
    Water   2 lb. 6 oz.
B   Purity C Bakers'
    Starch   3 oz.
    Water   7 oz.
C   Granulated Sugar   13 oz.
D   Lemon Juice   1 oz.

Ripe but firm greenings of average size, sliced in sixteenths, were used for developing this formula.

Bring A to boiling, stir in B, smoothly blended, and continue stirring as the mixture thickens. Add C and D. Turn the filling out to cool.

Fresh apple filling has found favor in such bakery and coffee shop items as Danish pockets, pastry slices, turnovers, and numerous forms of apple cake, both covered and open. Try this formula for your apple pastries. Add sliced and toasted almonds or filberts for the final tempting touch.

Blueberry Pie Filling
(Using 6 lb. dry frozen, unsugared berries, or fresh berries)

A	Blueberries	6 lb.
	Water	1 lb. 8 oz.
	Granulated Sugar	1 lb.
	Salt	¼ oz.
B	Purity C Bakers' Starch	5 oz.
	Water	8 oz.
C	Granulated Sugar	10 oz.
D	Lemon Juice	¾ oz.

Bring fruit and other ingredients grouped under A to 185–195°F., preferably in a steam-jacketed kettle. Some bakers prefer to bring the fruit to the boiling point, but care must be taken not to break and soften it excessively. Add the cold water and starch suspension, B, in a thin stream, with constant agitation. Cook until clear and thick (to about 185–190°F.). Add balance of sugar, lemon juice or citric acid, and butter, if used.

Lemon Chiffon Pie Fillings
Formula

		No. 1	No. 2
A	Water	32	32
	Sugar	12	12
	Salt	¼	¼
B	Purity C Bakers' Starch	5	5
	Water	8	11
C	Sugar	16	16
D	Lemon Juice	10½	10½
E	Egg Whites	16	13
F	Lemon Gratings	¼	¼

(Formula 1 will yield a fluffier filling and approximately 20% more pies than Formula 2.)

Bring A to boiling or at least to scalding heat (185–190°F.). Add a smooth suspension of B gradually with rapid stirring. Cook until clear and thick (185–190°F.). Dissolve C in the mixture, then blend in D.

Whip E to a soft peak, add hot mix gradually, and whip until light. Add F. The filling may be tinted a pale yellow with certified color, if desired. Fill into baked pie shells. Cool and decorate with a border of whipped cream.

Canned Cherry Pie Filling
(Based on water-packed, sour red cherries)

A	Sour Red Cherries (No. 10 Can)	6 lb. 8 oz.
	Granulated Sugar	10 oz.
	Salt	¼ oz.
B	Purity C Bakers' Starch	3¾ oz.
C	Granulated Sugar	1 lb.
D	Fresh Lemon Juice	½ oz.

Bring fruit and other ingredients grouped under A to 185–195°F., preferably in a steam-jacketed kettle. Some bakers prefer to bring the fruit to the boiling point, but care must be taken not to break and soften it excessively. Add the cold water and starch suspension, B, in a thin stream, with constant agitation. Cook until clear and thick (to about 185–190°F.). Add balance of sugar, lemon juice or citric acid, and butter, if used.

Frozen Cherry Pie Filling
(Using 4:1 sugar pack)

- A Cherry Juice* 4 lb.
- B Purity C Bakers'
  - Starch 5⅝ oz.
  - Water 12 oz.
- C Granulated Sugar 1 lb. 5 oz.
  - Salt ⅜ oz.
- D Lemon Juice 1 oz.
- E Drained Cherries 6 lb.

(For cooking, use same method as given above.)

Peach Chiffon Pie Filling

- A Egg Whites 1 lb. 4 oz.
- B Peach Filling used
  - hot from kettle 11 lb. 2 oz.
- C Fresh Lemon Juice 2 oz.
  - Certified Pink Color

Whip A to a soft peak, then add B gradually while whipping. Continue until mixture is light and glossy, with a fine, smooth texture. C may be adjusted or omitted, as desired. Fill into shells while warm, in order that the mixture may set up slightly. Yield: 16 eight-inch pies.

Frozen Peach Pie Filling
(4:1 sugar-packed fruit)

- A Frozen Peaches 10 lb.
  - Water 10 oz.

---

*Weight derived from draining of 10-lb. container of red cherries, packed 4 to 1 with sugar.

- B Purity C Bakers'
  - Starch 6¼ oz.
  - Water 12 oz.
- C Granulated Sugar 1 lb. 9 oz.
  - Salt ½ oz.
- D Fresh Lemon Juice 2¼ oz.

Half the fruit may be withheld from A, if desired, and added after cooking is completed. Bring A to boiling point, or at least to 190°F. Add a smooth mixture of B gradually while stirring, and cook until clear and thick (185–190°F.). Dissolve C and D.

Fresh Strawberry Pie Filling

- A Fresh Berries 7 lb. 8 oz.
  - Water 8 oz.
  - Granulated Sugar 1 lb. 8 oz.
  - Salt ¼ oz.
- B Purity C Bakers'
  - Starch 6¼ oz.
  - Water 12 oz.
- C Granulated Sugar 1 lb. 12½ oz.
  - Fresh Lemon Juice 1 oz.

Bring A to 185–190°F. Add smooth mixture of B, and cook until clear and thick, stirring well while cooking. Dissolve C.

Lemon Pie Filling

- A Water 44 oz.
  - Sugar 12½ oz.
  - Salt ¼ oz.
- B Purity C Bakers'
  - Starch 5⅝ oz.
  - Water 12 oz.
- C Sugar 12½ oz.
- D Fresh Yolks 7 oz.
- E Butter 1 oz.
- F Fresh Lemon Juice 7 oz.

Bring A to boiling. Add a smooth mixture of B gradually with rapid stirring. Cook until thick and clear, then dissolve C. Turn down the heat under the kettle. Stir a little of the hot mixture into the slightly beaten yolks, then stir these into the kettle mixture. If these increase the body of the mixture, no further heating is necessary. If they cool and thin the mixture, increase the heat slightly and cook until thickened to the previous consistency. Remove the kettle from the stove, blend in E and then stir F thoroughly into the mixture.

This filling may be poured into baked shells to cool, or it may be cooled in a suitable container, stirred slightly and filled into baked shells. Pies should be cool when topped with meringue.

Pie Meringue

A	Egg Whites	10	oz.
	Salt	1/8	oz.
B	Granulated Sugar	5½	oz.
	Water	14	oz.
C	Purity C Bakers' Starch	1	oz.
	Water	2	oz.
D	Granulated Sugar	16	oz.
	Corn Syrup or Honey	1	oz.
E	Vanilla	1/8	oz.

Have A ready, with a clean wire whip and bowl, free from oil or fat. Bring B just to boiling, stir in C and cook until clear. Dissolve D and boil to 216–218°F.

Have A whipped to a soft peak when the syrup is done. Add the syrup gradually while whipping and continue until the meringue will hold up in glossy peaks. Add vanilla.

Spread on pies and bake 10–15 minutes at 400–425°F.

Strawberry Whip Pie

A	Egg Whites	12	oz.
B	Fresh Strawberry Filling	6	lb.

Whip egg whites to a soft peak, on second speed of a three-speed machine. Use hot, freshly cooked filling, still at a temperature of 185–190°F. Add this gradually while whipping on second speed. Continue until the mixture is light enough to stand up in glossy peaks.

As this creamy, fluffy filling goes almost three times as far in pie shells as a true cream or fruit filling, the batches have been kept small.

Cherry Filling with Apple Sauce

A	Juice from 30 lb. can	11 lb.	4 oz.
	Water	27 lb.	
B	Purity C Bakers' Starch	3 lb.	6 oz.
	Water	6 lb.	
C	Granulated Sugar	10 lb.	8 oz.
	Salt		3 oz.
D	Apple Sauce	6 lb.	
	Corn Syrup	7 lb.	8 oz.
E	Drained Cherries	18 lb.	12 oz.

One No. 10 can of apple sauce to each 30-lb. can of frozen fruit is all right.

This formula contains 35% cherries.

Heat A to 175–180°F., add suspension B, heat to 185° until thick and clear. Shut off heat and add C, D and E in order named. Remove from kettle and cool.

Method for Milk Fillings

Any producer or would-be producer of milk fillings or custard cream items should become familiar with his own state and local board of health rulings on production and merchandising of these highly perishable foods. Such rulings may specify certain cooking procedures, as well as refrigeration, time limit for sales, and so on.

The amount of sugar considered appetizing in a milk filling is seldom high enough to interfere with cooking of the starch. Milk, sugar and salt are heated to scalding (slightly foamy appearance, or 180–190°F.) and the starch suspension is added gradually, with steady agitation. A temperature of 185–190°F. should be reached. In a vanilla cream, the eggs or yolks are incorporated at this point. For the finest, smoothest texture, thin the slightly beaten yolk with a little hot mixture, then incorporate it very carefully, with the heat reduced or shut off. Increase the heat, and cook to 175–180°F., stirring well. Add flavoring and pour into sanitary vessels to cool.

Another method, which unfortunately produces a duller, grainier cream, is to blend raw starch, cold milk and yolks, then add all together and cook as described. In cocoa powder fillings, the cocoa may be blended into the starch suspension. If chocolate liquor is used, it may be melted or grated, and added after the starch is cooked.

Butterscotch Pie Filling

Glossy, golden-brown, with the true fragrance of home-made butterscotch, this tempting filling may be used in a variety of combinations. Topped with whipped cream for pies, with nut meats or toasted cocoanut for tarts, it will add sales appeal to any pastry display.

A	Light Brown Sugar	1 lb.	14	oz.
	Butter		13	oz.
B	Liquid Milk	6 lb.	6	oz.
C	Salt		½	oz.
	Purity C Bakers' Starch		8	oz.
	Liquid Milk	1 lb.	1	oz.
D	Egg Yolks		9	oz.
E	Light Brown Sugar	1 lb.	4	oz.
	Vanilla Extract		½	oz.

Place A in copper kettle over low flame. Stir while melting and bubbling. When glossy and beginning to draw away from wall of kettle, butterscotch flavor will be well-developed. It may be heated longer for darker color, if watched carefully. Dissolve A in B, then follow method for Milk Filling.

Golden Vanilla Cream Fillings

Chill this rich custard cream in flaky pie shells lined with banana slices or pineapple shreds, top with whipped cream, and feature the products in bake-shop or restaurant. Use this filling for the daintier, finer types of eclairs, cream puffs, or Boston cream pies.

A	Liquid Milk	8 lb.	8	oz.
	Granulated Sugar	1 lb.	5	oz.
	Salt		¼	oz.
B	Purity C Bakers' Starch		8½	oz.
	Liquid Milk	1 lb.	1	oz.
C	Egg Yolks	1 lb.		
D	Vanilla Extract		1¼	oz.

(Follow method for Milk Fillings.)

Chef's Chocolate Cream Fillings

Dark, full-flavored, with a velvety texture, this filling is perfect for whipped cream or pecan pies. Not over-sweet, it is delicious in yellow layer cakes, individual pastries and tarts.

A	Liquid Milk	7 lb. 7	oz.
	Granulated Sugar	2 lb.	
	Salt	½	oz.
B	Purity C Bakers' Starch	7	oz.
	Liquid Milk	1 lb. 1	oz.
C	Bitter Chocolate Liquor	14	oz.
D	Egg Yolk (Fresh or frozen)	8	oz.
E	Butter	1½	oz.
F	Vanilla Extract	1	oz.

(Follow method for Milk Fillings.)

Light Vanilla Cream Fillings

Moderate in cost, fairly heavy in consistency, this cream filling is adapted to use in restaurant pastries of average quality. Its consistency is particularly good for the larger individual items, such as eclairs and cream puffs, and for heavy filled sheets and layers. Its color, flavor and texture can be improved by using yolks instead of whole eggs.

A	Liquid Milk	8 lb. 8	oz.
	Granulated Sugar	1 lb. 9	oz.
	Salt	¼	oz.
B	Purity C Bakers' Starch	9	oz.
	Liquid Milk	1 lb. 1	oz.
C	Whole Eggs or Yolks	10	oz.
D	Butter or Shortening	2	oz.
E	Vanilla Extract	¾	oz.

(Follow method for Milk Fillings.)

Lemon Chiffon Pie

Sugar	1 lb. 4	oz.
Borden's Powdered Lemon Juice	3	oz.
Salt	⅛	oz.

Mix dry. Add to the dry mix and then place on the fire:

Water	8	oz.
Butter or Vegetable Shortening	6	oz.

When the syrup is boiling, stir in vigorously:

Egg Yolk, slightly beaten	12	oz.

(Equals 18 Egg Yolks.)
Continue cooking until it thickens, which it should do almost immediately. Remove from fire and stir in:

Gelatin	¾	oz.

Previously soaked

until soft in Water	4	oz.

Have ready stiff meringue made with:

Egg White (Equals 6 Egg Whites)	8	oz.
Sugar	8	oz.

Whip meringue into the warm mix. Fill into baked shells while warm. Cool thoroughly (under a fan if necessary) until filler sets. Top with additional meringue. This makes an entirely new type of pie which offers a pleasing novelty. Owing to the lightness of the filling, only about 14 oz. are used per pie.

Sugarless Lemon Pie Filling

Powdered Gum Karaya	1	oz.
Water	12	oz.
Butter	¾	oz.
Egg Yolk	1	oz.

Lemons (Juice
 and Skin)   2 oz.
Saccharin   2 gr.

Add water gradually to the gum and then bring to a boil. Remove from the fire. Add the remaining liquid ingredients in the order given. Pour into a pan lined with pie crust which has been baked and cover with meringue.

Pie Crust Mix
Formula No. 1

Cake Flour	8 lb.
Low Gluten Bread Flour	2 lb.
Sugar	10 oz.
Salt	3 oz.
Baking Powder	2 oz.
Hydrogenated Shortening, High Melting Point	6 lb. 8 oz.

Sift the sugar, salt and 1 oz. of baking powder through the bread flour. Blend with the cake flour.

Have all the ingredients cold. Place in the mixer 2½ lb. of shortening. Into this blend 1 oz. of the baking powder. Add 2½ lb. of the flour blend and mix until it forms a pasty mass. Then add the remainder of the flour and work until smooth. Add the remainder of the shortening, half cut into small pieces about the size of peas and the rest somewhat larger.

For one 8½- to 9-in. pie tin, use 8 oz. of the mix. Add 3 to 4 oz. of water, which may be cold or boiling, as desired. Let stand in the refrigerator for 1 hour or so before rolling, if possible. This not only improves the rolling qualities but the color when baked.

No. 2

Flour (Soft cake)	8 lb.
Shortening (Soft)	4 lb. 8 oz.

Salt   3 oz.
Soyflake 8 oz.

Dissolve the salt and the Soyflake in 2 lb. 12 oz. cold water (amount of water variable). The dough should not be kept in the ice box over 2 days. Best results are obtained by making up the dough one day and using it next day. Care should be exercised to see that the dough is not over-mixed.

Pie Dough

Pastry Flour	12 lb.
High Grade Shortening	8 lb.
Soyflake	1 lb. 2 oz.
Milk Powder	12 oz.
Sugar	12 oz.
Salt	8 oz.
Water	6 lb.

Basic Sweet Dough with Soya Flour

A	Sugar	2 lb.
	Salt	4 oz.
	Shortening	2 lb. 4 oz.
	Mace	½ oz.
B	Whole Eggs	12
	Grated Lemons	2
C	Water	8 lb.
D	Soyflake Flour	1 lb.
	Bread Flour	12 lb.
E	Yeast	1 lb.
F	Pastry Flour	4 lb.

Mix A and add B and mix. Then pour in ¾ of C and add D and start mixing. Dissolve in the balance of water E and add F. Finish mixing.

Knead dough at low speed (in three speed machine) for from 5 to 8 minutes. Allow to proof for 2 hours and 30 minutes, at 78 to 80°F. Knock down, proof another 20 minutes, then to bench for make up. This mix can be made into various shapes and sizes, using a variety of fillings. Soyflake adds to the quality of the dough and retards staling considerably.

Commercial Danish Dough

Hard Wheat Flour	3	lb.
Pastry Flour	1½	lb.
Sugar	10	oz.
Yeast	6	oz.
Shortening	8	oz.
Salt	1	oz.
Whole Eggs	6	
Soyflake Flour*	6	oz.
Honey	2	oz.
Water (1¼ qt.)	2½	lb.
Milk Powder	2	oz.
Mace	¼	oz.
Pure Vanilla Extract	¼	oz.
Puff Paste Roll In	1½	lb.

Rich Coffee Cake

Milk	1	qt.
Yeast	3	oz.
Soyflake	2	oz.
Malt	1	oz.
Patent Flour	2 to 2½	lb.

Sponge, leave until ready (depending on shop temperature).

Sugar	12	oz.
Butter	5	oz.
Lard or Shortening	5	oz.
Yolks—3 Whole Eggs	½	lb.
Lemon Grating	1	
Patent Flour (Winter)	1½	lb.

Mix—After mixing well, leave dough.

Baking Powder Biscuit
(Exellent for Short-Cake)

Pastry Flour	3	lb.
Soyflake	4	oz.

*Special soyabean flour.

Baking Powder	3½	oz.
Shortening	12	oz.
Sugar	4	oz.
Milk	1	qt.
Eggs	3½	oz.

Beat the eggs well, add with milk. Salt.

Honey Muffins
or
Honey Bran Muffins

Sugar	12	oz.
Shortening	6	oz.
Soda	½	oz.
Eggs	7	oz.
Bran or Sunsoy Granules	8	oz.
Honey	4	oz.
Buttermilk	¾	qt.
Flour (Winter)	1½	lb.
Soyflake	8	oz.
Salt		

Crisp Pecan Wafers

Brown Sugar	1 lb.	14	oz.
Butter	1		lb.
Soda	½		oz.
Soyflake	6		oz.
Eggs	14		oz.
Pecan Meats (Ground)	1		lb.
Flour (Winter)	1 lb.	12	oz.

Salt to taste.

Drop on lightly greased pans, small drops as for vanilla wafers. Top with half of one pecan meat.

Cocoanut Bars

Cane Sugar	8	lb.
Lard Compound	4	lb.
Molasses	1	qt.
Water	3	qt.
Flour	12	lb.

Cocoanut	10	lb.
Soda	1	oz.
Soyflake	3	oz.

Run on greased pans with star tube.

Banana Muffins

Sugar		5	lb.
Shortening	2 lb.	8	oz.
Bananas		4	lb.
Salt		1	oz.
Soda		1½	oz.
Eggs		1	qt.
Milk		1½	qt.
Sunsoy Granules*		14	oz.
Ground Pecans		4	oz.
Cake Flour		7	lb.
Soyflake*		1	lb.

Ice with caramel icing and dip in Sunsoy Topping.

Almond Macaroons

Granulated Sugar	1	lb.
Powdered Sugar	1	lb.
Almond Paste	2	lb.
Soyflake*	4	oz.
Egg Whites	12	
Lemon Grating	1	

Layer Cake

Fine Granulated Sugar	3½	lb.
Hydrogenated Shortening	2	lb.
Whole Eggs	2	lb.
Water (or 2 lb. Fresh Milk)	4	lb.
Whole Milk (Powdered)	4	oz.
Pastry Flour	3½	lb.
Baking Powder	3	oz.
Soyflake (Dissolved in 8 oz. Water)	4	oz.
Salt		

Cream sugar and shortening, work in the eggs, cream up and add the Soyflake that has been dissolved in the 8 oz. water. Work in well as this helps to blend the mix.

Sift flour, powdered milk and baking powder three times (or add the powdered milk to the creamed mass) and add salt.

Add the water or milk and sifted flour alternately in 3 parts. Mix until smooth and bake in well greased pans. Flavor optional.

Pound Cake

A Standard Powdered

Sugar	3	lb.
Whole Eggs	1½	pt.
Salt	1½	oz.

Beat on second speed until light:

B Shortening (High Ratio)

Shortening (High Ratio)	1½	lb.
Cake Flour	1½	lb.
Soyflake	4	oz.
Skim Milk (Powdered)	3	oz.

Cream B for about 8 minutes and add carefully, but thoroughly to A.

C Water   1¾ pt.
   Vanilla Extract   ¼ oz.
Add C to A and B.

Add 1¼ lb. cake flour to the entire mixture and mix until smooth.

Bake for about 1 hour and 30 minutes at 340°F. This should be baked in wooden lined molds without a top.

Commercial Pound Cake

Cream until light:

Granulated Sugar		11	lb.
Butter and/or Shortening	6 lb.	12	oz.
Milk Powder		6	oz.
Salt		2½	oz.
Moderate Good Cake Flour	1 lb.		
Soyflake		8	oz.

---

* Special soyabean flour.

Flavor with Orange, Almond and Vanilla Extract.

While creaming the above mixture, add slowly the following:

Whole Eggs	7 lb.
Glycerin	6 oz.

Then mix in the following alternately to make a smooth batter:

Moderate Good Cake Flour	10	lb.
Baking Powder	½	oz.
Water	4 lb. 8	oz.

Total batter weight: 41 pounds 11 ounces.

Bake in wood lined pans in solid oven, with a little steam, at a temperature of 360°F.

It is highly important to have all the ingredients cool in the summer months as the batter temperature should not be above 68 to 70°F. to get the best results in volume and texture.

Pound Cake
(Approx. batter weight, 19 lb.)

Cake Flour	8	lb.
Soyflake	10	oz.
Shortening—Regular, Hi Emulsifying	2 lb. 8	oz.
Sugar	5	lb.
Salt	2½	oz.
Milk	4	oz.
Whole Eggs		lb.
Water	2 lb. 8	oz.
Baking Powder	1	oz.
Flavor	To suit	

Cream the Soyflake into the shortening and proceed as usual with the balance of ingredients.

White Cake with Soya Flour

Batch Weight

	25 lb.		50 lb.	
Step 1				
Flour	2 lb.	8 oz.	5 lb.	
Shortening	2 lb.	4 oz.	4 lb.	8 oz.
Step 2				
Flour	2 lb.		4 lb.	
Sunsoy *	1 lb.		2 lb.	
Sugar	7 lb.	10 oz.	15 lb.	4 oz.
Salt		2 oz.		4 oz.
Baking Powder		6 oz.		12 oz.
Milk Powder		12 oz.	1 lb.	8 oz.
Water	3 lb.		6 lb.	
Step 3				
Egg White	3 lb.	4 oz.	6 lb.	8 oz.
Step 4				
Water	2 lb.	2 oz.	4 lb.	4 oz.

Mix each step 2 minutes scraping down the bowl each time. Scale. Bake at 375°–400° F.

---

* Special soyabean flour.

Yellow Layer Cake

Batch Weight

	25 lb		50 lb	
Step 1				
Flour	2 lb.	8 oz.	5 lb.	
Shortening	2 lb.	4 oz.	4 lb.	8 oz.
Step 2				
Sugar	7 lb.		14 lb.	
Flour	2 lb.	8 oz.	5 lb.	
Sunsoy *	1 lb.		2 lb.	
Salt		2 oz.		4 oz.
Baking Powder		6 oz.		12 oz.
Milk Powder		12 oz.	1 lb.	8 oz.
Water	2 lb.	12 oz.	5 lb.	8 oz.
Step 3				
Eggs	2 lb.	12 oz.	5 lb.	8 oz.
Step 4				
Water	3 lb.		6 lb.	

Mix each step 2 minutes scraping down the bowl each time. Scale. Bake at 375°–400° F.

Fruit Cakes

	Golden Fruit Cake		White Fruit Cake		Spice Fruit Cake	
Flour	5	lb.	5	lb.	5	lb.
Sugar	3½	lb.	3½	lb.	—	
Brown Sugar	—		—		5	lb.
Molasses	2	lb.	—		1	lb.
Honey	1	lb.	1	lb.	—	
Butter	12	oz.	1	lb.	4	oz.
Shortening	3	lb.	3	lb.	3	lb.
Milk	4	oz.	8	oz.	3	lb.
Caramel Color	—		—		3	oz.
Whole Eggs	3	lb.	4	lb.	3	lb.
Salt	3	oz.	2	oz.	4	oz.
Cinnamon	1½	oz.	—		1	oz.
Mace	½	oz.	—		½	oz.
Cloves	1	oz.	—		½	oz.
Vanilla Extract	½	oz.	¼	oz.	¼	oz.
Almond Flavor	—		1	oz.	¼	oz.
Lemon Flavor	1	oz.	—		—	

* Special soyabean flour.

	Golden Fruit Cake	White Fruit Cake	Spice Fruit Cake
Rum Flavor	1 oz.	—	—
Baking Powder	5 oz.	5 oz.	5 oz.
Fruit Mixture:			
Seedless Raisins	6 lb.	7 lb.	6 lb.
Sliced Citron	3 lb.	3 lb.	1 lb.
Sliced Lemon Peel	¼ lb.	—	—
Sliced Orange Peel	½ lb.	3 lb.	—
Walnut, Pecan Pieces	3 lb.	—	1 lb.
Cherry Pieces	2 lb.	3 lb.	2 lb.
Chopped Dates	2 lb.	2 lb.	2 lb.
Water	2 lb.	—	—
Sugar Syrup	1 lb.	—	—
Almonds	—	2 lb.	—

Cream the shortening, add sugar, syrups, beat well, then add egg, milk, etc. Sift the flour with baking powder, salt and spices, and add to the creamed sugar, etc. Sift part of the flour over the fruit, add to the batter and add the nuts last. Place in loaf pans lined with greased paper. Cover tightly.

Bake with light steam for 1½ to 3 hours at 325 to 350°F.

Remove the covers at the end of the first hour and finish baking.

Light Fruit Cake

Sugar	2 lb.
Shortening	2 lb.

Salt   1 oz.
Flavor   ½ oz.
Cream together.
Eggs   2 lb.
Cream into the above mixture.
Flour   2 lb.
Baking Powder   ¼ oz.
Add and blend well.
Almonds   1½ lb.
White Raisins   2 lb.
Citron   2 lb.
Glace Cherries   2 lb.
Glace Pineapple   1½ lb.
Add to the mixture.
Scale into loaf tins and bake at low heat.

Devil's Food Cake

Batch Weight

	25 lb.		50 lb.	
Step 1				
Flour	3 lb.		6 lb.	
Shortening	2 lb.	4 oz.	4 lb.	8 oz.
Step 2				
Flour	1 lb.	4 oz.	2 lb.	8 oz.
Sunsoy *	1 lb.		2 lb.	
Salt		2 oz.		4 oz.

* Special soyabean flour.

Batch Weight

	25 lb.			50 lb.	
Step 2—Continued					
Soda		1½ oz.			3 oz.
Baking Powder		2½ oz.			5 oz.
Cocoa	1 lb.	4 oz.		2 lb.	8 oz.
Sugar	6 lb.	2 oz.		12 lb.	4 oz.
Water	3 lb.	12 oz.		7 lb.	8 oz.
Step 3					
Eggs	3 lb.			6 lb.	
Step 4					
Water	3 lb.			6 lb.	

Mix each step 2 minutes scraping down the bowl each time. Scale. Bake at 375°–400° F.

French Cheese Cake
- A Bakers' Pot
  - Cheese 1 lb. 10 oz.
  - Cake Flour 5 oz.
  - Salt ⅛ oz.
  - Granulated Sugar 5 oz.
  - Dried Milk 3 oz.
- B Egg Yolks 4 oz.
- C Melted Butter
  - or Shortening 4 oz.
- D Water 8 oz.
  - Lemon Juice 1½ oz.
  - Grated Lemon
    - Rind ⅛ oz.
- E Egg Whites 1 lb.
- F Granulated
  - Sugar 5 oz.
  - Purity C Bakers'
    - Starch 1½ oz.
  - Water 1 lb.
- G Granulated
  - Sugar 5 oz.

Blend A on low speed, add B, and mix 5 minutes. Incorporate C, then D. Prepare a cooked meringue by cooking F until clear and thick, dissolving G, and then beating this hot syrup into the whites, E, already whipped to a soft peak. Fold the meringue into the cheese batter, gently but evenly.

Prepare the pan with fat and crumbs. Fill the pan, set it in a second pan, and bake at 350°F.

Doughnuts
Formula No. 1

Granulated Sugar	2	lb.
Shortening	6	oz.
Salt	1½	oz.
Pastry Flour	7¼	lb.
Baking Powder	4	oz.
Egg Yolks	8	oz.
Whole Eggs	8	oz.
Liquid Milk (Variable)	4	lb.
Flavor	To taste	

Mix together at medium speed, varying the mixing time from 30 seconds to 4 minutes depending upon the ingredients, shop conditions, etc.

Use the formula as set up for machine-made doughnuts. If you cut down on the milk in the mix, you will get a dough which can be rolled and cut by hand.

Fry in shortening at 375 to 390°F.

Formula No. 2

4X Powdered Sugar	3	lb.
Shortening	8	oz.
Salt	2	oz.
Skim Milk Powder	2	oz.
Soda	½	oz.
Bread Flour	4	lb.
Pastry Flour	4	lb.
Baking Powder	6	oz.
Egg Yolks	2½	lb.
Liquid Milk (Variable)	4	lb.
Flavor	To taste	

Mix together at medium speed, varying the mixing time from 30 seconds to 4 minutes, depending upon the ingredients, shop conditions, etc.

Use the formula as set up for machine-made doughnuts. If you cut down on the milk in the mix, you will get a dough which can be rolled and cut by hand.

Fry in shortening at 375 to 390°F.

Formula No. 3

Sugar	2	lb.
Shortening	8	oz.
Salt	1	oz.
Soda	1	oz.
Baking Powder	2	oz.
Flour	8	lb.
Eggs	1	lb.
Buttermilk	5	lb.
Flavor	To taste	

Mix together at medium speed, varying the mixing time from 30 seconds to 4 minutes depending upon the ingredients, shop conditions, etc.

Fry in shortening at 375 to 390°F.

Chocolate Doughnuts

Sugar	6	lb.
Shortening	¾	lb.
Cocoa	1¼	lb.
Salt	3	oz.
Vanilla	2	oz.

Cream the above ingredients together.

Eggs	3	lb.
Milk	10	lb.
Bread Flour	5	lb.
Cake Flour	8	lb.
Baking Powder	½	lb.

Add the eggs, then the milk. Sift the dry ingredients together and add. Fry at 385°F.

Doughnut Glaze
Formula No. 1

Water	2½	lb.
Corn Syrup	8	oz.
Gelatin *	1	oz.

Dissolve the gelatin in the water. Add the corn syrup and bring to boil.

4X Powdered Sugar	10	lb.
Flavor	To taste	

Pour the above hot syrup over the sugar and mix until smooth. Stir in the flavor.

Dip the yeast raised doughnuts into the glaze as they come from the frying kettle. This glaze works best around 90 to 100°F. The glaze will set up while draining on the wire.

Formula No. 2

A Granulated Sugar	1	lb.
Agar	3	oz.
Water	1	qt.
Vanilla	If desired	

Mix sugar and agar, and boil for 3 minutes in 1 qt. of water.

B Mix smooth 10 lb. of powdered sugar with 1 qt. of hot water.

---

* The amount of gelatin may be varied depending upon its strength.

Pour hot A over B and mix thoroughly.

Doughnut Honey Glaze

4X Powdered Sugar	10 lb.
Honey	6 oz.
Hot water	1 lb.

Mix to a paste.

Gelatin *	3 oz.
Hot Water	1 lb.

Dissolve, and mix smooth. Bring to a temperature of 90 to 100°F. for use.

Dip the yeast raised doughnuts into the glaze, as they come from the frying kettle. Take the doughnuts from the glaze and allow them to drain on wire screens until the glaze sets up. Doughnuts may be covered with chopped nuts, coconut or sugar.

Doughnut Icing Glaze

Water	4 lb.
Tapioca Flour	1 oz.
Corn Starch	½ oz.
Agar-Agar	½ oz.

Bring to boil for 3 minutes. Place in a cake machine with the following:

Vanilla Extract	½ oz.
Salt	¼ oz.
Butter	2 oz.
Corn Syrup	1 lb.

Then add 10 lb. of icing sugar and mix well.

Use while warm.

Doughnut Icing

Water	1½ lb.
Corn Syrup	8 oz.

Bring the above to boil.

Gelatin *	½ oz.
Water	8 oz.

---

* The amount of gelatin may be varied depending upon its strength.

Dissolve the gelatin in hot water and add to the above syrup 4X Powdered Sugar 12 lb.

Add the above syrup slowly to the 4X sugar and beat until smooth.

Bring to a temperature of 90 to 100°F. for use.

Ice the doughnuts by dropping them into the icing made from the formula given above as soon as they come from the frying kettle. After complete immersion remove them and allow them to drain on a wire screen.

Lemon Powder for Pie Fillings

Corn Starch	75 lb.
Citric Acid	5 lb.
Powdered Sugar	20 lb.
Lemon Oil	6 oz.

Vanilla Powder

Powdered Mexican Vanilla Bean	50 lb.
Granulated Sugar	50 lb.
Bitter Almond Oil	½ oz.

Pie Filling Powder Base for All Flavors

Corn Starch	45 lb.
Tapioca Starch	15 lb.
Citric Acid	1 lb.
Gum Karaya	½ lb.
Salt	1 lb.
True Fruit Flavor	To suit
Imitation Fruit Flavor	To suit
Certified Food Color	To suit

This recipe, when properly mixed with sugar, will make the filling for pies. Cherry, strawberry, and raspberry flavors may be used. For lemon, a small percentage of powdered egg yolk will improve the taste of the product.

One pound of the above formula is dispersed in 1 qt. of warm water.

Then add this to a hot solution of 3 lb. of sugar in 3 qt. of water. Boil, and remove from fire as soon as it thickens.

Cream Filling for Pies

Milk	1 gal. or 8 lb.
Sugar	2 lb.
Cornstarch (High Grade)	8 oz.
Egg Yolks	9 oz.
Soyflake	4 oz.
Salt	1 oz.
Vanilla	1 oz.

Bring three quarters of the milk and half of the sugar to a boil. In the meantime mix the balance of milk and sugar with salt, cornstarch, Soyflake and well beaten eggs. When the milk solution boils, pour this mixture in, stirring briskly. Remove from the fire, cool at once and add vanilla. Care should be used in cooking the mixture thoroughly, and stirring briskly after the starch and the Soyflake have been added, otherwise the cream will become rough and lumpy.

Soyflake flour can be worked in very satisfactorily in proportion of 4 oz. to 1 gal. of liquid into most cream fillings.

Cream Filling for Ice Cream Wafers

Icing Sugar	225 lb.
Cocoanut Butter (M.P. 92°)	125 lb.
Powdered Milk	25 lb.
Soyflake	15 lb.

Cream sugar and 125 lb. cocoanut butter until light, add powdered milk and Soyflake, and continue creaming until desired consistency is reached.

Cream Icing and Filling

Sugar	5 lb.
Cocoanut Butter (M.P. 92°)	2 lb.
Glucose (43° Bé.)	1 lb.
Marshmallow	2 lb.
Salt	½ oz.
Soyflake	10 oz.
Water	10 oz.

Cream sugar, cocoanut butter, salt, glucose, and water until fluffy; add Soyflake, mix thoroughly; add marshmallow, and continue mixing until smooth.

Improved Cake Icing

Glyceryl Monostearate (S 928) improves cake icings when used as follows:

CREAM ICING

(*Formerly used*)		(*Now used*)
100 lb.	Sugar	100 lb.
13 lb.	Shortening	......
....	Glyceryl Monostearate (30% Water dispersion)	5 lb.
2½ lb.	Powdered Milk	......
5 oz.	Salt	5 oz.
5 oz.	Vanilla	5 oz.
13 lb.	Water	13 lb.

NOTE: When using a Glyceryl Monostearate S-928 dispersion in icings, it is best to make up the icing with hot water. This formula gives an icing which creams up better, stays soft longer and prevents adhesion to the wrapper.

Cream Puff Paste

Melt the shortening, then heat on full flame until rapid stirring will not stop the boil:

Water	8 oz.
Shortening	10 oz.
Salt	¼ oz.

Mix, stir rapidly until thoroughly and evenly mixed:

Bread Flour	8 oz.

Turn off the heat; place in a mixing bowl and mix at medium speed for 30 to 45 seconds to aid cooling to approximately 170°F. Add gradually, 1½ to 3 oz. at a time, to make a total of 1 lb. egg at 70 to 75°F.

After each addition of egg, the mixture should be stirred until the eggs have been taken up and sufficient body developed for the paste to clean the side of the bowl.

Mix together; add, and stir in:

Ammonium Carbonate	1/16 oz.
Milk	1 oz.

If necessary, add 2 or 3 additional oz. of milk to make a paste which is glossy and yet has enough body to hold its shape when deposited on a paper-lined pan. The temperature of the paste should be approximately 100°F. Bake at 425°F.

Boiled Meringue Icing

Egg Whites	2 pt.
Meringue Powder	8 oz.
Water	2 pt.
Color and Flavor	To suit

Beat to a mazette.

Powdered Agar-Agar	2¾	oz.
Sugar	12	lb.
Corn Syrup	3	lb.
Cream of Tartar	1	oz.
Water	2	qt.

Mix the sugar with the agar. Boil the water. Add cream of tartar. Then add the sugar-agar mixture slowly, then the corn syrup and cook at least 10 minutes. Add this slowly to the above mazette and beat for 10 minutes at high speed.

Light Meringue Icing

Salt	¼ oz.
Vanilla Flavor	To suit
Egg Whites	10 lb.

Beat until stiff.

Sugar	32 lb.
Corn Syrup	3 lb.
Water	10 pt.

Heat to 240°F.

Add the hot syrup to the beaten egg whites and beat to a good consistency. Chopped nuts or fruit juices may be added.

Sugarless Meringue

Egg White	1
Saccharin	¼ gr.
Lemon Extract	¼ tsp.
Salt	To taste

To the partly beaten egg white, add the other ingredients in the order given, until stiff. Spread over surface and bake in a moderately hot oven.

Billowy Marshmallow

Gelatin	1
Cold Water	8
Egg White	15
Icing Sugar	65
Standardized Invert Sugar	10
Flavor and Color	As desired

Soak the gelatin in the cold water. Heat the soaked gelatin carefully until fluid (about 140°F.). Add the fluid gelatin to the egg whites, 35 parts of sugar and the standardized invert sugar. Whip this mixture until it is fairly stiff. Just before finishing the whipping, add the remaining 30 parts of sugar along with the flavor and color. Beat until it is of the consistency of mashmallow.

For banana flavor, use about 4 oz. of ripe banana powder containing dry milk solids for each 10-lb. batch. Incorporate it in any way convenient, preferably first blended with about an equal weight of sugar.

Marshmallow Fluff

Corn Syrup	3½ lb.
Sugar	3½ lb.
Water	14 oz.

Boil to 230°F. Then remove from fire.

Corn Syrup	2½ lb.
Invert Sugar Syrup	1 lb.

Add to the above in a clean beating bowl:

Water	13 oz.
Egg Albumen	4½ oz.
Vanilla Extract	½ oz.
Salt	¼ oz.
Cream of Tartar	¼ oz.

Dissolve. Add to above. Then beat until stiff on medium speed. Fill containers for storage.

This fluff spreads easily. It never gets tough or rubbery.

Egg White Substitute

Skim Milk	5500 l.
Whey	1000 l.
Slaked Lime	23.2 kg.

Mix and concentrate by vacuum evaporation to about ⅕ of original volume. Then spray dry, starting with 160°F. and going down to 70°F.

Yeast Cake Improver

One pound of Glyceryl Monostearate is dispersed in 6 pounds of hot water and this is further diluted with 4 pounds of cold water, thus producing a 10% dispersion. This dispersion is added, when cold, to 400 pounds of yeast containing 70–72% moisture. One half to three-quarters pound of sulfonated olive oil is also added to the yeast. The resultant product has better keeping qualities, improved color immediately after mixing, and also after aging, improved cutting qualities, no coring, etc., and appears to be a better dispersion.

Improving the Quality of Baked Goods

In the last few years widespread investigations concerning edible emulsifying agents have been undertaken by the baking industry with a view to improving the quality of bakery products in many respects and to reduce the amount of shortening.

In the forefront of the materials studied were the mono- and diglycerides, and, in particular, Glyceryl Mono Stearate. Several forms of this material were studied, and it has become apparent that the particular formulation known as Glyceryl Mono Stearate (S-928) is exceptionally well adapted.

Three advantages accrue to the baker in whole or in part as a result of using Glyceryl Monostearate (S-928).

1. Improved Quality
2. Replacement of Shortening
3. Economic Savings

The use of S-928 is of definite advantage in:

a. Cakes
b. Bread
c. Icings

Glyceryl Monostearate (S-928) is generally applied or used by the baker in the form of a water dispersion. The standard 30% dispersion is made by heating 70 parts of water almost to boiling, followed by the addition of 30 parts of S-928, with constant stirring until the dispersion is cool. S-928, which is supplied in the form of fine beads, is readily dispersible under such conditions, and forms a stable creamy dispersion of a consistency somewhat akin to shortening. Such a standard 30% dispersion should be used the same day that it is made unless refrigerated storage space is available. If it is kept in refrigerated storage it will remain in first class condition for a week or more. A convenient vessel which may be used in making this dispersion is the usual type of cake mixer.

Cakes

Mono- and di-glycerides are used in making cakes, sweet pastries and icings. They promote emulsification between the fatty and aqueous ingredients. The high degree to which the fat may be dispersed, and the consequent mechanical strength of the emulsion, permit the baker to use a greater amount of water in the dough, without causing falling during baking. A softer, smoother and stronger product is obtained. The greater amount of water that can be added permits higher ratios of sugar to flour, and a sweeter product is obtained. The ratio of sugar to flour may be as high as 140:100 by weight compared to ordinary shortenings of ratio 100:100.

In bakeries which are particularly concerned with the maximum improvement in quality characteristics of their product, it is often the practice to add from 5% to 10% of the standard 30% dispersion * (this 5-10% is based on the flour content) to the other ingredients in the cake mix and then to follow the usual baking procedures with little or no modifications.

Although the cost of the cake is slightly increased, the product is outstanding in texture, aroma and, in particular, its shelf life or retained freshness and tenderness is remarkably increased.

On the other hand, some bakeries which are particularly concerned with savings of shortening and, of course, with substantial financial savings, are cutting out as much as 50% of their usual shortening content. In place of the 50 parts of shortening which have been removed, they will add 25 parts of the standard 30% dispersion * and make up the remaining 25 parts by the addition of flour and water. This produces a cake which is approximately equal to that made by the usual production methods with the normal amount of shortening.

However, it is generally advocated to replace only a smaller percentage of the shortening. In this case there is a definite improvement in the product with a saving in costs.

---

* Glyceryl Monostearate (S-928) 30
Boiling Water 70

Baking Test in Which Glyceryl Monostearate S-928 Was Used in the Production of Layer and Loaf Cakes

Formula	(Control) Cake No. 1A	Cake No. 2A	Cake No. 3A
Flour	400 g.	400 g.	400 g.
Sugar	400 g.	400 g.	400 g.
Shortening	200 g.	140 g.	140 g.
Standard 30% Dispersion*	...	60 g.	60 g.
Eggs	280 g.	280 g.	280 g.
Milk	280 cc.	280 cc.	280 cc.
Baking Powder	20 g.	20 g.	20 g.
Salt	6 g.	6 g.	6 g.
Vanilla	6 cc.	6 cc.	6 cc.
Specific Gravity of Batter	0.8355	0.6944	0.6744
Specific Volume of Batter	1.1968	1.4400	1.4827

*Glyceryl Monostearate (S-928) 30.
Boiling Water 70.

Cake Scores

	Layers			Loaves		
	No. 1A	No. 2A	No. 3A	No. 1A	No. 2A	No. 3A
Symmetry	4.37	4.37	4.45	4.37	4.37	4.37
Bloom	4.37	4.07	4.07	4.37	4.07	4.07
Color of crust	4.37	4.07	4.07	4.37	4.07	4.07
Volume	4.37	4.54	4.69	4.37	4.54	4.54
Consistency of Crust	4.37	4.22	4.22	4.37	4.22	4.27
Color of Crumb	8.75	8.44	8.59	8.75	8.44	8.59
Grain	8.75	8.90	9.06	8.75	8.75	8.90
Texture	13.13	14.06	14.06	13.13	13.59	14.06
Aroma	8.75	8.75	8.75	8.75	8.75	8.75
Flavor	17.52	18.12	18.12	17.52	17.82	18.12
Eating Quality	8.75	9.22	9.22	8.75	8.90	9.06
Total Score	87.50	88.76	89.30	87.50	87.52	88.80
Specific Gravity of Cake	0.2933	0.2733	0.2500	0.3200	0.2766	0.2533
Specific Volume of Cake	3.4094	3.6589	4.0000	3.1250	3.6153	3.9478
pH Value	8.90	8.83	8.80	8.61	8.45	8.45

Color Readings of Crumb

	Layers			Loaves		
	No. 1A	No. 2A	No. 3A	No. 1A	No. 2A	No. 3A
Red	14.8%	14.3%	14.3%	14.8%	14.3%	14.3%
Yellow	34.8	35.3	32.5	34.8	35.3	32.5
Black	0.2	0.5	0.5	0.2	0.5	0.5
White	50.2	49.7	52.7	50.2	49.7	52.7

Baking Test for Pound Cake with Glyceryl Monostearate (S-928)

Formula	Control Cake No. 1	Cake No. 2	Cake No. 3
Flour	350 g.	350 g.	350 g.
Sugar	350 g.	350 g.	350 g.
Shortening	200 g.	140 g.	140 g.
Standard 30% Dispersion	...	60 g.	60 g.
Eggs	220 g.	220 g.	220 g.
Milk	180 cc.	180 cc.	180 cc.
Baking Powder	5.25 g.	5.25 g.	5.25 g.
Salt	12.5 g.	12.5 g.	12.5 g.
Vanilla	3.5 cc.	3.5 cc.	3.5 cc.
Specific Gravity of Batter	0.7555	0.6722	0.6855
Specific Volume of Batter	1.3236	1.4876	1.4587

Pound Cake Scores

	Cake No. 1	Cake No. 2	Cake No. 3
Symmetry	4.37	4.37	4.37
Bloom	4.37	4.37	4.37
Color of Crust	4.37	4.37	4.37
Volume	4.37	4.54	4.54
Consistency of Crust	4.37	4.07	4.07
Color of Crumb	8.75	8.90	8.90
Grain	8.75	8.75	8.75
Texture	13.13	13.59	13.59
Aroma	8.75	8.75	8.75
Flavor	17.52	18.12	18.12
Eating Quality	8.75	8.90	8.90
Total Score	87.50	88.73	88.73
Specific Gravity of Cake	0.3600	0.3133	0.3133
Specific Volume of Cake	2.7777	3.1918	3.1918
pH Value	7.55	7.60	7.58

Color Readings of Crumb

	Cake No. 1	Cake No. 2	Cake No. 3
Red	13.5%	14.5%	14.5%
Yellow	34.2	36.3	36.5
Black	0.8	1.0	1.0
White	51.5	48.2	48.0

Although the proper use of the standard 30% dispersion of Glyceryl Mono Stearate (S-928) in the production of cakes may result in lower costs, its usage is particularly recommended because the cake so produced is outstanding as to texture, keeping qualities, aroma, and flavor.

Bread

Excellent bread of the sponge-dough type is readily made by the use of 2% shortening plus 2% of the standard 30% dispersion.* The dispersion should be added in the dough stage. (Percentages are based on the amount of flour used.) All other ingredients may be added in the usual way.

Bread so produced will have good volume with remarkable freshness and softness on keeping, and will have tender crust and improved texture. If the straight-dough method is used, it is desirable to give the dough at least two hours fermentation time.

In general, it may be stated that if the baker has as his sole object the improvement of his product notwithstanding any economic factors involved, he should retain his present formula but add 2% of the standard 30% dispersion * as indicated. If it is important to reduce the costs, a few trials with varying reduced amounts of shortening will indicate the possibilities in this regard. (Naturally, in each case 2% of the dispersion * is added.)

Recently instead of the dispersion, Glyceryl Mono Stearate (S-928) has been added directly to the sponge. In this application, 2% of shortening and 0.6% of S-928, based on the total weight of the flour, are added. This S-928 should be added to compensate for the water used in the standard dispersion method. The sponge takes the usual mixing time, about 5 minutes, with the usual fermenta-

---

* Glyceryl Monostearate (S-928) 30
  Boiling Water 70

tion time of 4 to 5 hours. The sponge on going back to the mixer should be less sticky and more pliable. It has been found in this method that the fine beads of S-928 seem to disappear in the sponge and dough and no evidence of them can be found in the resultant bread.

Greater tenderness of the sponge will be noticed and greater capacity to retain moisture is likewise apparent. Every favorable quality will be evidenced as to whiteness of crumb, volume of the loaf, flavor, aroma and increased and prolonged softness.

Icings

It is interesting to compare a typical formulation for a cream icing according to standard baking procedure, with one in which the standard 30% dispersion * of S-928 has been incorporated, omitting all shortening. Icings so produced cream better, stay soft longer and exhibit a surface hardness which almost totally prevents sticking to the wrapper. This last factor is of particular interest to many bakeries. The only variation in method of manufacture of such an icing lies in the recommended use of hot water instead of the cold water used in the old formula.

* Glyceryl Monostearate (S–928) 30
  Boiling Water 70

Cream Icings

	Formerly used	Now used
Sugar	200 lb.	200 lb.
Shortening	25 lb.	......
30% Dispersion *	......	10 lb.
Milk Powder	5 lb.	......
Salt	10 oz.	10 oz.
Vanilla	10 oz.	10 oz.
Water	25 lb.	25 lb.

In the manufacture of cup cakes, it has been reported that where ordinarily 27 lb. of shortening are used per 450 lb. of cake dough, with the use of slightly more than 1% Glyceryl Monostearate (S-928), it has been possible to cut this amount back one-fifth. However, such an increase in volume has been obtained that it has been possible to save 3 oz. per dozen cup cakes, getting better volume than previously obtained. The usual 30% dispersion is made up as needed.

It has also been reported that 20% of shortening can be replaced in chocolate cakes and 10% in yellow cakes.

White Bread Containing Soybean Flour

	Batch Weight					
	25 lb.		50 lb.		100 lb.	
	Sponge	Dough	Sponge	Dough	Sponge	Dough
Flour	8 lb. 8 oz.	5 lb. 11 oz.	17 lb.	11 lb. 6 oz.	34 lb.	22 lb. 12 oz.
Yeast	4 oz.		8 oz.		1 lb.	
Yeast Food	1 oz.		1¼ oz.		2½ oz.	
Sunsoy [1]	8 oz.		1 lb.		2 lb.	
Sugar		6 oz.		12 oz.		1 lb. 8 oz.
Salt		4 oz.		8 oz.		1 lb.
Shortening		4 oz.		8 oz.		1 lb.
Dry Milk Solids		8 oz.		1 lb.		2 lb.
Water	5 lb. 3 oz.	3 lb. 7 oz.	10 lb. 6 oz.	6 lb. 14¾ oz.	21 lb.	13 lb. 9½ oz.

[1] Sunsoy (Soya flour) may be replaced by Soyflake if desired.

Mix sponge 4 to 5 minutes at low speed, fermentation time 4½ hours at 80°F. Mix doughs, set approximately 15–20 minutes at 80°F., scale and proof for 10 minutes at 80°F., pan and proof for 70 minutes at 90°F., saturated humidity. Bake for 30 minutes at 450°F.

Bread Improver
Lard	1–2 lb.
Glyceryl Monostearate	2 lb.
Wheat Germ Oil	2 oz.
Commercial Lecithin	1 oz.

The above is mixed with 100 lb. flour to give better mixing and a whiter bread of finer texture, which does not dry out quickly.

Irish Soda Bread
Soft Flour	12 lb.
Milk	3 lb. 12 oz.
Salt	3 oz.
Soda	3½ oz.
Shortening	8 oz.
Sugar	4 oz.
Cream of Tartar	7 oz.
Raisins	2 lb. 8 oz.

Mix to a smooth dough, taking care not to get the mixture too stiff. When nearly mixed, add the raisins. In mixing, be sure that the milk is cold. Adjust according to the strength of the flour.

Shape dough into flat disks, brush over with milk and egg wash. Cut a cross on top before baking at about 400°F.

If scaled at 17¾ oz. for each loaf, this batch makes approximately 17 loaves.

Vitamin Mixture for Flour Enrichment
Thiamine	380 mg.
Riboflavin	230 mg.
Niacin	2740 mg.
Iron	2400 mg.
Starch	To make 1 oz.

½ oz. of the above is added to 100 lb. of flour.

Noodles
Durum Flour	4 lb. 8 oz.
Special X Flour (or Soyflake)	1 lb.
Eggs	3½ oz.
Water	1½ lb.
Salt	⅛ oz.

Mix all ingredients together well. This makes a very stiff dough.

Wheat Cake or Flannel Cake
Pastry Flour	2 lb.
Soyflake	3 oz.
Baking Powder	2 oz.
Sugar	2 oz.
Melted Butter	2 oz.
Salt	

Enough milk to make a thin batter. Bake on a medium hot griddle.

Bakers' Pan Grease
Formula No. 1
Soyflake	1
Shortening	7
Regular flour	3

Cream well together.

No. 2
Soyflake	5
Shortening	5
Salad Oil	10

Whip the shortening and Soyflake flour up together, then add the oil slowly. This Soyflake pan grease is fine for greasing pans for bread, cup cakes, pecan rolls, high sugar content or any other bakery product.

Yeast Fermentation Defoamer
0.9 kg. crude lanolin is used per 100 kg. yeast content of a batch.

Fudge Bars

Corn Syrup	50
Granulated Sugar	40
Standardized Invert Sugar	10
Edible Oil	5
Unsweetened Evaporated Milk	30
Basic Casting Fondant	100
Salt	1
Frappe	20
Full Fat Soya Flour	20
Flavoring	As desired

The casting fondant is made with 80 parts of granulated sugar, 20 parts of corn syrup and 10 parts of standardized invert sugar, with water to dissolve the sugar. Heat all together, stirring occasionally, until the batch boils, then wash down all grains that adhere to the inside of the kettle and heat rapidly to 242 to 246°F. Cool to 130 to 115°F. Beat into fondant.

The standard frappe is made with equal weights of corn syrup and standardized invert sugar, with albumen in the proportion of 1½ oz. dissolved in 3 oz. of water for 10 lb. of the sugars. Heat the corn syrup to 245°F. Turn off the heat. Add the invert sugar, stirring until melted. Beat in a marshmallow beater, adding the albumen solution. Beat until light.

To make the fudge, place the corn syrup, sugar, invert sugar, edible oil and evaporated milk in a kettle. Mix well. Continue to stir while cooking to a medium ball (approximately 245°F.). Turn off the heat. Add the fondant and salt, mixing thoroughly, then add the frappe, soya flour and flavor and mix well.

Spread the batch on heavy waxed or oiled paper. When it has set, it may be scored or marked into squares.

Nougat Fruit Nut Bar

Granulated Sugar	13
Shortening (Preferably Part Butter)	4
Standardized Invert Sugar	4
Salt	¼
Flavor (Vanilla and Maple)	As desired
Whole Eggs	4
Liquid Milk	1½
Flour	10
Dates or Other Fruits	4
Nuts	4

Cream the sugar, shortening, standardized invert sugar, salt and flavor until smooth. While creaming, add the eggs slowly and cream very little. Add the milk to the creamed mixture and mix in until distributed. Then add the flour. Mix until smooth.

Add the fruit and chopped nuts Mix in until distributed. The nuts may be pecans, walnuts or whatever is desired.

Deposit the dough about an inch deep on flour-greased sheet pans. Bake at about 360°F.

After baking cut into rectangular pieces as when making chocolate brownies. Sprinkle with powdered sugar.

Santo Domingo Nougat

Egg Albumen	10 oz.
Tapioca Flour	10 oz.
Water	1 lb. 8 oz.
Molasses	10 lb.
Granulated Sugar	30 lb.
Standardized Invert Syrup	5 lb.
Corn Syrup	10 lb.

Coconut Butter (M.P. 86 to 88°F.)	3 lb.
Powdered Whole Milk	4 lb.
Powdered Salt	2 oz.
Shredded Citron	5 lb.

Dissolve the albumen in the water, add the tapioca flour and 5 lb. of molasses. Beat until quite light. Transfer the whipped batch to an upright beating machine.

Boil 5 lb. of sugar, the corn syrup, enzyme-converted corn syrup and molasses with enough water to dissolve the sugar. Bring to 250°F. Add 1 gal. of this syrup to the beaten portion and beat for several minutes. Gradually add the balance of the batch and beat until quite light and grain is visible. Add the citron, coconut butter, powdered milk and salt, mixing well.

Spread the batch on oiled and dusted slabs. It may be cut into bars or formed into kisses or other shapes by passing the batch through a rolled cream center machine.

Short Nougat

Sugar	50 lb.
Corn Syrup	50 lb.
Full-Fat Soya Flour	7–12 lb.
Egg Albumen	2 lb.
Salt	8 oz.
Flavoring	As desired

Place 5 lb. of corn syrup in a nougat beater with the egg, having the beater in operation. Heat the sugar and the remaining corn syrup to the desired degree, usually about 270°F., and pour this slowly into the egg mixture in the beater. Beat to the desired lightness.

Adjust the beater to slow speed and add the soya flour and salt. Mix well. Flavor and finish as usual.

Low-fat soya flour can be used if desired. This formula is suggested to increase the nutritive balance of the candy through introduction of soya flour.

Fruit Bars

Granulated Sugar	20 lb.	
Shortening	10 lb.	
Milk Powder	1 lb.	4 oz.
Soda		5 oz.
Salt		5 oz.
Eggs	5 lb.	
Molasses	6 lb.	14 oz.
Water	5 lb.	
Seedless Raisins	12 lb.	
Oatmeal	1 lb.	8 oz.
Cake Flour	32 lb.	8 oz.
Cinnamon		7 oz.
Ginger		3 oz.
Allspice		5 oz.

Mix together the sugar, shortening, milk powder, soda and salt. Add the molasses and water, stirring into the mix. Add the oatmeal and the seedless raisins, then the flour and spices, folding the latter in.

These can be run on a cookie machine, using a special four-hole die.

Scale 13½ oz. to a strip of 18. Bake at 380°F.

To run on the cookie machine, remove the cutter arm and use a special die which is made of a piece of flat steel the same size as the regular die, with four small holes spaced so that they will drop the material into the pans. These pans are 3x30 in., with a 1-in. sidewall. It takes about nine strokes to run out a full-length bar, as the pans pass under the die. When the full-length bar has been obtained, a short stop must be made to cut it off before continuing with the next pan.

Bars are slightly flattened in the pan before baking. After baking they are dumped on a conveyor belt while warm and stacked four high, to be cut with a knife by hand. They are then ready for wrapping.

Candied Fruit Peel

Wash either 4 oranges, or 6 lemons, or 2 grapefruits, or 6 tangerines and remove the peelings. Cut these in 1/4 to 1/2 in. strips and cover with cold water and then heat to boiling. Allow to boil till tender, which may take 10 or 15 minutes. Drain well on a screen. Make a sugar solution by adding 2 cups of granulated sugar to 1 cup of water. Cook this syrup solution until it reaches a temperature of 238°F. or 240°F. Next add the cooked peel and boil slowly, with stirring for 20–30 minutes, until a greater part of the syrup has been absorbed. Allow the peel to stand overnight in the sugar solution and next day, drain thoroughly, roll them in granulated sugar and dry finally in an oven at 240°–250°F. until each piece is firm.

Butterscotch Topping

Granulated Sugar	15 lb.	
Clear Syrup	28 lb. 8	oz.
Water	44 lb.	
Butter	8 lb.	
Egg Yolks	12 lb.	
Cornstarch	6 lb.	
Salt	1½ oz.	
Vanilla	1 lb.	

Heat the granulated sugar and corn syrup with 2 lb. of water to 280°F. Add the butter to this. Blend together the egg yolks, cornstarch and 6 lb. of water. Bring to a boil 36 lb. of water containing the salt, and add the sugar and cornstarch mixtures. Cook until clear. Blend in the vanilla.

Cream Fondant for Chocolate Dipping

Take ½ the white of a large fresh egg and beat it till it is quite stiff. This usually takes 3–5 minutes. Now, take ½ lb. of XXXX confectioner's sugar and add to the beaten white of the egg in small quantities. Do not add more than ½ teaspoonful at a time and continue beating. When nearly all the XXXX sugar has been added, add the flavoring, e.g., a few drops of peppermint oil, vanilla extract, maple, or any other flavor. Finally add the rest of the sugar and knead and mix thoroughly with the hands. Set away for 2–3 hours before use and then break up in small portions and dip in chocolate.

Candy Coatings (Glazes)

Formula No. 1

Edible White Shellac	1 lb.
Pure Alcohol	7 pt.
Ethyl Acetate	1 pt.

No. 2

Gum Benzoin	1 lb.
Pure Alcohol	8 pt.

No. 3

Edible Shellac	1 lb.
Gum Sandarac	1 lb.

Pure Alcohol	8	pt.
Soluble Brown Dye No. 4	1	oz.
Gum Mastic	4	oz.
Edible Shellac	1	lb.
Brown Dye	¼	oz.
Gum Sandarac	1	lb.
Alcohol	8	pt.

Sugarless Chocolate Bars

Bitter Chocolate	20
Skimmed Milk Powder	50
Fatless Soybean Flour	30
Vanillin	Enough to flavor
Saccharin	Enough to sweeten

Sugarless Mapleine Syrup

Powdered Gum Karaya	1	oz.
Water	32	oz.
Saccharin	3	gr.
Mapleine Extract	9	cc.
Caramel Color	Enough to suit	
Salt	To taste	

Whip the gum up with the cold water to a thick liquid. Then add flavor, sweetener and color.

Compound Table Syrups
Formula No. 1
Cane Sugar and Maple Sugar Blends

Sugar Syrup	85	pt.
Vermont Maple Syrup	15	pt.

No. 2
Corn Syrup and Cane Sugar Blend

Corn Syrup (39° Bé.)	50	pt.
Sugar Syrup	50	pt.

No. 3
Cane Sugar and Molasses Blend

Sugar Syrup	25	pt.
New Orleans Molasses	25	pt.

Improved Shortening

Propylene Glycol Monostearate	100
Lecithin	5
Water	400

Heat to a boil and stir until uniform. Add 5–10% of the above to ordinary shortening; mix it in uniformly.

Stabilized Whipped Cream
Formula No. 1

(a) 40% Cream	1	gal.
(b) High Grade Gelatin	1½	oz.
Hot Water (200°F.)	1	lb.
Granulated Sugar	10	oz.

Dissolve (b). Add while hot to (a) stirring constantly. Place in refrigerator overnight. Whip cream the following day. Then add 10 oz. granulated sugar. Flavor to taste.

This cream may be whipped either with a regular egg beater or on an air whip machine. If possible whip in a cool place.

Be sure the product for which this cream is intended is thoroughly cooled before using. When finished hold whipped cream products under refrigeration to prevent deterioration.

No. 2

Water	1½	pt.
Sugar	1¾	lb.
Corn Starch	1	oz.
Agar-Agar	½	oz.
Egg Whites	1	pt.

Boil the water, sugar, corn starch and agar-agar for 3 minutes. Whip the egg whites with medium speed. When the whites are up to the full volume pour the syrup slowly over the whipping whites and whip until cool.

Blend with 2 qt. whipped cream.

No. 3

Sugar	27	oz.
Corn Starch	8	oz.
Gelatin	1	oz.
Agar-Agar	1	oz.
Salt	½	oz.
Vanilla Flavor	If desired	

The above ingredients are well blended, added slowly to 4 pints of boiling water, and heated to 190°F. The mixture is cooled to room temperature until it forms a paste. One gallon of chilled cream (45°F., 38% fat) is then whipped at second speed until it holds the marks of the beater. One pint of the above paste is added to the cream, and the mixture heated for ½ to 1 minute. This whipped cream will not weep and will hold its shape.

Imitation Whipped Cream

This is made without cream for filling and covering cakes, filling sandwich sponge cookies and small sponge cakes.

Shortening (Preferably Half Butter and Part Plastic Coconut Fat)	25
Sugar (Icing)	15
Nulomoline	20
Milk (Evaporated Basis)	32½
Prepared Marshmallow	7½
Flavor (Vanilla)	As desired

The fats are creamed until thoroughly mixed and slightly lightened. The powdered sugar and Nulomoline are added, and the mixture whipped lightly. During the whipping, the evaporated milk is added gradually. Just before the whipping is finished, the prepared marshmallow and flavor are whipped in.

Whipped Cream

To make whipped cream from ordinary top of the bottle cream, add 1 teaspoonful of baking soda per pint. Stir vigorously until the desired consistency is obtained.

Fortified Whipped Cream
(For bakers)

Cold Water	5	qt.
Meringue Powder	6	oz.
Sugar	4	lb.
Salt	1	oz.
Starch	14	oz.
Gelatin	¾	oz.
Vanilla Extract	1	oz.
Heavy Cream	1	qt.

In a whipping machine put 1 qt. of water, the meringue powder, and 3 lb. of sugar and whip to just peak (not stiff). Put 3 qt. of water, the remaining sugar and the salt into a kettle and bring to a boil. Dissolve the starch and gelatin in the remaining water, add to the boiling mass, and stir until it is thick and clear. Blend the two mixtures carefully with a wire whip and put in the refrigerator until needed. When ready to use, put the mixture into a clean bowl and smooth down with a wire beater. Do not beat. Bring the whipping cream up to about three-fourths stiff, pour it over the boiled mixture, and fold together only until the cream is well incorporated and the mass is smooth. This should make topping enough for 30 or 40 nine-inch pies.

Frosted Chocolate Malted

Fluid Ice Cream Mix	4 qt.
Fluid Whole Milk	4 qt.
Malted Milk Powder	1 lb.
Chocolate Syrup	1 qt.

Dissolve the malted milk powder in the fluid fresh milk, flavor with the chocolate syrup and chill to about 50°F. Combine the flavored milk with the liquid ice cream mix and freeze to the consistency of frozen custard. Serve in glasses or paper containers with long handle spoon.

It is advisable to use an ice cream mix stabilized with gelatin.

Soybean Milk

Infants' Soy Milk Formula No. 1

Cleaned Whole Soybeans	100 lb.
Sugar (Calculated as Dextrose)	55 lb.
Oil	30 lb.
Salt	6 lb.
Water to make:	
Before filtering a volume of	70 gal.
At finish a volume of	80 gal.

Start by soaking the beans in water at room temperature for about 12 hours, keeping well covered. Then grind finely in a manner similar to wet-grinding of corn or any other grain. A stream of water should be fed into the mill along with the beans to avoid clogging and to form a slurry.

Add enough water to give a volume of 70 gal. preparatory to heating in a glass-lined, stainless steel or other sanitary, corrosion-resistant metal vat. Bring to temperature of 130°F. with continuous agitation, and filter to remove granular or coarse portions of beans. Boil in the same type of vat or cooker for 45 minutes, with constant agitation.

To a conveniently handled portion of this hot milk add the oil (soya, corn or cottonseed) and whip into an emulsion.

Add this emulsion, with the sugar and salt, to the hot milk and continue boiling and stirring for 15 minutes. Add water to bring the volume up to 80 gal. Cool quickly and bottle.

No. 2

Sugar	40
Soya Flour	125
Lactose	30
Peanut Oil	20
Dextrin	20
Liquid Egg Yolk	50
Calcium Lactate	6
Salt	2

Moldproofing Cheese

After the cheese has been salted by the dry salting method, it is washed and brushed on all surfaces with a 5% solution of sodium propionate in water. This solution is applied for the first two days when the cheese is placed in the curing room.

Custard Powder

Corn Flour (St. Vincent)	300	lb.
Arrowroot Starch	20–30	lb.
Vanilla	6	oz.
Nutmeg Essence	1½	dr.
Pure Food Color	35	dr.

The above mixture is to be used at the rate of 1½ oz. per pint of milk. The smoothness of the product is increased by the amount of corn flour used.

Cocoanut Flavor Pudding Powder
Granulated Sugar	40 lb.
Corn Starch	38 lb.
Skimmed Milk Powder	10 lb.
Gelatin Powdered	2 lb.
Cocoanut Fine Cut	10 lb.
Vanillin	2 oz.

Imitation Flavors

Apricot
Linalyl Formate	1½ oz.
Amyl Valerianate	½ oz.
Oenanthic Ether	¾ oz.
Aldehyde C_{14}	⅛ oz.
Benzaldehyde	¼ oz.
Peach Flavor	8 oz.
Glycerin	1 pt.
Alcohol	67 oz.
Water	34 oz.

Banana
Amyl Acetate	3 oz.
Butyl Butyrate	⅛ oz.
Isobutyl Ketone	¼ oz.
Ethyl Benzoate	⅛ oz.
Orange Oil	¼ oz.
Benzyl Valerianate	⅛ oz.
Ceylon Cinnamon Oil	15 min.
Mace Oil	30 min.
Heliotropin	¼ oz.
Glycerin	52 oz.
Water	5 pt.
Alcohol	3 pt.

Rum Essence
Rum Ether	200
Ethyl Acetate	40
Tincture of Cinnamon	10
Tincture of Catechu	10
Tincture of Vanillin	10
Ethyl Formate	75
Tincture of Angelica Root	2
Tincture of Peruvian Bark	15
Orange Flower Water	100
Woodruff Essence	30
Butyric Ether	20
Alcohol (90%)	650
Rum	1000

Raspberry
Isobutyl Acetate	420
Isoamyl Acetate	275
Ethyl Acetate	200
Ethyl Formate	35
Benzyl Benzoate	25
Bromelia	15
Vanillin	10
Linalool	10
Eugenol	6
Benzyl Acetate	2
Geraniol	1
Ionone	1

Dilute with 2 voumes of alcohol to make the essence.

Pineapple
Ethyl Butyrate	190
Isoamyl Isovalerate	810

Dilute with 5 volumes of alcohol to make the essence.

Peach
Cyclohexyl Butyrate	600
Ethyl Cinnamate	140
Benzyl Butyrate	140
Isoamyl Butyrate	70
Isobutyl Salicylate	20
γ-Undecalactone	20
Geranyl Formate	10

Golden Ginger Ale Flavor for Beverages
Oleoresin of Ginger	10	oz.
Citral	1	oz.
Lemon 10-fold	3	cc.
Orange 10-fold	3	cc.
Lime Extract (2 oz./gal.)	1	gal.
Alcohol	2.5	gal.

Caramel	3	oz.
Water q.s.	5	gal.

Rock and Rye Whiskey Essence

Grain Fusel Oil Rectified	340
Green Wine Lees Oil	12
Peru Balsam	12
Jamaica Rum Essence	12
Vanillin	6
Ethyl Acetate	12
Coumarin	15
Raisin Wine Essence	580
Peach Essence	8
Bitter Orange Extract	50
Cinnamon Oil	2.5
Cloves Oil	2.5

Artificial Cinnamon Oil

Cinnamic Aldehyde	96.0
Eugenic Acid	4.0
Ceylon Cinnamon Oil	0.2

Artificial Cinnamon Sugar

Artificial Cinnamon Oil	4
Sugar (Powdered)	96

Imitation Clove Oil
(For pickling)

Eugenol	6
Dipentene	3
Eugenol Acetate	1

Vanilla Concentrate

Vanillin	1	lb.
Coumarin	3	oz.
Alcohol	3	pt.
Glycerin	3	pt.
Caramel	1	pt.
Water	To make 1	gal.

For use add to 1 pt. of the above concentrate 5 pt. 6 oz. of water and mix well.

Vanilla Sugar

Powdered Vanillin, U.S.P.	7	oz.
Powdered Coumarin, U.S.P.	1	oz.
Confectioners' Sugar XXXXXX	5	lb.
Corn Starch	1	oz.

Use in place of liquid vanilla flavoring. One teaspoonful flavors 1 qt. of liquid.

Flavor Emulsions

Flavor emulsions for food products have been successfully made using Glyceryl Monostearate (S-928). Both fluid and paste emulsions can be made. A general formula for a fluid emulsion is:

Lemon Oil U.S.P.	10
Glyceryl Monostearate (S-928)	12
Water	75
Preservative	3

The S-928 is dispersed in the water at 170°F. with continuous stirring. When completely dispersed the temperature is lowered to 120°F. The lemon oil is then added together with the preservative and when completely dispersed the temperature is lowered rapidly. These operations may be handled in a jacketed kettle fitted for steam and cold water.

Pastes may be made in the same way using less water.

Almond Oil Emulsion

Gum Tragacanth	1/3	lb.
Propylene Glycol	2	lb.
Bitter Almond Oil	1 3/4	lb.
Water	12	pt.

Add the oil to the propylene glycol and mix. Then add the gum and the water, and mix to a uniform emulsion.

Lemon or Orange Oil Emulsion
Gum Tragacanth ½ lb.
Propylene Glycol 3 lb.
Lemon or Orange Oil 2½ lb.
Water 17 pt.

Add the oil to the propylene glycol and mix. Then add the gum and the water, and mix to a uniform emulsion.

Mace or Cinnamon Oil Emulsion
Gum Tragacanth ¾ lb.
Propylene Glycol 3 lb.
Oil of Mace or
  Cinnamon 2¼ lb.
Water 17 pt.

Add the oil to the propylene glycol. Then add the gum and the water and mix to a uniform emulsion.

Spicy Flavors

Spice Meat Extract
Mace Extract ¾
Clove Extract ⅛
Thyme Extract ¾
Basil Extract ¾
Shallott Extract 7
Celery Extract 4

Spice Fish Extract
Onion Extract 4
Pepper Extract 1
Pimenta Extract 1
Ginger Extract ⅛
Clove Extract ⅛
Bay Laurel Extract ¼

Spice Sausage Extract
Savory Extract 1
Pimenta Extract 2
Pepper Extract 5
Onion Extract 6
Bay Laurel Extract ⅛

Spice Smoked Meat Extract
Savory Extract 1
Cardamom Extract ½
Basil Extract 2
Pepper Extract 4
Garlic Extract 8

Spice Pickle Extract
Celery Extract 1½
Clove Extract ¾
Dill Extract 1¼
Pimenta Extract 2½
Capsicum Extract 3
Pepper Extract 4

Spice Curry Extract
Tumeric Extract 2
Ginger Extract 2
Pepper Extract 4

Spice Tarragon Extract
Tarragon Extract 6
Celery Extract 1½

Spice Mixed Pickle Extract
Tarragon Extract ½
Dill Extract 1
Pimenta Extract 2
Capsicum Extract 2
Caper Extract 3
Shallott Extract 3

Spice Mustard Pickle Extract
Dill Extract 1
Pepper Extract ¾
Celery Extract ½
Capsicum Extract ½
Onion Extract ½
Cardamom Extract ½
Volatile Mustard Oil 1/30

Spicy culinary extracts are generally manufactured by dissolving ½ to 1½ oz. of the essential (spice) oil in about 12 oz. of pure alcohol and then adding 2 to 3 oz. of water. Extracts such as garlic,

FOOD PRODUCTS

onion, celery, etc., are best manufactured by allowing the latter to soak in dilute alcohol, and then drawing off the clear liquid for use as a flavor.

Spice salts are made by simply rubbing the salt and the spice extracts together and drying them in the air.

Spice Vinegar Flavors

Tarragon Vinegar

Tarragon Oil	1
Cognac Flavor	1/8
Alcohol	16

Fruit Vinegar

Cognac Flavor	1/8
Strawberry Flavor Essence	1
Raspberry Flavor Essence	3
Ethyl Acetate	1/2
Alcohol	16

Mustard Vinegar

Mustard Oil Volatile	1/8
Pimenta Oil	1/4
Peppermint Oil	1/30
Tarragon Oil	1
Alcohol	16

Wine Vinegar

Cognac Flavor	1/4
Celery Oil	1/2
Ethyl Acetate	3/4
Alcohol	16

Malt Vinegar

Cognac Flavor	1/8
Ethyl Acetate	1/4
Cardamom Oil	1/30
Alcohol	16

Universal Vinegar Essence

Tarragon Oil	1
Celery Oil	1/2
Thyme Oil	1/2
Pimenta Oil	1/8
Clove Oil	1/30
Alcohol	16

Celery Vinegar

Tarragon Oil	1/2
Celery Seed Oil	1 1/2
Alcohol	16

Artificial Lemon Syrup

Sugar	2 lb.
Citric Acid	2 oz.
Concentrated Essence of Lemon	2 dr.
Essence of Almonds	20 drops
Hot Water	2 pt.

The citric acid is dissolved in the hot water, the sugar is added, and finally the essence of lemon and almonds. The mixture is stirred thoroughly, covered, and allowed to cool. Two tablespoons of the syrup are recommended in a glass of cold water to make a refreshing lemonade.

Frozen Desserts

Improved Ice Cream Manufacture

In the manufacture of ice cream, the butter fat content and the amount of total solids are set by law. By virtue of this, most good ice creams are more or less also standardized in respect to other requirements. Hence overrun, which is the percent volume increase in the mix due to incorporation of air and which can also be called the whipping ability of the cream, ranges between 85% and 90%.

Stabilizers are used in ice cream to:

1. Produce a set or gel structure to retain the air in the ice cream.
2. Protect the ice cream from losing volume in handling, shock, etc.
3. Act as protective colloids and help to disperse the butter fat.
4. Prevent the formation of ice crystals.
5. Help to give body to the ice cream.

Stabilizers, most commonly used, are gelatin and sodium alginate. The latter is sold under the trade name of Dariloid. Gelatin, if used, should be a high bloom grade, for larger amounts may be necessary if low test gelatins are utilized. Other stabilizers are methyl cellulose about which very little is known to date, and vegetable gums.

Whipping agents: Egg yolk is the most commonly used agent to give improved whipping properties to the cream, but it is on the wane in progressive plants. It is very expensive to use and most plants are getting away from its use altogether.

Glyceryl Monostearate (S-928) can supplant egg yolk where the latter is used. One illustration of such use may be gained from the following:

Ice cream made with 0.35% gelatin and 0.5% egg yolk having an overrun of 135% (a specialty ice cream of the cheaper grade) was made with 0.35% gelatin and 0.2% S-928. When finished the latter cream had 135% to 140% overrun finer texture, and was equal and even superior to the former in all respects. The economy of this use is self-evident. Hence S-928 can be thought of as being a whipping agent, or an agent to produce overrun.

Another example is the replacement of a certain percent of gelatin to obtain set and overrun. Here 0.62% gelatin was formerly used. (An abnormally high amount). Gelatin was reduced to 0.35% and 0.15% S-928 was added. The resulting cream was finer in all respects.

It is thus seen that Glyceryl Monostearate (S-928) contributes to the set of ice cream. By virtue of its dispersing action it gives a finer texture, dispersing the butter fat more finely and more uniformly, and reduces the tendency for ice crystals to form, which give a grainy texture to the ice cream.

Glyceryl Monostearate (S-928) stabilizes the emulsion and maintains the volume of the ice cream, reduces losses in dipping when packaged from bulk and prevents too fast melting. It also gives a dry ice cream, that is, when frozen it is not wet and runny but stiff and dry.

Note: Don't use more than 0.45% gelatin, and not more than 0.35% Dariloid, with Glyceryl Monostearate (S-928). Dariloid gels too fast and will sometimes cause a set in the holding tank or on the cooling coils.

Don't use egg yolk when using S-928, otherwise too much overrun will result.

Improved Ice Cream Formula
Gelatin (225 lb. test)   20   lb.
Glyceryl Mono-
    stearate (S-928)   7¼ lb.

Sweetened Condensed
  Milk   160   gal.
  Cream 40%   140   gal.
  Milk   230   gal.

Mix at 160°F. for 45 minutes. Homogenize at 3200 lb. per square inch and cool at 38°F. Hold for 12 hours just below 40°F. and freeze.

If super heated condensed milk is used and an increased set occurs, reduce the gelatin content slightly, and operate at slightly higher temperature.

Powdered Ice-Cream Mix
(Army Quartermaster Corps Specifications)

Composition of a powder meeting the specifications:

  Fat   27.79
  Milk-Solids-Not-Fat   28.31
  Sugar   40.65
  Stabilizer   1.0
  Moisture   2.25

A. Eight combinations of cream and skim or whole milk to make 100 parts of powder:

(1) Cream (50% fat)   55.58
     Skim milk   286.45
(2) Cream (40% fat)   69.45
     Skim milk   272.58
(3) Cream (30% fat)   92.65
     Skim milk   249.38
(4) Cream (20% fat) 138.80
     Skim milk   203.23
(5) Cream (50% fat)   32.69
     Whole milk (3.7%
      fat)   309.34
(6) Cream (40% fat)   41.69
     Whole milk (3.7%
      fat)   300.34
(7) Cream (30% fat)   57.49
     Whole milk (3.7%
      fat)   284.54
(8) Cream (20% fat)   92.70
     Whole milk (3.7%
      fat)   249.33

To reconstitute the ice-cream powder for freezing, add 1.717 lb. of water to each lb. of powder. (This is at the same rate as the Army specifies, i.e., 7 pt. or 7.3 lb. of water to 4.25 lb. of powder.)

This makes a mixture of the following composition:

  Fat   10.22
  Milk-Solids-Not-Fat   10.42
  Sugar   14.96
  Stabilizer   0.37
  Water   64.03

B. Amounts of sugar to add to the powder before and after drying (depending on the proportion of total sugar wanted in the liquid mix before drying)

Proportion of sugar wanted in the liquid mix	Sugar to add before drying lb.	Sugar to add after drying (per pound of powder) lb.
All	40.65	—
One-half	20.33	0.2551
Two-fifths	16.26	0.3222
Three-tenths	12.20	0.3975
One-quarter	10.16	0.4389
One-fifth	8.13	0.4825
One-tenth	4.07	0.5762

C. To substitute corn sugar or corn-sirup solids for ¼ of the sucrose, omit 10.16 lb. of sucrose, and add:

10.16 lb. corn sugar (anhydrous)
    or
11.06 lb. corn sugar (hydrated)
    or
10.53 lb. corn sirup solids
    or
12.67 lb. corn sirup
    or
12.40 lb. high equivalent sirup

D. To obtain approximately 1.0% whole egg in the finished ice cream add:
10.20 lb. liquid whole egg
or
2.76 lb. powdered whole egg

To obtain approximately 0.5% egg yolk in the finished ice cream add:
3.11 lb. liquid egg yolk
or
1.38 lb. powdered yolk

NOTE: The addition of eggs will increase the weight of the powder to over 100 lb. by an amount equal to the weight of egg solids, and will reduce the percentage of other ingredients slightly.

Ice Cream Mix
Formula No. 1
U.S. Patent 2,395,587

Cream	826.6
Skim Milk	1562.6
Skim Milk Powder	151.8
Sugar	420.0
Glyceryl Monostearate	3.0
Sugared Egg Yolk (40%)	27.0

No. 2

Cream (30%)	424.0
Fluid Milk	367.0
Powdered Skim Milk	42.0
Sugar	120.0
Anhydrous Dextrose	40.0
Gelatin	3.5
Egg Yolk	3.5
Flavor	To suit

Heat to 130°F. and homogenize.

Ice Cream Stabilizer
Formula No. 1

Locust Bean Gum	16
Glyceryl Monostearate (S-928)	10

No. 2

Gelatin (225 lb. test)	2
Glyceryl Monostearate (S-928)	1

Ice Milk 5%

Cream (20% fat)	25
Condensed Skim Milk	25
Whole Milk	25
Granulated Sugar	15
Gelatin	½
Water	9½

Dissolve the gelatin in the water. Combine all liquid ingredients in a pasteurizing tank and heat to 150°F. for 30 minutes. Homogenize at 2500 lb. pressure per square inch. Cool to about 40°F. Age for 12 hours. Freeze in an ice cream freezer. Place in containers and handle the same as ice cream.

Rhubarb Ice

Granulated Cane Sugar	12 lb.
Corn Syrup Solids or Dextrose	6 lb.
Water	14 qt.
Rhubarb Sauce	2 qt.
Locust Bean Gum	2 oz.
Pectin	1 oz.
Strawberry Color	½ oz.
Lemon juice *	1 qt.

Combine the cane sugar, corn syrup solids or dextrose, locust bean bum and pectin, mixing thoroughly. Heat the water to 180°F. Sift the dry mixture into the hot water and stir vigorously until it is completely dissolved. Cool this mix to about 50°F. Add the rhubarb sauce, lemon juice or citric acid solution and color by stirring until well mixed.

Put the combined mixture in a freezer and start whipping and

---
* Or 2 oz. of 50% citric acid solution.

freezing until an overrun of about 40% is obtained. Remove from the freezer to the containers and place in a hardening room at about −15°F.

In case citric acid solution is used instead of lemon juice it is necessary to increase the water to 15 qt.

If the rhubarb sauce is of a pale color it is advisable to increase the amount of the color from 0.5 to 0.75 oz.

Low Cost Sugar-Fruit-Pectin Jellies

Manufacture of Liquid Pectin

Dried Apple Skins or Apple Pomace	200 lb.
Water	250 gal.

Add water to a 350-gal. wood tank with cover. Bring the water to a boil. Add the dried skins. Now boil vigorously, having live steam coming through several perforated pipes extending up and down the tank. Allow to boil for about 40 minutes. Transfer the cooked batch to a hydraulic press and collect the liquid. Approximately 180 to 190 gal. of liquid pectin juice results. It may read from 3 to 4° Brix. Concentrate this juice to 9 to 10° Brix. This liquid pectin product can be used in jellies, jams and preserves.

Acid Coagulant for All Jellies Except Imitation Jellies

Tartaric Acid	48 oz.
Citric Acid	16 oz.
Water Enough to make a total of	1 gal. acid solution

Use 2 to 4 oz. of this liquid acid for each 30-lb. pail of jelly. If the jelly is too firm or leathery use less acid. If the jelly is not firm enough use more acid, provided you do not make the jelly too tart.

Apple-Raspberry Jelly
Formula No. 1

Liquid Pectin	50 gal.
Cane Sugar	600 lb.
Red Raspberry Juice	4 gal.
Black Raspberry Juice	5 gal.

Add the 50 gal. of pectin juice to the 600 lb. of sugar. Heat with stirring to 219°F., or, if a heavier product is desired, heat to a higher temperature. Close the steam valve and now add the 9 gal. of fruit juices. Mix the contents. Allow the jelly to cool down to around 150°F. Pour the required amount of liquid acid into 30-lb. pails, and then let the jelly run in quickly.

No. 2

Liquid Pectin	45 gal.
Cane Sugar	375 lb.
True Fruit Raspberry Flavor	6 oz.
Red Raspberry Juice	4 gal.
Black Raspberry Juice	4 gal.

This jelly is to be made in the same way as described for Formula No. 1.

Raspberry-Pectin Jelly

Water	30 gal.
No. 100 Powdered Pectin	5 lb.
Cane Sugar	260 lb.
Corn Syrup	140 lb.
Amaranth Color (4 oz. Color Dissolved in 1 Gal. Water)	12 oz.
True Fruit Raspberry Flavor	20 oz.

Mix the pectin with the sugar

and add gradually this mixture with stirring into the boiling water until the solution is complete. Add the balance of the sugar and heat to 214°F. Now add the corn syrup and the color, and heat to 220°F. Close the steam valve and then add the flavor. Cool the jelly and then fill it into 30-lb. pails, with the required amount of liquid acid already added.

Imitation Jelly

Concentrated Liquid Pectin	55 gal.
Corn Syrup (42° Bé)	85 gal.
Red Food Color	3 pt.
Sodium Benzoate (Dissolved in 1 qt. of Warm Water)	2 lb.
Acid	Sufficient to make the jelly set
Flavor	As desired

To the warm solution of pectin add 85 gal. of corn syrup. Mix with an agitator. Now add the color, the flavor and the sodium benzoate, and, when the solution is thoroughly mixed, run into pails or barrels. The temperature of the batch should be around 150°F. If pail goods are desired (30 lb. of jelly to a pail) add 4 oz. of phosphoric acid solution to the pail and then fill it with the jelly. In about 1 minute the jelly will set. If barrels are desired, fill the barrel with the jelly first, and then add the required amount of phosphoric acid, with vigorous up-and-down motion with a paddle. In about 1 to 2 minutes the jelly should be firm. If a barrel holds 51 gal. add 68 oz. of the prepared acid.

The jelly pails should be closed the day following the filling. If the jelly is low in solids it may ferment or develop surface molds.

Pure Orange Jelly

Water	85 lb.
100-Grade Citrus Pectin	18 oz.
Concentrated Orange Juice *	14 lb. 8 oz.
Sugar	100 lb.
Citric Acid Solution	20 fl. oz.

Put the water into the kettle and turn on the steam. When it has reached a temperature of 160 to 180°F., add the pectin, thoroughly mixed with 8 times its weight of granulated sugar. Allow the mixture to come to a brisk boil, stir occasionally. Boil vigorously for about ½ minute.

Add the fruit juice and the balance of the sugar. Heat quickly to 219°–220°F. Turn off the steam and immediately add the acid solution. Mix thoroughly.

Draw off the hot jelly and fill it into containers immediately. Filling and capping should be finished while the jelly is still above 190°F., or the capped jars must be sterilized.

The yield should be approximately 170 lb. of finished jelly with 65% soluble solids.

Cranberry Jelly

Cranberry Juice	41 lb.
100-Grade Citrus Pectin	8 oz.
Sugar	50 lb.
Fruit Acid Solution	2 fl. oz.

To prepare the juice, boil 45 lb. of cranberries with just enough water to prevent burning, until

---

* 72% soluble solids.

thoroughly pulped. Cool to 100 to 110°F., and add ½ oz. of pectinase, previously made into a smooth paste with cold water (Pectinol M, Soluble, made by Rohm and Haas, is recommended). The temperature must be that specified. Stir the pectinase paste thoroughly into the batch, and allow to stand overnight. Allow the batch to cool during this period. On the next day reheat the batch to boiling. Strain the pulp through a sieve or cloth to remove the seeds, skins and pulp. By straining through a fairly close-weave muslin a sparkling clear juice can be obtained. The yield is high and should amount to about 41 lb. of cranberry juice.

In making the jelly, heat the juice to 180°F. in a kettle. Mix the pectin thoroughly with 4 lb. of granulated sugar. Add this mixture to the juice. Continue to stir while heating to boiling. Boil vigorously for about ½ minute.

Add the remainder of the sugar. Heat rapidly to 219 to 220°F. Turn off the steam. Immediately add the fruit acid solution and mix. The fruit acid solution is made by dissolving 1 lb. of citric acid or ½ lb. of tartaric acid (crystals or powder) in 1 pt. of hot water.

Draw off the hot jelly promptly and fill into containers as quickly as possible. Filling and capping should be completed while the product is still above 190°F.

The yield should be approximately 81 lb. of the finished jelly. The product is slow setting. It should have the same fine texture and consistency as other fruit juice jellies.

Artificial Honey

Clarified Sugar	10 lb.
Pure Honey	3 lb.
Water	3 pt.
Cream of Tartar	1 dr.
Essence of Peppermint	10 drops

The sugar is dissolved in the warm water, and mixed with the honey and cream of tartar dissolved in the balance of the water. The mixture is brought to the boiling point, well stirred; the scum is removed, and the product is allowed to cool.

Salad Dressing

A heavy-bodied stable emulsion of egg yolk, oil, spices, salt and sugar furnishes color, flavor, food value and richness of appearance to a salad dressing. A starch paste, usually consisting of a special starch or starches cooked in a mixture of vinegar and water, contributes a tart flavor, body and bulk and a fine, smooth texture. The following suggestions represent good commercial practice, to be varied to suit manufacturing and market conditions.

Starch Paste

Water	16 lb. 8 oz.
Vinegar (5% Acid)	10 lb.
Special Starch	3 lb. 8 oz.
Granulated Cane Sugar	3 lb.

The vinegar and 10 lb. of water are brought to about 160°F. in a steam jacketed kettle. Meanwhile, blend the starch to a smooth, free-flowing suspension with the remaining water.

Add the starch suspension only

as rapidly as the agitator will disperse it. Continue agitation, at about 30 rotations per minute, until the paste is clear and thick.

As soon as cooking is complete, add the sugar and mix thoroughly. If desired, 3 lb. of corn sirup or corn sugar may replace 2 lb. of cane sugar.

Cool quickly, at least to 80°F., with slow agitation. It is not wise to prepare starch pastes too far in advance. Best practice is to cool to 80°F. and use immediately.

Emulsion Base

Frozen Egg Yolk (10% Sugar)	1 lb. 8 oz.
Granulated Cane Sugar	3 lb.
Salt	1 lb. 4 oz.
Dry Mustard	4 oz.
Water	12 oz.
Vinegar (5% Acid)	4 oz.
Salad Oil *	10 lb.

Place the defrosted yolks in the bowl of the mixer with the sugar, salt and mustard. Blend. Add about one-third of the water and whip until light.

Begin adding the oil and water or vinegar alternately. First add, during 30 seconds, oil in a thin stream while whipping at high speed until a uniform mass is obtained. Next add a small amount of water, then more oil, working in each addition well. The quantity of oil added at one time may be gradually increased. Add the water or vinegar in small amounts so that there is some left to finish the process. The temperature for emulsifying should be 43 to 45°F.

* Preferably cottonseed or corn oil.

This is the emulsion base suggested for a salad dressing with 20% oil in the product.

Final Blend

Emulsion Base	17 lb.
Starch Paste	33 lb.

Blend the emulsion and the paste in a power mixer until a fine, glossy, uniform mass is obtained. Overbeating will cause loss of body or the incorporation of large air bubbles with consequent danger of rancidity through oxidation or slack-fill through settling.

Russian Salad Dressing

Mayonnaise	25 lb.
or	
Salad Dressing	25 lb.
Drained Chili Sauce	2 gal.
Tabasco Sauce	1/6 oz.
Minced Green Pepper	1 lb.

Mix the ingredients.

Sauerkraut-Tomato Juice Cocktail

Sauerkraut Juice	1 pt.
Tomato Juice	3 pt.
Onion Salt	1/3 oz.
Worcestershire Sauce	1 1/2 oz.
Salt	3/4 oz.
Pepper	3/4 oz.

Mix the ingredients. Fill into bottles and pasteurize.

Peanut Butter Spreads

The spreadability of peanut butter varies with the kind of peanuts used, as Spanish and Runner peanuts presumably make oilier and softer peanut butter than Virginia peanuts. The ease of spreading is also influenced by the fineness of grinding, as very coarsely or very finely ground peanut butter does not spread as well as a moderately fine ground butter. But in any case the normal stickiness makes spreading difficult.

The spreadability may be increased by the addition of glyceryl monostearate or lecithin. Ingredients added to peanut butter to produce firmness or to increase the spreadability tend to dilute the peanut flavor. This is not true in all cases; certain flavoring materials added in minute quantities, like salt, accentuate the peanut flavor.

If water in a considerable quantity is added to peanut butter to aid spreadability, the product will spoil unless it is sterilized, used immediately, or stored under refrigeration. Heating usually causes the oil to rise to the surface, destroys the spreadability, and impairs the appearance.

The formulas given below are for products to be used immediately or to be held in cold storage. It was found that all of them could be held at 34°F. for 2 weeks or more with very little change.

Formula No. 1
Orange Spread

Peanut Butter	12.4
Sucrose Syrup (65%)	11.1
Citric Acid Solution (0.7%)	1.2
Orange Oil	To suit

No. 2
Chocolate Spread

Peanut Butter	11.1
Chocolate Syrup	9.9
Water	3.7

No. 3
Raisin Spread

Peanut Butter	9.9
Sucrose Syrup (65%)	6.2
Raisins (Ground)	5.0
Citric Acid Solution (0.7%)	3.7
Cloves	

No. 4
Cherry Spread

Peanut Butter	12.4
Sucrose Syrup (65%)	5.0
Maraschino Cherries (Ground)	2.5
Cherry Juice	3.7
Citric Acid Solution (0.7%)	1.2

No. 5
Pickle Spread

Peanut Butter	12.4
Sweet Pickle (Ground)	5.0
Pickle Juice	3.7
Water	3.7

No. 6
Olive-Pimiento Spread

Peanut Butter	12.4
Olives (Ground)	3.7
Pimiento (Ground)	1.2
Olive Juice	3.7
Water	3.7

Flavored Peanut Butter

A number of formulas have been developed for making firm, flavored peanut butter. In using these the amount of mixing must be carefully controlled to attain desired firmness and prevent oil separation. Excessive mixing seems to mash the peanut particles, press out oil, and prevent firming, while insufficient mixing does not produce a homogenous product. After preparing and packaging, the product should be stored under moderate refrigeration (50 to 60°F.) for 1 day or 2 to allow the oil to equalize.

Formula No. 1
Orange Flavor

Roasted Peanuts	7 lb. 2.4 oz.
Salt	3.5 oz.

Dextrose	2 lb.
Powdered Sugar	8 oz.
Hydrogenated Oil	1.3 oz.
Glycerin	0.8 oz.
Orange Oil	0.4 oz.

Grind the peanuts, add the salt and the hydrogenated oil; hold the mixture at 150°F. until the oil is melted; add the glycerin, then the dextrose and the orange oil; mix thoroughly; pack into molds or containers while hot.

No. 2
Chocolate Flavor

Roasted Peanuts	7 lb.
Salt	3.2 oz.
Powdered Sugar	8 oz.
Dextrose	2 lb.
Cocoa	3.2 oz.
Glycerin	1.6 oz.
Vanilla	0.7 oz.

Grind the peanuts and the salt; add the glycerin, then the other ingredients and mix thoroughly; pack into molds or containers while hot.

No. 3
Raisin and Other Fruit Flavors

Roasted Peanuts	5 lb. 14.1 oz.
Salt	1.9 oz.
Dextrose	2 lb.
Raisins (Chopped)	2 lb.
Cloves	trace

Mix the salt and the dextrose with the peanuts thoroughly; grind to the desired fineness; mix the raisins into the peanut butter and pack into molds or containers while hot. Chopped dehydrated figs, dates, prunes, and candied or glazed fruits may be substituted for the raisins.

Ripe Olive Sandwich Spreads
Formula No. 1

Ripe Olives (Chopped)	1 gal.
Liverwurst	1 gal.
Ground Boiled Ham	2 gal.

No. 2

Ripe Olives (Chopped)	1 gal.
Flaked Shrimps	1 gal.
Mayonnaise	½ gal.
Minced Onions	2 oz.

No. 3

Hard-Boiled Eggs	12
Mayonnaise	1 cup
Chopped Olives	1 cup

No. 4

American Cheese (Grated)	1 gal.
Mayonnaise	2 oz.
Chopped Olives	1 gal.

Table Mustard

Mustard Flour	1 lb.
Vinegar	1 qt.
Jamaica Pepper	1 dr.

Boil the vinegar and pour it over the mustard. Stir until all lumps are gone. Add the pepper and let stand several days, well covered. Add salt to suit. Force through a colander, if necessary.

French Mustard

The same as table mustard with the addition of 8 oz. of sugar and a trace of powdered cloves.

Sauces
Sea Food Sauce

Catsup	4 lb.
Lemon Juice	6 oz.
Tabasco Sauce	½ tsp.
Celery Salt	3 oz.
Salt	1½ oz.
Pepper	½ oz.
Worcestershire Sauce	6 oz.
Dry Mustard	1 oz.
Minced Parsley	1½ oz.
Minced Onion	1½ oz.

Mix the ingredients thoroughly and keep in a cold place.

Mustard Sauce

Table Mustard	20 lb.
Onions	2 lb.
Garlic	2 lb.
Sugar	20 lb.
Salt	7 lb.
Sodium Alginate	6 oz.
Ground Mace	4 oz.
Vinegar (24 Grain)	25 gal.
Yellow (Certified Color)	As desired

Soak the gum in 2 gal. of water for 24 hours.

Soak the mustard and mace in 2 gal. of water for 24 hours. Peel and chop the onions and garlic. Rub mustard and gum through a fine sieve.

Place all ingredients in a steam-jacketed kettle and boil gently for 15 minutes. Add the yellow color. Pass the batch through a fine sieve. Fill the container at a temperature higher than 180°F.

Chili Sauce

Ripe Tomatoes (Finely Chopped)	5 lb.
Vinegar	2 pt.
Garlic	1 oz.
Red Pepper	1 dr.
Salt	2 oz.
Lemon Juice	5 oz.

Boil for 1 hour. Then force through a colander and bottle while warm. Cork tightly.

Barbecue Sauce

Tomato Catsup or (Strained) Chili Sauce	8 cups
Meat Stock	8 cups
Worcestershire Sauce	1 cup
Black Pepper	½ tsp.
Cayenne Pepper	⅛ tsp.
Distilled Vinegar	1 cup
Salt to taste	3 to 4 tsp.

Combine all the ingredients and heat to boiling. Fill in sterile jars and process 2–2¼ hours in a hot water bath, or 45 minutes at 10 lb. pressure. Pint jars will require only 2 hours in the water bath, while the quarts will take 2¼ hours.

Pepper-Onion Relish

Bermuda Onions (Finely Chopped)	1 qt.
Sweet Red Peppers (Finely Chopped)	2 cups
Green Peppers (Finely Chopped)	2 cups
Granulated Sugar	¾ cup
Cider Vinegar (45 grain)	1 qt.
Salt	4 tsp.

Mix all the ingredients together and bring the mixture slowly to the boiling point. Cook 15 or 20 minutes until slightly thickened, then pour into hot sterilized jars and seal tightly at once.

Vinegar in Form of Tablets
Russian Patent No. 46121

Sodium Acetate	70
Powdered Sugar	3.15
Dehydrated Lemon Extract	26.85

Dye may be added if desired.

The sodium acetate and sugar are mixed and dried at 80–100°F.; after that the powdered lemon extract is added. For use, dissolve the tablets in water.

Pickling Vinegar Essence

Pimento Oil	½ fl. oz.
Nutmeg Oil	30 min.
Clove Oil	90 min.

Tincture of Capsicum	½ fl. oz.	
Acetic Acid (BP)	20 fl. oz.	

This formula makes a concentrated liquid for making pickling vinegar. One teaspoonful of this essence is mixed with each quart of vinegar to spice it.

Stuffed Olives

California green olives (pitted) can be stuffed with pieces of onions, nuts, or pimientos and place-packed in small glass jars. The brine for these products can be made as follows:

High-Grade Salt	28 lb.
Edible Lactic Acid	1 pt.
Water	43 gal.

The brine should be heated to about 160°F. and poured hot into the jars containing the olives. The caps are screwed on and as the contents cool, a vacuum is formed. This prevents the growth of yeast films on the surface of the brine. This type of pack will keep without being cooked.

Manufacture of Olive Oil
(California Method)

Almost any variety of olives will do. The Mission and Manzanillo varieties are popular. Usually, the riper the olives are, the better will be the quantity and quality of oil. The olives should not be stored after they have been harvested, since long storage will lower the quality of the oil.

The olives are first warmed in hot water to a temperature of about 150°F. and then they are ground, crushed, and put into press bags. The press bags are very strong cocoa-fiber bags equipped with a flap which can be tucked in after the bag is filled with ground olives.

The bags of crushed olives are stacked in a hydraulic press and are pressed at a pressure of more than 25 tons per square foot of bag until juice and oil cease to run from the bags.

The oil is immediately separated from the juice by means of a supercentrifuge.

Fresh olive oil must be filtered very well and freed from all particles of olives and juice if it is to keep well. Preliminary filtering is done with canvas cloth, but final filtering is done with special paper or cellulose pads. Filter clay is used to facilitate filtering. Very dry oil has a lower clouding temperature, more stable color, and greater resistance to rancidity.

The equipment used to make olive oil should be acid resistant. Metal contamination is usually detrimental to the quality of the oil.

The oil is usually stored in acid-resistant tanks for a few months to allow emulsified oil to coalesce, and the oil is refiltered then and bottled.

Oil, thus carefully made, is called California Virgin Oil and does not need refining.

Beef-Soya Sausage

		% Weight (approx.)
Beef Meat (30% fat)	7½ lb.	37½
Soya Grit	1½ lb.	7½
Dry Rusk	3 lb.	15
Water (¾ gal.)	7½ lb.	37½

Seasoning (approx.) 6 oz. 2
Approximately 20 lb. chopping.

Stock Seasoning

Salt	1 lb. 0 oz.
White Pepper, Ground	4 oz.
Nutmeg, Ground	1 oz.
Ginger, Ground	1 oz.
Sage, Ground	1 oz.

An analysis of such a sausage would probably give: 57% water, 11.5% protein, 13% fat, 13% carbohydrate. There would also be small amounts of calcium and iron, with a useful quantity of vitamin B_1, and the calorific value would be in the region of 1,000 cal. per pound.

Pork-Soya Sausage

Lean Pork	7 lb.
Back Fat	3 lb.
Dry Rusk	3 lb.
Soya Grits	1 lb.
Water	6 lb.
Seasoning	6 oz.

Stock Seasoning

Salt	1 lb. 4 oz.
White Pepper	4 oz.
Mace	4 oz.
Ginger	2 oz.
Sage	1 oz.

The method of manufacture is as follows:

The soya grit and dry rusk are mixed and added to the water. The lean meat is placed in a bowl chopper, the machine started, the seasoning added, followed by the soaked binder and finally the cubed backfat. Chopping is continued to give a fairly fine texture and the meat is filled into hog casings.

Under normal storage conditions there would be a slight loss of weight owing to water evaporation.

Meat and Vegetable Dehydration
U. S. Patent 2,354,495

Solid food stuffs, which may be in the raw state, or else partly or wholly cooked, are placed in a perforated receptacle within a heated bath of a chemical compound maintained at a temperature at or above the vaporization of water. The compound comprising the bath is of edible nature, which will become impregnated in the food stuff as the water content of the food treated, is vaporized. Upon rehydration of the food stuff the compound is such that it will form a colloidal dispersion with the water adsorbed by the food stuff during rehydration. The dehydration is continued for such period of time until the foodstuff is entirely free from water, or until its residual water content has been reduced to any desired point.

Formula No. 1

Lean pork is cut into 1-in. cubes and precooked. The cubes are immersed in a bath of molten glycerol monostearate at a temperature from about 218°F. to 228°F., for ¾ to 1½ half hours, or for such a time as is sufficient to decrease the moisture content of the meat to that desired, i.e., 1% to 10%.

No. 2

A quantity of precooked pea beans is immersed in a molten bath of propylene glycerol monopalmitate at a temperature from about 220°F. to 230°F. for 25 to 45 minutes, or for a time sufficient to decrease the moisture content of the beans to the desired percentage; 40

minutes of treatment as set forth being sufficient to remove about 95% of the water content.

No. 3

Raw lean beef is diced into 1-in. cubes and then subjected to a bath of molten glycerol monooleate maintained at a temperature of about 218°F. to 228°F. for 2 hours, for practically complete dehydration.

Improving Dehydrated Sweet Potatoes

Cooking dehydrated sweet potatoes in about a 0.1% solution of either citric or tartaric acid instead of plain water noticeably improves the appearance of the cooked product.

Stuffing for Fowls

Rice (Cooked)	6 gal.
Ripe Olives (Pitted)	1½ gal.
Onions (Chopped)	⅓ gal.
Sage (Ground)	1 oz.
Thyme	⅛ oz.
Parsley (Chopped)	¼ gal.
Butter (Molten)	¼ gal.
Salt	1 oz.
Pepper	⅛ oz.

Detecting Cold Storage Eggs

By dipping eggs in lamp black, one can tell immediately whether they are freshly laid or cold storage eggs.

The test depends upon the fact that storage eggs are treated with an oil to preserve them. If it is a cold storage egg, the lamp black will cling readily to the outer shell, while the amount of lamp black adhering to a fresh egg is said to be negligible.

Preventing Mold Growth in Cold Storage Rooms

First give the inside surfaces of the cooler a thorough washing down to remove existing surface growth. Follow this with a fungicidal wash or spray. A common agent used to kill vegetative growth of molds is a sodium hypochlorite or calcium hypochlorite solution containing 200 p.p.m. of available chlorine. If products, coming into the storage room, are likely to be heavily contaminated with mold spores, it may be necessary to use solutions containing as much as 400 p.p.m., in order to make certain that all the spores adhering to the inner surfaces of the cooler are killed.

After existing growth has been destroyed, the cooler can be kept reasonably free from mold growth by periodically washing down with 10% solutions of chloride of lime, or an equivalent preparation. Chloride of lime has a corrosive effect upon the metal surfaces of the cooling coils.

Methods of Treating and Handling Fruits, Flavors, Nuts, and Colors

Fruits

Store frozen fruits a little above zero.

Do not thaw and refreeze fruits.

Thaw only the amount needed for the day.

Thaw at 32–40°F.

Pasteurize strawberries and raspberries 30 minutes at 145°F.

Dip oranges in 75–100 p.p.m. chlorine solution before juicing.

Flavors and Flavor Extracts

Pasteurize at 145°F. for 30 minutes those extracts that will not be injured.

Use only good grade of raw materials.

Use sterile utensils and avoid human contact.

Nuts

Buy from reliable concerns where it is known that sanitary measures are taken to protect the quality of the meats.

Use only clean, sterile choppers or grinders.

Handle the meats with clean sterile scoops or sterile rubber gloves.

The best method of treating nut meats is to dip them in a 25% sucrose (sugar) solution at 180°F. for 30 seconds, followed by gas oven drying for 2½ minutes at 250°C. The addition of 1% salt to the hot sugar solution was reported to have improved the flavor.

A 50–75% boiling sucrose solution dip plus the 1% salt, followed by drying in a hot air oven is also used.

Treated nut meats are best stored in glassine bags at room temperature with a relative humidity of 42–50%. When nut meats are stored in refrigerators they are likely to be less crisp.

Colors

Dissolve the dry powder in a 45% sugar solution and heat to 180°F. or higher. The bottle in which the liquid is to be kept should be thoroughly cleaned and treated with boiling water to kill mold spores. The liquid color, while still hot, should be added to the bottle. Cork the bottle tightly.

Mix the dry color with finely pulverized sugar in the proportion 1 part color to 4 parts sugar. This may be added to the mix at the freezer and satisfactory color results will be obtained.

Revivifying Dry Popcorn

Popcorn which pops poorly because of having become too dry may be restored to good popping condition by the following method:

Put 40 pounds of corn into a 10-gallon can. Add 1 to 3 lb. of water, according to the dryness of the corn, as indicated by the way it pops. If its popping yield is less than one-third the normal yield of the variety, add 3 lb. of water; if it is two-thirds normal, add only 1 lb. For intermediate degrees of popping add intermediate quantities of water. For different varieties of popcorn the normal popping yield varies from 15 to about 30 vol.

Put on the cover, using a rubber, and clamp it down tightly. Shake thoroughly. Let it stand 2 days or longer before popping.

The poor popping quality of some lots of corn is due to other causes than lack of moisture. In such cases the popping cannot be improved by adding water.

The above method of restoring poppability is applicable only to small quantities of corn. When it becomes necessary to increase the moisture content of a large quantity of popcorn, it should be done by storing the corn in a cool, damp place, as, for example, in a shed outside during winter, or in a basement room during summer.

Rhubarb Wine

Run 32 lb. rhubarb through a meat chopper and strain the juice into a vat. Add 6 gal. of water and allow to stand 2 days. The solution is then strained and let stand for 1 to 2 days, after which the clear liquid is siphoned off into a suitable keg, and 24 lb. of sugar are added. Boil up 2 lb. of raisins in a little water and add them together with 1 lb. of sugar coloring. A small amount of gelatin is also added as a clearing agent. Let ferment for about 14 days, or until fermentation is complete. Fill up the keg with water and let stand for 5 months before tapping.

Chapter VIII

INK AND ALLIED PRODUCTS

Waterproof Ruling Ink
1) Dye Solution (2%)   500 cc.
   Distilled Water   1500 cc.
2) Carnauba Wax
   Emulsion   1000 cc.
   Distilled Water   1000 cc.

Mix the two solutions together to make a total of 4000 cc. If the ink is too strong, dilute with No. 2 solution. The color of the formula may be varied by the use of different dyes such as:

Red—Acid Red (Color Index No. 31)

Blue—Xylene Cyanole (Color Index No. 715)

Green—Acid Green L (Color Index No. 666)

In the use of these waterproof ruling inks, at times the pens have a tendency to clog up due to coagulation of the ink. Possibly such coagulation occurs in the felts or feeding yarns due to the presence of the wax in the waterproofing agent rather than to any precipitating action of the dye or pigment.

The difficulty is easily overcome by spraying the pen points and sometimes the feeding wicks and felts with alcohol or with weak ammonia water. These agents serve to cut the wax and also increase the flowing properties of the dye so that the work goes on again normally. This spraying may be done with an ordinary DeVilbiss atomizer such as is used for throat spraying.

Safety (Check) Paper Ink
U. S. Patent 2,380,195

Benzidine Sulfate	40
p-Diphenyl Disulfonic Acid	30
Ethyl Cellulose	4
Dimethyl Phthalate	100

Ink from Old Used Mimeograph Paper

Take two sheets of the old used mimeograph paper and cut it into strips, and place it in a wide mouth bottle. Add 10 cc. of denatured ethyl alcohol, 2 cc. of liquid soap, 100 cc. of warm water, and 2 drops of cresol. Shake well to let the liquid dissolve off the dye of the paper, then filter through glass wool. This forms a fairly thick ink, suitable for using with a steel pen.

Spirit Inks

The following formula makes a satisfactory, rapid drying ink which can be applied to paper or Cellophane from rubber rolls:

Orange Shellac	1 lb.
Dye	1 lb.
Denatured Alcohol	1 gal.

The dye should be dissolved hot. The following colors* are satisfactory:

Luxol Fast Yellow G
Luxol Fast Yellow T
Rotalin Yellow G

*Luxol and Rotalin are registered trade-marks of E. I. du Pont de Nemours & Co., Inc.

Luxol Fast Orange GS
Luxol Fast Orange R
Rotalin Orange R
Luxol Fast Brown G
Luxol Fast Brown K
Luxol Fast Brown R
Rotalin Chocolate
Luxol Fast Red B
Luxol Fast Red BB
Luxol Fast Scarlet C
Rotalin Brilliant Red B
Rotalin Red B Extra Conc.
Rotalin Red S Conc.
Rotalin Red Y
Rotalin Violet Conc.
Rotalin Violet NB Conc.
Luxol Fast Blue AR
Luxol Fast Blue G
Luxol Fast Blue MBS (Pat.)
Rotalin Blue B Conc.
Rotalin Brilliant Blue 2B Conc.
Luxol Brilliant Green BL
Luxol Fast Green B
Rotalin Green B Conc.
Rotalin Green Y Conc.
Luxol Fast Black L
Rotalin Black RM

Hectograph Ink
Formula No. 1

Methyl Violet	1
Water	8
Glycerin	1
Alcohol	¼

The methyl violet is dissolved in the water and the glycerin added. After gently warming for an hour, the mixture is allowed to cool, and the alcohol is added. For black ink, 2 oz. of negrosine are used in place of the methyl violet.

No. 2
U. S. Patent 2,382,796

Crystal Violet Dye	40 –60
Nitrocellulose (½ sec.)	0.1– 3.0
Cellosolve	10 –15
Ethyl Acetate	12 –18
Denatured Alcohol	19.5–25.0

Mix until uniform.

Hectograph Composition
(For colored copying)
U. S. Patent 2,195,926

Gelatin	30
Water	45
Sodium Lactate	100
Glycerin	250

Typewriter Ribbon and Rubber-Stamp Ink

Aniline	½
Alcohol	4
Glycerin	7
Water	4

Typewriter Ribbon Ink for Plastic Printing
U. S. Patent 2,382,861

Carbitol Acetate*	58.8
Carbon Black	3.0
Methyl Violet Base	38.2

Indestructible Ink

Lampblack	1
Potash Water Glass	12
Aqua Ammonia	1
Distilled Water	38

Glass Etching Ink

Glycerin	240 cc.
Methanol	215 cc.
Lead Borate	105 g.
Silver Oxide	350 g.

Grind in a ball mill to smooth consistency. Evaporate the methanol. The ink must be fired on glass.

Ceramic Stenciling Ink
U. S. Patent 2,318,124

Copaiba Resin	32

* Acidified to a pH less than 3.03.

INK AND ALLIED PRODUCTS

Venice Turpentine	16
Molasses	4
Dammar	4
Dibutyl Phthalate	1/32

To this may be added some pulverized, vitrifiable enamel frit and powdered color.

Metal Marking (Etching) Ink
U. S. Patent 2,377,593

Molybdic Acid (Powdered)	437 g.
Mercuric Chloride	23 g.
Copper Sulfate	100 g.
Antimony Trichloride	585 g.
Hydrochloric Acid (Conc.)	2450 cc.
Nitric Acid (Conc.)	530 cc
Sulfuric Acid (Conc.)	80 cc.
Water	360 cc.

Bakelite Stamping Ink

Clear Printing Ink	60 cc.
Fortifying Acid*	10 cc.
Filler†	15 cc.

Marking Ink

Logwood Extract	8	oz.
Bichromate of Potash	1	oz.
Hydrochloric Acid	3/4	oz.
Dextrin	4	oz.
Water	1/2	gal.

The logwood is boiled in the water, the acid and potash added, and finally the dextrin.

Thermoplastic Solid Printing Ink
U. S. Patent 2,322,445

Carbon Black	10

---

*Fortifying Acid

Hydrochloric Acid	2000 cc.
Antimony Trichloride	300 g.
Bismuth Chloride	300 g.
Platin Nig.	400 g.

† Filler

Glyceryl Phthalate	5 g.
Methanol	10 cc.

Gilsonite	15
Candelilla Wax	45

Melt together and stir till uniform. Apply by fusing.

Thermofluid Printing Ink
U. S. Patent 2,351,585

Hard Cumar	25–40%
Gilsonite	15–30%
Hydrogenated Soybean Oil	30–45%
Carbon Black	To suit

Steam-Setting Printing Ink
U. S. Patent 2,390,102

Lampblack	2.6
Carbon Black	10.1
Petrolatum	3.5
Dipropylene Glycol	28.9
Vinsol Resin	25.9
Diethylene Glycol	21.9
Iron Blue	3.0
Methyl Violet	1.3
Talc	2.8

Dampener for Lithographic Printing Plates
U. S. Patent 2,395,654

Water	1 gal.
Aerosol (Wetting Agent)	5 cc.
Sodium Dichromate	40 g.
Gum Arabic	2 oz.

Mix until uniform.
This is applied to ink-repellent parts of lithographic printing plates.

Ink Eradicator

1: Hydrochloric Acid	1/8 oz.
Water	1 pt.
2: Chlorinated Soda Solution	2/3 pt.
Water	1 pt.

Solution 1 is applied to the ink, followed by 2, and the stain removed with clean blotting paper.

Secret Writing Detector

Many of the so-called "sympathetic" inks used for secret writing can be detected and brought out by the use of the following solution:

Potassium Iodide	4.0 g.
Iodine	0.1 g.
Sodium Chloride	5.5 g.
Aluminum Chloride	2.3 g.
Glycerin	3.5 g.
Ethyl Alcohol (95%)	5 cc.
Water	25 cc.

Apply the above solution to the suspected paper with a wad of cotton and the secret ink will usually be brought out. Only in very exceptional cases does the above formula fail to show up "sympathetic" inks.

Cleaner for Ruling Pens Using India Ink

Ethyl Alcohol	50 cc.
Conc. Ammonia	10 cc.
Water	40 cc.

Soak the pen in the solution and wipe it off, repeat if necessary.

Chapter IX

INSECTICIDES, FUNGICIDES AND WEED KILLERS

Household Insecticides

German Cockroach Poison

Sodium Fluoride	1
Borax	1

Control of Bedbugs

Cresol U.S.P.	1
Methyl Salicylate	2
Kerosene	97

Mix together and use as a spray.

Control of Ants

Warm Water	0.5	pt.
Tartaric Acid	0.85	g.

Dissolve and add

Sugar	1.0	lb.
Honey	3.0	fl. oz.
Glycerin	45	cc.

Stir in well, and heat to bring the mixture slowly to boiling. Withdraw from fire and add a solution of

Thallium Sulfate	1.7	g.
Warm Water	0.5	pt.

Stir very thoroughly, and place in shallow cans at the infested areas.

Control of Silverfish

Rolled Oats	86
Powdered Sugar	5
Salt	2
Sodium Fluoride	7

Grind to a very fine consistency. Spread near book shelves, wood work, radiators, etc., and replace when covered with dust.

Body Louse Insecticide
Formula No. 1
(British Army)

Cresylic Acid	2.0
Powdered Derris Root	14.3
Powdered Naphthalene	50.0
China Clay	33.7

Mix together until uniform.

No. 2

Diphenylamine	25
Talc	75

No. 3

Mineral Oil	5.0
Aerosol OT	0.6
Isobornyl Thiocyanate	5.0
Diglycol Laurate S	5.0
Water	84.4
Perfume	To suit

Sticky Fly Paper

Castor Oil	5
Powdered Rosin	8

The two ingredients are mixed and heated until the rosin is dissolved. Do not boil. This material, once prepared, can be stored indefinitely in cans. Just before using heat the mixture so that it can be applied while hot. Spread on paper in as thin a coat as possible. A coated paper is preferred with a hard finish.

Mothproofing
Formula No. 1

Magnesium Silicofluoride	0.8
Ethanolamine Silicofluoride	0.6
Talc	98.6

No. 2
U. S. Patent 2,350,814

Hexachlorethane	95–98%
p-Dichlorbenzol	2– 5%

No. 3

Sodium Aluminum Silicofluoride	1	g.
Pulverized Gum Arabic	½	g.
Warm Water	150	cc.

DDT Preparations

Non-Penetrating DDT Spray
(For citrus trees)

DDT	4
Aluminum Stearate	2
Kerosene	94

Warm together and mix until uniform.

DDT Housefly Spray
Formula No. 1

DDT	100	g.
Benzol	140	cc.
Dibutyl Phthalate	140	cc.
Triton NE	12	g.
Water	1	l.

No. 2

DDT	14	lb.
Deodorized Kerosene	44	gal.
Perfume	3	oz.

No. 3
Make up a stock solution as follows:

50 parts of DDT is dissolved in
110 parts of xylene by stirring occasionally for 3 hours.
Then add
40 parts of Glycox 1400 or Emulphor DDT
Stir until uniform.

For use in spraying or painting:
10 parts of the stock solution and
90 parts of water
are mixed very well to produce a good emulsion.

DDT Emulsion Screen Coating

Dilute 166 g. of 30% DDT solution with 274 g. of mineral spirits. Then, dissolve 30 g. of Vistac #2 (Synthetic resin) in this solution. When the solution is complete, add 25 g. of Advawet #33 (Wetting agent).

To form an emulsion, simply pour the above solution into an equal part of water. The emulsion may be sprayed or painted on screens, doors, jambs, etc. and, when the liquids have evaporated, a thin transparent resinous film is deposited. This film holds the DDT and permits it to be effective for as long as 4 months on exposure to the elements.

DDT Insecticide Emulsion Concentrate
Formula No. 1

Xylol	65
DDT	25
Emulsifier (Diglycol Laurate)	10

One part of the above is mixed vigorously with 11 parts of water before use.

No. 2

DDT	53.8
Water	40.4
Nonaethylene Glycol Monooleate	5.4
Methyl Cellulose (400 cps.)	0.4

The above is diluted with cold water before use.

No. 3

DDT	20–35
Xylol	70–55
Emulsifier (Nonaethylene Glycol Laurate)	10

DDT Solvent

No. 2 Diesel Oil	85
Lubricating Oil	10
Gasoline	5

10% of DDT can be dissolved in above by stirring in the sun.

Aerosol Insecticide
Formula No. 1

Freon—12	85
Pyrethrum (20% Solution)	2
Cyclohexanone	5
Lubricating Oil (S.A.E. 30)	5
DDT	3

This is packed in sealed cans or bombs.

No. 2

DDT	5
Cyclohexanone	5
Dichlorodifluoro Methane	90

No. 3

Pyrethrum Extract (20%)	2
DDT	3
APS-202 (Petroleum Solvent)	12
Freon 12	83

DDT Delousing Preparations

The stock solution for making the requisite delousing spray for humans consists of:

DDT	6.0
Benzyl Benzoate	68.0
Benzocaine	12.0
Wetting Agent	14.0

This is diluted with 5 parts of water by volume and used within 24 hours.

About two-thirds of an ounce is required per person. A bath should not be taken for at least twenty-four hours after application of the spray to the body. When applying to the scalp, the eyes must be protected. For head lice, not more than 0.5 oz. is usually needed for complete deinfestation.

Two Army uses for a 10% DDT powder may also be adopted for civilian use. For head lice the powder is dusted lightly into the hair and rubbed in with the fingertips. For crab lice apply the powder thoroughly to all regions of the body having a moderate growth of hair. Do not bathe for at least 24 hours. Repeat the application after 1 and 2 weeks. The DDT powder does not have ovicidal properties.

DDT Powder

DDT	10
Pyrophyllite	90

It is reported that milling the DDT with diluents prevents some difficulties. In addition to the above base, a variety of talcs, clays and soapstone have been used.

Quick Acting DDT Insecticide

DDT	5
Lethane A-70	5
Pyrophyllite	90

All of the above should be ground to finest particle size to get best results.

Wettable DDT Powder
Formula No. 1

DDT	75
Nonaethylene Glycol Monooleate	10
Water*	15

Melt DDT and dissolve "nona" in it. Add water at 70–80°C. Stir slowly until a thick paste is formed.

* Half of the water may be replaced by mineral oil or paraffin wax if desired.

No. 2
DDT	90
Santocel 45	8
Emulgor A	2

Airplane Interior DDT Insecticide
(Wettable DDT powder)
DDT (Purified Micronized)	90.5
Silica Aerogel	6.0
Polyvinyl Alcohol	2.0
Naphthalene Formaldehyde Sulfonate	1.0
Dibutylphenylphenol Disulfonate	0.5

Mosquito and General Insect Repellent
Formula No. 1
An effective general pest repellent is the mixture obtained from 1 part of dimethyl phthalate, one part of Rutger's 612 (3-ethyl-1,3-hexanediol), and one part of indalone. The mixture should be applied to the skin as a thin film, which is effective for 2 to 3 hours.

No. 2
U. S. Patent 2,356,801
Dimethyl Phthalate	33⅓–80
2-Ethyl-1,3-Hexanediol	33⅓–10
n-Butyl Mesityl Oxide	33⅓–10

No. 3
Dimethyl Phthalate	6
Indalone	2
2-Ethylhexanediol-1,3	2

No. 4
2-Phenyl Cyclohexanol	7
2-Cyclohexylcyclohexanol	3

No. 5
Dimethyl Phthalate	6
Benzyl Benzoate	2
Indalone	2

No. 6
Saponified Cresol	15
Pinene	8
Water	77

No. 7
U. S. Patent 2,352,746
Cyclohexyl 2-Ethyl Hexoate	25.0
Stearic Acid	20.0
Potassium Hydroxide	0.67
Water	54.33

Mosquito Repellents
Formula No. 1
Citronella Oil	2
Spirits of Camphor	1
Cedar Oil	1

No. 2
Castor Oil	4
Pennyroyal Oil	2
Citronella Oil	1
Camphor	1
Pine Tar	2

No. 3
Thyme Oil	1
Concentrated Extract of Pyrethrum in Mineral Oil	2
Castor Oil	4

No. 4
Turmeric	5
Mustard Oil	95

No. 5
Dimethyl Phthalate	80
Citronella Oil	10
Cedar Oil	10

Mix well and spray in the room to repel mosquitos and other insects.

No. 6
Trichloracetylchlorethylamide	7.50
Calcium Chloride	1.25
Magnesium Chloride	1.25
Alcohol	65.00
Water	25.00

No. 7
Cinnamyl Alcohol	10
Alcohol	90
Calcium Chloride	10

INSECTICIDES, FUNGICIDES AND WEED KILLERS

No. 8
Presinol or Mipax

Cinnamyl Alcohol	100
Alcohol	894
Calcium Chloride	60
Magnesium Chloride	40
Geraniol	½
Water	106

Mosquito Attractants

Tributyrin, lactic acid or methyl caproate in small percentages attract mosquitos.

Exhaled human air adsorbed on charcoal also is an attractant. Warm moist air is another attractant.

Insect Bite Lotion

Zinc Oxide	25.0
Talc	25.0
Bentonite or Kaolin	5.0
Camphor	5.0
Menthol	0.5
Water	30.0
Alcohol (95%)	30.0

Midge Repellent
Formula No. 1

Dimethyl Phthalate	100 cc.
Diglycol Stearate S	5 g.
Oleic Acid	27 cc.
Triethanolamine	9 cc.
Water	100 cc.
Perfume	To suit

Heat to about 60°C. and mix until uniformly emulsified.

No. 2

Lanette Wax SX	5.0
Triethanolamine	9.0
Oleic Acid	27.0
Dimethylphthalate	100.0
Water	100.0

The emulsion, smeared on the exposed parts, will ward off midges for at least two hours. It is not injurious to the skin; there may be some slight tingling when it is first applied and afterwards when washing the face. The emulsion should not be allowed to get into the eyes.

Stable Bordeaux Mixture

Dissolve 35 g. of copper acetate in 2 l. of water. This is solution I. The equivalent amount of copper sulfate may also be used.

Disperse 35 g. of a good grade of hydrate of lime in 2 l. of water. Add 1 g. of dispersing agent, such as Tamol NNO, and 1 g. of wetting agent, such as the alkyl naphthalene sodium sulfate type. Any wetting agent which is stable in alkaline solution may be used. This is solution II.

When ready for use pour solution I slowly into solution II with stirring. This produces a deep blue colloidal suspension which remains stable for many days.

Fatted Calcium Arsenate Dust
(Gralit)

Calcium Arsenate	29.3
Tallow	70.6
Alkali Green Dye	0.1

Rotenone Insecticides

Dusts on the basis of rotenone usually should contain at least 0.5% rotenone. To mix a 0.5-percent dust, use 12½ lb. of the rotenone root powder containing 4% of rotenone and 87½ lb. of talc, sulfur, or other diluent. If the root powder contains 5% of rotenone, use 10 lb. of it and 90 lb. of the diluent. For smaller quantities, weigh by ounces instead of pounds.

To prepare a spray for bean beetle control, use 3¼ lb. of rotenone root which contains 4% of rotenone, or 2½ lb. of rotenone root which contains 5% of rotenone, to 100 gal. of water. For smaller quantities use 1 oz. of 4-percent powder to 2 gal. or 1 oz. of 5-percent powder to 2½ gal. This mixture will contain approximately 0.015% of rotenone.

A spray mixture consisting of derris powder, pyrethrum extract, and sulfonated castor oil with water has been found effective against red spiders, thrips (except the gladiolus thrips) on certain flowering plants, the cyclamen mite on chrysanthemums, aphids, cucumber beetles, tarnished plant bugs, certain species of leaf rollers, and leaf tiers. The spray is made up according to the following formula:

	For small quantities	For large quantities
Rotenone - Containing Root Powder (Containing 4% of Rotenone)	1 tbsp.	1 lb.
Pyrethrum Extract (Alcoholic Extract, Containing 2% of Pyrethrins)	4 tsp.	2 qt.
Sulfonated Castor Oil	2 tsp.	1 qt.
Water	1 gal.	50 gal.

In preparing this spray, add the sulfonated castor oil to the water. Next add a small quantity of this oil-and-water mixture to the derris or cube powder to make a uniform paste. Then stir the paste slowly into the remainder of the oil-and-water mixture. Finally add the pyrethrum extract to this mixture in case it is intended for the control of thrips or the cyclamen mite. For either red spiders or whiteflies, the pyrethrum may be omitted. A proprietary spreader-sticker, such as sodium oleyl sulfate plus synthetic resinous base, may be substituted for the sulfonated castor oil in the above formula, since the oil may at times injure the petals of open flowers and also the foliage of some plants. This material is used at the rate of ¾ tsp. per gallon, or 1½ pt. per 100 gallons, of spray mixture.

Derris Diluents

Tests of mixtures of 3 parts by weight of various diluents to 1 part of derris containing 5% of rotenone, showed that the toxicity of derris dust to the housefly is influenced greatly by the kind of carrier or diluent used. The percentage killed with some of the diluents tested was as follows: precipitated sulfur 93.5, cupric sulfide 90.9, dusting sulfur 89.5, sodium chloride 85.4, talc 41.0, paraformaldehyde 33.3, calcium carbonate 22.8, kaolin 11.6, and manganese dioxide 11.5%.

Naphthalene Plant Bactericide

Naphthalene	25.00
Glue	>0.30
Ammonium Sulfate	<0.15
Bentonite	3
Water	71.55
Preservative (Moldex)	0.1

Velsicol 1068 Insecticide Emulsion Base

Velsicol 1068	50
Kerosene	30
Glycox 1300	20

The above is diluted with an equal amount of water and mixed thoroughly to give a stable emulsion.

This is effective against grasshoppers when sprayed at rate of 1 lb. of Velsicol per acre.

Nicotine Spray

Add 1 tbsp. of a 40% solution of nicotine sulfate (Black Leaf 40) to 1 gal. of water. To this solution add 1 tsp. of soap and 5 g. of a good wetting agent of the alkyl naphthalene sodium sulfate type. Stir until everything is in solution and then spray. This solution remains stable over a long period of time. The addition of the wetting agent increases the efficiency of the nicotine sulfate.

Lethane Insecticide Emulsion

Lethane 384 Special	10
White Mineral Oil	10
Diglycol Laurate	5
Water	75

This emulsion forms an insecticide which is very effective against lice.

Gammexane * Insecticides
Formula No. 1

Crude Gammexane	20
Gypsum (Powdered)	80

No. 2

Solutions of 5% or more Gammexane in mineral oil, xylene, carbon tetrachloride or Dekalin are used. These solutions may be diluted with kerosene. Emulsions may be made from these solutions with sulfonated castor oil or waste sulfite lye (Goulac) and water.

Gammexane does not give an immediate "knock-down" of the insects. For this purpose some pyrethrum extract should be added.

Fungicides
Formula No. 1

Copper Sulfate	8 lb.
Lime	6 lb.
Water	100 gal.

* γ-Benzene hexachloride.

No. 2

Copper Sulfate	4 lb.
Lime	4 lb.
Water	50 gal.

No. 3

Copper Sulfate	4 lb.
Lime	2 lb.
Water	100 gal.

Formula 2 gives the strongest solution, and formula 3 is the mildest.

Fungicide for Mildew Control
Formula No. 1

Disodium Ethylene Bisdithiocarbamate (Solution)	2 qt.
Zinc Sulfate	1 lb.
Hydrated Lime	½ lb.
Water	100 gal.

This fungicide is particularly recommended for potatoes, celery and tomatoes.

No. 2
Plant Fungus Control

Basic Copper Chloride	26.0
Powdered Sulfite Lye	6.6
Methyl Cellulose	3.0
Precipitated Chalk	64.4

Use 1 pound per 99 pounds of water.

Apple Fungicide

Ferric Dimethyldithiocarbamate	½ lb.
Flotation Sulfur	5 lb.
Water	100 gal.

Cedar Rust Control

When it is desired to use a fungicide to control cedar rust on apples, Fermate, an organic sulfur compound, may be used in the pink, petal fall and first cover sprays at the rate of 1 lb. to 100 gal. of water together with 3 lb. hydrated lime. An alternate recommendation which will probably give as good cedar rust control, better scab control and cost somewhat less is Fermate ½ lb., flotation sulfur paste 8 lb., and lime 3 lb.

Spray for Brown Rot on Citrus Fruits

Copper Sulfate	1 lb.
Zinc Sulfate	5 lb.
Slaked Lime	4 lb.
Water	100 gal.

Spreader-Sticker for Agricultural Sprays

Formula No. 1

Casein	¼ lb.
Wheat Flour	2 lb.
Spray Mixture *	100 gal.

No. 2

Use ¼ lb. high protein soybean flour per 100 gal. of spray.

No. 3

Casein	10
Calcium Carbonate	16
Calcium Hydroxide	69
Calcium Oxide	5

Insecticides for Animals and Plants

Winter Control of Cattle Lice

Dusts containing as low as ¼ and ½% nicotine in a sulfur base give excellent control. If nicotine is not available, a dust made up of phenothiazine 1 part, sodium fluosilicate 2 parts, and sulfur, flour or another carrier 5 parts, is suggested as next best choice. Concentrated nicotine dusts rather than liquid nicotine sulfate are recommended as sources of nicotine. Dusts should be applied by shaker-top can or by hand and rubbed in thoroughly, particularly on the under side of the animal.

A dust composed of 1 part cube or derris (5% rotenone) to 10 parts wettable sulfur gives a very good control of all four species of cattle lice. Finely ground sabadilla seed 1 part to 10 parts wettable sulfur controls all species of cattle lice. Ground yam bean seed 1 part to 10 parts wettable sulfur is also effective.

Cattle Grub Control

The most effective dust for control of cattle grubs consists of ground cube or derris 1 part, plus double-ground cream tripoli earth 2 parts by weight, containing at least 1.5% of rotenone. Mixtures with micronized volcanic ash or pyrophyllite (90% through a 325-mesh screen) are somewhat less effective.

A high degree of control of the grubs is obtained by means of a spray containing 7.5 lb. of cube or derris powder (5% rotenone) per 100 gal. of water, applied to the backs of the animals at a 400–410 lb. nozzle pressure. With a pressure of 400 lb. or more, sprays containing cube powder and either wettable sulfur or a wetting agent are less effective than those containing only cube powder. A satisfactory wash is composed of ground cube or derris 12 oz., and granular laundry soap 4 oz. per gallon of warm water. A suitable dip contains ground cube or derris 10 lb., and a wetting agent such as sodium lauryl sulfate 2 oz. per 100 gallons of water.

---

* Spray mixture means that the stated quantities of casein and flour are added to 100 gallons of Bordeaux mixture fungicide spray, for example, to provide sticking and spreading qualities.

DDT Cattle Spray

Velsicol AR	44
DDT	10
Pine Oil	6
Water	40
Glycox 1300 (Emulsifier)	6

Control for Cattle Ticks

DDT	5
Rosin	47
Hercolyn	33
Dibutyl Phthalate	15

Melt together and mix well. Apply warm to the ears of the cattle. The treatment should be renewed every 3–6 weeks.

Goat Louse Dip

DDT	1 lb.
Pine Oil	5 lb.

Mix until dissolved and then stir into 60 gal. water.

Dog Flea Powder

Thanite	4–5
Talc	96–95

Mix well until uniform. Use ½ oz. of above for small dogs; 1 oz. for medium size dogs; 1½ oz. for large dogs.

Rub in well and allow to remain on the dogs.

Chicken Louse Powder
Formula No. 1

Sodium Fluoride	9.0
Sulfur	10.0
Nicotine	0.2
Talc	80.8

No. 2

Nicotine	1
Napthalene	½
Sulfur	10
Talc	88½

Chicken Roost Spray
(To kill lice)

Nicotine	5
Ethylene Diamine	5
Water	90

Killing Insects on Bulbs and Corms

Immersion of plants, corms, or bulbs in heated water, maintained at a constant temperature ranging from 110° to as high as 120°F. for the period of treatment, is a method used in the elimination of a number of pests, including the gladiolus thrips, aphids, and mealybugs on gladiolus corms, the larvae of bulb flies and mites in narcissus and other bulbs, and the cyclamen mite in crowns and distorted growths of some ornamental plants.

The treatments for these pests vary, and publications dealing with each should be consulted for specific recommendations. Small quantities of bulbs or plants can be treated in a laundry tub or similar container, provided that an accurate thermometer is available for checking the temperature. In carrying out the treatment, fill the vessel three-fourths or more full of water, using sufficient hot water to bring the temperature up to the desired point. Submerge the plants or bulbs in screen boxes or loose net bags and add hot water to maintain the desired water temperature, as it is lowered by the cooling effect of material being treated or by radiation. After the bulbs or plants have been warmed to the desired temperature in the bath, less additional hot water will be required to maintain the temperature. If the treatment can be carried out in a warm room and the tanks kept covered, the temperature will be more easily maintained. During the entire treating process the water should be stirred with a paddle, frequently enough, to maintain a uniform temperature throughout the container. Free circulation of the water should not be blocked by the treatment of too many plants or bulbs at one time. The duration of treatment is calculated from the time the temperature is brought up to the desired point after the plant material has been placed in the water.

The treatments required to control some common pests are:

The cyclamen mite and broad mite, 15 minutes at 110°F., except 20 minutes for large clumps of delphinium or gerbera and for trays of loosely placed strawberry plants.

Bulb mites on tuberoses, narcissus, and other bulbs, 1 hour at 110°F.

Bulb flies in narcissus and amaryllis, 1½ hours at 111°F.

The grape mealybug on gladiolus corms, 30 minutes at 116°F.

The gladiolus thrips on gladiolus corms, 30 minutes at 112°F.

The boxwood leaf miner on boxwood, 5 minutes at 120°F. during late fall and early spring.

Concentrated Insecticide Paste
U.S. Patent 2,369,855

Rotenone Root Powder	20
Wood Flour	20
Pine Oil	20
Aerosol	5
Naphtha	35

The above is mixed with water before use.

Termite Control

Pentachlorphenol	5
Fuel Oil	95

Dissolve with stirring. Mix well into soil to act as a termite repellent.

Anti-Termite Impregnant for Wood

Pentachlorophenol	5
Soft Asphalt	5
Varsol Solvent	90

Control of Cabbage Maggots and Chinch Bugs

Mix about 10% of a good oil emulsifier, such as the condensation of ethylene oxide with fatty acids (Glycox 1300), an oleic acid triethanolamine soap or other oil emulsifiers which may be obtained on the market, with β,β'-dichlorethyl ether until in solution. Avoid unnecessary breathing of the fumes of dichlorethyl ether. Use 1–2 tbsp. of this mixture per gallon of water. Spray or pour on the surface of the ground which is infected. It may be washed into the ground with water. Dichlorethyl ether is heavier than water and sinks into the ground.

Potato Psyllid and Flea Beetle Control
Formula No. 1

Sulfur	3%
Cryolite	1%

No. 2

Sulfur	3%
Basic Copper Arsenate	1%

No. 3

Sulfur	2½%
DDT	5%

Corn Earworm Insecticides
Formula No. 1
Pyrethrum Extract
  (20% Pyrethrin)   1¼ fl. oz.
White Mineral Oil   1 gal.
No. 2
Dichloroethyl Ether
  or Styrene Di-
  bromide   2½ fl. oz.
White Mineral Oil   1 gal.
No. 3
Dichlorethyl Ether or
  Ethylene Dichloride   2
White Oil   100

Corn Borer Dust
Nicotine Sulfate   40
Talc   60

Pea Aphid Spray
Derris (4% Rotenone)   3 lb.
Sodium Lauryl Sulfate   4 oz.
Water   100 gal.

Mexican Bean Beetle Control

To keep the beetle in check, spraying or dusting must be done when the worms are very young, and spray or dust must be applied to the underside of the leaves where the worms feed.

Spraying gives better control than dusting. The recommended spray, made into a thin paste and poured into the spray tank, is comprised of:

Cryolite   3 lb.
Water   50 gal.

For smaller amounts:

Cryolite   3–6 tbsp.
Water   1 gal.

Generally about 100 gal. of spray per acre will be required when plants are small, 150–250 gal. per acre for thorough wetting of full-grown or large bean vines.

Grasshopper Control
Velsicol 1068   50
Kerosene or Light Solvent   30
Glycox 1300   20

This formula is used as a stock solution and then diluted 1 to 1 with water to give a concentrated stable emulsion for spraying from the air. The same formula can be used with orchard spray equipment where 200 to 300 gal. of spray may be used for 5 acres. Of Velsicol 1068, ½ to 1 lb. has been used per acre, giving almost 100% kill of grasshoppers and many other insects.

Cotton Flea Hopper Insecticide

	Formula No. 1	No. 2
DDT	2.5–10	5
Pyrophillite	97.5–90	—
Sulfur	—	95

Grape-Bud Beetle Insecticide
Formula No. 1
DDT   1–2 lb.
Xylol   4–8 lb.
Nonaethylene
  Glycol Oleate S   ½–1 lb.
Water   100 gal.
No. 2
Kerosene   100 gal.
DDT   35 lb.
Water   100 gal.
Casein Spreader   4 oz.

Codling Moth Spray
Formula No. 1
DDT   3 lb.
Wettable Sulfur   6 lb.
Slaked Lime   6 lb.
Water   100 gal.
No. 2
DDT   4 lb.
Pyrophyllite   4 lb.
Water   100 gal.

No. 3
Derris (Powdered)	2½ lb.
Summer (Mineral) Oil	1 qt.
Water	100 gal.

Peach-Tree Borer Spray
Propylene Dichloride	8
Fish Oil Soap	1

Mix well and then stir in until emulsified

Water	16

Dilute above with water, to suit, before use.

Blueberry Thrips Control
Kerosene	12 pt.
Water	24 gal.

Soap	¾ lb.

Apply 1 gal. per square yard.

Vine Moth Control (Nirosan Dust)
Formula No. 1
Tetranitrocarbazol	10
Tallow	90

Warm, mix, cool and powder.

No. 2
Tetranitrocarbazol	25
Powdered Sulfite Lye	10
Polyglycol Monostearate	1
Precipitated Chalk	20
China Clay	44

Use 1 pound per 99 pounds of water.

Sprays for Ornamental Plants
For red spiders:
Derris or Cube Powder (4% Rotenone)	1 tbsp.
White Oil Emulsion (83% Oil)	4 tsp.
Water	1 gal.

For mealybugs and scale insects:
Nicotine Sulfate Solution (40% Nicotine)	1½ tsp.
White Oil Emulsion	3 tbsp.
Water	1 gal.

For newly hatched scale insects on hardy shrubs and also against lacebugs:
White Oil Emulsion (83% Oil)	1 cupful (or ½ pint).
Soap Flakes	1½ cupfuls.
Nicotine Sulfate Solution	4 tsp.
Water	3¼ gal.

Horticultural Lice Control Spray
Nicotine (95–98%)	3
Polyglycol Monostearate	4
Water	93

Protecting Stored Seed from Insects
Formula No. 1
Use 1 oz. magnesium oxide per bushel of seed.

No. 2
DDT	3
Talc	97

Use ½ oz. per bushel of seed. This formula should not be used for food for humans or livestock.

Colorado Beetle Spray Base
Calcium Arsenate	93.90

Sulfite Lye (Powdered) 6.00
Green Water Soluble Dye 0.01
Add 4 lb. of above to 96 lb. of water.

Lucerne Snout Beetle Control

Soybean Meal	100
Sugar	15
Sodium Silicofluoride	6

Press into pellets.

Slug Killer

	For small quantities	For large quantities
Calcium Arsenate	1 oz.	1 lb.
Metaldehyde	½ oz.	½ lb.
Bran	1 lb.	16 lb.
Molasses	2 tsp.	1 pt.
Water	1 pt.	2 gal.

Pine Sawfly Control

Lead Arsenate	16 lb.
Fish Oil Spreader	1 gal.
Water	100 gal.

Apply at high pressure, upward between the trees, 100 gallons per acre.

Control of Sandflies and Midges

Paint or spray mosquito screens with a 5% solution of DDT in kerosene.

Field Control of Chiggers

	Formula No. 1	No. 2
Hexachlorocyclohexane	3	25
Fuel Oil	97	—
Talc	—	75

Formula No. 1 is applied at a concentration of 25 gal. per acre. No. 2 is used at 40 lb. per acre.

Japanese Beetle Lure
Formula No. 1

Caproic Acid	8
Phenyl Ethyl Butyrate	1
Eugenol	1

This is 2.8 times as successful as the standard lure of geraniol (9) and eugenol (1).

No. 2

Anethol	90
Pimenta Leaf Oil	10

Fire Ant Spray

DDT	¼–1	lb.
No. 2 Fuel Oil	1	gal.
Glyceryl Mono Oleate	1.3	fl. oz.
Water	To make 100	gal.

Spray on trunk and larger lower limbs of trees.

Mound Ant Control

This insect is controlled by treating its mounds with 2 oz. of a solution containing 2.0% of pyrethins and 33.0% vegetable oil soap, in 10 gal. of water. Ten gallons of this mixture treats a mound two and one-half feet in diameter and the effect of the mixture on the ants is immediate.

Treatment of Insect Infected Wheat
British Patent 553,633

Add 1–2 grams sodium chlorite per sack of wheat.

Fumigants and Disinfectants

Fumigant for Stored Products

Ethylene Dichloride	3
Carbon Tetrachloride	1

This is recommended by the U. S. Department of Agriculture for use in ridding stored products of insect infestation. Five qt. or 14 lb. of the mixture are recommended for each 1000 cu. ft. of space to be fumigated at temperatures of 65°F. or above. The fumigant is poured into shallow pans placed above the products to be fumigated and left for 24 hours in a gastight space.

Shelled Corn Crib Fumigants

The following mixtures seem to be most toxic for shelled corn in steel bins: ethylene dichloride-carbon tetrachloride Methyl bromide, 10%; propylene dichloride-carbon tetrachloride Methyl bromide, 10%; and Ethide-carbon tetrachloride and chloropicrin-carbon tetrachloride. Satisfactory mixtures, if sufficient dosage is used are: ethylene dichloride-carbon tetrachloride; carbon bisulfide; carbon tetrachloride-carbon bisulfide, and similar solvent mixtures. Carbon bisulfide gives excellent control at dosage levels of 1½ gal. (15.75 lb.) or more per 1000 bushels. Mixtures of carbon tetrachloride and carbon bisulfide (80-20) give good kills at a dosage of 5 gal. (63.7) or more per 1000 bushels.

Narcissus Bulb Fly Fumigation

Methyl bromide (3 lb./1000 cu. ft. air) is used for 4 hours at 21°C.

Fumigant for Dairies and Factories

In practical tests methyl bromide gives excellent control of cockroaches, cheese skippers, cheese mites, rats, mice and other minor pests of dairy plants and cold-storage warehouses. The fumigant has no adverse effects on dairy products or factory equipment.

Citrus Fruit Disinfectant
U. S. Patent 2,374,209

The fruit is passed through a tank containing

Soap	0.20
Sodium Carbonate	0.50
Sodium Ortho-phenylphenate	0.15
Tetrasodium Pyrophosphate	1.00
Water	98.15

Insect Fumigant
U. S. Patent 2,362,472

Methyl Bromide	70
Methyl Isopropyl Ketone	30

Tomato Seed Disinfection

Ethyl Mercury Phosphate	1
Water	24,000

Treat for 5 minutes at 43–80°F. and then centrifuge.

Wheat Seed Disinfectant
(Tritisan)

Pentachloronitrobenzene	20
Tallow	80

Herbicides

Poison Ivy and Poison Oak Eradicator

Dissolve 1 lb. of ammonium sulfamate in one gal. of water. Add 50 g. of a good wetting agent, such as Deceresol OT, Nekal NS, or one of the alkyl naphthalene sodium sulfate type, and stir until dissolved. Spray on a bright, hot, sunny day; within three days the poison ivy will turn yellow and die.

Herbicide (Weed Killer)
U. S. Patent 2,370,349

Arsenic Oxide	2	lb.
Caustic Soda	⅕	lb.

INSECTICIDES, FUNGICIDES AND WEED KILLERS

Sodium Chlorate	3	lb.
Sodium Pentachloro-phenate	3	lb.
Water	100	gal.

Weed Killer
Formula No. 1

An effective weed killer for dandelions, poison ivy, and other broad leaf weeds, is obtained by dissolving 1 g. of 2,4-dichlorophenoxyacetic acid and about one-half level teaspoon of Dreft, to serve as a wetting agent, in 1 qt. of water. The solution should be applied with a sprayer or an ordinary garden sprinkler.

No. 2
U. S. Patent 2,396,513

2-4 Dichloro-phenoxyacetic acid	5
Sulfonated Castor Oil	95

Mix until dissolved and then add

Water	945

No. 3

2-4 Dichloro-phenoxyacetic acid (Ammonium Salt)	0.2–0.4
Water	To make 100

Apply to the lawn, only, with a sprayer or sprinkling can.

No. 4

Sodium Chlorate	70
Calcium Chloride	14½
Magnesium Chloride	14½
Water Soluble Dye	To suit

No. 5

Ferrous Sulfate	75
Ammonium Sulfate	15
Fine Sand	10

Drain Plant Root Destroyer
Canadian Patent 426,319

A dry mixture of

Caustic Soda	100
Copper Sulfate	8
Ammonium Sulphate	½

For use put in drain with hot water.

Rat Control

Kind of poison	Poison Oz.	Bait Lbs.
Red Squill	16	9
Thallium Sulfate	1	4
Micronized Arsenic Trioxide	1	2
Barium Carbonate	16	5
Zinc Phosphide	1	6
Antu (Alpha Naphthyl Thiourea)	1/6	10

Formula No. 1

Ground Bread Crumbs	8
Ground Bacon	1

After grinding the bacon stir in the selected powdered poison for 10 minutes to insure a thorough mixture; then add the bread crumbs, thoroughly mixing for 5 minutes, in a power mixer if available. This bait should be used within a few days after preparation.

No. 2

Ground Bread Crumbs	35
Peanut Butter	5
Blackstrap Molasses	5

Add the selected powdered poison to the peanut butter and stir into a mixture of the other ingredients for 10 minutes or more. This bait will remain acceptable to rats for several weeks.

No. 3

Ground Bread Crumbs	40
Peanut Butter	4
Cottonseed Oil	1

Add the selected powdered poison to the oil and stir into a mixture of the other ingredients for 10 minutes or more. This is a permanent bait that can be kept for a long time if stored in a tightly closed container.

No. 4

Ground Bread Crumbs	10 lb.
Freshly Ground Hamburger	10 lb.
Glycerin	10 oz.

To prepare as a fresh bait, stir the desired powdered poison into the hamburger, then add to a thorough mixture of the other ingredients, stirring for 10 minutes or more.

To prepare as a dry permanent bait use any of the poisons, except barium carbonate and zinc phosphide, and after thoroughly mixing let it dry completely in an oven gently heated. Put aside in a tight container in a dry place. Before using moisten with a little cold water.

No. 5

Ground Bread Crumbs	4 lb.
Cream Cheese	1 lb.
Mineral Oil	1 oz.

Add the selected powdered poison to the oil and stir thoroughly into a mixture of the other ingredients, or first mix well the poison with the cheese, stirring this into the mixture of bread crumbs and oil. This bait will keep 2 months or more, and though it may become rancid, it will remain acceptable to rats.

Rolled oats of the grade used for feeding poultry may be substituted for bread crumbs in any of these formulas, though they are not so attractive to rats, or a mixture of bread crumbs and corn meal may be used.

Often on farms or where rats for some time have been eating stored feeds, one of these can well be employed as a bait material. When poison is mixed with this bait and it is exposed where rats can get it, the clean feed should, whenever possible, be removed from reach or access to it shut off. In follow-up work the operator must seek the type of bait material that will be most acceptable under the prevailing conditions.

No. 6

Wheat	1000
Thallium Sulfate	20

No. 7

p-Dimethylamino-phenyldiazosulfonicacid	6.0
Bran	93.4
Red Dye	0.6

No. 8

Dry Bread Crumbs	24
Glycerin	2
Meat (Ground)	20
Powdered Red Squill	5

Mix ingredients thoroughly.

Leave bait out for 3 days. If, after 1 week, live rats are still noticed, wait 2 weeks and pre-bait for several days with the same mixture, but without the red squill. Then use red squill mixture.

Chapter X

LEATHER TREATING PREPARATIONS

Hide Depilatory

Caustic Soda	1– 2
Sodium Sulfide	2– 4
Water	97–94

Put the hides in a drum containing the above mixture and work them for 16–22 hours.

Oropon Type Bate

Ammonium Chloride	65
Wood Fibre	31
Dry Pancreas	3½

Chrome Tan

Sodium Bichromate	1,000 lb.
Sulfuric Acid (66° Bé.)	530 lb.
Anhydrous Sodium Bisulfite	970 lb.
Water To make	500 gal.

Vegetable Tan

Dry Quebracho Extract	10,000 lb.
Water	1,000 gal.
Anhydrous Sodium Bisulfite	400 lb.
(in 250 gal. water)	

Leather Dubbing

Tallow	50
Neatsfoot Oil	40
Paraffin Wax	9
Aluminum Stearate	1

Heat and mix until uniform.

Mildew Preventive for Leather Book Bindings
Formula No. 1

Make a 2% to 5% solution (not more than 5%) of copper sulfate. Immerse a soft towel or cloth in this solution. Remove the cloth and thoroughly wring out. Then hang out to dry. When thoroughly dried, it can be used to rub leather bound books. One treatment of the cloth will easily take care of scores or a hundred volumes, and the leather will not be marked by the chemical.

No. 2

Thymol	10 g.
Mercuric Bichloride	4 g.
Ether	200 cc.
Benzene	400 cc.

Apply with absorbent cotton or brush.

Leather Fungicide Protection
Formula No. 1

Salicylanilide	2.2
Isopropyl Alcohol	25.0
Paraffin Wax	33.0
Stoddard Solvent	39.8

Warm gently and mix until dissolved. Impregnate the leather, squeeze, drain and dry.

No. 2

Perchloroethylene	67.0
Mineral Oil	10.0
Neatsfoot Oil	10.0
Cyclohexanone	10.0
Paranitrophenol	1.5
Pentachlorophenol	1.5

Increasing Durability of Leather Soles

Sole leather should be immersed for 30 minutes in a solution of not less than 60% by weight of a non-volatile base, preferably solvent naphtha, not less than 25% by weight of a fatty oil with a viscosity of at least 2,500 Saybolt units at 100°F. The balance of the oil base should consist of a mineral oil, preferably naphthenic-base oils, of a viscosity such that the overall viscosity of the oil base, without solvent, exceeds 1,800 Saybolt units at 100°F.

Cleaning and Dressing Leather Belts

Keep belts as clean as possible at all times for best results. If machine bearings are throwing oil or grease, these substances will get on the belt, reducing its life and pulling power.

If the leak cannot be stopped at the source, the installation of deflectors or throwing discs will be helpful.

A small amount of oil on a belt can sometimes be removed by ordinary wiping. If this does not do the job, give it a thorough scrubbing with a solution of carbon tetrachloride and unleaded gasoline, using a stiff jute brush and working in the direction of lap joints so as not to lift them but rather lay them down, or remove the belt and soak it for five or six hours in a degreasing solution consisting of one part carbon tetrachloride to three parts of unleaded gasoline. If carbon tetrachloride is not available, the belt can be soaked in any of the cleaning fluids used by dry cleaning establishments. Due to the fire hazard, the soaking of the belt should be done in the open or where ventilation is good. After removing from the bath, allow the belt to dry thoroughly.

Leather belting should always be redressed after cleaning.

Use a belt dressing approved by the belt manufacturer and designed to supply the necessary currier's oils which were lost in use or during cleaning.

When pulley faces begin to polish, it is a sign that dressing is needed on the belt. Under normal conditions, dress belts every three to 6 months.

If cemented laps show signs of opening, stick them down immediately.

Artificial Leather Dope

Nitrocellulose	10.0
Polyvinyl Methyl Ether	20.0
Methyl Acetate	58.0

Leather Conditioning Agents

Combinations of fish oil and degras together with water make a suitable conditioning agent for leather.

Iron Free Oxidized Fish Oil	30–50
Common Degras	15–30
Water	10–35

The oxidized fish oil and common degras are heated under constant stirring until they melt, then they are cooled to a slightly pasty consistency and the necessary water is added in a slow stream under constant stirring. Some sulfonated fish oil or sulfonated higher alcohol might be added to make the emulsion stable. The ready material has to be of a tannish brown color and no water is allowed to separate.

Inexpensive Leather Conditioner
For Keeping Leather Smooth and Soft

Fish Oil With Low Stearine Content	40
Pale Paraffin Oil (100 @ 100)	40
Oleyl Alcohol	20

Some oil soluble yellow dye may be added as well as a few drops of mirbane oil to cover the odor of the fish oil.

The leather has to be cleaned thoroughly before applying the conditioning agent. Then a rag should be dipped in the agent and applied to the leather. No excess should be used. The appearance of the treated leather shall be soft but not greasy.

Preparation of Sulfonated Castor Oil
(For leather finishing)

The preparation of the sulfonated oil must be carried out under carefully controlled conditions. To a given amount of a light colored practically neutral castor oil, 20–40% highly concentrated sulfuric acid (98–100%) is slowly added under constant stirring and cooling. The temperature should be held around 90°F. and shall never go above 100°F. After all of the acid has been added, the mixing of the sulfonates continues for ½–1 hour. Following this procedure, the material is allowed to stand. During this time, the castor oil reacts with enough sulfuric acid to become water soluble. This will take a few hours and depends on the size of the batch and the temperature conditions. The material shall be tested in a test tube ½ filled with distilled water. The sulfonation might be considered finished as soon as the sulfonated castor oil is entirely soluble in distilled water. The content of the test tube appears then as a clear and transparent homogeneous solution.

Whenever this stage is reached, the sulfonated oil is ready for the washing process. The washing is carried out with water in such a way that the sulfonate is mixed into the water during which process it forms a white dough-like mass. The water separates by standing. The clear oil is then neutralized with any suitable medium, e.g., caustic solution. The neutralization is finished when the oil gets water soluble or forms a white milky emulsion with water and the pH is around 6.5.

The sulfonated castor oil prepared as given above can be applied widely in the leather finishing industry. It serves as a softener when applied by itself. It also may be added to the tanning solution. In this case, it will accelerate the absorption of the tanning agent by the leather. The amounts of sulfonated castor oil recommended for sole leather is 1–2% of the dry weight of the leather itself. The sulfonated castor oil may also be used in combination with other fats, e.g., neatsfoot oil for greasing of fine leather. In this case, it possibly should be neutralized slightly farther than recommended above.

Dyeing Gloves

The following formulas for the production of these shades are calculated per 100 lb., dry weight, crusted chrome tanned skeepskins.

The stock is wet back at 130°F. with 2 lb. ammonia 28% and 1 lb. of Nacconol NR. It is drummed for 45 minutes, drained, rinsed and then dyed with the indicated amounts of dyestuffs.

Dyeing is carried out by drumming for 45 minutes at 130°F., then exhausting slowly with 2½ to 5 lb. formic acid, added in three portions at 10-minute intervals. After the last addition of formic acid, the stock is drained, rinsed and finished in the usual manner.

Rum Frappe:
National Resorcine
  Brown R   1⅝ oz.
National Nigrosine
  12525 J Powder   3/16 oz.
National Para
  Bordeaux FS   ⅛ oz.
National Naccotan
  A Powder   1 lb.
Frappe Cocoa:
National Alizarol
  Brown EB   6 lb. 15/16 oz.

National Resorcine
  Brown R   2 lb.
National Para
  Blue BB   9⅝ oz.
National Fast
  Red S Conc.   5 7/16 oz.
Blue Cedar:
National Buffol
  Black NBR   2 lb. 6⅜ oz.
National Wool
  Green S   8 15/16 oz.
National Acid
  Green L Extra   5⅛ oz.
National Resorcine
  Brown R   1 15/16 oz.
Horizon Gold:
National Para
  Yellow CW   3 lb. 3 3/16 oz.
National Resorcine
  Brown R   3 lb. 3 3/16 oz.
National Induline
  B Extra   1 15/16 oz.

Wool Sheepskin Preservative
Salt   99.5
Zinc Chloride   0.5

Chapter XI

LUBRICANTS AND OILS

Tire Rim Lubricant

Boiling Water	60.0
Triethanolamine	7.5
Add slowly	
Stearic Acid	32.5
Stir until cool.	

Pipe Thread Seal and Lubricant
U. S. Patent 2,324,729

	%
Aluminum Powder (Fine)	0.5–10
Barium Sulfate	10 –80
Blown Castor Oil	10 –89.5

Valve Lubricant
U. S. Patent 2,393,800

Stearamide	1–2
Glyceryl Monoricinoleate	4–3

Wire and Bolt Drawing Lubricant
U. S. Patent 2,319,393

Calcium Myristate	10
Acrawax C	5–7

Extreme-Pressure Lubricant for Cold Rolling
U. S. Patent 2,391,631

Tritolyl Phosphate	15
Lorol Phosphate	1½
Sulfonated Castor Oil	3
Kerosene	20
Mineral Oil (50–200 sec. S. S. U. @ 100°F.)	1960–4960

High Quality Soluble Oil

Dibutyl Tartrate	50
Castile Soap	30
Trichlorethylene	20

The above can be diluted with vegetable or mineral oil; when water is mixed in, a stable emulsion results.

Cutting Oil Base
U. S. Patent 2,393,927

Potassium Rosin Soap	45	lb.
Glyceryl Monoleate	35	lb.
Red Oil Soap	5	lb.
Water	7½	%

Cutting Oil

91 Oil (Sunoco)	11	gal.
Twitchell Base #262	15	lb.

This will form a very stable emulsion on dilution with water.

Glass Grinding Fluid

Turpentine	45.0	cc.
Ether	22.5	cc.
Camphor	31.0	g.

To be used with powdered emery for grinding glass.

For smoothing edges a sheet of emery cloth moistened with the above solution may be used.

Plane surfaces should be ground on thick plate glass.

For grinding glass stoppers use coarse emery, turn in one direction, finish with fine emery.

Dental Grinding Coolant

Diethylene Glycol Mono Oleate	20.0
Paraffin Oil	20.0
Wintergreen Oil	0.5
Distilled Water	100.0

Drill Lubricant
U. S. Patent 2,408,385

	%
Soda Base Grease	45–52
Paraffin Wax (143–145°F.)	38–43
Turpentine	8–12
Aluminum Stearate	¼–½

Lubricant for Split Dies
U. S. Patent 2,334,076
Formula No. 1

Calcium Resinate	2– 3
Acrawax C	2– 4
Hexachlorbenzene	10–12
Graphite	5– 6

No. 2

Hydrated Lime	18
Lime Base Petroleum Grease	5
Calcium Resinate	2

Electroforming Mold Parting (Release) Medium

Air-Spun Graphite	300–400 g.
Beeswax	100 g.
Rosin	25 g.
Trichlorethylene	1 l.

Water-Soluble Lubricant
(German Torpedo Oil)

Triethanolamine Caproate	20–40
Diethylene Glycol	80–60

Gasoline-Insoluble Plug Valve Lubricant
U. S. Patent 2,321,384

Calcium Soap of Blown Castor Oil	5–15
Calcium Soap of Beeswax	30–40
Fused Lead Stearate	40–70
Aniline	2– 4

Solvent-Insoluble Lubricant
U. S. Patent 2,382,860

Glycerin	75
Mica (180 mesh)	20
Starch	5

Nitration-Resistant Lubricant
U. S. Patent 2,335,331

	%
Polyisobutylene (M. W. 12000–20000)	10–15
White Mineral Oil	65–80
Light Petrolatum	15–30
Paraffin Wax	<10

Nitration-Resistant Packing
Acid resistant blue African asbestos is impregnated with the above.

Packing Lubricant

Heavy Petrolatum	92
Vistanex Medium	8

Incorporate Vistanex at 240°F. with stirring. Apply lubricant to braided cotton, asbestos, rayon or other fiber packing in kettle at 240°F. The packing is drained, cooled, graphited and calendered in the customary manner.

Stuffing Box Lubricant

Potassium Soap	16
Glycerin	4
Water	1

Rubber Lubricant

Glycerin	40 cc.
Diglycol Stearate S	80 g.
Water	400 cc.
Ammonia	Enough to neutralize

Mold Lubricant for Rubber Goods
U. S. Patent 2,388,153

	%
Gum Arabic	0.1 –0.4
Pine Oil	0.3 –0.9
Turkey-Red Oil	1.2 –2.1
Trisodium Phosphate	0.34–0.4
Water	To make 100

Slide Rule Lubricant

Tapioca or Rice Flour	3
Zinc Stearate Powder	1

Optical Instrument Lubricant (Zeiss)

Aluminum Stearate	18.33
Spindle Oil	81.67

Warm together and mix until dissolved.

Fine Instrument Lubricant
U. S. Patent 2,409,443

Tricresyl Phosphate	10
Ethylene Glycol Monobenzyl Ether	10

Clock Lubricant
U. S. Patent 2,409,444

	%
Tricresyl Phosphate	40–60
Dibutyl Phthalate	25–15
Triethylene Glycol Di-2-Ethylbutyrate	35–25

Drawer, Window and Door Lubricant

Paraffin Wax (M. P. 140–145°F.)	20.0
Orthodichlorobenzene	12.0
V. M. & P. Naphtha	67.5
Methyl Salicylate	0.5

Melt the wax in the orthodichlorobenzene and then add naphtha and finally the methyl salicylate.

Lubricating Mastic

Graphite Flake	100
Asbestos Fiber	100
Medium Heavy Grease	100

Goldbeaters' Lubricant
U. S. Patent 2,391,653

Sodium Stearate	6½	oz.
Alcohol	3	qt.

Dissolve above and add

Calcium Sulfate (Dehydrated)	12 lb.

Then evaporate until dry.

Wire Rope Lubricant and Preservative

Acrawax C is used in a compound to lubricate and preserve steel wire rope. This compound is incorporated into the rope when the strands of wire are twisted, thereby securing a thorough coating on every strand of wire. This gives the rope pliability through internal lubrication, supplies sufficient effect for external lubrication, and serves as a preservative. A typical formulation consists of a combination of mineral oil, asphalt, and blown animal oil or fish oil.

When Acrawax C is added to such a compound several effects are noticed— a) higher softening point, b) lower viscosity. For example, a test on a viscosimeter at 210°F. gives these results:

Before adding Acrawax C: viscosity 273 seconds
After adding Acrawax C(1%): viscosity 172 seconds.

The higher softening point characteristic is particularly important when the wire ropes are used in hot climates. The lowering of the viscosity of the blend is of value and, in addition, better cold weather characteristics are obobserved.

Waterproof Rope Lubricant
U. S. Patent 2,199,695

Calcium Stearate	12–14
Paraffin Wax	35

Oleic Acid	8– 9
Lubricating Oil	35
Wool Fat	4

Water Removable Textile Lubricant
British Patent 553,562

Mineral Oil	70% or more
Diglycol Stearate	30% or less

Rayon Lubricant
U. S. Patent 2,176,510

Refined Mineral Oil	5
Olive Oil	2
Sulfonated Spermaceti	3
Water	90

Belt Dressing Compound

Cumar Resin (M. P. 60–75°F.)	60
Rubber	40

Thin with benzol.

Hydraulic Pressure Fluid
U. S. Patent 2,402,754

Tetradecanol	10.00
Di-(Ethoxydiethylene Glycol) Phthalate	25.00
Trimethylene Glycol	20.00
Carbitol	34.85
Butyl Carbitol	10.00
Diamylamine Phosphate	0.15

Hydraulic Brake Fluid
Formula No. 1

Isopropyl Alcohol	50
Castor Oil	50

Secondary butyl alcohol can be substituted for isopropyl alcohol.

No. 2
U. S. Patent 2,200,494

Castor Oil	31.0
Polypropylene Glycol	67.0
Caustic Soda (36% Solution)	0.2

Heat to about 200°C. for 1¾ hours, then at 210°C. for ½ hour. Then add:

Butyl Alcohol	0.5

No. 3

Castor Oil	45
Propylene Glycol	15
Butyl Cellosolve	15
Isopropyl Alcohol	24
Phosphoric Acid	¼
Cresylic Acid	¼
Triethanolamine	½

Penetrating Oil

Secondary Butyl Alcohol	15
Kerosene	25
Viscosity Blending Oil	60

Fat and Oil Oxidation Inhibitor

Toloquinone	0.02
α-Tocopherol	0.02

Use 0.04% of the above on the weight of oil or fat.

Antifoaming Oil Mixture

Crude Petroleum	3 fl. oz.
Toluol	To make 1 qt.

Antifoaming Additive for Lubricants
U. S. Patent 2,377,654

The addition of 0.2 to 0.8% of potassium oleate to heavy oils, such as gear lubricants, serves to reduce or prevent frothing and foaming when the oil is subjected to violent agitation. The soap does not impair the lubricating qualities of the oil.

Non-Foaming Gear Lubricant
U. S. Patent 2,377,654

Refined Mineral Lubricating Oil	92.0
Refined Sulfurized Sperm Oil	7.5
Potassium Oleate	0.5

Lubricating Oil Filter
U. S. Patent 2,195,272

Mineral Wool	43.40
Water	52.60
Ammonia	1.30
Formaldehyde	1.25
Casein	1.72

Mix, form into shape and bake.

Salt Water Protective Grease
(For protecting iron-work on ships)
U. S. Patent 2,383,148

Dark Green Petrolatum	85.0
Lithium Stearate	11.0
Aluminum Tristearate	2.2
Lead Oleate	0.5

Heat together and mix until uniform, then add:

Tributyl Phosphite	0.5

Chapter XII

CONSTRUCTION MATERIALS

Wood Preservatives
For Pressure Impregnation
Formula No. 1
Nuodex Copper*
(10½% Cu)   10½
Fuel Oil   89½

No. 2
Nuodex Copper
(10½% Cu)   10½
Creosote   89½

Nuodex Copper 10½% should be heated to 200–250°F. before dissolving in carriers.

For Field Application to Surfaces
Not to Be Painted
(Fence Posts, etc.)
Formula No. 1
Nuodex Copper (8% Cu)   25
Fuel Oil   75

No. 2
Nuodex Copper (8% Cu)   25
Waste Crank-Case Oil   50
Kerosene   25

This mixture should be allowed to stand for a short time to settle out the solid impurities and water in the crank-case oil.

No. 3
U. S. Patent 2,384,026
7 to 145 parts of ferrous sulfate; 6 to 26 parts of orthoarsenic acid; 0.4 to 4 parts of sulfuric acid, and 0.1 of potassium bromide.

---

* Copper naphthenate.

For Surfaces Not to Be Painted
(Where the odor of cresote or oiliness is not desired)
Nuodex Copper (8% Cu)   25
Kerosene (or Mineral
Spirits)   75

For Surfaces to Be Painted
Nuodex Copper (8% Cu)   29
Linseed Oil   35
Nuodex Cobalt (6% Co)   ½
Petroleum Naphtha   36

Soluble Dyes for Wood Impregnation with Dimethylol Urea

The following dyes are satisfactory for coloring wood throughout in conjunction with the application of dimethylol urea. This process is carried out by evacuating the wood, then applying a solution of resin and dye under pressure. A pH of 8–9 must be maintained.

    Pontacyl Light Yellow 3G Conc. 150%*
    Du Pont Quinoline Yellow P Extra Conc.
    Du Pont Orange II Conc. for Lakes
    Du Pont Orange RO for Lakes
    Du Pont Resorcin Brown 5G Conc. 200%
    Du Pont Resorcin Brown 3R

---

* Chromacyl, Pontacyl and Pontamine are registered trade-marks of E. I. du Pont de Nemours and Co., Inc.

Pontacyl Fast Red AS Extra Conc.
Du Pont Crocein Scarlet N Extra
*Pontamine Fast Red 8BL Conc. 125%
Du Pont Anthraquinone Violet R
Pontacyl Violet 4BL Conc. 125%
Du Pont Anthraquinone Blue SWF Conc. 150%
Pontacyl Wool Blue BL Conc. 200%
Pontacyl Dark Green B Conc. 175%
Du Pont Anthraquinone Green G
Pontamine Green 2GB Extra Conc. 150%

Pontacyl Black BX	90
Du Pont Crocein Scarlet N Extra	10

Chromacyl Black W

Wood Bleach
U. S. Patent 2,397,193

A	Borax	1
	Caustic Soda	2
	Sodium Silicate (42° Bé.)	1
	Water	20
B	Hydrogen Peroxide (30%)	20

Mix together before use, apply to the wood surface and allow to dry.

Detecting Heartwood and Sapwood in Douglas Fir Lumber

Prepare a solution of:

Ferric Chloride	83 g.
Distilled Water	417 g.

Mix thoroughly.

The borings of the fir lumber to be tested are dipped into a small amount of the ferric chloride solution. Then on a paper towel, they are dried quickly.

The *heartwood* gets a greenish-brownish black, the *sapwood* a light greenish iridescent color.

Cork Substitute

Glue	100
Glycerin	75
Glucose	75
Ground Peanut Hulls	100
Water	350
Saponin	1–2

Warm to 60°C. and beat vigorously to incorporate air into the mix. Pour into crown caps or on a smooth surface to form sheets, from which discs can be punched after cooling.

Building Brick
U. S. Patent 2,302,988

Blast Furnace Slag (Ground)	50
Lime Sludge	25
Hydraulic Cement	14
Calcium Chloride	4
Calcium Stearate	2

Wet and press to shape and dry at 38°C. for 10 hours.

Dark Gray Concrete

Cement	12 sacks (1504 lb.)
Sand (¼ In. Diameter to 100 Mesh)	3200 lb.
Gravel (¼ to ¾ In. Diameter)	3200 lb.
Water	96 gal.
Continental A or Witco #1 Pellets (Carbon Black)	½ bag (12½ lb.)

Light Weight Cement
Portland Cement	35
Wood Fibers	10
Burned Rock (Porous)	30

Weatherproof Cement
Portland Cement	95
Sodium Silicate (Powdered)	3
Calcium Lignosulfonate	2

Cement Improver
British Patent 560,258
Van Dyke Brown	1 lb.
Soda Ash	30–60 lb.

The above is dissolved in 40–80 gal. of water mixed with the cement.

Floor Composition and Road Marker
U. S. Patent 2,393,525
Marble Dust	71
Amorphous Petroleum Wax	5
Rosin	12
Castor Oil	4
Titanium Dioxide	3
Wood Flour	5

Warm and mix until uniform. Apply hot. This composition expands and contracts about the same way as a road surface does.

Sanitary (Fungous Proof) Floor Surface
Portland Cement	7
Sand and Gravel (3/8 in.)	6
Malachite Ore (Powdered)	5
Salt	1

The above is made into a grout with water and applied. The copper compound in the malachite prevents or retards growth of microorganisms.

Concrete Floor Hardener
Zinc Fluosilicate	1/2 lb.
Magnesium Fluosilicate	1½–2 lb.
Water	1 gal.

Interior Wall Plaster
U. S. Patent 2,407,739
Portland Cement	100 lb.
Hydrated Lime	10 lb.
Brick Dust	20 lb.
Sand and Marble Dust	10 lb.
Water	10–15 gal.

Building Board
German Patent 745,907
Slag	20–30
Gypsum	42–45
Clay	16–18
Cellulose Fibers	24–30
Sodium Silicate (28–30° Bé.)	11–16
Iron Oxide	5–8

Mix well; form into sheets and subject to a pressure of 2500–3000 lb.

Coating Asbestos Cement Sheet
U. S. Patent 2,347,684
Kaolin	50
Titanium Dioxide	50
Cryolite	10
Sodium Silicate	220
Water	30

Blend in a ball mill; apply and heat to 850°F.

The soda to silicate ratio should be 1 to 2.6.

Crinkle Finish Enamel (Ceramic)
U. S. Patent 2,359,283
Borax	15.4
Soda Ash	6.5
Silica	23.8
Feldspar	33.2
Cryolite	10.5
Fluorspar	4.3
Antimony Oxide	8.5

Fuse the above and then add:
Sodium Nitrate 3.2
After the enamel slip is applied fire in an oxidizing atmosphere.

Preventing Scumming in Antimony Free Enamels

A satisfactory formula is:

Antimony Free Frit	100	lb.
Clay	7½	lb.
Opacifier	1	lb.
Sodium Nitrite	1½	oz.
Potassium Carbonate	4	oz.
Zinc Oxide	1	oz.

Opaque Titanium Enamel (Frit Compositions)

Feldspar	4.5
Quartz	25.3
Soda Ash	24.2
Sodium Nitrate	2.4
Whiting	3.4
Zinc Oxide	19.5
Cryolite	2.6
Sodium Antimonate	.8
Titanium Oxide	17.2

Synthetic Mica

Alumina	11.6
Magnesia	32.6
Kieselguhr	30.7
Potassium Silicofluoride	25.1

Grind finely, mix well and form into pellets 20 mm. in diameter and 6 mm. thick.

Place a few pellets in an alundum crucible which is placed in a cylindrical larger carborundum form.* The space between (about 1½ in.) is packed with fused alumina sand. The top of the form and the lid of the crucible have holes for introducing the pellets. The whole is placed in a rig for inserting into a gas-fired furnace and to withdraw it when necessary.

First the crucible and form is heated to 900°C. in an electric heater and then put into the gas furnace and heated to 1500°C. As the pellets melt more are added until the crucible is nearly full of molten material. Allow to cool slowly, without vibration (0.1–0.03°C. per minute) until about 1320°C., when solidification occurs. Then cool rapidly to 900°C., take out of gas furnace into the electric heater. Allow to cool gradually then remove the form and break the crucible. When cold, cleave or split mica into sheets.

Refractory Lining for Foundry Crucibles

Silica Sand #80	80
Silica Flour	15
Bentonite	5
Water	6–8

Mix thoroughly and ram into place. Bake at least 4 hours at 400°F.

Cement Coating for Steel Reaction Chambers

Commercial Furnace Cement	60
Silica Foundry Sand	30
Asbestos	1
Water Glass	10
Water	1½

The water glass should be a sodium silicate having a soda to silica ratio of 1 to 3.22. Only a furnace cement having a silicate binder is to be used. The material is sprayed on metal surfaces with a cement gun after the metal has been roughened by sand blasting. The sprayed coat is brushed with a

---
* This has a separate top, bottom and cylindrical walls.

mixture containing 5 lb. furnace cement, 11.8 lb. water glass, and about 3 lb. of water. Thorough drying and careful baking are required, at temperatures up to 900°F.

Refractory Lining for Melting Furnaces
Formula No. 1

Water Glass	13.5
Silicon Carbide Firesand	100
Water	8

The water glass, which should be a sodium silicate with 1 to 2.40 soda to silica ratio, is diluted in the approximate amount of water stated, and then mixed into the silicon carbide; stirring is continued until free from lumps.

No. 2

Fireclay or Kaolin	10
Silicon Carbide Firesand	70
Ground Mica	4
Ground Quartz	16
Water Glass	4
Water	10

The ground quartz and mica should be of a grade to pass a 40-mesh screen. The water glass required is a sodium silicate with 1 to 2.40 soda to silica ratio. The dry ingredients are first mixed thoroughly, then the sodium silicate (water glass) is diluted with the water and stirred into the mixture.

Protective Coating for Refractories

Raw Fireclay	25
Calcined Fireclay	75
Water Glass	18

The water glass should be a sodium silicate containing soda and silica in the ratio 1 to 2.40 (52° Baumé). The product may be painted or troweled on the desired refractory surfaces, sufficient water being used to give the proper consistencies for the respective purposes. The coating is particularly useful in the arch and bridge walls of boiler furnaces, potters' kilns, welding and malleable iron furnaces, oil burners, and foundry cupolas.

Steatite Insulation Body
U.S. Patent 2,382,137

Ball Clay	10
Whiting	6
Barium Carbonate	24–35
Zinc Oxide	1–10
Boric Acid	1–10
Talc	58–29

Machinable Ceramic Insulation
U.S. Patent 2,391,376

Titanium Dioxide	10
Mica	24
Sodium Potassium Borate	10
Water	4½

Mix well and heat to 1100–1400°F.; press to desired form and allow to cool.

High Dielectric Ceramic
Formula No. 1

Rutile	69
Titanium Dioxide	10
Lanthanum Oxide Hydrate	10
Zirconium Hydrate	10
Beryllium Carbonate	1

No. 2

Rutile-Oxide	97
Lanthanum Oxide Hydrate	1
Zirconium Hydrate	2

The dielectric constants (K) of these two materials are 90 and 105 respectively with temperature coefficient of -7×10^{-4} per degree Centigrade. A formula for a ceramic with a K of 405 and a positive temperature coefficient of 1.8×10^{-3} is:

Titanium Dioxide	70
Rutile	10
Titanium Peroxide	13
Lanthanum Oxide Hydrate	7

A high permeability material with high electrical conductivity is made as follows:

Ferric Oxide (Fe_2O_3)	80
Soapstone	15
Magnesium Carbonate	5

This material has a higher conductivity in the center than the outside. The conductivity of the entire piece can be raised by baking in nitrogen, thus reducing the oxidation.

In producing these various ceramic materials, the raw materials in finely pulverized form are weighed, mixed and tipped into revolving drums. Each drum contains about 1/3 flintstone, 1/3 mixture and 1/3 water. The quantity of water is measured so that about 2.5 l. is allowed for each 2 kg. of mixture. Water and mixture are milled for about 100 hours. Interior walls of the drum may be lined with porcelain or flintstone.

After the milling process is completed the mixture is passed through a fine sieve into a vat fitted with a propeller-shaped whirl. The whirl is kept in constant motion and the mixture pumped into a filter press at a pressure of 8 atm.

Cakes of substance emerging from the press are passed along to a roller device and reduced to small pieces which are once more passed through different sieves of varying mesh. It can then be pressed again into cakes and once more finely divided in the chopper.

The final dimensions of the grains depend to a great extent on the nature of the objects to be made and also on the nature of the matrix used. The powder is then subjected to the ordinary processes of pottery.

For those masses with a high dielectric constant and low loss angle, the process of the ordinary tunnel kiln will serve satisfactorily. According to the size and strength of the objects, they will remain at a temperature of up to 1,400°C. for periods of from 10 to 75 hours.

For adding the quality of permeability the operations are exactly the same. According to the value desired, the firing takes place in a reducing or an oxidizing atmosphere. Shut off the air or use hydrogen for a reducing atmosphere.

Tungsten Cement for High Temperature Conduction

Nitrocellulose	1⅓	g.
Butyl Acetate	30	cc.
Tungsten Powder	75	g.

High Temperature Insulating Coating

Aluminum Nitrate	70	g.
Water	140	cc.
Alundum	300	g.

High Temperature Insulation Spray

Methanol	900	cc.
Butyl Acetate	300	cc.
Diethyl Oxalate	150	cc.
Alundum	750	g.
Binder *	300	cc.

* The binder is a 5% solution of nitrocellulose in butyl acetate.

Thermal Insulation
(Non-inflammable, oil- and water-proof)
U.S. Patent 2,389,460

A Mineral Wool	10.7
Asbestos	5.0
B Methyl Ethyl Ketone	24.8
Solvesso No. 1	49.7
Vinylite	3.6
Tritolyl Phosphate	6.2

Mix B until dissolved, then mix into A. Put into forms and dry at 200–300°F. for 12–16 hours or at room temperature for several days.

Fireproof Building Insulation
U.S. Patent 2,364,344

A Ammonium Stearate Paste (25%)	1 gal.
Corn Starch	1 oz.
Ammonium Aluminum Sulfate	1 oz.
B Portland Cement	94 lb.
Vermiculite	Up to 12 volmes of cement
Water	7–14 gal.

Use 0.6 lb. of A to 94 lb. Portland cement.

Insulating Mortar
U.S. Patent 2,364,344

Ammonium Stearate Paste (28%)	1 gal.
Corn Starch	1 oz.
Ammonium Aluminum Sulfate	1 oz.

The mortar body itself consists of about 94 lb. of Portland cement and a vermiculite aggregate of up to 12 volumes to 1 volume of Portland cement, and 7 to 14 gal. of water.

Ceramic Tile and Insulation
U.S. Patent 2,356,214

Magnesite	15
Florida Clay	25
Talc	26
Bentonite	4
Asbestos	15
Silica	5
Dextrin	3
Portland Cement	7
Calcium Chloride Solution	5½
Water	To suit

Mix, mold and dry.

Heat Insulating Compound
U.S. Patent 2,364,344

This compound consists of a paste and a strong, light-weight and fire-proof mortar body, these being employed in the proportions of 0.6 lb. of paste to 94 lb. of Portland cement. The paste consists of:

Ammonium Stearate (28%)	1 gal.
Corn Starch	1 oz.
Ammonium Aluminum Sulfate	1 oz.

The required mortar body may be made from about 94 lb. of Portland cement, and a vermiculite aggregate of up to 12 volumes to 1 volume of the Portland cement, and 7 to 14 gal. of water.

Molded Soapstone

Ground Soapstone	100
Water Glass	15
Water	25

The grade of sodium silicate (water glass) should be one in which the ratio of soda to silica is 1 to 2.40. After diluting in the water the sodium silicate is mixed thoroughly with the ground soapstone. The material can be molded to a variety of shapes, and after air-drying can be baked just under 212°F. The product withstands temperatures up to red heat without warping or cracking.

Chapter XIII

METALS AND THEIR TREATMENT

Coloring Anodized Aluminum

The following dyes may be applied to anodized aluminum from a water bath. The dye concentration used will depend on the intensity of shade desired:

    Pontachrome * Yellow SW Conc. 150%
    Pontachrome Yellow 3RN
    Pontachrome Yellow GS
    Pontacyl * Light Yellow GX
    Pontacyl Light Yellow GG Conc. 125%
    Pontacyl Light Yellow 3G Conc. 150%
    Du Pont Quinoline Yellow Conc.
    Du Pont Tartrazine Conc.
    Du Pont Orange II Conc.
    Pontachrome Fast Red E Conc.
    Pontachrome Red B
    Pontacyl Rubine R Extra Conc. 125%
    Pontacyl Light Red BL Conc. 175%
    Lithosol * Rubine BLM Powder
    Lithosol Scarlet 3BI
    Du Pont Anthraquinone Blue BN
    Du Pont Anthraquinone Blue WSA
    Du Pont Anthroquine Blue B
    Du Pont Indigotine Conc.

* Pontachrome, Pontacyl and Lithosol are registered trade-marks of E. I. du Pont de Nemours & Co., Inc.

    Du Pont Anthraquinone Green G
    Du Pont Naphthol Green B Extra Conc. 125%
    Pontacyl Blue Black RC
    Du Pont Nigrosine WSJ Crystals

Bluing Steel and Iron

The metal is cleaned with a potassium bichromate sulfuric acid mixture, then washed with ammonium hydroxide and rubbed dry. Apply ammonium polysulfide until the desired depth of color is obtained, allowing the object to dry after each application and rubbing briskly with a soft clean cloth. The result is a deep blue which may be made very nearly black by repeated applications. Rubbing with boiled linseed oil will deepen this color. The finish thus obtained is very resistant to oxidation.

Black Finish on Steel
Formula No. 1

Caustic Soda	70
Sodium Nitrite	29
Sodium Chlorate	1

Use 7¼–8 lb. of the mixture per gallon water and bring to a boil. Cleaned pieces are immersed for 10–20 minutes.

Black Finish for Silverware

Platinum Chloride	2.3
Water	1000.0

Dissolve and add:
Alcohol   500.0

Clean the silverware in a caustic soda solution, immerse in the above mentioned solution, remove, drain and heat lightly over a flame. Repeat until the desired black color is obtained.

Rustproof Blackening of Stainless Steel

The usual cleaning of parts to be treated precedes the blackening. All scale, grease, oil or other foreign substances are removed. Meanwhile a molten solution of dichromates, preferably sodium dichromate, should be raised to a temperature between 730° and 750°F. Naturally, treatment time will vary depending on the number and size of the parts and the grade of stainless from which they are made. Stabilization occurs in about ½ hour. The parts are then removed from the hot bath, allowed to cool to room temperature and rinsed in hot water. The salts used are readily soluble and it is simple to remove them. Dry with air.

Pretreatment of Stainless Steel for Electroplating

Clean the parts in an alkaline cleaner. Dip for 30 seconds in a solution of:

$NiCl_2 \cdot 6H_2O$ 16 oz.
HCl 6N   To make 1 gal.

Rinse and plate.

Fast Nickel Plating

Nickel sulfate, 200 g. per l.; nickel chloride, 175 g. per l.; boric acid, 40 g. per l. Optimum operating conditions for a semi-bright plate are: Temp., 115°F.; pH, 1.5. At these conditions a current density of 11 amp. per sq. dm. (100 amp. per sq. ft.) is readily usable, resulting in 0.001-in. deposit thickness in 11 min. Twice this current density can be used without burning. The cathode current efficiency at these conditions is 98%.

Silver Plating Steel

Etch the steel for 5 minutes in a 20% sulfuric acid at 200 amp. per square foot, rinse, plate with a minimum thickness of 0.00005 in. (0.00012 cm.) of nickel, rinse, plate with a minimum of 0.0001 in. (0.00025 cm.) of copper from an acid copper bath, rinse, dip 10 to 15 seconds in a solution containing 50 g. per liter sodium cyanide, silver strike, silver plate.

The strike composition found best contains:

Silver Cyanide	4.25 g.
Sodium Cyanide	75.0 g.
Sodium Carbonate	17.5 g.
Water	To make 1 l.

Black Nickel Plating

Nickel Sulfate	10 oz.
Nickel Ammonium Sulfate	6 oz.
Zinc Sulfate	5 oz.
Sodium Sulfocyanide	2 oz.
Water	To make 1 gal.

Use at 1.6 amp. per square foot at 80–90°F.

Indium Plating

Potassium Cyanide	140–160 g.
Indium	15–30 g.
Potassium Hydroxide	30–40 g.
Dextrose	20–30 g.
Water	To make 1 l.

Use at room temperature at 15–30 amp. per square foot with a steel anode.

High-Speed Copper Plating

Bath Composition:

Copper Fluoborate	450 g.
Fluoboric Acid	30 g.
Boric Acid	30 g.
Water	To make 1 l.

Operating Conditions:

pH	Not over 0.6
Temperature	80 to 140° F.
Cathode Current Density	125 to 150 amp. per square foot
Anodes	Rolled Annealed or Electrolytic Copper
Anode Current Density	Same as the Cathode Current Density

Cadmium Plating Solution

Bath Composition:

Cadmium Fluoborate	240 g.
Fluoboric Acid	15 g.
Boric Acid	20 g.
Ammonium Fluoborate	60 g.
Water	To make 1 l.

Operating Conditions:

pH	4.6
Temperature	75 to 100° F.
Cathode Current Density	40 amp. per square foot
Anodes	Pure Cadmium
Anode Current Density	20 amp. per square foot

Lead Plating

Bath Composition:

Lead Fluoborate	400 g.
Fluoboric Acid	40 g.
Boric Acid	40 g.
Water	To make 1 l.

Operating Conditions:

pH	1.0 to 1.5
Temperature	75 to 100° F.
Cathode Current Density	30 to 70 amp. per square foot
Anodes	Chemical Pure Lead
Anode Current Density	10 to 30 amp. per square foot

Tin Plating

Bath Composition:

Stannous Fluoborate	200 g.
Fluoboric Acid	50 g.
Boric Acid	25 g.
Glue	6 g.
Water	To make 1 l.

Operating Conditions:
Temperature	75 to 100° F.
Cathode Current Density	50 amp. per square foot
Anodes	Pure Tin
Anode Current Density	25 amp. per square foot

High-Speed Zinc Plating

Bath Composition:
Zinc Fluoborate	300 g.
Ammonium Chloride	27 g.
Ammonium Fluoborate	35 g.
Water	To make 1 l.

Operating Conditions:
pH	3.5 to 4.0
Temperature	130° F.
Cathode Current Density	Up to 800 amp. per square foot
Anodes	Cast, Pure Zinc
Anode Current Density	One-half of the Cathode Current Density

---

Plating on Plastics

It is of primary importance that the surface to be plated is properly prepared.

Slight Roughening or De-glazing Operation

This procedure has two fundamental purposes: (1) to roughen the surface slightly, or to remove glaze from the plastic article as it comes from the molding operation and (2) to remove the *flash* or feather-edge from the molded piece. This operation is usually performed by wet-tumbling in a mixture of moderately coarse pumice and water, blasting with 220-mesh grain aluminum oxide, or etching the part chemically. The surface preparation procedure used depends also upon the size, shape and number of pieces to be treated and the chemical structure of the plastic. If the parts are small, the wet-tumbling operation is employed, the tumbling barrels rotating at 40 to 50 rotations per minute for 1 to 5 hours. With casein products, kerosene is substituted for water because of the swelling effect of the water on casein plastics. Larger pieces are de-glazed with the aluminum oxide blast in the usual type of blast cabinet.

Chemical roughening methods must be used with extreme care so that the etchant will not cause too severe an etching action. Plastics of the phenol formaldehyde and urea formaldehyde types as well as the cellulose esters can be treated in the following acid etch provided the excess acid is immediately rinsed away and the parts are immersed in a neutralizing solution such as a 10% sodium carbonate solution:

Sulfuric Acid (66° Bé.)	256 cc.
Nitric Acid (40° Bé.)	128 cc.
Hydrochloric Acid (Sp. Gr. 1.2)	1 cc.
Water	32 cc.

Treatment for 1 to 5 minutes in

a 5% to 10% sodium hydroxide solution may be used for etching plastics of the cellulose ester group, also organic reagents, such as a 10% acetone solution.

Cleaning the Surface

The surface of the plastic must be freed from any greasy film, finger marks or contamination following the roughening operation, otherwise a non-continuous, non-adherent and mottled silver film may result. Mild proprietary cleaners such as those used for cleaning non-ferrous metals suffice in the majority of cases. Carbon tetrachloride or solutions of trisodium phosphate, or wetting agents, or mild caustic soda solution may also be used. However, the chemical structure of the material must be considered, in order to avoid chemical reaction between the cleaning agent and the plastic surface. Elevated temperatures may be employed in cleaning, provided these temperatures do not approach the distortion point of the plastic. Hand scrubbing the part with a stiff brush will aid the cleaning operation. The plastic piece should be thoroughly rinsed and not allowed to dry prior to the application of an adherent silver film.

Sensitizing Treatment

Perhaps the most important step in the preparation of the plastic surface to receive a continuous and adherent metallic silver film that is capable of carrying current in the subsequent electroplating operation is the so-called sensitizing treatment. The function of this treatment of plastic surfaces is subject to discussion.

Sensitizing Solution:
Stannous Chloride 10 g.
Hydrochloric Acid
 (Conc.) 40 cc.
Water 1000 cc.

Immerse, with agitation, for 1–2 minutes.

Silver Plating Solutions:
Ammoniacal Silver Nitrate Solution:
Silver Nitrate (C.P.) 60 g.
Distilled Water 1000 cc.
Ammonium Hydroxide
 (28%) 60 cc.

Reducing Solution:
Formaldehyde (40% by Vol.) 65 cc.
Distilled Water 1000 cc.

The concentrated ammonium hydroxide is added after the silver nitrate has completely dissolved in the water. The precipitate first formed will redissolve when the entire amount of ammonium hydroxide is added, and the resulting solution will contain the proper amount of free ammonia. This is important since a relatively small excess of ammonia will prevent any deposition of silver, while too little ammonia will result in an excess of precipitated oxide which is undesirable and must be removed by filtration.

The properly prepared plastic parts, if in bulk, are *bonded* or silvered by placing them in an inclined, rubber-lined container, rotating at 4 to 5 rotations per minute with enough water to completely cover the pieces. The container usually possesses ribs on its inner wall to insure against mass rolling of the parts and incomplete silvering. The proper volume of the ammoniacal silver nitrate solution is then added and an equal volume of the formadlehyde reducing solution. The quantity of the silver solution needed depends upon the total area of the work being bonded. Larger pieces, not suitable for bulk handling, are treated on racks, which can be raised and lowered to bring about the necessary agitation.

The bonding treatment continues until all silver has been precipitated from the ammoniacal silver nitrate solution. This is best determined by removing a sample of the bonding mixture and testing for silver with a 10% sodium chloride solution. The bonding operation usually requires from 20 to 25 minutes to produce a silver film capable of carrying current to receive the intermediate electrodeposit. One such treatment deposits an approximately 0.000006 in. (0.15 millimicron) thick film of silver, as determined by both direct and indirect methods, to be discussed presently.

After a suitable silver film has been applied, as determined by test with an ohmmeter or with a flashlight bulb connected in series with two test prods and a dry cell, the plastic parts are rinsed thoroughly and allowed to dry. On plastics with high water absorption, and on large, intricate parts, a second or even a third silvering treatment may be necessary to obtain sufficient conductivity for subsequent electroplating. Fortunately, the freshly silvered surface, upon immediate immersion in a fresh portion of bonding mixture, is receptive to an additional silver deposit and thus a thick enough coating can be obtained. A thorough drying of the parts after bonding is recommended before electrodepositing the intermediate layer of metal.

Intermediate Electrodeposited Layer of Metal

When the parts are dry, they are usually copper plated in an acid copper electrolyte. It was recommended to apply first the copper in a solution with less free sulfuric acid than in the conventional bath, so that the relatively thin silver film will not be harmed in any way, and then continuing the deposition of copper in the usual acid bath. The recommended formulas are as follows:

Initial Bath:
Copper Sulfate	40–100 g.
Sulfuric Acid	2.5 g.
Water	To make 1 l.
pH	1.8–3.4

Second Bath:
Copper Sulfate	200–300 g.
Sulfuric Acid	15–40 g.
Water	To make 1 l.
pH	0.9–1.1

The thickness of copper deposited may range from 0.0005 in. to 0.010 in. (0.01 to 0.25 mm.) depending upon the design of the piece and the operations which follow, such as bright-dipping, buffing, tumbling, etc.

Metals other than silver and lead may be deposited directly upon the silver bond coating but copper is recommended as the intermediate layer, except in special cases.

METALS AND THEIR TREATMENT

Outer Layer of Metal

The outer layer can be any metal which can be electrodeposited from presently used electrolytes using the usually suitable technics.

Metallographic Etchants for Magnesium

Formula No. 1

Ethylene Glycol	75 cc.
Distilled Water	24 cc.
Nitric Acid	1 cc.

No. 2

Alcohol	100 cc.
Picric Acid	4 g.
Orthophosphoric Acid	0.7 cc.

No. 3

Acetic Acid (10% solution)

Solutions for Polishing Stainless Steel Specimens

Solutions Requiring Low Voltage

(All solutions are given in *weight* percentages unless otherwise indicated; current in amperes per square inch; temperatures in °F.)

Solution		Conditions
Phosphoric Acid	100	1 amp. 100–200° 5 minutes
Phosphoric Acid	80	1–12 amp. 100–220° 5 minutes
Water	20	
Phosphoric Acid	75	3 amp. 100–220° 5 minutes
Water	25	
Pyrophosphoric Acid	400 g.	2 amp. 100–220° 10 minutes
Alcohol To make	1 l.	
Sulfuric Acid Aqueous Solution	75–100	20 amp. 150° 5 minutes
Chromic Acid	43	0.4 amp. 100° 1 hour
Water	57	
Phosphoric Acid	60	5 amp. 175° 3 minutes
Sulfuric Acid	20	
Water	20	
Phosphoric Acid	65	3.5 amp. 200° 5 minutes
Sulfuric Acid	15	
Water	20	
Phosphoric Acid	63	0.35 amp. 120° 1 hour
Sulfuric Acid	15	
Water	22	
Phosphoric Acid	30	0.2 amp. 120° 5 minutes
Sulfuric Acid	60	
Water	10	
Phosphoric Acid	30	1.8 amp. 120° 2 minutes
Sulfuric Acid	60	
Water	10	
Phosphoric Acid	45	0.5–3.5 amp. 85–300° 5 minutes
Sulfuric Acid	40	
Water	15	
Phosphoric Acid	14	No current 120° 5 minutes
Sulfuric Acid	50	
Nitric Acid	36	
Phosphoric Acid	63	3 amp. 120° 1 hour
Sulfuric Acid	15	
Chromic Acid	10	
Water	12	
Phosphoric Acid	67	3 amp. 120° 1 hour
Sulfuric Acid	20	
Chromic Acid	4	
Water	9	
Phosphoric Acid	15	4 amp. 120° 30 minutes
Sulfuric Acid	60	
Chromic Acid	10	
Water	15	
Phosphoric Acid	67	1.7 amp. 120° 1 hour
Sulfuric Acid	20	
Chromic Acid	2	
Water	11	
Phosphoric Acid	44	3 amp. 120° 1 hour
Sulfuric Acid	40	
Chromic Acid	6	
Water	10	
Phosphoric Acid	13	7 amp. 120° 5 minutes
Sulfuric Acid	16	
Glycerol *	56	
Water	15	
Phosphoric Acid	42	0.1–0.6 amp. 200–300° 8–15 minutes
Glycerol	47	
Water	11	
Phosphoric Acid	40	0.5 amp. 212° 10 minutes
Glycerol, Tallow, Benzoic Acid, Picric Acid, or Inhibitors	50	
Water	10	

---

* Other organic compounds, which are high boiling, soluble in phosphoric acid, and contain one or more hydroxyl groups may be added.

Solution		Conditions
Phosphoric Acid	50	3–4 amp.
Sulfuric Acid	40	160°
Solution A (Containing 7 lb. Glycine Dissolved in 1 gal. Hot Water)	10	2–3 minutes
Phosphoric Acid	70	0.7–3.5 amp.
Chromic Acid	10	100°
Water	20	5 minutes
Phosphoric Acid	60	1.8 amp.
Arsenic Acid	15	120–160°
Water	20	40 minutes
Sulfuric Acid	48	3.5 amp.
Glycerol	32	130°
Water	20	5 minutes
Sulfuric Acid	58	1–2 amp.
Glycerol	20	70–100°
Water	22	5 minutes
Sulfuric Acid	73	1.5 amp.
Glycerol or Toluene Sulfonates	20	85°
Water	7	0.5–5 minutes
Sulfuric Acid	50	7 amp.
Glycerol	40	180°
Water	10	8 minutes
Sulfuric Acid	60	14 amp.
Ammoniated-glycyrrhizin	0.01	165°
Methanol	0.55	2 minutes
Water	40	
Sulfuric Acid	15	0.5–40 amp.
Citric Acid	60	125–250°
Water	25	0.5–3 minutes

Add 4% of the total volume of methyl alcohol, or other alcohols, such as, ethyl, propyl, butyl, glycerol, glycol.

Solution		Conditions
Sulfuric Acid	20	0.5–1.5 amp.
Citric Acid	55	180–190°
Water	25	5 minutes
Sulfuric Acid	15	0.5–40 amp.
Citric Acid	55	120–250°
Water	30	0.5–3 minutes
Sulfuric Acid	10	0.3–0.7 amp.
Arsenic Acid	77	140°
Water	13	20–90 minutes
Sulfuric Acid	25	0.3–0.7 amp.
Arsenic Acid	69	140°
Water	6	20–90 min.
Sulfuric Acid	10	0.7 amp.
Arsenic Acid	77	140°
Water	13	60 minutes
Sulfuric Acid	36	0.3–0.7 amp.
Arsenic Acid	47	140°
Water	17	60 minutes
Sulfuric Acid	55	0.5–2 amp.
Hydrofluoric Acid	7	70°
Water	38	0.5–4 minutes
Sulfuric Acid	69	2 amp.
Hydrofluoric Acid	10	70°
Hydrogen Peroxide	21	5 minutes
Sulfuric Acid	73	3.5 amp.
Hydrofluoric Acid	13	70°
Water	14	5 minutes
Sulfuric Acid	73	1 amp.
Hydrofluoric Acid	7	140°
Water	20	5 minutes
Sulfuric Acid	40	Specimen as cathode
Hydrochloric Acid	29.5	170°
Titanium Tetrachloride	5.5	
Nitric Acid	0.5	2–5 minutes
Water	24.5	
Sulfuric Acid	17	1 amp.
Calcium Phosphate (Phosphoric Acid Added)		75° 1.5 minutes as anode
Sodium Sulfate Added to Saturation		1.5 minutes as cathode
Arsenic Acid	84	2.5 amp.
Chromic Acid	10	140°
Water	6	10 minutes

An aqueous acid electrolytic bath containing any of the following ions that form easily soluble salts with the steel may also be used: oxalate, bisulfite, chloride, nitrate, bromide, cyanate, sulfite, acetate, fluoride, iodide, perchlorate, nitrate, chlorate, thiocyanate, bromate, hydrosulfide, and cyanide. Current density sufficient to remove solid anodic oxidation products from the surface. 50–100°

Solutions Requiring High Voltage

(All solutions are given in *volume* percentages; current in amperes per square inch; temperatures in °F.)

Solution		Conditions
Perchloric Acid	75	2–4 amp.
Water	25	50–85°
		5 minutes
Perchloric Acid	32	0.4 amp.
Acetic Anhydride	64	85° max.
Water	4	5 minutes

Solution		Conditions
Perchloric Acid	19	0.6–4 amp.
Acetic Anhydride	76	85° max.
Water	5	4–10 minutes
Perchloric Acid	34	0.5 amp.
Acetic Anhydride	66	85° max.
		5 minutes
Perchloric Acid	20	13–40 amp.
Ethyl Alcohol	80	95° max.
		10–15 sec.
Perchloric Acid	66	30 amp.
Ethyl Alcohol	80	85° max.
Water	14	30 sec.
Nitric Acid	34	10 amp.
Methyl Alcohol	66	90–100°
		1 minute

Strip Coating

The formula for an excellent strip coating material used in masking the terminal wire of a resistor which is coated with a ceramic insulating coating by baking at a temperature of 100°C. is:

Acrawax C	100
Cellosolve Ricinoleate	22
Ethyl Cellulose (N 100)	10

This strip coating successfully operates under this severe temperature condition and is rapidly stripped off after the operation.

Stripping Copper, Brass or Zinc from Iron

Immerse the object at room temperature in a solution of:

Chromic Acid	500 g.
Sulfuric Acid	50 g.
Water	To make 1 l.

The iron base will not be attacked by this solution. An hour is usually sufficient to dissolve thin layers of surface metals. This solution should not be used on zinc base objects.

Stripping Copper from Iron or Zinc Alloys

Use a lime-sulfur solution, such as is used for spraying of orchards, or make a sodium sulfide solution as follows:

Sodium Hydroxide	100 g.
Sulfur	170 g.
Water	1 l.

Keep for ½ hour at 80°C.

Immerse the object in this hot solution for 10 minutes to 3 hours, depending upon the thickness of the copper to be removed, or overnight in a cold solution. This treatment changes the copper to copper sulfide. The object should be washed vigorously to remove as much as possible of the copper sulfide and then should be immersed in a warm sodium cyanide solution which will dissolve the rest of the copper sulfide. If there are any thick spots where the copper remains, repeat the treatment.

Stripping Metallic Coatings

Coating	Basis Metal	Solution	
Cadium	Steel	Hydrochloric Acid	12 oz.
		Antimony Trioxide	2 oz.
		Water	½ pt.

Simple immersion at room temperature. Rapid strip, but smut of antimony is left.

Ammonium Nitrate	1 lb.
Water	To make 1 gal.

Stripping Metallic Coatings—*Continued*

Coating	Basis Metal	Solution	
		Simple immersion at room temperature, leaves clean, smut-free surface.	
Copper	Non-ferrous	Sodium Sulfide	28 oz.
		Sulfur	2 oz.
		Water To make	1 gal.
		Alternate immersion and brushing required to remove loose sulfide, followed by a 10% sodium cyanide dip. (Formula suggested by Bell Telephone Labs.)	
	Zinc	Sodium Sulfide	1 lb.
		Water To make	1 gal.
		Make the work the anode at 2 volts in a solution at room temperature.	
	Steel	Sodium Cyanide	1 lb.
		Water	1 gal.
		Make the work the anode at 6 volts in a room temperature solution.	
Chromium	Steel or Nickel	Sodium Hydroxide	6 oz.
		Water To make	1 gal.
		Reverse current at 6 volts.	
	Brass, Copper or Nickel	Sulfuric Acid	1 gal.
		Glycerin	1 oz.
		Water	1 pt.
		Make the parts the anode at 6 volts. Copper sulfate crystals may be added instead of glycerin in ratio of 4 oz. per gallon. The solution will remove nickel undercoats.	
		Hydrochloric Acid	1 pt.
		Water To make	1 gal.
		Use concentrated acid at room temperature; diluted acid at 125° F. Nickel undercoats will become passive and must be reactivated before rechroming.	
Gold	Copper Alloys, Nickel Alloys, Ferrous Metals	Sodium Cyanide	2 oz.
		Water	1 pt.
		Hydrogen Peroxide	$\frac{1}{2}$ fl. oz.

Stripping Metallic Coatings—*Continued*

Coating	Basis Metal	Solution	
		Use the solution by the pint and strip only a few pieces at a time to prevent violent gassing.	
Nickel	Steel	Sulfuric Acid	1 gal.
		Glycerin	1 oz.
		Water	1 pt.
		Make the parts the anode at 6 volts. Copper sulfate crystals may be added instead of glycerin in ratio of 4 oz. per gallon.	
	Brass or Copper	Fuming Nitric Acid, Hydrochloric Acid	2 oz.
		Water	1 gal.
		This is an excellent nickel strip but extreme precautions must be taken to prevent dilution of the acid and subsequent attack on the underlying metal.	
		Use gas carbon cathodes and reverse current.	
Silver	Brass or Nickel Silver	Sulfuric Acid	19 parts by vol.
		Nitric Acid	1 part by vol.
		Simple immersion in water-free solution at 180° F.	
	White Metal	Sodium Cyanide	4 oz.
		Water	To make 1 gal.
		Reverse the current at 4 volts.	
Tin	Steel, Copper or Brass	Hydrochloric Acid	12 oz.
		Antimony Trioxide	2 oz.
		Water	1 gal.
		Simple immersion at room temperature.	
	Brass, Bronze or Copper	Ferric Chloride	10-14 oz./gal.
		Copper Sulfate	18-21 oz./gal.
		Acetic Acid, 56%	40-60 fl. oz./gal.
		Reactivate by additions of hydrogen peroxide to oxidize the reduced iron.	

NOTE: Navy Aeronautical Specification PC-12 recommends removal of tin coatings in an alkaline cleaner rather than in an acid bath.

Stripping Metallic Coatings—Continued

Coating	Basis Metal	Solution	
Zinc	Steel	Hydrochloric Acid	1 gal.
		Antimony Trioxide	3 oz.
		Simple immersion at room temperature.	
	Brass	Hydrochloric Acid	15 oz.
		Water	To make 1 gal.
		Simple immersion at room temperature.	

GENERAL NOTE: The majority of plated coatings, such as brass, bronze, cadmium, copper, zinc, gold and silver, are soluble in cyanide and can be stripped electrolytically if made the anode at 6 volts in the following solution:

Sodium Cyanide		12 oz.
Sodium Hydroxide		2 oz.
Water		To make 1 gal.

Stripping Oxide Films from Aluminum
U.S. Patent 2,353,786
Immerse in

Sulfuric Acid	8.2
Phosphoric Acid	5.2
Chromic Acid	2.0
Water	84.6

Rinse in clear cold water, then dip into hot water.

Cleaning Magnesium Welds

Immerse in a 5 to 10% caustic soda solution at 150°F. for 5 minutes, rinse thoroughly in cold water, immerse in 20% chromic acid at 150°F. for 2 minutes; rinse thoroughly in cold running water; dry by air blast.

Aluminum Cleaner
U.S. Patent 2,409,271

Sodium Sulfate	20
Nitric Acid	10
Water	70

Pickling (Rust Removing) Solution
U.S. Patent 2,360,509

Phosphoric Acid	7–10
Sodium Bisulfate	3–5
Sodium Acid Phosphate	5–7
Sulfonated Mineral Oil	0.1–0.5
Water	80–100

Pickling Bath Inhibitor
Canadian Patent 434,365

Di-o-Xenylthiourea	5–10
Sodium Decylbenzene Sulfonate	1.2–10
Glauber's Salt	75–85
Bone Glue	3–10

Cleaning Steel Tools and Equipment

A polish for cleaning steel so as to bring back the original luster of machinery, tools or equipment can be made as follows:

Oxalic Acid	1	oz.
Boiling Water	1	pt.
Pumice Stone (Finely Powdered)	1	oz.
Ammonia	1½	oz.

Place in a bottle. Moisten a piece of felt with this solution and rub the steel briskly. Be sure to shake the bottle well before using.

Metal Cleaner
U. S. Patent 2,386,789

Solvent (⅔ Ethylene Dichloride, ⅓ Propylene Dichloride)	20.0
Soap (Vegetable Oil Soap)	10.0
Penetrant (Tar Acid Oil Containing Not Less Than 50% Tar Acid)	33.0
Coupling Agent (Butyl Alcohol)	8.0
Antifoaming Agent (Denatured Alcohol)	6.0
Preservative (Finely Powdered Rosin)	1.0
Blending and Thinning Agent (Sodium Bichromate)	2.0
Water (Sufficient to dissolve the soap)	20.0

Radiator Rust Remover
Formula No. 1

Sodium Bisulfate	8
Oxalic Acid	8

Dissolve in water and add to radiator. Run motor 1–2 hours. Drain and flush well.

No. 2

Potassium Dichromate	10	oz.
Sodium Hydroxide	2	g.

Rust Remover for Iron and Steel

Ammonium Citrate	15	g.
Water	1	pt.

Mix this solution thoroughly and immerse the iron or steel objects to be cleaned therein. Let stand ½ to 3 hours depending on the depth of the rust, remove, wash in boiling water, dry thoroughly, and oil to prevent further rusting.

Rustproofing Steel

Alcohol 70% containing either 0.15% of sodium bicarbonate or 1% of liquor ammon. fort. B.P. is satisfactory; also aqueous solutions of liquor ammon. fort. B.P. 1%, sodium nitrite 1%, potassium hydroxide 0.1% or 1%. All these solutions prevent rust formation for well over 12 months.

Rust Inbibitors for Steel
(Apply prior to painting)

Formula	No. 1	No. 2
Phosphoric Acid	7.00	12 gal.
Sodium Dichromate	0.75	—
Caramel Coloring	0.50	—
Cellosolve	0.50	6 gal.
Water	91.25	160 gal.
Chromium Sulfate	—	25 lb.
Areskap (Wetting Agent)	—	1½ gal.

No. 3
U. S. Patent 2,374,565

Petrolatum (140–150° F. M. P.)	50–70
Petrolatum (150–165° F. M. P.)	0–15
Cylinder Stock	10–25
Viscous Pale Oil	5–15
Wool Fat	1–8
Aluminum Stearate	0.4–1.0
Butyl Stearate	1–5

U. S. Patent 2,400,573

	No. 4	No. 5	No. 6
Diphenyl Phosphate	5	—	—
Mineral Oil	95	99.75	99.75

Tritolyl
  Phosphate — 0.25 —
Triisobutyl
  Phosphate — — 0.25

Steel sheets are dipped into any of the mixtures (No. 4, No. 5, No. 6) heated to 180°C. and then drained.

Slushing (Metal Protective) Grease
U. S. Patent 2,359,946

Microcrystalline Wax	15.00
Paraffin Wax	10.00
V. M. & P. Naphtha	74.85
Aluminum Stearate	0.15

Protective Treatment for Tinplate

Trisodium Phosphate	20
Sodium Dichromate	8
Sodium Hydroxide	20
Wetanol (Wetting Agent)	3
Water	1000

Immerse the tinplate in the mixture for 10–60 seconds at 70–80°C., drain and rinse with water.

Protecting Aluminum Against Corrosion
U. S. Patent 2,364,964

A process for protecting an aluminum or an aluminum alloy article against corrosion comprises immersing the article to be protected in a bath consisting of approximately 30 g. of sodium aluminate, 10 g. of sodium silicate and 8 g. of caustic soda per liter of solution having a pH between 8.5 and 12 and then passing between the article which serves as an electrode in the bath and another electrode therein an electric current having an initial potential greater than 30 volts.

Corrosion Protection of Magnesium Alloys

Before using any one of the protective treatments, the metal surface must be thoroughly cleaned. All grease, oil, oxide film, casting skin, paint, or welding flux must be removed. Grease and oil may be removed by the use of an organic solvent, or by vapor degreasing. When liquid solvents are used, the work should be given a final rinse in a clean, unused solvent.

Oxide films, casting skin, and previously applied paint or other finish may be removed by abrasive equipment. On sheets and extrusions, sanding, wirebrushing, or scrubbing with steel wool will give a satisfactory cleaning. Rough unmachined castings may be sandblasted or cleaned in a nitric-sulfuric acid pickle. Machined parts are best cleaned in a chromic acid bath.

Old paint is removed by scraping, sanding, or with an alkaline liquid paint remover, such as caustic soda solution. If a solvent-type paint remover is used, it should be followed by a wash with a wax-free solvent. Sandblasting will remove old paint coatings from heavy parts.

Welding flux is removed with hot water and a wire brush. Where the areas cannot be reached with a brush, a high-velocity jet of water may be used.

A joint specification issued by the Army and Navy requires that all chemicals used in cleaning and treating magnesium be free from salts of heavy metals. A final cleaning in a strong alkali cleaner, similar to those used for steel, is demanded just prior to the surface treatment.

The nitric-sulfuric acid pickle, used for the cleaning of sand castings, is used with sandblasted parts and with those showing surface oxidation also. It must be used before machining, however, as an appreciable amount of metal is dissolved. In using the pickle, the work is immersed in it for a period of 10 seconds, then rinsed thoroughly in running water, and finally dipped into hot water to speed drying.

Composition of the nitric-sulfuric acid pickle is:
    Sulfuric Acid (concentrated) 2 parts by volume
    Nitric Acid (concentrated) 8 parts by volume
    Water 90 parts by volume

The chromic acid bath for use with machined parts is composed of:
    Chromic Oxide (Cr_2O_3) 1½ lb.
    Water 1 gal.

Parts to be cleaned are immersed in the bath for about 15 minutes at room temperature, or for 1 minute at 200°F.

Chrome-Pickle Treatment

In using the chrome-pickle treatment, the work is immersed for from ½ to 2 minutes in a bath of the following composition:
    Sodium Dichromate   1.5 lb.
    Nitric Acid   1.5 pt.
    Magnesium Sulfate   0.2 lb.
    Water   To make 1.0 gal.

The bath is used at room temperature. The magnesium pieces are then exposed to the air for 5 seconds and washed in cold running water, followed by a dip in hot water to assist in drying. Welded sections should be allowed to remain in the solution somewhat longer, and agitation is used during immersion.

When removed from the wash water, the work has a red-and-yellow iridescence. If it should show a bright yellow coloration, an excess of nitric acid in the bath is indicated.

Sealed Chrome-Pickle Treatment

For the sealed chrome-pickle treatment, parts from the chrome-pickle treatment are immersed for 30 minutes in a boiling solution composed of:
    Potassium or Sodium Dichromate   10 to 20
    Magnesium or Calcium Fluoride   0.25
    Water   To make 100

Hot and cold water rinses follow the chemical treatment. Paint immediately after drying.

Dichromate Treatment

In applying the dichromate treatment, the parts to be treated are immersed for 10 minutes in a 15 to 20% solution of hydrofluoric acid by weight, used at room temperature, then rinsed thoroughly in cold running water, and dipped into hot water to speed drying.

A 5% solution of sodium acid fluoride, potassium acid fluoride, or ammonium acid fluoride may replace the hydrofluoric acid bath, provided the work has not been sandblasted. The substitute bath requires 15-minute immersions at room temperature.

Galvanic Anodizing

When using the galvanic anodizing treatment, the work is given the hydrofluoric acid dip as in the preceding method, rinsed thoroughly in cold water, then suspended in a steel tank or a tank containing steel cathode plates, in the following solution:

Ammonium Sulfate	4	oz.
Sodium Dichromate	4	oz.
Ammonia	1/3	fl. oz.
Water	To make 1	gal.

The magnesium parts are externally connected to the steel by electrical connectors, and a current not exceeding 10 amp. per square foot of anode is passed, and maintained for about 30 minutes. To obtain a satisfactory coating, the treatment should be about 70 amp.-minutes or more.

Either the dichromate or galvanic anodizing treatment may be applied to work previously chrome-pickled, or to assemblies containing brass, bronze, cadmium or steel. Aluminum inserts are rapidly attacked by the hydrofluoric acid dip, so the acid fluoride baths must be substituted for it when parts of that metal are in the assembly.

Chrome-Sulfate Treatment

The cleaned parts are immersed in a bath at 150 to 160°F. for 10 to 20 minutes for the chrome-sulfate process. The bath is composed of:

Magnesium Sulfate	8.0 oz.
Potassium Dichromate	5.3 oz.
Water	To make 1.0 gal.

A cold-water and a hot-water rinse follow. For small scale operations, the bath may be used at room temperature, increasing the time to 30 to 60 minutes. Glass or earthenware tanks may be used, or iron or aluminum containers, although in the latter case the work dare not be allowed to touch the metal of the tank.

Chrome-Alum Treatment

Treatment in a boiling chrome-alum bath for from 2 to 15 minutes produces a smooth, adherent, compact film, brown to black in color. The bath consists of:

Chrome-Potash Alum	4.0 oz.
Sodium Dichromate	13.3 oz.
Water	To make 1.0 gal.

Development of the black color indicates completion of the treatment.

Modified Alkali-Chromate Treatment

For the modified alkali-chromate treatment, the cleaned parts are held for 5 minutes in a 15 to 20% solution of hydrofluoric acid, washed thoroughly, then placed in a boiling bath composed of:

Ammonium Sulfate	4	oz.
Sodium Dichromate	4	oz.
Ammonia (Conc.)	1/3	fl. oz.
Water	To make 1	gal.

After 45 minutes in the bath the parts are removed, washed well in cold running water, and placed in a boiling solution of arsenous oxide, 1 1/3 oz. per gallon. The work is held here for 5 minutes or more, then removed and washed with cold and hot water.

Borax Treatment

A 30-minute treatment in a boiling solution of borax and sodium bicarbonate will produce a light gray film on cleaned magnesium pieces, especially the magnesium-manganese and certain wrought alloys. Because of its light color, this film may be dyed to produce a decorative finish.

Caustic-Pressure Treatment

When painting cannot be applied to the treated work, the caustic-pressure process may be considered for the treatment. The pieces are subjected to a pressure of about 225 lb. per square inch in an autoclave with a caustic soda solution of 13.3 oz. per gallon, for 3 to 5 minutes. This film also is suitable for dyeing, and a dye bath may follow after thoroughly washing the caustic-treated work, or certain dyes may be used in the bath.

Tarnishproof Finish for Silver
U. S. Patent 2,400,784

Treat silver with a 2% water solution of

Diethylenetriamine	103
Oleic Acid	282

Warm together and mix. Such a coating will give protection about a month.

Dental Amalgam Alloy

Silver	50.00
Tin	33.33
Gold	16.67

Anti-Friction Aluminum Alloy
Russian Patent 57,946

	%
Nickel	2–3
Iron	< 1
Silicon	< 1
Aluminum	To make 100

Non-Oxidizing Bursting Disc Alloy
U. S. Patent 2,353,254

Silver	96
Copper	4

Printing Type Alloy
German Patent 741,743

Aluminum	4%
Magnesium	0.01–0.3%
Zinc	To make 100%

Non-Poisonous Buckshot
U. S. Patent 2,193,664

Zinc	98
Sodium	2

Melt and at 450–590° allow to fall down a 130 ft. shot-tower.

Galvanizing Alloy
U. S. Patent 2,360,784

	%
Lead	0.5 – 1
Cadmium	0.2 –0.5
Antimony	0.05–0.2
Zinc	To make 100

Porous Oil-Absorbing Metal
U. S. Patent 2,182,741

Copper	87
Tin	10
Soybean Meal	1
Graphite	1
Borax	1

Press together and heat to 480–590°C. and then to 835–850°C. This will absorb 16–18% (vol.) of oil.

Platinum Substitute
(For dental work)
U. S. Patent 2,198,400

Silver	56
Palladium	27
Copper	14
Gold	2
Zinc	1

Light Porous Metal-Like Composition
U. S. Patent 2,394,993
Formula No. 1
Shellac	20
Alcohol	80

Mix until dissolved then add

Aluminum Powder	20

Mix and heat slowly to 140°F.

No. 2
Phthalic Anhydride	100
Glycerin	46

Heat to about 200°F. until the reaction is completed, then add

Aluminum Powder	20

Cigarette Lighter Flint
U. S. Patent 2,389,198
Misch-Metal	300
Iron	75–100
Copper	5
Silver, Nickel and Chromium	2–8

Spherical Metal Powders
Metal Filings, chips, or irregular shaped particles (40–100 mesh)	12
Calcium Oxide (325 mesh)	1

The mixture is heated above the melting point of the metal in a hydrogen atmosphere where the metal particles assume a spherical shape. They are prevented from fusing together by the calcium oxide. The metal powder can be removed from the calcium oxide by screening, or leaching with water.

EXAMPLES
Metal	Time	Temperature
Aluminum	15 minutes	1900°F.
Silver	20 minutes	2100°F.
Copper	20 minutes	2200°F.

Tempering Steel (Very Hard)
Water	4
Flour	1
Salt	2

Mix to a smooth paste, then heat the steel until a coating adheres when it is dipped in the mixture. Continue to heat the steel to a cherry red color, then cool in cold distilled water. This forms a very hard temper for certain types of steel.

Anti-Carburizing Composition
British Patent 570,971
Formula No. 1
Fire Clay	70
Cement	5
Soft Soap	10
Graphite	5
Linseed Oil	10

No. 2
Abopon (sodium borophosphate) has been proven satisfactory as a coating on steel during hardening to prevent decarburization. When used in vertical type furnaces, the price difference between Abopon and the borax usually used is too great to warrant its use. However, in oven type furnaces, Abopon will pay for itself, in that it apparently has weaker fluxing action on tile cement linings of furnaces than borax. When borax is used it falls off the steel in the oven type furnace on the bottom of the furnace, causing softening of the tile cement.

Foundry Core Wash
Formula No. 1
Silica Flour	64.0
Bentonite	1.5
Dextrin	3.0
Sodium Benzoate	0.2
Water	31.3

Mix the dry ingredients and add

water slowly, mixing well, before use.

No. 2
(For gray iron castings)

	%
Silica	65–68
Iron Oxide	2
Alumina	6
Tellurium	23–25

This mixture is used to prevent shrinks and to produce a local chill.

Casting Foundry Core
Formula No. 1

Albany Sand	300	lb.
Rosin (Powdered)	5½	pt.
Water	22	lb.

No. 2

Silica Sand	15 parts by volume
Truline Binder	1 part by volume
Water	6% by weight

No. 3

Bentonite	3
Dextrin	6
Silica Flour	91

Mix dry and, before using, mix with water to give a putty-like consistency.

Foundry-Core Binder
German Patent 743,092

Sulfite Liquor	95–85
Urea	5–15

Foundry Core Crack Filler

Bentonite	3
Dextrin	3
Silica Flour	94

Wetting Agent for Dental Castings

Sulfonated Castor Oil	10
Propyl Alcohol	90

Magnesium Molding Sand Inhibitors
Formula No. 1

	%
Bentonite	4.0
Sulfur	0.7
Boric Acid	1.0
Diethylene Glycol	0–1
Sodium Fluoride	3.0
Water	2.3–3.7

No. 2

	%
Bentonite	4
Sulfur	2–2.8
Boric Acid	< 1.2
Diethylene Glycol	1–1.5
Water	2–2.5

Foundry Parting Compound
U. S. Patent 2,358,157

Fly ash alone or with 2% of stearic acid, aluminum stearate or coconut oil is dusted on the patterns and the core boxes to aid removal.

Non-Corrosive Soldering Flux
Formula No. 1
(For tin cans)

Rosin WW Grade	2 qt.
Gum Turpentine	2 qt.
Amyl Acetate	25 cc.

No. 2
U. S. Patent 2,361,867

Pale Rosin	15 lb.
Denatured Alcohol	5 gal.

Mix until dissolved. Mix 3 gal. of the solution with 1 gal. levulinic acid (98–99%).

General Soldering Flux
U. S. Patent 2,379,234

Zinc Chloride	74.8
Ammonium Chloride	19.3
Sodium Fluoride	0.9
Potassium Bromide	5.0

Stainless Steel Soldering Flux

A	Boric Acid	5 parts
	Potassium Fluoride	5 parts
B	Hydrochloric Acid	5 parts
	Water	5 parts

Make a thin paste of A and B and apply to the joint.

Soft Soldering Flux
U. S. Patent 2,179,258

Sodium Fluoride	1
Ammonium Chloride	20
Zinc Chloride	79

Silver Soldering Flux

Potassium Borofluoride	50–75
Boric Acid	50–25

Apply with water as a thin paste.

Welding Flux for Magnesium Alloys
U. S. Patent 2,396,604
Formula No. 1

Calcium Chloride	25–50
Potassium Chloride	20–65
Sodium Chloride	5–40

No. 2
British Patent 567,725

Lithium Chloride	25
Potassium Chloride	35–20
Sodium Chloride	30–45
Potassium Fluoride	10

Aluminum Welding Flux
Formula No. 1

Borax	50.0
Potassium Chloride	25.0
Sodium Chloride	25.0
Titanium Dioxide	2.5
Sodium Bisulfite	11.3

No. 2

Potassium Chloride	60
Sodium Chloride	12
Potassium Bisulfate	4
Lithuim Chloride	20

Mix with water and apply as a thin paste.

Arc-Welding Flux
U. S. Patent 2,360,716

	%
Silica Sand	30–65
Cryolite	2–15
Manganese Dioxide	10–25
Silico Manganese	10–25

Electric Welding Flux
U. S. Patent 2,194,200

	%
Alumina	42–60
Silica	58–40
Calcium Fluoride	4
Sodium Fluoride	7

Heat to 1200° and grind.

Aluminum Melting and Degassing Flux

Sodium Chloride	40
Potassium Chloride	35
Potassium Borofluoride	15
Sodium Fluoride, or Sodium Silicofluoride	10

Flux for Melting Tin Bronze

The most suitable flux contains 34% borax, 50% sand and 20% cupric oxide. (Copper mill scale is as satisfactory as technically pure copper oxide.) This flux does not attack the melting pots severely, is only slightly fuming at 2200 to 2400°F., and can easily be thickened for removal from the melt so slag inclusions in the metal are rarely found. The improved soundness and mechanical properties of the metal indicate that this flux also has a considerable degassing action. Pre-fusing the flux is preferable to using a dried mixture. The addition of the flux in two stages gives no advantage over adding it all with the charge. The time allowed for reaction between the metal and flux is restricted by the necessity of obtaining the correct casting temperature and avoiding over-oxidation.

Dried sand or fritted mixtures of 3 sand to 1 borax or 3 sand to 1 sodium fluoride are the most satisfactory thickeners for this flux.

White Metal (Pewter) Welding Composition
U. S. Patent 2,333,989

Tin	60.6
Zinc	34.8
Aluminum	1.5
Paraffin Wax	1.5
Stearic Acid	1.5

Aluminum Solder
Formula No. 1
Belgian Patent 447,781

	%
Tin	22–45
Cadmium	20–50
Silver	10–20

No. 2
U. S. Patent 2,364,402

	%
Tin	30–45
Cadmium	35–50
Silver	10–20

No. 3
Canadian Patent 442,421

	%
Tin	56–92
Cadmium	2–24
Zinc	3–16
Copper	1½– 5
Silver	1½– 3

No. 4
British Patent 571,208

Tin	30–45
Cadmium	35–50
Silver	5–20

Aluminum Foil Solder
British Patent 555,720

Bismuth	500–550
Lead	60–120
Tin	180–200
Cadmium	80
Silver	50

Wiping Solder
U. S. Patent 2,372,745
Formula No. 1

Tin	30
Silver	1¼
Antimony	2
Lead	To make 100

No. 2
U. S. Patent 2,351,477

Antimony	0.5
Arsenic	0.1
Tin	13.0
Bismuth	23.0
Lead	63.4

Brazing Solder
U. S. Patent 2,355,067

	%
Silver	10–15
Cadmium	10–15
Copper	55–70
Zinc	7–12
Sodium	0.05– 1
Phosphorus	0.04–1.5

Solder for Zinc
German Patent 744,790

	%
Tin	3–12.7
Phosphorus	0.001– 0.3
Cadmium	3–10
Lead	To make 100

Low Melting Solder
Formula No. 1

Cadmium	85
Zinc	15

No. 2

Cadmium	62
Lead	18
Zinc	20

No. 3

Tin	50
Lead	32
Cadmium	18

Soft (Berzelit) Solder
Formula No. 1

Cadmium	75
Zinc	15
Lead	10

No. 2

Cadmium	17
Zinc	1
Lead	82

No. 3
British Patent 555,844

	%
Tin	11–62
Silver	0.2–1.3
Lead	To make 100

Silver Solder
U. S. Patent 2,374,183

	%
Silver	40–50
Copper	15–19
Zinc	14–20
Cadmium	18–24
Tin and Lead	0.1– 5

Bonding Cast Aluminum to Steel

The steel part that is to have aluminum bonded to it is first degreased, and pickled in 30% hydrochloric acid solution. Then it is tinned in a bath of molten tin using a flux composed of 25% ammonium chloride and 75% zinc chloride. When molten aluminum is poured against the tinned surface, fusion takes place between the aluminum and tin resulting in an alloy bond.

Porous Metal Filters

*Spherical Copper Powder (40–60 mesh)	85–95
Powdered Tin (–325 mesh)	5–15

The copper powder is first given a coating of castor oil by washing it with a 5% solution of castor oil in alcohol and drying. Add the tin, and mix thoroughly. The tin will adhere to the castor oil on the copper particles, and will produce a very even distribution of tin on each particle. Next the mixture is poured into a hollow graphite block, and heated in a hydrogen atmosphere to 1500°F. The resulting slug will have the appearance of brass, and the copper particles will be firmly brazed together. The porosity can be controlled by variations in the particle size of the copper. Higher tin contents will produce stronger slugs.

Removing Carbide Tips From Tool Shanks

A solution of 80% nitric acid in 20% water at 175 to 195°F. is very useful for removing brazed carbide inserts from steel shanks. The carbide is not attacked by the solution, neither is the shank. The latter is clean and can be retipped without machining. It takes from 30 to 90 minutes depending on the size of the tool. Some precautions are necessary:

Preheat the tool to 200°F.; otherwise violent foaming will occur.

Place the tool so the tip falls out as soon as it is loose.

If the tool is removed for inspection, wash and reheat it before putting it back into the hot acid.

*See Spherical Metal Powders.

Immerse only a little beyond the tip.

The operator should wear goggles, acidproof gloves and apron, and keep some commercial ammonia handy to dash on any accidental splash reaching the person. Fumes should be carried off through a hood. Handle acid in stainless steel breakers.

Chemical Sharpening of Files

Files that are not too badly damaged may be sharpened by immersing them in an acid solution. Prior to attempting to resharpen a file, it should be thoroughly cleansed of oil and other particles of dirt, preferably by the use of a solvent. The files are then resharpened by immersing them for 20 to 25 minutes at room temperature in a solution composed of the following:

Sulfuric Acid	7 oz.
Copper Sulfate	2 oz.
Borax	2 oz.
Water	1 pt.

When the files are removed, they will be covered with a sludge. This should be removed by a vigorous wire brushing.

Flotation Reagent
(For Molybdenum Ore)
Formula No. 1

Pale Neutral Mineral Oil	1.000 lb.
Pine Oil	0.028 lb.
Syntex (Sulfated Glyceride)	0.050 lb.

The above quantity is used per ton of ore.

No. 2
(For Mercury Antimony Fluorite Ore)

Soda Ash	0.20 kg.
Copper Sulphate	0.75 kg.
Butyl Xanthate	0.15 kg.
Pine Oil	0.03–0.04 kg.

This quantity is used per ton ore.

Chapter XIV

PAINT, VARNISH, ENAMEL, LACQUER AND OTHER COATINGS

Exterior House Paints

Outside White House Paint Formula

Titanium Dioxide	116 lb.
Leaded Zinc (35%)	463 lb.
Basic Carbonate White Lead	116 lb.
Magnesium Silicate	270 lb.
Refined Linseed Oil	59 gal.
Q-Bodied Linseed Oil	6.5 gal.
Petroleum Spirits	6.5 gal.
Combination Drier	2.5 gal.

White house paint should not be tinted.

House Paint Base for Tints

Titanium Dioxide (Chalk-Resistant)	179 lb.
Leaded Zinc (35%)	452 lb.
Magnesium Silicate	272 lb.
Raw Linseed Oil	59¼ gal.
Q-Bodied Linseed Oil	5¾ gal.
Petroleum Spirits	7 gal.
Combination Drier	3 gal.

This is a good house paint base for tints. It can be tinted with oil colors.

White Primer Undercoat

A primer-undercoat formula which has good sealing properties for use on new wood or old porous surfaces is given below.

Titanium Barium Pigment	354 lb.
Basic Carbonate White Lead	354 lb.
5X Asbestine	100 lb.
3X Asbestine	135 lb.
Litharge	9 lb.
Raw Linseed Oil	28 gal.
Litharge-Treated Linseed Oil Viscosity U-W. 70% N.V.	14 gal.
Ester Gum-Dehydrated Castor Oil Varnish (5-Gal.)	7 gal.
Petroleum Spirits	32 gal.
Cobalt Naphthenate Drier (6%)	4 fl. oz.

Brick and Stucco Paints

Brick and Stucco White Paint

Rutile Titanium Dioxide	116 lb.
Asbestine	270 lb.
Co-Fumed Leaded Zinc (35%)	463 lb.
Basic Carbonate White Lead	116 lb.
Pale Refined Linseed Oil	32 gal.
*Cut Z-4 Linseed Oil	19.75 gal.
Petroleum Spirits	31.5 gal.
Cooked Paint-Type Drier	1.125 gal.

Brick and Stucco Paint Base for Tints

Chalk-Resisting Titanium Dioxide	179 lb.
3X Asbestine	272 lb.

* 80% solids by weight in petroleum spirits.

PAINT, VARNISH, ENAMEL, LACQUER AND OTHER COATINGS

Co-Fumed Leaded Zinc (35%)	452	lb.
Raw Linseed Oil	29½	gal.
Cut Z-4 Linseed Oil	25	gal.
Petroleum Spirits	18¼	gal.
Paint Drier	1⅝	gal.

Exterior Trim and Trellis Paints

Black Trim and Trellis Paint (Oil Type)
Formula No. 1

Carbon Black	21	lb.
Litharge	4	lb.
Z-2 to Z-3 Dehydrated Castor Oil	21	gal.
Ester Gum* Varnish (50 Gal.)	55	gal.
Petroleum Spirits	21	gal.
Lead Naphthenate Drier (8%)	1	gal.
Manganese Naphthenate Drier (2%)	4	pt.
Cobalt Naphthenate Drier (2%)	6	pt.

No. 2 (Alkyd Type)

Carbon Black	22	lb.
Litharge	4	lb.
Raw Linseed Oil	10	gal.
Long Oil Alkyd	77	gal.
Petroleum Spirits	9	gal.
Lead Naphthenate	1	gal.
Manganese Naphthenate	1⅜	gal.
Cobalt Naphthenate	½	gal.

No. 3 (Green) (Alkyd Type)

Medium Chrome Green	110
Falkovar NS	8
Zinc Oxide (Acicular)	5
Falkyd Solution AA-6	250

* 80% solids by weight in petroleum spirits.

Grind and Thin:

Mineral Spirits	169
Falkyd Sol. AA-6	160
Nuodex Manganese (6%)	1
Nuodex Lead (24%)	7
Pine Oil	3

No. 4
Green Trim Enamel—Utilizing Raw Linseed Oil

Medium Chrome Green	110
Zinc Oxide (Acicular)	5
Falkyd Sol. A-5-D	237

Grind and Thin:

Falkyd Sol. A-5-D	168
Raw Linseed Oil	90
Mineral Spirits	115
Nuodex Manganese (6%)	1.4
Nuodex Lead (24%)	8.8

No. 5 (Brown) (Alkyd Type)

Falkovar NS	2
Falkyd Sol. A-5-D	113
Iron Oxide Brown	200
Mineral Spirits	52
Nuodex Lead (24%)	5

Grind and Thin:

Falkyd Sol. A-5-D	402
Raw Linseed Oil	40
Mineral Spirits	148
Nuodex Cobalt (6%)	2

General Utility Exterior
Green Maintenance Paint

1	Chrome Green Medium	300
	Asbestol Filler	50
	Falkovar YY	328
	Raw Linseed Oil	100
	#10 Mineral Spirits	33
2	#10 Mineral Spirits	127
	Nuodex Cobalt (6%)	4
	Nuodex Manganese (6%)	4
	Nuodex Lead (24%)	10

Grind 1 and thin with 2.

Wagon, Tractor and Implement Paints

Synthetic Primer, Rust-Inhibiting

No. 1

1.
- ZV-3391 Glyptal Solution — 20.0
- Petroleum Spirits — 22.0
- Ester Gum — 3.5
- Zinc Chromate — 3.4
- Red Iron Oxide — 16.6
- Magnesium Silicate — 27.5

2.
- Heavy Bodied Blown Oil — 6.0
- Lead Drier (24%) — 0.7
- Cobalt Drier (6%) — 0.3

Grind 1 in a pebble mill and add 2.

No. 2

1.
- ZV-3391 Glyptal Solution — 20.0
- Ester Gum (50% in Petroleum Spirits) — 7.0
- Lead Drier (24%) — 0.7
- Zinc Chromate — 3.4
- Red Iron Oxide — 16.6
- Magnesium Silicate — 17.5

2.
- Celite No. 266 or Equivalent — 10.0
- Heavy Bodied Blown Oil — 6.0
- Petroleum Spirits — 18.5
- Cobalt Drier (6%) — 0.3

Grind 1 in a roll mill and add 2.

No. 3

1.
- ZV-3391 Glyptal Solution — 19.1
- Ester Gum (50% in Petroleum Spirits) — 6.8
- Lead Drier (24%) — 0.7
- Yellow Iron Oxide — 9.0
- Lampblack — 1.0
- Magnesium Silicate — 23.0
- Cobalt Drier (6%) — 0.3

2.
- Celite No. 266 or Equivalent — 10.0
- Heavy Bodied Blown Oil — 6.0
- Petroleum Spirits — 24.2

Grind 1 in a roll mill and add 2. Mix the Celite with the ground paste before adding the blown oil and petroleum spirits.

No. 4

- ZV-3550 Glyptal Solution — 36.0
- Petroleum Spirits — 16.50
- Cobalt Drier (6%) — 0.03
- Lead Drier (24%) — 0.47
- Zinc Chromate — 3.30
- Red Iron Oxide — 16.70
- Magnesium Silicate — 27.00

Enamel Gloss White

1.
- ZV-3202 Glyptal Solution — 15.0
- Petroleum Spirits — 8.0
- Rutile Titanium Oxide — 25.0

2.
- ZV-3202 Glyptal Solution — 44.0
- Petroleum Spirits — 7.0
- Lead Drier (24%) — 0.8
- Cobalt Drier (6%) — 0.2

Grind 1 in a pebble mill and add 2

Enamel Gloss Gray

1.
- ZV-3550 Glyptal Solution — 15.0
- Petroleum Spirits — 8.0
- Rutile Titanium Oxide — 10.0
- Rutile Titanium-Calcium — 9.8
- Yellow Iron Oxide — 3.0
- Lampblack — 0.2

2.
- ZV-3550 Glyptal Solution — 39.0
- Petroleum Spirits — 14.0
- Lead Drier (24%) — 0.8
- Cobalt Drier (6%) — 0.2

Grind 1 in a pebble mill and add 2.

Black Machinery Enamel

1.
- ZV-3391 Glyptal Solution — 12.0
- Petroleum Spirits — 6.0
- Carbon Black — 2.0
- Lead Drier (24%) — 1.3

PAINT, VARNISH, ENAMEL, LACQUER AND OTHER COATINGS

2
- ZV-3391 Glyptal Solution 65.5
- Petroleum Spirits 13.0
- Cobalt Drier (6%) 0.2

Grind 1 in a roll mill and add 2.

Light Gray Machinery Enamel

1
- ZV-3320 Glyptal Solution 15.0
- VM&P Naphtha 8.3
- Rutile Titanium Oxide 6.4
- Lithopone 14.6
- Magnesium Silicate 10.7
- Carbon Black 0.2
- Lampblack 0.1

2
- ZV-3320 Glyptal Solution 30.0
- ZV-3209 Glyptal Solution 9.0
- Xylol 5.0
- Lead Drier (24%) 0.5
- Cobalt Drier (6%) 0.2

Grind 1 in a pebble mill and add 2.

Olive Drab Lusterless

ZV-3550 Glyptal Solution	34.0
Petroleum Spirits	22.5
Cobalt Drier (6%)	0.1
Lead Drier (24%)	0.4
Yellow Iron Oxide	9.0
Lampblack	1.0
Magnesium Silicate	33.0

Olive Drab Gloss Enamel

1
- ZV-3180 Glyptal Solution 17.0
- Petroleum Spirits 5.8
- Medium Chrome Yellow 17.5
- Red Lead (97%) 8.5
- Zinc Oxide 6.0
- Lampblack 3.0
- Magnesium Silicate 15.0

2
- ZV-3209 or ZV-3263 Glyptal Solution 27.0
- Cobalt Drier (6%) 0.2

Work 1 to a paste in a roll mill and add to 2.

Black Lusterless Enamel

1
- ZV-3320 Glyptal Solution 23.0
- Velvet Lampblack 5.0
- Magnesium Silicate 31.0
- Petroleum Spirits 20.0
- Lead Naphthenate Drier (24%) 0.8

2
- ZV-3320 Glyptal Solution 15.0
- Cobalt Naphthenate Drier (6%) 0.3
- Anti-Skinning Agent (National Aniline A.S.A.) 0.1
- Xylol 4.8

Grind 1 in a pebble mill and add to 2.

Exterior Metal Paints

Synthetic Primers for Ferrous Metals

	Formula No. 1	No. 2
ZV-3550 Glyptal Solution	36.0	....
ZV-3560 Glyptal Solution	....	30.0
Petroleum Spirits	16.5	22.0
Cobalt Drier (6%)	0.1	0.3
Lead Drier (24%)	0.4	0.7
Zinc Chromate	3.3	3.3
Red Iron Oxide	16.5	16.5
Zinc Oxide	2.5	2.5
Lampblack	0.5	0.5
Ground Limestone	12.0	12.0
Magnesium Silicate	12.2	12.2

No. 3
Rust-Inhibiting Primer

ZV-3550 Glyptal Solution	36.0
Petroleum Spirits	16.50

Cobalt Drier (6%)	0.03
Lead Drier (24%)	0.47
Zinc Chromate	3.30
Red Iron Oxide	16.70
Magnesium Silicate	27.00

Multiple Pigment Red Lead Paints

Formula No. 1

A red lead multiple pigment paint from which good service has been obtained on structural steel is:

Red Lead (97% Grade)	1001.0
Iron Oxide 78% Fe_2O_3	331.0
Aluminum Stearate	2.6
Raw Linseed Oil	326.0
Spar Varnish*	143.0
Thinner and Drier†	103.0

No. 2

U.S. Maritime Commission Specification 52MC18 (March 2, 1943)

A red lead multiple pigment paint which will dry hard in 9 hours and set to touch in not more than 4 hours.

Red Lead (97% Grade)	818.0
Basic Carbonate White Lead	86.0
Indian Red (Federal Spec. TT-I-511, Type 1)	50.0
Magnesium Silicate, (52MC11)	186.0
Raw Linseed Oil	160.0
Pale Heat Bodied Linseed Oil (R to S Body)	127.0
Spar Varnish (52MC7)‡	211.0
Thinner (52T9, Grade 1) and Drier §	86.0

* A 30-gal. length estergum-China wood oil varnish meeting the requirements of Federal Specification TT-V-121a.
† Drier: 0.3% lead, 0.2% manganese and 0.005% cobalt based on oil content.
‡ An ester gum spar varnish vehicle of 35-gal. length may be used.
§ Drier: 0.2% lead, 0.02% manganese, 0.005% cobalt based on the oil content.

No. 3

A red lead multiple pigment paint used by a southern railroad for priming steel surfaces conforms to the following formula:

Red Lead (95% Grade) *	738.5
Diatomaceous Silica	98.5
Magnesium Silicate	147.7
Linseed Oil †	375.0
Thinner and Drier ‡	203.2

No. 4

Red Lead 95%	620.0
Magnesium Silicate	399.7
Raw Linseed Oil	322.0
Heat Bodied Linseed Oil Z2	45.7
Mineral Spirits and Drier §	183.3

No. 5

This red lead multiple pigment paint contains leaded zinc oxide. The dehydrated castor oil varnish vehicle helps to reduce the drying time.

Red Lead (97% Grade)	634.0
Leaded Zinc (35%)	176.0
Diatomaceous Silica	35.2
Magnesium Silicate	176.0
Mica (Water-ground)	70.5
Raw Linseed Oil	203.0
Ester gum Varnish §§	263.0
Mineral Spirits	43.9
Lead Manganese Drier	27.3

* The addition of aluminum stearate up to 0.3% by weight of pigment is suggested to maintain suspension of the pigment in the vehicle.
† NOTE: In general a vehicle meeting the requirements described for Federal Specification TT-P-86, Amend. 1, Sept. 1, 1943, meets the requirements called for in this formulation.
‡ Drier: 0.3% lead, 0.02% manganese, 0.01% cobalt based on the oil content.
§ Drier: 0.3% lead, 0.02% manganese, 0.02% cobalt based on the oil content.
§§ A 25-gal. length dehydrated castor oil varnish vehicle containing 45% non-volatile matter.

PAINT, VARNISH, ENAMEL, LACQUER AND OTHER COATINGS

No. 6
A red lead-iron oxide-linseed oil paint.

Red Lead*	696
Iron Oxide	232
Magnesium Silicate	232
Raw Linseed Oil	349
Linseed Oil (Z-2 Body)	26
Thinner and Drier †	169

No. 7

Red Lead 97% Grade	634.5
Iron Oxide	211.5
Magnesium Silicate	158.5
Diatomaceous Silica	53.0
Raw Linseed Oil	185.6
Linseed Oil Z-2	185.6
Lead Drier (16%)	9.1
Manganese Drier (4%)	3.4
Mineral Spirits	184.9

No. 8
This red lead-iron oxide paint will dry moderately fast due to its varnish contents.

Red Lead (97% Grade)	595.5
Iron Oxide	198.5
Magnesium Silicate	148.8
Diatomaceous Silica	49.6
Raw Linseed Oil	174.9
Linseed Oil Z-2	174.9
Congo Varnish ‡	71.7
Lead Drier (16%)	8.8

---

* The addition of aluminum stearate up to 0.3% by weight of pigment is suggested to maintain suspension of the pigment in the vehicle.

† Drier: 0.3% lead, 0.02% manganese, 0.02% cobalt based on oil content.

‡ A 25-gal. congo varnish, viscosity B–C (G–H), weight per gallon 7.35 lb. containing 50% solids is recommended. Heat 100 lb. of fine melt congo resin and 200 lb. of kettle bodied linseed oil (Z viscosity) to 530°F. (time 30 minutes). Let cool to 450°F. with 296 lb. of mineral spirits, after which add 2.1 lb. lead naphthenate (24%), 1.3 lb. cobalt naphthenate (6%) and 9.7 lb. of manganese naphthenate (6%). This varnish contains 33.3% linseed oil by weight.

Manganese Drier (4%)	3.5
Mineral Spirits	149.2

No. 9
A red lead multiple pigment paint formulated similarly to the below has been used extensively on iron and steel surfaces.

Red Lead (95% Grade)	523.0
Carbon Black	4.9
Inert	439.6
Treated Linseed Oil	340.5
Spar Varnish	68.1
Thinner and Drier †	158.8

Rustproofing Coating

U.S. Patent 2,366,486

Graphite	2½	oz.
Glycerin	2	oz.
Alcohol	1	oz.
Manganese Resinate	7	g.
Liquid Japan	220	g.

Marine Paints

Anti-Corrosive Shipbottom Paint

Zinc Oxide	186
Venetian Red	93
Silica	93
WG or N Rosin	145
Coal-Tar Naphtha	380
Coal Tar	47½
Manganese Linoleate	129

Anti-Fouling Shipbottom Paint
No. 1

Leaded Zinc Oxide	12.0
Red Iron Oxide	6.7
Kieselguhr	6.3
Talc	5.0
Mercuric Oxide	6.8
Flaked Copper	8.7

---

† Drier: 0.3% lead, 0.02% manganese, 0.02% cobalt based on the oil content.

Xylol	6.1
Benzine	2.9
Color Lacquer *	45.5
Xylol	8.2

No. 2
(For Submarines)

Leaded Zinc Oxide	12.4
Titanium Dioxide	5.4
Black Iron Oxide	10.5
Paris Blue	1.5
Kieselguhr	4.3
Flaked Copper	8.7
Mercuric Oxide	6.3
Color Lacquer *	40.4
Xylol	7.6
Benzine	2.9

No. 3

Phenolic-Ester Gum Varnish	22.8
Mercurous Arsenite	3.1
Cuprous Oxide	12.3
Talc	19.6
Red Iron Oxide	19.6
Benzine	11.3
Solvent Naphtha	11.3

No. 4

Rosin-Maleic Ester	5.5
Paraffin Pitch	8.0
Mercuric Oxide	6.0
Ester Gum	4.0
Cumarone Resin	1.9
Rosin	4.0
Cuprous Oxide	18.0
Leaded Zinc Oxide	18.5
Milori Blue	1.4
Benzine	20.3
Solvent Naphtha	7.4

---

* Color Lacquer

Rosin	51.7
Cupric Hydroxide	2.1
Copper Naphthenate	10.6
Coal Tar Pitch	12.8
Xylol	7.1
Solvent Naphtha	6.4
Tetralin	4.3
Benzine	4.3
Triethanolamine	0.7

Anti-Fouling Copperbottom Paint for Wooden Vessels

WG or N Rosin	330
Hydrogenated Methyl Abietate	165
Coal Tar Naphtha	98
Mineral Spirits	110
Cuprous Oxide	660
Diatomaceous Silica	110

Black Hull Paint

Carbon Black	36¾	lb.
Litharge	7	lb.
Asbestine	138	lb.
40-Gallon Pentalyn Ester Linseed Oil Varnish	19	gal.
Linseed Oil	51¼	gal.
Mineral Spirits	22	gal.
Lead Naphthenate (8%)	2	gal.
Manganese Naphthenate (2%)	7	pt.
Cobalt Naphthenate (2%)	7	pt.

Light-Gray Hull Paint

XX Zinc Oxide	389	lb.
Barium Base Titanox Pigment	198	lb.
Asbestine	72	lb.
Linseed Oil	18	gal.
40-Gallon Pentalyn Ester Linseed Oil Varnish	44	gal.
Mineral Spirits	16	gal.
Lead Naphthenate (8%)	2	gal.
Manganese Naphthenate (2%)	5	pt.
Cobalt Naphthenate (2%)	3	pt.

Fire-Resisting Canvas Preservative Paint

Antimony Oxide	187	lb.

Titanium Dioxide	47 lb.
Zinc Borate	47 lb.
Aluminum Stearate	1 lb. 12 oz.
Chlorinated Paraffin (42%)	16 gal. 2 pt.
Alkyd Resin (50% Solids)	15 gal. 7½ pt.
Chlorinated Paraffin (60%)	5 gal. 6 pt.
Varnolene	50 gal.
Lead Naphthenate (8%)	6½ pt.
Manganese Naphthenate (2%)	3¼ pt.
Cobalt Naphthenate (2%)	3¼ pt.

Marine Interior Flat Paint

Rutile Titanium Dioxide	382 lb.
Titanium Calcium Pigment	109 lb.
Zinc Oxide	169 lb.
Magnesium Silicate	85 lb.
Antimony Oxide	100 lb.
Alkyd Resin Solution (52-R-13)	28.37 gal.
Petroleum Spirits	23.25 gal.
Heavy Petroleum Spirits	22.85 gal.
Lead Naphthenate Drier (24%)	2.5 pt.
Cobalt Naphthenate Drier (6%)	1.0 pt.

Can be tinted with color, ground in oil, or varnish.

Marine Interior Enamel

Titanium Dioxide	220 lb.
Zinc Oxide	38 lb.
Titanium Calcium Pigment	106 lb.
Soya Lecithin	8 lb.
Alkyd Resin Solution (52-R-13)	68.75 gal.
Dipentene	2.94 gal.
Petroleum Spirits	17.23 gal.
Lead Naphthenate Drier (8%)	0.83 gal.
Cobalt Naphthenate Drier (2%)	0.29 gal.

Can be tinted with colors ground in oil or varnish.

Light Green Fire-Retardant Paint (Semi-Gloss)

Titanium Dioxide	250
Titanox-Calcium	235
Zinc Oxide	170
Magnesium Silicate A	90
Antimony Oxide	100
Aluminum Stearate	8.5
Chromium Oxide Green	2
Grinding Varnish	222
Petroleum Spirits I	281
Lead Naphthenate	7.8
Cobalt Naphthenate	1.0
Manganese Naphthenate	0.5

Waterproof Awning Paint

Crude Beeswax	1 lb.
Rosin	1 lb.
Non-Drying Vegetable Oil	0.5 gal.
Outdoor House Paint	1.25 gal.
Volatile Mineral Spirits (Painters' Naphtha)	3.75 gal.

Beeswax and rosin are melted in heated vegetable oil. The mixture is added to the paint with stirring; when cooled sufficiently, it is diluted with naphtha. Allow it to dry a week between coats. Do not fold for storage until thoroughly dry.

Moistureproof Coating
Araclor 5460	96
Paraffin Wax	4

Emulsion Paints

Gray Water Paint (Exterior) (For Plaster)
Mowolith D32 *	15
Mowolith D	10
Lithopone	15
Titanium White	10
Talcum	5
Chalk	9
Greenish Umber	1
Water	35

Red Emulsion Paint (Exterior) (For Wood)
Mowolith D32	25
Mowolith D	25
Red Iron Oxide	25
Talc	3
Water	22

Brown Emulsion Wall Paint
Mowolith D32	20
Ochre, Dark	20
Titanium White	8
Talc	8
Greenish Umber	4
Water	40

Cream-Colored Emulsion Paint (Interior) (For Wood)
Mowolith D32	24
Methyl Cellulose (4% Solution)	5
Lithopone	24
Chalk	23½
Light Ochre	½
Water	23

* Mowolith D is Mowolith D32 (see) without tricresyl phosphate and dibutyl phthalate.

Polyvinyl Emulsion Coating
Polyvinyl Acetate	38.7
Polyvinyl Alcohol	1.9
Tricresyl Phosphate	11.6
Dibutyl Phthalate	3.3
Water	36.8

Emulsion Paint (Interior)
Falkene 10 (Resin)	115.0
Mineral Spirits	40.0
Caustic Soda Solution (50%)	4.0
Unadjusted Calgon Solution (18%)	1.0
Carbitol	7.5
Ammonia	6.0
Enamel Grade Lithopone	440.0
Methyl Cellulose Solution (2% 400 Cp.)	50.0
Surfex	50.0
China Clay	20.0
Water	20.0

Gloss Emulsion Paint (Varnish Enamel)
Varnish (25-Gal. Length)	100.0
Titanium Dioxide	To suit
Emulsifier *	8.0
Water	5–10

Grind together until uniform. The above is thinned with an equal amount of water for use.

Polyvinyl Acetate Emulsion Paint

Formula No. 1
Copolymer of Vinyl Acetate and Vinyl Chloride	38.0
Dioctyl Phthalate	7.7
Dibutyl Phthalate	1.2
Methanol	3.0
Water	To make 100.0

* Emulsifier consists of:
Nonaethylene Glycol Monolaurate	2–5
Morpholine	3–6

No. 2

Polyvinyl Acetate	38.7
Polyvinyl Alcohol	1.9
Tricresyl Phosphate	11.6
Dibutyl Phthalate	7.7
Alcohol	3.3
Water	36.8

Emulsion Paint Base
Formula No. 1

Alkyd Resin	44.00
Casein	3.50
Montan Wax Acids	2.50
Lead-Manganese Napthenate	1.36
Borax	0.56
Ammonia	1.60
Water	46.54

Prepare the above before use and work in the desired pigment.

No. 2

Rosin	50.0
Montan Wax Fatty Acids	8.0
Potassium Hydroxide	0.5
Ammonia	8.0
Toluene	6.0
Turpentine	4.0
Water	250.0

Prepare the above before use and work in desired pigment.

Oleoresinous Water Paint

Mix thoroughhly:

Low Acid Ester Gum	65.0
Refined Linseed Oil (Acid Value 9–12)	215.0

Heat to 300°F. within 25 minutes.

When dissolved, allow to cool. Add

Monosulph	22

Stir well, add:

Lead Naphthenate (24%)	15
Cobalt Naphthenate (6%)	4

Stir well, add:

Emulphor AG Oil Soluble	7

Mix well, add the following solution:

Tylose S-25	5

Wet with:

Hot Water	30

Chill and add:

Caustic Soda (10% Solution)	31

Stir well and add:

Casein BI (Casein Co. of America)	10

This mixture must be stirred well to dissolve before adding to the foregoing mixture of oils and driers.

Mix well and add the following paste:

Titanox A 168 LO	400
Asbestine X	600
Water	500

Mix and add:

Emulphor ELA	10

Mix very well and grind twice on a three roller mill. For brushing, thin 1:1 with water. This paint possesses excellent stability as paste and in the diluted form.

Water Emulsion Paint
Formula No. 1

Titanox-A No. 168 LO	400.0 lb.
Magnesium Silicate —Low Oil Absorption	600.0 lb.
Varnish (40-Gal.)*	35.0 gal.
Sulfonated Castor Oil (75%)	2.5 gal.
Casein (Muriatic)	10.0 lb.
Sodium Hydroxide Solution (10%)	3.5 gal.
Methyl Cellulose	5.0 lb.
Lead Naphthenate (24%)	1.5 gal.

*A 40-gal. solution of ester gum in high-acid linseed oil.

Cobalt Naphthenate (6%)	0.5 gal.
Water	27.0 gal.

No. 2

Titanox-RC-HT	900.0 lb.
Diatomaceous Silica	70.0 lb.
Varnish (40-Gal.) *	33.0 gal.
Sulfonated Castor Oil (75% Non-Volatile)	5.0 gal.
Emulsifying Agent †	3.5 gal.
Sodium Hydroxide (10%)	3.5 gal.
Starch Solution (25%)	26.5 gal.
Water	27.0 gal.
Lead Naphthenate (24%)	1.5 gal.
Cobalt Naphthenate (6%)	0.5 gal.

No. 3

Titanox-RC-HT	17.0 gal.
Titanox-RA 10 LO	1.5 gal.
Diatomaceous Silica	0.5 gal.
Varnish (10-Gal.)‡	750.0 lb.
Vacuum Bodied Linseed Oil (Vis.-600 Poises)	100.0 lb.
Linseed Oil Fatty Acids	150.0 lb.
Sulfonated Castor Oil (75%)	26.0 gal.
Emulsifying Agent †	5.0 gal.
Sodium Hydroxide Solution (10%)	2.0 gal.
Methyl Cellulose Solution (3%)	5.0 gal.
Water	3.5 gal.
Lead Napthenate (24%)	3.5 gal.
Cobalt Naphthenate (6%)	40.0 gal.

* A 40-gal. solution of ester gum in high-acid linseed oil.
† Octyl alcohol—polyethylene oxide condensation product.
‡ A 10-gal. solution of ester gum in refined linseed oil.

No. 4

Titanox-A No. 168 LO	400.0 lb.
Magnesium Silicate —low oil absorption	600.0 lb.
Emulsifiable Varnish	41.0 gal.
Lead Naphthenate (24%)	1.5 gal.
Cobalt Naphthenate (6%)	0.5 gal.
Emulsifiable Varnish:	
Ester Gum	100.0 lb.
Grinders' Linseed Oil (Acid No. 9–12)	40.0 gal.
Sodium Hydroxide Solution (10%)	1.5 gal.
Anhydrous Sulfonated Castor Oil	11.0 gal.
Soya Lecithin	7.5 lb.

Heat the ester gum and linseed oil to 300°F. in approximately 25 minutes. After the resin is completely dissolved, remove from the fire and add the 10% sodium hydroxide solution slowly. Some foaming will occur; however, it may be beaten down without difficulty. To insure complete removal of the water, the temperature should not be allowed to drop below 240°F. before foaming ceases. Allow to cool to 220°F. then add the anhydrous sulfonated castor oil and the soya lecithin.

All of these formulations, with the exception of No. 4, may be thinned with half their volume of water.

No. 4 contains no water but will tolerate a 100% thinning with either water or other thinners.

Tinting of these border-line emulsion paints is best accomplished with dry colors which have been ground in a similar emulsified vehicle, although some of the common colors-in-oil are suitable. Due to the fact that some colors-in-oil are incompatible, however, it is best to proceed with caution. Water colors may be stirred into the emulsion paint even after complete thinning and are a good means of producing light tints. This method should not be used to produce dark tints, as the large addition which is necessary will have a detrimental effect upon the washability of the paint film.

The preparation of these formulas requires no unusual equipment. The pigments may be either wet down with the varnish and driers and the remaining ingredients added step-wise as they appear in the formula, or they may be added to the composite vehicle after it has been preemulsified by any ordinary means of mixing. When the first procedure is followed, the paste which is formed before the addition of water is quite stiff, but the slight difficulty which might be experienced in mixing is more than off-set by the fact that the consistency of the paste and the wetting action of the emulsifying agents will produce excellent dispersion without grinding. In fact, this dispersing action is so effective that finished white pastes have been tinted to uniform dark shades by merely mixing in dry color with a pony mixer.

Exterior Varnishes

Formula No. 1
Water-Resisting Spar Varnish
ZV-3173 Glyptal Solution	66.0
Low Body Oxidized Linseed Oil	22.0
Hi-Flash Naphtha	10.5
Lead Drier (24%)	1.0
Cobalt Drier (6%)	0.5

No. 2
Glyceryl-Phthalate Spar Varnish
ZV-3202 Glyptal Solution	75.0
Petroleum Spirits	23.7
Dipentene	1.0
Cobalt Drier (6%)	0.3

No. 3
Spar Varnish
Piccolyte S-115	100.0	lb.
Castung 103	40.0	gal.
Apco-#10 Mineral Spirits	60.0	gal.
Nuodex Cobalt (6%)	0.75	gal.
Nuodex Lead (24%)	1.5	gal.

Heat the Castung 103 to 600°F. Hold for 45 minutes. Reheat to 580°F., hold for body. Cool to 450°F., thin, add drier.

No. 4
Clear Varnish
(For Trucks and Floors)
Falkyd Solution A-3	770
Union Solvent 3 or Equivalent	230
Nuodex Lead (24%)	3¾
Nuodex Cobalt (6%)	1¾

No. 5
Chemical-Resistant Spar Varnish
Dyphene 13080	100
Varnish Makers' Linseed Oil	75
China Wood Oil	200
Mineral Spirits	250
Dipentene	50
Lead Naphthenate (24%)	5
Cobalt Naphthenate (5%)	2

Heat Dyphene 13080 and linseed oil to 560°F. Hold 10 minutes. Add China wood oil. Heat to 460°F. Hold about 30 minutes for required viscosity. Cool to 400°F. and thin with mineral spirits and dipentene. Add driers last.

Flat Wall Finishes

On new wood, plaster or composition board it is the usual practice to apply one coat of sealer, one coat of undercoat and one coat of flat wall finishes. The following formulae may be tinted with oil colors to give any desired light tints.

Sealer

Titanox RCHT	200 lb.
Whiting	200 lb.
Asbestine	100 lb.
Aluminum Stearate	5 lb.
Modified Phenolic-Dehydrated Castor Oil Varnish (25-Gal) (Viscosity E, Solids 50%)	70 gal.
Mineral Spirits	15 gal.
Lead Naphthenate Drier (16%)	½ gal.
Cobalt Naphthenate Drier (6%)	⅜ gal.

Undercoat
Formula No. 1

Titanox C	300	lb.
Whiting	125	lb.
Asbestine	75	lb.
Zinc Stearate	10	lb.
Ester Gum-Linseed Oil Varnish (45-Gal.) (Viscosity P-R, Solids 40%)	23	gal.
Gloss Oil (65% Solids)	4	gal.
Mineral Spirits	6½	gal.
Japan Drier	1½	gal.

Yield: 55 Gal.

No. 2

Rutile Titanium Calcium Pigment	720	lb.
Calcium Carbonate	174	lb.

Pentaerythritol Esterified Rosin Varnish (⅔ Dehydrated Castor-⅓ Linseed) (25-Gal.) — 49 gal.
Petroleum Spirits — 16 gal.
Lead Naphthenate Drier (8%) — 1 gal.
Cobalt Naphthenate Drier (2%) — 1 gal.

Flat Wall Paint
Formula No. 1

1 {
Aluminum Stearate	5
Titanox RCHT	700
Surfex	450
Celite 110 or Equivalent	150
Castung 403-Z5	200
Piccolyte S-115 Sol.	120
Apco–18	228

2 {
Apco-#10 Mineral Spirits	81
Nuodex Cobalt (6%)	1.5
Nuodex Lead (24%)	3.5

Grind 1 and thin with 2.

No. 2

Titanox C	800	lb.
Natural Whiting	200	lb
Litharge	10	lb.
Aluminum Stearate	5	lb.
Blown Soya Bean Oil (Z-6)	8	gal.
Kettle Bodied Linseed Oil (Z-2)	8	gal.
Refined Linseed Oil	7½	gal.
Ester Gum Solution (60% Solids)	9½	gal.
Kerosene	10	gal.
Mineral Spirits	43¼	gal.
Cobalt Naphthenate (6%)	¼	gal.

Yield: 125 Gal.

No. 3

Mowolith D32 (see)	15
Methyl Cellulose (4% Solution)	5

Chalk	15
Lithopone	30
Talc	5
Water	30

Interior Semi-Gloss Paint

Surfaces to be finished with semi-gloss should be prepared with the same type of primer and undercoat as given under flat wall paints. The following semi-gloss formulae can be tinted with colored pigments ground in oil or varnish to produce all desired shades:

Formula No. 1

	Titanox RCHT	500
	Surfex	300
	Ponolith LRJ	200
1	Castung 403-Z5	200
	Piccolyte S-115 (60%)	120
	Apco-#10 Mineral Spirits	98
2	Apco-#10 Mineral Spirits	146
	Nuodex Cobalt (6%)	2
	Nuodex Lead (24%)	4½

Grind and thin with 2.

No. 2

Titanium Calcium Pigment	600
Lithopone	400
Zinc Oxide	50
Transparent Litho Oil No. 2/0	160
Dyal 15002	135
Debloomed Kerosene	40
Mineral Spirits	150
Lead Naphthenate (24%)	19
Manganese Naphthenate (6%)	¼
Cobalt Naphthenate (6%)	1

Grind the pigment in Litho Oil and Dyal 15002. Thin the paste with the remainder of ingredients.

Interior Gloss Paints

Surfaces to be finished with interior gloss should be prepared by applying one coat of primer and one coat of undercoat as shown under flat wall paints. The following interior gloss paint formulae can be tinted with colored pigments ground in oil or varnish to produce any desired shade.

Formula No. 1

	Ponolith LRJ	450
	TiPure R-110	100
	Surfex	150
1	Castung 403-Z3	116
	Castung 103	117
	Piccolyte S-115 Solution *	240
2	Apco-#10 Mineral Spirits	65
	Nuodex Cobalt (6%)	2
	Nuodex Lead (24%)	4.5

Grind 1 and thin 2.

No. 2

	TiPure LOCR or equivalent	150	lb.
1	Titanox C or equivalent	400	lb.
	Surfex	200	lb.
	Thinned SKA	37	gal.
	Mineral Spirits	8	gal.
	1000 Vis. Falkovar Blown Soya	2	gal.
	Falkovar K Heavy	10	gal.
2	Ester Gum Cut (10 lb.)	21	gal.
	Mineral Spirits	5	gal.
	Nuodex Cobalt (6%)	½	gal.
	Nuodex Lead (24%)	1	gal.

Grind 1 and thin with 2.

No. 3

Mowolith D32	15

---

* Piccolyte S-115 Solution
   Piccolyte S-115   100 lb.
   Apco-#10 Mineral Spirits   65 lb

Mowolith D (See)	15
Kaolin	10
Lithopone (50% Zinc Sulfide)	10
Talc	3
Titanium White (50% Barium Sulfate)	5
Water	40

Quick-Dry Enamels

Interior Quick-Dry Enamels

Interior quick dry enamels are commonly used for interior trim, woodwork, furniture, toys, etc. Surfaces should be prepared with one coat of primer and one coat of undercoat before applying the finish coat of enamel. Formulae for primer and undercoat are given under flat wall finishes.

Non-Yellowing White Enamel
Formula No. 1

1	TiPure Locr or equivalent	288
	Blk Label 15 Kadox Zinc Oxide	48
	Falkyd Solution B–41	226
	Nuodex Lead (24%)	8
2	Falkyd Solution B–41	340
	Solvesso 3 or equivalent	134
	Nuodex Cobalt (6%)	1.5
	Nuodex Lead (24%)	4.0

Grind 1 and thin with 2.

No. 2

1	Falkovar NS	4
	TiPure Locr	288
	Blk Label 15 Kadox Zinc Oxide	32
	Falkyd Solution B–5	263
	Nuodex Lead (24%)	3
2	Falkyd Solution B–3	147
	Mineral Spirits	159
	Nuodex Cobalt (6%)	3
	Nuodex Lead (24%)	6

Grind 1 and thin with 2.

No. 3

1	ZV–3202 Glyptal Solution	15.0
	Petroleum Spirits	8.0
	Rutile Titanium Oxide	25.0
2	ZV–3202 Glyptal Solution	44.0
	Petroleum Spirits	7.0
	Lead Drier (24%)	0.8
	Cobalt Drier (6%)	0.2

Grind 1 in a pebble mill and add 2.

No. 4

Titanium Dioxide	240
Zinc Oxide	60
Dyal 15001	450
Lead Naphthenate (24%)	3¾
Manganese Naphthenate (6%)	1
Cobalt Naphthenate (6%)	2
Mineral Spirits	220

Grind pigment in 160 lb. Dyal 15001. Thin the paste with remainder of ingredients.

No. 5

1	ZV–3202 Glyptal Solution	15.0
	Petroleum Spirits	8.0
	Rutile Titanium Oxide	25.0
2	ZV–3202 Glyptal Solution	44.0
	Petroleum Spirits	7.0
	Lead Drier (24%)	0.8
	Cobalt Drier (6%)	0.2

Grind 1 in a pebble mill and add 2.

Colored Quick-Dry Enamels

Four Hour Black Enamel

Coresin Black or equivalent	90
Falkyd Sol. A–5–D	425
Mineral Spirits	175
Nuodex Cobalt (6%)	4

PAINT, VARNISH, ENAMEL, LACQUER AND OTHER COATINGS

Nuodex Manganese (6%)	2	
Nuodex Lead (24%)	10	

Fast Air Dry Red Enamel

Toluidine Red	100
Falkyd Sol. J–203	700
Xylol	130
Nuodex Lead (24%)	9
Nuodex Manganese (6%)	5
Nuodex Cobalt (6%)	2¼

Olive Drab

1	ZV–3550 Glyptal Solution	16.0
	Petroleum Spirits	8.0
	Yellow Iron Oxide	9.0
	Lampblack	0.7
	Limestone	6.3

2	ZV–3550 Glyptal Solution	52.0
	Petroleum Spirits	7.0
	Lead Drier (24%)	0.8
	Cobalt Drier (6%)	0.2

Grind 1 in a pebble mill and add 2.

Four Hour Green Enamel

1	Hilton Davis HD–50A–254 Yellow	81.3
	Hilton Davis HD–20A–331 Blue	18.7
2	*Falkyd Solution A–5–D (Reduced)	244.0
	Nuodex Cobalt (6%)	0.2
	Nuodex Lead (24%)	0.5

Mix 1 and add slowly 2.

	Pebble Mill Grind	Black	Red	White
1	ZV–3263 Glyptal Solution	15.0	15.0	15.0
	Terpene Resin	9.5	9.0	6.5
	Petroleum Spirits	10.0	10.0	10.0
	Lead Drier (24%)	1.0	1.0	21.0
	Carbon Black	3.0	....	....
	Toluidine Red	....	8.0	....
	Titanium Oxide (Y-CR or AA)	....	....	26.0
2	ZV–3263 Glyptal Solution	35.0	32.0	26.0
	Heavy Bodied Blown Oil	9.5	9.0	6.5
	Petroleum Spirits	16.7	15.7	8.8
	Cobalt Drier (6%)	0.3	0.3	0.2

Grind 1 in a pebble mill and add 2.

Heavy bodied blown linseed oil provides better drying than a similar soya oil, but may cost more. Heat bodied oils may be used, but they are not as compatible as blown oils with Glyptal resins.

If the bodied blown oil is added as a part of the mill base, the enamels may sag badly. The above enamels brush and flow very well without sagging.

When the grinding is done with a roll or stone mill, the terpene resin should be added as a cold cut or as a cooked varnish in the oil.

---

* Falkyd Sol. A–5–D	61½ gal.
Apco–18	30¾ gal.
Nuodex Cobalt (6%)	2¼ gal.
Nuodex Lead (24%)	6 gal.

Quick-Drying Camouflage Enamel

Neutral Gray

1.
ZV-3391 Glyptal Solution	28.0
VM&P Naphtha	13.4
Rutile Titanium Oxide	6.0
Yellow Iron Oxide	0.5
Lampblack	0.5
Barytes	10.0
Magnesium Silicate	26.0

2.
ZV-3391 Glyptal Solution	15.0
Cobalt Drier (6%)	0.1
Lead Drier (24%)	0.5

Grind 1 in a pebble mill and add 2.

Olive Drab

ZV-3263 Glyptal Solution	40.0
Dark Orange Iron Oxide	8.5
Red Iron Oxide	0.7
Phthalocyanine Blue	0.8
Magnesium Silicate	20.0
Celite 266 or Equivalent	10.0
VM&P Naphtha	19.0
Lead Drier (24%)	0.75
Cobalt Drier (6%)	0.2
Anti-Skinning Agent	0.05

Sea Gray

ZV-3263 Glyptal Solution	40.00
VM&P Naphtha	19.00
Lead Drier (24%)	0.75
Cobalt Drier (6%)	0.20
Anti-Skinning Agent (National Aniline)	0.05
Rutile Titanium Oxide	6.00
Lampblack	1.50
Magnesium Silicate	22.50
Celite No. 266 or Equivalent	10.00
Red Iron Oxide	Tint

Gray Enamel

1.
ZV-3550 Glyptal Solution	15.0
Petroleum Spirits	8.0
Rutile Titanium Oxide	10.0
Rutile Titanium Calcium	9.8
Yellow Iron Oxide	3.0
Lampblack	0.2

2.
ZV-3550 Glyptal Solution	39.0
Petroleum Spirits	14.0
Lead Drier (24%)	0.8
Cobalt Drier (6%)	0.2

Grind 1 in a pebble mill and add 2.

Floor Enamels

It is advisable to treat new wood floors with a sealer before painting. The following is a type of sealer for this purpose.

Clear Liquid Wood Sealer

Formula	No. 1	No. 2
ZV-3375 Glyptal Solution	66.0	45.0
ZV-3290 Glyptal Solution	....	20.0
Xylol	3.0	....
Petroleum Spirits	30.5	34.5
Cobalt Drier (6%)	0.5	0.5

Gray Floor Enamel

Lithopone	170.0 lb.
Rutile Titanium Calcium Pigment	82.0 lb.
Lead Titanate	62.0 lb.
Modified Phenolic Linseed-Dehydrated Castor Oil Varnish (80% Castor Oil, 20% Linseed Oil) (25-Gal.)	89.0 gal.
Lead Naphthenate Drier (8%)	1.5 gal.

PAINT, VARNISH, ENAMEL, LACQUER AND OTHER COATINGS

Manganese Naphthenate
Drier (2%)   1.5 gal.
Cobalt Naphthenate
Drier (2%)   1.5 gal.

Light Oak Floor Enamel

1 { Falkovar NS   10.0
    Ferrite Yellow   200.0
    Falkyd Solution A–3   400.0
    Falkyd Solution A–3   50.0
    Falkote 285 Cut–50%
      Solids in Xylol   150.0
2 { Mineral Spirits   70.0
    Xylol   35.0
    Nuodex Cobalt (6%)   3.0
    Nuodex Lead (24%)   7.5

Grind 1 and thin with 2.

Black Oil-Resisting Floor and Deck Paint

Lampblack   28.0
Magnesium Silicate I   240.0
Phenolic Mixing Varnish   492.0
Aromatic Petroleum
  Naphtha III   20.0
Petroleum Spirits I   104.0
Lead Drier   16.0
Cobalt Drier   14.0
Manganese Drier   3.0
Anti-Skinning Agent of Volatile Type As Required.

Floor Enamel Vehicle

BR–11544 Resin   100.0
Linseed Oil (Z1)   79.0
Cicoil   81.0
Dehydrated Castor
  Oil (Z3)   79.0
Mineral Spirits   286.0

Interior Varnishes

Interior Floor and Trim Varnish (Quick Dry)

Dyphenite 13133   135.0
Dehydrol   130.0
China Wood Oil   100.0
Mineral Spirits   375.0
Lead Naphthenate (24%)   5.0
Cobalt Naphthenate
  (6%)   2.0

Heat Dyphenite 13133 and China wood oil to 525°F. Add 65 parts of Dehydrol. Heat to 565°F. Add the remaining 65 parts of Dehydrol. Heat to 525°F. and hold about 20 minutes for viscosity. Cool to 400°F. and thin. Add driers last.

Mixing Varnish

Piccolyte S–115   100.0   lb.
Castung 403–Z3   25.0   gal.
Apco–#10 Mineral
  Spirits   45.0   gal.
Nuodex Cobalt (6%)   ⅜   gal.
Nuodex Lead (24%)   ¾   gal.

Heat resin and oil to 585°F. Hold for body. Cool to 450°F., thin and add drier.

Quick-Dry Varnish for Four Hour Enamels

1 { Titanox AA or
      equivalent   250.0   lb.
    Kadox Blk Label
    #15 Zinc Oxide   10.0   lb.
    Falkyd Solution
      A–5–D   33.5   gal.
2 { Falkyd Solution
      A–5–D   24.75   gal.
    Falkyd Solution B–3   29.0   gal.
    Mineral Spirits   30.0   gal.
    Nuodex Cobalt (6%)   3.5   lb.
    Nuodex Lead (24%)   4.0   lb.

Grind 1 and thin with 2.

Short Oil Blending Varnish

BR–11544 Resin   100.0
Linseed Oil (Z1)   39.0
Cicoil   40.0
Dehydrated Castor
  Oil (Z3)   40.0
Mineral Spirits   182.0

Heat the resin and linseed oil to

560°F. in 35 minutes and hold at this temperature for 15 minutes. Check with the Cicoil (470°F.). Heat to 540°F. in 10 minutes and add the dehydrated castor oil (490°F.). Cool to 465°F. in 20 minutes and thin.

Tall Oil Varnish
Undistilled Tall Oil	1050.0
Maleic Anhydride	37.0
Polypentek	194.0
Mineral Spirits	252.0

The tall oil and maleic anhydride are charged into a 200-gallon, gas-fired, stainless steel kettle equipped with a turbine type agitator, condenser, receiver, and bottom outlet. The furnace setting is equipped with a centrifugal blower for external cooling.

The mixture is heated with good agitation to 150°C. in 2 hours and held at 150° for 1 hour to form the maleic adduct. The Polypentek is then added and the temperature increased to 275°C. in 2 hours. The charge is cooked at 275°C. for 8 hours, cooled with the help of the blower to 150°C., diluted with the mineral spirits, and discharged into suitable containers.

The mixture is thinned further to 50% solids with a mixture of equal parts of mineral spirits and turpentine.

When 0.05% Co and 0.05% Mn, as naphthenates are added and a 24 hour ageing period permitted, the resulting films dry tack free on glass in 7 hours.

Electrical Insulating Varnish
Whale Oil	10.0
Toluol	30.0
Sulfur	1.0

Apply and bake at 170°C. for 1 hour.

Varnish for the Impregnation of Fiber, Paper or Pulp Products

This very simple mixture can either be cured in an oven at temperatures up to 270°F. or cured by air drying. The following ingredients are used and are mixed by stirring until a homogeneous mixture is obtained.

Poly Pale Resin	40.0
Linseed Oil	5.0
Xylol	55.0

Phenolic Mixing Varnish
Phenolic Resin	100.0
Tung Oil	195.0
Driers	none
Petroleum Spirits to Give 60% ±1% Non-Volatile Matter.	

Ready-Mixed Aluminum Vehicles
Falkote 420 ES (60% sol.)	166.0
Falkovar K Medium	80.0
Mineral Spirits	58.0
Nuodex Cobalt (6%)	0.2

Lacquers

Wood Lacquer for Pianos, Furniture, etc.
Nitrocellulose (high viscosity)	1.5
Film Scrap Nitrocellulose	19.0
Tricresyl Phosphate	2.0
Butyl Acetate	20.0
Butanol	7.0
Fatty Acid—Pentaerythritol Ester	2.0
Toluol	14.0
Cyclohexanol Resin	5.0
Benzol	9.0
Benzine	20.5

Gray Finish Coat for Wood and Metal

Zinc Sulfide	5.8
Lampblack	1.6
Yellow Iron Oxide	1.7
Blanc Fixe	1.5
Talc	4.4
Nitrocellulose	12.5
Cyclohexanol Resin	6.0
Dibutyl Phthalate	8.5
Solvents and Diluent	58.0

Red Primer

Red Iron Oxide	11.0
Blanc Fixe	8.8
Talc	14.0
Nitrocellulose (medium viscosity)	11.5
1,3 Butyleneglycol Phthalate	7.5
Soft Butyl Urethane Formaldehyde	1.9
Methanolethyl Acetate-Butanol	6.2
Ethyl Acetate	7.1
Butyl Acetate	7.0
Butanol	1.7
Tricresyl Phosphate	2.8
Dibutyl Phthalate	4.0
Xylol	16.5

Metal Lacquer

Acryloid B72 (Resin)	50.0
Dibutyl Phthalate	1.0
Xylol	28.0

If necessary thin with L.C. Xylol for spray application.

Clear Polishing Lacquer for Furniture

Nitrocellulose	21.6
Cyclohexanone Resin	4.0
Urea-Formaldehyde plus Adipictrimethylolpropane Ester added before condensation	10.0

Dibutyl Acetate	4.4
Butyl Acetate	19.0
Ethyl Formate	13.1
Xylol	12.4
Benzine	1.7
Ethyl Alcohol	9.0
Ethyl Glycol	4.0
Methanol	2.8

Clear Lacquer

Alcohol	40.0
Ethocel (20 cps.)	10.0
Durez #219 Resin	5.0
Rosin	2.0
Castor Oil	2.5

Stir until dissolved and add:

Sovasol #1 (Mineral Spirits)	80.0

Cloth Waterproofing Lacquer
U. S. Patent 2,188,901

Nitrocellulose	12.7
Butyl Acetyl Ricinoleate	15.3
Glyceryl Sebacate	12.7
Ethyl Acetate	22.9
Alcohol ⎫ Toluol ⎭	To make 100.0

The wet surface of treated fabric is dusted with powdered mica, brushed and dried at about 82°C.

Engine Gray Lacquer

1	ZV–1420 Glyptal Solution	7.0
	Cellosolve Acetate	5.0
	Rutile Titanium Oxide	5.0
	Yellow Iron Oxide	3.0
	Lampblack	2.0
2	Half-Second Nitrocellulose Solution (25%)	48.0
	Lacquer Thinner	12.5
	ZV–3352 Glyptal Solution	17.5

Grind 1 in a pebble mill and add 2.

White Cellulose Acetate Lacquers

Formula No.	1	2	3	4	5	6	7	8
Cellulose Acetate	7.5	7.5	7.5	7.5	7.5	7.5	7.5	7.5
Santicizer M–17	3.75	3.75	3.75	3.75	3.75	3.75	3.75	3.75
Titanium Dioxide	4.0	4.0	4.0	4.0	8.0	8.0	8.0	8.0
Rezyl 14	1.5	3.0	...	...	1.5	3.00	...	...
Bakelite XR–4357	...	...	1.5	3.00	...	...	1.5	3.0

Formula No.	9	10	11	12	13	14	15	16
Cellulose Acetate	7.5	7.5	7.5	7.5	7.5	7.5	7.5	7.5
Santicizer M–17	3.75	3.75	3.75	3.75	3.75	3.75	3.75	3.75
Zinc Sulfide	5.0	5.0	10.00	10.00	...	...	...	...
Zinc Oxide	...	...	...	...	6.0	6.0	12.0	12.0
Rezyl 14	1.5	3.0	1.5	3.0	1.5	3.0	1.5	3.0

Medium Green Lacquer

1.
- ZV–3352 Glyptal Solution   12.5
- Xylol   10.0
- Butyl Alcohol   3.0
- Cellosolve Acetate   4.0
- Chrome Oxide   8.5
- Purple Iron Oxide   2.0
- Red Iron Oxide   0.4
- Yellow Iron Oxide   1.7
- Phthalocyanine Blue   0.6
- Celite 266 or Equivalent   5.0
- Magnesium Silicate   7.0
- Santocel 45–F   0.8

2.
- Half-Second-Nitrocellulose Solution (25%)   28.0
- Cellosolve Acetate   3.0
- Lacquer Thinner   9.0
- 1247 Glyptal Solution   4.5

Grind 1 in a pebble mill and add 2.

Olive Drab Lacquer

1.
- ZV–3352 Glyptal Solution   12.5
- Xylol   10.0
- Butyl Alcohol   3.0
- Cellosolve Acetate   4.0
- Yellow Iron Oxide   8.6
- Red Iron Oxide   2.0
- Medium Chrome Yellow   1.7
- Phthalocyanine Blue   0.7
- Magnesium Silicate   7.0
- Celite 266 or Equivalent   5.0
- Santocel 45–F   1.0

2.
- Half-Second-Nitrocellulose Solution (25%)   28.0
- Cellosolve Acetate   3.0
- Lacquer Thinner   9.0
- 1247 Glyptal Solution   4.5

Grind 1 in a pebble mill and add 2.

Cellulose Ether Coatings
(Water Soluble)
U. S. Patent 2,362,761

Formula No.	1	2	3	4	5	6	7	8
Ethylene Glycol Borate (Aquaresin)	2.5	...	...	1.0	...	...	1.0	...
Glycol Borate condensate	...	...	1.5	...	...	2.5	...	...
Glycol Borate (Bori-Borate)	...	2.5	...	...	2.0	...	...	2.0
Methyl Cellulose	5.0	5.0	...	...	...	5.0	3.0	5.0
Cellulose Ether of Sodium Glycollate	...	...	...	4.0	...	...	...	...
Water Soluble Ethyl Cellulose	...	...	4.0	...	...	...	...	...

PAINT, VARNISH, ENAMEL, LACQUER AND OTHER COATINGS

Formula No.	1	2	3	4	5	6	7	8
Water Soluble Hydroxy Ethyl Cellulose	...	...	...	...	4.0	...	...	...
Glycerol	...	...	...	1.0	...	...	...	1.0
Dextrose	...	...	...	...	1.0	...	...	...
Sodium Caseinate	...	...	1.5	...	...	...	1.0	1.0
Water	92.5	92.5	93.0	94.0	93.0	92.5	95.0	91.0

The above are prepared by heating the water to 70°C., stirring in the cellulose ether and cooling the resulting mixture to 2°C. The resulting solution is then passed through a colloid mill. The other ingredients are then dissolved in the cellulose ether solution. The cellulose ethers used are of the low-viscosity water-soluble type.

The compositions below are coating compositions designed for substantially rigid surfaces. They are adapted for use on plaster or concrete surfaces, iron structural members, black iron sheets, and the like in locations not subject to the direct action of water.

Formula No.	9	10	11	12	13
Methyl Cellulose	6.0	5.0	5.0	3.0	5.0
Glycerol Bori-Borate (Aquaresin GB)	2.0	1.0	2.5	1.0	2.0
Sodium Tetraborate	0.5	0.4	...	0.5	0.5
Whiting	1.0	...	...	...	1.0
Asbestine	0.5	0.2	...	...	...
Zinc Oxide	...	1.4	...	...	1.0
Sodium Caseinate	2.0	1.0	...	1.0	2.5
Water	78.0	81.0	92.5	94.5	88.0
Ethyl Alcohol	10.0	10.0	...	...	...

Cellolyn Lacquer Base		Plasticizer:	
Nitrocellulose (RS½-Sec. Dry Basis)	40.0	Processed Castor Oil (Baker's No. 15)	13.0
Cellolyn 102	40.0	Dibutyl Phthalate	7.0

Nitrocellulose Furniture Lacquers

Formula No.	1	2	3	4	5	6	7	8	9	10	11
Nitrocellulose (RS ½ Sec.)	1.0	...	...	...	1.0	...	1.0	...	1.0	...	...
Nitrocellulose (RS ¼ Sec.)	...	1.0	...	...	...	1.0	...	1.0	...	1.0	1.0
Nitrocellulose (RS 30–35 Cp.)	...	...	1.0	...	...	...	...	...	...	...	...
Nitrocellulose (RS 18–25 Cp.)	...	...	...	1.0	...	...	...	...	...	...	...
Aroplaz 905	...	...	...	...	0.9	...	1.2	...	2.0	2.0	2.0
Glyptal 2477	0.6	0.6	0.6	0.6	...	0.9	...	1.0	...	...	...
Lewisol 33	0.4	0.4	0.4	0.4	0.6	0.6	0.8	...	1.0	1.0	...
Melmac 245-8	...	...	...	...	...	...	...	1.0	...	...	1.0

Cellulose Acetate Lacquers

Solids	Formula No. 1	2	3	4
Cellulose Acetate				
Type used	FM-6	WH-1	LH-1	PH-1 or LH-1
Amount used	6.4	6.4	10.0	16.0
Triphenyl Phosphate	0.7	0.7	6.0	....
Santicizer 10	....	....	5.0	....
Santicizer M-17	....	....	....	2.0
Diethyl Phthalate	....	....	....	2.0
Solvents				
Acetone	32.5	....	....	68.0
Methyl Ethyl Ketone	37.6	....	....	....
Methyl Acetate	....	74.9	65.0	....
Ethanol	....	....	10.0	12.0
Butanol	....	....	4.0	....
Diacetone Alcohol	9.2	10.0	....	....
Toluene	18.6	8.0	....	....

Clear Lacquer Base Compositions

Formula No.	1	2	3	4	5	6	7	8
Nitrocellulose, (RS ½-Sec.)	1.0	1.0	1.0	1.0	1.0	1.0	1.0	1.0
Rezyl	2.0	...	...	...	3.0	2.0	2.0	...
Syntex 16	...	2.0	...	...	...	...	...	...
Beetle 227-8	...	0.2	0.3	...	...	...	...	...
Aroplaz 905	...	...	3.0	3.0	...	...	...	...
Melmac 245-8	...	...	...	0.3	0.3	...	...	...
Aroplaz 1130	...	...	...	...	...	1.0	2.0	...
Glyptal 2477	...	...	...	...	...	...	...	2.0
Rezyl 387-5	...	...	...	...	...	...	...	1.0

Lacquer Thinners

Butyl Acetate	20.0
Butyl Alcohol	10.0
Ethyl Acetate	10.0
Petroleum Naphtha	15.0
Toluol	45.0

Thinner for Furniture Lacquer (High Solid Content)

Standard Solvent

Butyl Acetate	35.0
Butanol	15.0
Toluene	50.0

High-Solvency Solvent

Methyl Ethyl Ketone	25.0
Butyl Acetate	20.0
Ethyl Acetate	10.0
Butanol	10.0
Ethanol	5.0
Toluene	30.0

Nitrocellulose Lacquer Thinners

Solvent Ingredients	Per Cent of Ingredients (By Wt.)		
	Standard Mixture	High-Solvency Mixture	Hot-Spray Mixture
Butyl Acetate	35.0	20.0	37.5
Butanol	15.0	10.0	12.5
Ethyl Acetate	....	10.0	....
Methyl Ethyl Ketone	....	30.0	....
Denatured Alcohol	....	....	10.0
Toluene	50.0	30.0	....
Xylene	....	....	40.0

PAINT, VARNISH, ENAMEL, LACQUER AND OTHER COATINGS

Paint and Varnish Removers
Formula No. 1
Non-Inflammable

Trichlorethylene	30.0 cc.
Methyl Ethyl Ketone	35.0 cc.
Xylene	30.0 cc.
Microcrystalline Wax	5.0 g.

No. 2
U. S. Patent 2,398,242

Water	12.0
Monoethanolamine	10.0
Monoethanolamine Oleate	13.0
Kerosene	10.0
1-Nitropropane	36.0
Isopropyl Alcohol	19.0

No. 3

Acetone	67.0 cc.
Ethylene Dichloride	20.0 cc.
Lactic Acid (Sp.Gr.1.2)	3.0 cc.
Water	10.0 cc.
Paraffin Wax	1.0 g.
Cellulose Acetate (High Visc.)	3.0 g.
Sulfonated Castor Oil	3.0 g.
Diamylamine Phosphate	1.0 g.

No. 4
(Army Air Forces #14119)

Methyl Ethyl Ketone	87.0 cc.
Lactic Acid	3.0 cc.
Water	10.0 cc.
Paraffin Wax	1.5 g.
Cellulose Acetate (Med. Visc.)	4.0 g.
Wetting Agent (Aerosol or Nacconal)	7.0 g.

No. 5
U. S. Patent 2,393,798

Nitropropane	15.0
Methyl Amyl Ketone	15.0
Oleic Acid	10.0
Triethanolamine	4.3
Pine Oil	2.0
Water	30.0

No. 6
U. S. Patent 2,346,622

2-Chloro-2-Butene	40.4
Ethyl Methyl Ketone	21.0
Acetone	18.5
Methanol	16.7
Paraffin Wax	3.4

Special Paints, Coatings and Compounds
Waterproof Emulsion Wax Coating
U. S. Patent 2,371,473

Carnauba Wax	25.0
Pentaerythritol Ester of N Wood Rosin	5.0
Beeswax	15.0
Ceresin Wax	15.0
Stearic Acid	8.0
Triethanolamine	4.3

Above are melted together at about 100°C. When these ingredients are thoroughly mixed, they are cut with 26 parts of turpentine. The turpentine solution is slowly diluted with 24 parts of naphtha. The temperature of this solution is held at 90°C. while 65 parts of water, also at 90°C. are added.

Gasolineproof Coating
(Water Soluble)
U. S. Patent 2,357,275

Dextrin	60.0
Glycerin	30.0
Sodium Nitrate	10.0

Waterproof Label Glaze

Clear Pliolite Resin	70.0 lb.
Mineral Spirits	50.0 gal.

Mix the resin into the spirits with high-speed agitation.

White Stencil Paint

Hexone	2.5
Denatured Alcohol	47.5
Vinylite Resin A Y A F	12.5
Titanium Dioxide	12.5
Whiting	25.0

Grind all the ingredients in a pebble mill. Use 50% by volume of butanol and 50% by volume denatured alcohol as a thinner. The amount of dilution required is dependent on the spray equipment.

Acidproof Tank Lining

Pitch	75.0
Plaster of Paris	9.0
Yellow Ochre	9.0
Beeswax	15.0
Litharge	3.0

Melt the pitch and beeswax over an electric burner or a very low flame and add all the other ingredients. Apply this mixture while hot with a stiff brush, then let cool.

Asphalt Clear Sealer

Accroides Gum	96.0 lb.
Denatured Alcohol	20.0 gal.
Butanol	4.0 gal.

Cold cut the gum in the solvent by agitation.

Knot Sealer (Varnish)

BV-9700, 60% Solids (Bakelite)	5.0
Polyvinyl Butyral XYHL Low Viscosity	0.5
Denatured Alcohol (95%)	9.5

The polyvinyl butyral resin should first be dissolved completely in the alcohol by stirring, followed by the addition of BV-9700 with thorough mixing. This low-cost sealer weighs approximately 7.5 lb. per gallon and the raw materials are available in quantity. It should be well brushed over the unprimed knot and surrounding area to insure complete coverage. A gallon of this product will cover about 500 sq. ft. of surface and will dry, set to touch, in about 10–15 minutes. However, the sealer should be allowed to dry overnight. Then the coats of paint may be applied in the usual manner. The sealer is satisfactory when used under regular outdoor house paints.

Mosquitoproof Paint

Paint	1.0 gal.
Citronella Oil	4–8.0 oz.

Protective Coating for Methyl Methacrylate Sheet

Polyvinyl Alcohol	1.49 lb.
Water	0.93 lb.
Alcohol	0.09 lb.
Aerosol (100%)	0.008 lb.
Methyl Orange Dye	0.003 lb.
Glycerin	0.026 gal.

Apply by dipping and hang until dry, using a heated room with exhaust fan. When desired, this coating is stripped off. If it does not strip easily, spray with water and allow latter to soak in for about ½ hour.

Welding Spatter and Cleaning Shield Mixture

U. S. Patent 2,343,158

Asbestos	75.0
Bentonite	9.0
Borax	8.0
Graphite	8.0
Water	To suit

Apply as a paint-like slurry on seams to be welded.

Thermocolor Paints

Heat-sensitive metallic salts or pigments are a convenient method for measuring approximate temperatures, on large surfaces such as boilers, dryers, furnaces, etc. Compounds of cadmium, cobalt, nickel, copper and manganese, in combination with other pigments and a suitable binding medium are useful for this purpose. Standard crayons or marking colors have been calibrated against time and temperature. Marking paints in which the color has been dispersed in urea-formaldehyde resin solution have been prepared. A few examples are tabulated below.

Color Changes in Thermocolor Paints

Code	Color Change	Time in Minutes	Time vs. Temperature				
			10	30	60	90	120
F36b	yellow—red brown		300	290	280	270	260
F214	purple—blue		150	140	137	133	130
F217	green—brown		230	220	210	200	195
F318	lt. red—lt. blue		72	65	62	60	58
F318	lt. blue—beige	centigrades	155	145	135	130	125
F320	grey-green—lt. blue		72	65	62	60	58
F320	lt. blue—olive gn.		155	145	135	130	125
F320	olive gn.—brown		230	220	210	200	195
F333	yellow—violet		120	110	105	103	100
F334	lt. green—blue		63	60	55	52	50
F335	red—blue		40	38	36	34	33

Composition of the Paints

Code	Formula	
F36b	Ferrite yellow	7
	Plastopal *	3½
F214	$CoNH_4PO_4.H_2O$	10
	Plastopal	2.3
F217	$CuSO_4.3Cu(OH)_2.H_2O$	7
F318	$MgNH_4PO_4.6H_2O$	3.75
	$CoNH_4PO_4.6H_2O$	1.25
	$Pb(OH)_2$	2.5
	Plastopal	2.2
F320	$MgNH_4PO_4.6H_2O$	4.5
	$CoNH_4PO_4.6H_2O$	1.5
	$Pb(OH)_2$	4.2
	$CuSO_4.3Cu(OH)_2.H_2O$	2.8
	Plastopal	4.5
F333	$NiCl_2.2C_6H_{12}N_4.2H_2O$	4.0
	TiO_2	2.0
	Plastopal	2.5
F334	$NiBr_2.2C_6H_{12}N_4.10H_2O$	4.0
	TiO_2	2.0
	Plastopal	2.5
F335	$CoCl_2.2C_6H_{12}N_4.10H_2O$	4.0
	TiO_2	2.0
	Plastopal	2.0

* Plastopal is a 50% solution of ureaformaldehyde resin in butyl alcohol. To obtain spraying or brushing consistency, the products are thinned with ethyl alcohol, usually 100 parts color paste to 60 or 80 parts alcohol.

Wallpaper Remover
U.S. Patent 2,317,505

Butyl Carbitol	40–60
Sodium Beta Ethyl-hexyl Sulfate	60–40

A 2–5% solution of the above in water is used.

X-Ray Protective Coating
U.S. Patent 2,315,061

Lead Oxide	78.0
Barium Sulfate	14.0
Bismuth Oxychloride	8.0
Water	To make a paste
Rubber Latex (55%)	300.0

This is molded or coated.

Coating for Inside of Petroleum Tanks
U.S. Patent 2,367,376

Pyroxylin	3–5
Acetone	12–15
Ethyl Acetate	40–50
Benzol	10–20
Triacetin	¼
Pigment	2–3
Blown Castor Oil	15–25

Gasolineproof Coating for Concrete

Butyl Acetate	50.0
Toluol	35.0
Alcohol	15.0
Polyvinylacetal	7–15

Coatings for Plastics
No. 1

Titanium Dioxide	9.6
Zinc Oxide	2.4
Vinylite Resin XYSG	6.0
Bakelite BV9700	10.0
Raw Castor Oil	1.0
Flexol 3GH	1.0
Butanol } Isopropanol } Solvesso No. 1 }	72.0

Pigments and resins are dispersed in the solvents and thinned further to spraying consistency.

No. 2

Titanium Dioxide	10.50
Antimony Oxide	1.15
Vinylite Resin VYHH	17.50
Flexol DOP	1.35
Hexone } Toluene }	69.5

Pigments and resins are dispersed in the solvents and thinned further to spraying consistency.

Plastic Coating for Iron Nails
Canadian Patent 427,632

Cumar Resin (CX)	86.0
Gilsonite	155.0
Soybean Oil	2–8
Naphtha	475.0
Asbestine	125.0
Iron Oxide Red	75—90

Mix until uniform.

Nails are coated with the above and dried. They do not rust and are difficult to pull out after insertion.

Chemical-Resistant Paints
Formula No. 1
Gray Primer

Talc	17.8
Zinc Dust	10.0
Chlorinated Rubber	24.1
Xylol	48.1

Finish Coat

Talc	15.0
Silcar	22.0
Standoil	4.0
Chlorinated Rubber	15.0
Xylol	44.0

No. 2

Titanium Dioxide	24.0
Chlorinated Rubber	20.0
Solvent Naphtha	14.0
Plasticizer	13.0
Toluol	29.0

No. 3

Titanium Dioxide	12.0
Blanc Fixe	8.0
Lampblack	0.5
Chlorinated Polyvinyl Chloride	11.4
Chlorinated Diphenyl Resin	10.5
Chlorinated Diphenyl	3.5
Tall Oil	3.0
Xylol	35.0
Alcohol	2.1
Butyl Acetate	6.8
Butanol	4.5
Propylene Glycol	2.7

Caulking Compound

Nuodex Cobalt 6%	6.0
Falkomast	172.0
Titanium Dioxide	19.25
Short Asbestos Fiber	108.0
Marble Dust (85%–200 mesh)	434.0
Falkomast	264.0

Saran Coating

Saran Latex (57% Solids)	100.0
Dibutyl Phthalate (60% Emulsion)	23.8
Hydroxyethyl Cellulose (5% Solution)	1.7

Black Wrinkle Finish—Medium Wrinkle

1:
- Superba Black — 12.0
- Asbestine 3X — 150.0
- Falkyd Solution W-200 — 460.0

2:
- Toluol — 60.0
- Petrolene — 163.0
- Nuodex Cobalt (6%) or — 14.0
- Nuodex Manganese (6%) — 14–28 (for Minimum Skinning)

Grind 1 and mix with 2.

Yellow Fluorescent Pigment
British Patent 568,445

Zinc Oxide	95–60
Vanadium Pentoxide	5–40

Heat together at 780°C.

Fungus Treating Coatings
No. 1

Sodium Thiosulfate	40.0
Copper Sulfate	10.0
Sodium Carbonate	5.0
Acetone	200.0
Diglycol Laurate	20.0
Water	50.0

No. 2

Vinegar	200.0
400 Mesh Copper	60.0
Methyl Cellulose	20.0

No. 3

Copper Oxide	15.0
Cellulose Acetate Cement	400.0

Add acetone to dilute. Apply paint in a thin film.

Fungusproof Wax Coating

Dissolve 1% phenylmercuric stearate in melted microcrystalline wax.

Red Barn Paint

Literite WS	3.0
Spanish Red Oxide	200.0
Asbestol Filler	75.0
Falkovar YY	450.0
Raw Linseed Oil	100.0
#10 Mineral Spirits	135.0
Nuodex Cobalt (6%)	5.0
Nuodex Manganese (6%)	6.0
Nuodex Lead (24%)	7.0

Superior Quality Whitewash

Slaked Lime	10.0	lb.
Liquid Hide Glue (LePages')	0.25	lb.
Warm Water	5.0	gal.

Pulverized Salt	0.5 lb.
Carbolic Acid (Phenol: 88%)	10 drops
Pulverized Portland Cement	0.5 lb.
Laundry Liquid Blueing	1.0 oz.

Mix thoroughly all the ingredients and apply with a large whitewash brush, brushing out well. Then let dry.

Powdered Cold Water Paint

Titanium Dioxide	7.9
Calcium Carbonate	18.3
Hydrated Lime	2.1
Clay	16.7
Casein	6.8
Preservative (Moldex)	0.5
Linseed Oil	4.1

Before use add:

Water	43.6

Whitewash

Casein	5.0 lb.
Lime Paste *	8.0 gal.
Trisodium Phosphate	3.0 lb.

Kalsomine

Whiting	75.0
China Clay	21.0
Glue	3.0
Phenol	1.0

White Pigment Composition for Fire-Retardant Paint

Formula No. 1

Barium Sulfate	21.9
Aluminum Silicate	3.2
Antimony Oxide	0.5
Zinc Oxide	24.8
Titanium Dioxide	49.6

* Lime Paste

Quicklime	25.0 lb.
Water	10.0 gal.

After thoroughly slaking strain through a screen to remove lumps.

No. 2

Barium Sulfate	21.9
Aluminum Silicate	2.6
Antimony Oxide	0.4
Zinc Oxide	25.5
Titanium Dioxide	49.6

No. 3

Barium Sulfate	24.0
Aluminum Silicate	1.2
Antimony Oxide	49.3
Titanium Dioxide	25.5

Concrete Floor Treatment

Paraffin Wax	4.0
Turpentine	1.0
Naphtha	16.0

Warm together until dissolved. Apply warm.

Antiseptic Barnyard Whitewash

Slaked Lime	7.0 lb.
Glue	6.0 oz.
Water	2.5 gal.

Mix until glue has dissolved. Then add

Formaldehyde	6.0 oz.

Acidproofing Laboratory Tables

A

Ferrous Sulfate	20.0 g.
Copper Sulfate	20.0 g.
Potassium Permanganate	40.0 g.
Water	500.0 cc.

B

Aniline	60.0 cc.
Hydrochloric Acid	90.0 cc.
Water	500.0 cc.

Apply one coat of Solution A and allow at least twelve hours to dry. Apply second coat, again allowing twelve hours. After the second coat is completely dry, apply two coats of Solution B, allowing twelve hours between each coat. When the second coat is thoroughly dry, add one coat of boiling hot linseed oil with a cloth.

Quick-Drying Red Primer for Wood
No. 1

Zinc Oxide	2.5
Iron Red	7.5
Talc	10.0
Vinoflex MP 400 (Copolymer of 75 pt. Polyvinylchloride and 25 pt. Polyvinylisobutyl Ether)	9.6
Tricresyl Phosphate	3.2
Alkydal ST (42% Glyceryl Phthalate—58% Castor Oil)	9.6
Ethyl Acetate	9.6
Butyl Acetate	8.0
Toluol	40.0

No. 2

Red Iron Oxide	16.3
Blanc Fixe	12.7
Kaolin	12.7
White Lead	2.2
Alftalate (Long-Oil Phthalic Alkyd Resin)	10.3
Stand Oil	1.1
Albertol (Rosin Modified Phenolic—75% Rosin)	1.6
Cobalt-Lead-Manganese Driers	2.9
Alcohol	20.2
Turpentine	20.0

Improving Lithographic Varnish

Acrawax C, 0.75%, incorporated by means of a roller mill, is used in lithographic varnishes such as are applied to metal cans, in order to impart improved water resistance. In addition, slip and lubricating effect is obtained which facilitates the stamping and forming of the can.

Crankcase Insulating Coating
U.S. Patent 2,350,559

Vermiculite, Powdered	22.0 oz.
Pre-Vulcanized Latex (30%)	0.5 gal.
Water	To suit

Black Shingle Stain and Preservative

Powdered Graphite	3
Linseed Oil	7

Walnut Furniture Stain

Asphalt	20 cc.
Kerosene or Gasoline	80 cc.

Apply to wood to be stained and remove excess by wiping. The depth of the color produced depends on time of contact.

Wood Stains—Fast to Light

Dissolve the following amounts of dyes in methyl alcohol to form 10 gallons of solution.

Maple

Luxol Fast Orange GS	3	oz.
Luxol Fast Brown K	1¼	oz.
Luxol Fast Black L	¼	oz.

Cherry

Luxol Fast Red B	11	oz.
Luxol Fast Black L	½	oz.

Light Oak

Luxol Fast Orange GS	4	oz.
Luxol Fast Brown K	1½	oz.
Luxol Fast Black L	½	oz.

Dark Oak

Luxol Fast Orange GS	2¾	oz.
Luxol Fast Brown K	2¾	oz.
Luxol Fast Black L	1	oz.

Red Mahogany

Luxol Fast Brown K	6	oz.
Luxol Fast Red B	4¼	oz.

Luxol Fast Black L   2½ oz.
    Brown Mahogany
Luxol Fast Orange GS ½ oz.
Luxol Fast Brown K   3 oz.
Luxol Fast Red B   1½ oz.
Luxol Fast Black L   1¼ oz.
    Light Walnut
Luxol Fast Orange GS 1½ oz.

Luxol Fast Brown K   7 oz.
Luxol Fast Red B   ½ oz.
Luxol Fast Black L   3 oz.
    Dark Walnut
Luxol Fast Orange GS 1 oz.
Luxol Fast Brown K   10 oz.
Luxol Fast Red B   ¾ oz.
Luxol Fast Black L   4 oz.

Chapter XV

PAPER

Paper Finish

Clay	300 lb.
Water	20 gal.
Talc	18 lb.
Wax Emulsion *	12 gal.
Casein Glue †	25 gal.

The three ingredients are boiled together, until the wax is emulsified, and sufficient water is added to bring the volume to 50 gal.

Another formula, for making a wax emulsion for flint papers, is as follows:

Laundry Soap	7 lb.
Carnauba Wax	50 lb.
Water	12½ gal.

Boil with live steam till thoroughly emulsified (generally for 4 to 5 hours). Cool to 35°C. and add 2 lb. 26° ammonia. Make up to 50 gal. with cold water. The emulsion should be allowed to stand, before using, as it seems to improve with age.

A beeswax emulsion suitable for friction-calendered papers is made as follows:

---

* The wax emulsion is made up as follows:

Carnauba Wax	50 lb.
Water	50 gal.
Castile Soap	12 lb.

† The casein glue may be made up as follows:

Casein	100 lb.
Water	50 gal.
Borax	17 lb.
Ammonia (26 Bé.)	1 qt.

Yellow Beeswax	150 lb.
Castile Soap	28 lb.
Water	25 gal.

Dissolve the soap in water and add the wax. Melt and stir till emulsified and smooth. Add water to make 150 gal.

In the manufacture of coated boxboard, a wax emulsion is sometimes added to the casein coating mixture to make the coating more flexible and to improve the bending and folding properties of the coated board. Such an emulsion is usually made with Japan wax.

Japan Wax	75 lb.
Stearic Acid	17 lb.
Water	25 gal.

Heat till the waxes are melted. Cool slightly and add 10 lb. of borax dissolved in 10 gal. of water. Boil till a smooth emulsion is formed, cool and add 1 qt. of 26° ammonia and make up to 75 gal.

Paper Sizing
U.S. Patent 2,320,771

	%
Animal Glue	3.5–8.5
Borax	0.5
Bentonite	0.25–1.25
Water	To make 100

Use at 32–46°C.

Transparentizing Paper
U.S. Patent 2,383,660

Impregnate the paper with the following mixture:

Triethyleneglycol Ester
of Hydrogenated Rosin 37½
Ethyl Acetate 62½
Squeeze out the excess and dry.

Improved Drawing Surface
U.S. Patent 2,386,626

A transparent cellulose sheet is coated with the following suspension:

Ground Glass	3.54
Cellulose Acetate	5.80
Gelatin	0.26
Sodium Sulfate	0.60
Methyl Cellosolve	14.00
Water	0.80
Acetone	69.50
Methanol	3.50
Acetic Acid	2.00

Allow to dry and then leach out the sodium sulfate with water and dry.

Paper Cap Die Cutting Lubricant

A 2% aqueous dispersion of Diglycol Stearate S, applied as hot as possible, is excellent for the dry-die forming of paper caps. The dispersion can successfully replace high-grade soap. When soap is used, the dies have to be replaced every five or six weeks; when the Diglycol Stearate dispersion is used instead of soap, the dies stand up more than 5 months.

Greaseproofing Composition
U.S. Patent 2,367,678

Glue	16–30
Water	40–67
Sodium Nitrate	⅔–6½
Gelatin	½–5
Glycerin	12½–20
Acetic Acid	3–15
Hexamethylene Tetramine	½–3

Waterproof Coating for Paper

Polystyrene	10
Carbon Tetrachloride	90

Mix until dissolved. Dip the paper in the mixture; allow to drain and dry.

Waterproof Coating for Paper Cartons
Formula No. 1

A hot melt blend of 5% Vistac, 0.5% Acrawax C and 94.5% asphalt (softening point 160°F.) has been successfully substituted for micro-crystalline waxes in materials such as ordnance wrap. The coat is reported to be only ½ of that when using micro-crystalline wax.

Destroying Paper Mill Foam

To kill the foam 0.1 to 1.0% diglycol laurate based on the protein present has proven successful. In other cases, mills have used 0.002% diglycol laurate based on the total solids, or 0.0008% on the contents of the coating bath, including water. The amount needed will vary widely in individual cases.

Inhibiting Bacterial Slime

For inhibiting the deposition and development of bacterial slime formations and mild growths as well as killing algae in paper mill water and stock systems 2,2'-dihydroxy-5,5'-dichloro-diphenylmethane is dissolved as follows:

2,2'-dihydroxy-5,5'-dichloro-diphenylmethane	6
Caustic Soda (76% flakes)	1
Water	24

Mix together. The heat of solution of the caustic soda is sufficient to cause its reaction with 2,2'-

dihydroxy-5,5'-dichloro-diphenyl-methane, forming the soluble sodium salt. This solution may then be diluted to a convenient concentration for application to the system in any proportion necessary for control of the micro-organisms encountered. Usually 3% of the above product is used based on the solid material (paper pulp, etc.) that is to be protected.

Chapter XVI

PHOTOGRAPHY

Metal-Pyro Developer

Stock Solution A
- Water (at 125°F.)   16 oz.
- Metol   1 oz.
- Sodium Bisulfite   ¼ oz.
- Pyro   ¼ oz.
- Potassium Bromide   60 gr.
- Cold Water to make   32 oz.

Stock Solution B
- Water   32 oz.
- Sodium Sulfite (Desiccated)   5 oz.

Stock Solution C
- Water   32 oz.
- Sodium Carbonate (Desiccated)   2½ oz.

This modified pyro developer is a big improvement on the standard three-solution pyro developer so far as the keeping qualities are concerned. Several negatives can be developed in the same tray solution before discoloration becomes objectionable and tank solutions can be kept for several days, if covered with a floating paraffin lid.

For tray use take 1 ounce of each solution, A, B, and C, to 8 ounces of water and develop for 7 to 9 minutes at 65°F. For tanks use twice the amount of water and develop for 10 to 12 minutes.

Developers for Low-Temperature Processing

	Kodak D–8		Kodak D–82 + Caustic	
	Metric	Avoirdupois	Metric	Avoirdupois
Elon	....	....	14.0 g.	200 gr.
Hydroquinone	45.0 g.	1½ oz.	14.0 g.	200 gr.
Sodium Sulfite	90.0 g.	3 oz.	52.5 g.	1¾ oz.
Sodium Hydroxide	37.5 g.	1¼ oz.	17.6 g.	250 gr.
Potassium Bromide	30.0 g.	1 oz.	8.8 g.	125 gr.
Benzotriazole	....	....	0.2 g.	3 gr.
Water To make	1,000.0 cc.	1 gal.	1,000.0 cc.	32 oz.

For Use down to 30°F.: D–8 2 parts, water 1 part. Use D–82 + caustic undiluted.

For Use down to +5°F.: D–8 2 parts, water 1 part, ethylene glycol 1 part. D–82 + caustic 3 parts, ethylene glycol 1 part.

The glycol should be added previous to storage at low temperatures.

Amidol-Catechol Developer
(Kodak SD–22)

Solution A		Metric	Avoirdupois
Sodium Bisulfite		100 g.	3 oz. 145 gr.
Amidol		40 g.	1 oz. 145 gr.
Catechol		40 g.	1 oz. 145 gr.
Benzotriazole		2 g.	30 gr.
Water	To make	1 l.	32 oz.

Solution B

Sodium Hydroxide		120 g.	4 oz.
Potassium Bromide		20 g.	290 gr.
Potassium Iodide		4 g.	60 gr.
Water	To make	1 l.	32 oz.

For Use down to $+30°F.$: Solution A 1 part, Solution B 1 part, water 2 parts.
For Use down to $+5°F.$: Solution A 1 part, Solution B 1 part, ethylene glycol 1 part, water 1 part.
For Use down to $-40°F.$: Solution A 1 part, Solution B 1 part, ethylene glycol 2 parts.

The glycol may be divided and added to each of these solutions previous to storage at low temperatures. Combine Solutions A and B only immediately before use since the mixed developer oxidizes rapidly. Solution A may also deteriorate on keeping and should be kept well-stoppered and as cool as possible.

Fine-Grain Developer
Formula No. 1

Pyro		2.5 g.
p-Phenylenediamine		10.0 g.
Sodium Sulfite		100.0 g.
Water	To make	1 l.

No. 2

Metol	10 g.
Sodium Sulfite	100 g.
Water	1000 g.

Development of Brown or Blue-Black Tone

Development in a solution composed of 40 g. of sodium sulfite, 6 g. of glycin, 6 g. of hydroquinone, 30 g. of sodium carbonate, and 2 g. of potassium bromide in 1,000 cc. of water yields a tone which, after long exposure and short development, changes from warm black through brown to brownish red and red. With a developer consisting of 25 g. of hydroquinone, 70 g. of sodium sulfite, 90 g. of potassium carbonate, and 2 g. of potassium bromide in 1,000 cc. of water, more reddish tones will be obtained with increasing dilution. Blue-black images are obtained with a solution consisting of 3 g. of metol (Elon, Scadol), 40 g. of sodium sulfite, 12 g. of hydroquinone, 75 g. of sodium carbonate, and 0.8 g. of potassium bromide in 1,000 cc. of water; with small additions of Bellaton (nitrobenzimidazole), e.g., 2 to 5 cc. of 1% solution to 100 cc. of developer, a pronounced blue tone is obtained.

Developing Old Printing Paper

Metol	3 g.
Sodium Sulfite	30 g.
Hydroquinone	3 g.
Borax	40 g.
Water	1000 cc.
Potassium Bromide (10% Solution)	5–50 cc.

Succinic Acid In Photography

Succinic acid may be successfully substituted for acetic acid in the short-stop and fixing solutions and in all other photographic solutions. It has two outstanding advantages over acetic acid. It is a solid, and it is odorless. The odor of acetic acid is offensive to many people, and the fact that it is a liquid makes it difficult to handle under some circumstances.

Succinic acid is conveniently prepared as a 4% stock solution in water. Four parts of this solution are equivalent to one part of 28% acetic acid in any photographic formula.

Sometimes, for reasons that are

not always clear, it works better than acetic acid. The following variation of a familiar intensifier formula works exceptionally well:

Intensifier

Potassium Ferricyanide	2.5% Solution
Uranyl Nitrate	2.5% Solution
Succinic Acid	4% Solution

Mix one part of each at the time of use. The only problem is to avoid overintensification as it works rather rapidly. Some may prefer to dilute the intensifier with at least an equal volume of water.

Recovering Silver from Photographic Film

Place the cut film in a washing machine, the regular laundry type will do, and add sufficient solution containing ½ lb. pepsin (U.S.P.) to 250 gal. of 21/N hydrochloric acid. The solution should be heated to 110°F. and the wash wheel containing 200 lb. of the film should be rotated for 30 minutes. The pH of the solution should be maintained at 2.0–2.5.

After the silver and the emulsion have been removed, the precipitated silver is allowed to settle. The liquid is decanted after the pepsin has been exhausted. The black sludge is dried, and then heated to eliminate the organic matter.

To this residue is added 2 parts of soda ash to 1 part of silver and the whole mass is placed in a crucible and melted. After completing the melt, the charge is poured. The pure silver button is separated from the slag by hammering to remove the last traces of slag.

Photographic (Light) Filter

Films of cellulose acetate are soaked at 20–50°C. in

Dye	0.14	g.
Glycerin	1	cc.
Caustic Soda (38° Bé.)	2	cc.
Water	10	cc.
Sodium Hyposulphite	½	g.
Water	40	cc.

Heat this solution at 50°C. for 30 min. and add:

Water	150	cc.

Indanthrene dyes that are used are:

Dark Blue BO
Brilliant Green B
Brilliant Green TT
Brilliant Violet RR
Golden Yellow GK
Chloroindanthrene GCDN
Brown R
Brilliant Pink

Chapter XVII

PLASTICS, RUBBER, RESINS AND WAXES

Identification Tests for Plastic Raw Materials

The material to be identified may be in one of three forms:
(1) As a pure constituent, or raw material.
(2) As a processed article, either of a pure constituent, or modified by the addition of a plasticizer, or of a thermoset resin.
(3) As a solution, as a varnish in a mixed solvent.

A direct examination of the material can only be made in the first case, and in the others the plastic base may have to be separated from the plasticizer or solvent by extraction.

The following table applies to the identification of plastics and plastic raw materials in their simple form, as the analysis of complex articles or solutions requires more or less elaborate chemical technique.

Identification of Raw Material by Heating Tests

Material	Test	Result
Formvar (Polyvinyl formal)	Heat in test-tube	Melts. Some discoloration, slight charring. Fishy odor and smell of formaldehyde.
Alvar (Polyvinal acetal)	Heat in test-tube	As Formvar, but no smell of acetaldehyde, which can be detected by Schiff's reagent.
Polyvinyl Chloride	Heat in test-tube	Browns immediately, turning black. Little melting. Copious evolution of HCl.
Mixed Vinyl-chloride Acetate Polymer	Heat in test-tube	Same as for Polyvinyl Chloride.
Polystyrene	Heat in test-tube	Melts to a clear liquid which boils. Slight discoloration. Characteristic smell of monomer.
Methyl Methacrylate	Heat in test-tube	Does not melt or char appreciably. Decomposes and monomer distils off.
Bakelite	Heat in test-tube and in flame	Presence of wood flour causes much charring and evolution of smoke, which disguises characteristic odor. Without wood flour phenol and formaldehyde can be detected.
Urea Formaldehyde	Heat in flame	Strong smell of formaldehyde and ammonia. Much charring, but highly non-inflammable.
	Heat in test-tube	Little smell of formaldehyde. Smell of ammonia and pyridine.
Thiourea-formaldehyde	Heat in test-tube	Pronounced smell of H_2S and ammonia.

Material	Test	Result
Casein	Heat in flame	Chars readily. Smell of burning protein. More inflammable than urea formaldehyde.
Cellulose Acetate	Heat in flame	Melts and chars. Pungent smell of burning cellulose and acetic acid.
Ethyl Cellulose	Heat in flame	Chars. Readily melts, with smell of burning cellulose and oil.
Cellulose Acetobutyrate	Heat in flame	Chars. Readily melts, with characteristic smell of butyric acid.

Plastic Compositions

Black Polyvinyl Plastic

Formula	No. 1	No. 2	No. 3	No. 4
Polyvinyl Chloride	100.0	100.0	100.0	100.0
Tricresyl Phosphate	50.0	55.0	60.0	70.0
Lead Oleate	2.2	2.2	2.2	2.2
Basic Lead Carbonate	2.2	2.2	2.2	2.2
Carnauba Wax	1.5	1.5	1.5	1.5
Witco #1 (Carbon Black)	—	10.0	20.0	40.0

Mold at 307°F. for 10 minutes at 1700 lb. per square inch pressure.

Plastic Molding Composition

Formula No. 1
Canadian Patent 425,002

Cellulose Fibers Pulp	40
Graphite (Powdered)	40
Soap	6
Rosin	10
Alum	4

Air dry and then mold under high pressure and heat.

No. 2
Canadian Patent 424,936

Hydrolyzed Wood Pulp	100
Rosin	16
Magnesium Stearate	½
Water	2

	No. 3	No. 4
Blood (Dried)	40	40
Wood Flour	38	—
Water	20	20
Zinc Stearate	2	2
Lignin	—	31
Carbon Black	—	2
Furfuramide	—	5

No. 5

Waste Sulfite Liquor (50% Lignin)	65
Calcium Oxide	35

Mix to a thick paste and mold cold under moderate pressure.

No. 6

Resin (Bakelite BR-1922)	1.0
Lignin	3.0
Water	0.3
Carbon Black	2.0
Zinc Stearate	0.5

No. 7

Casein	50
Slate Powder	40
Furfural	5
Rosin Ester Gum	5
Basic Calcium Lignosulfonate	5

The furfural is dissolved in the ester then the other materials are added.

Mold at 2500 lb. per square inch at 212°F. for 30 minutes.

No. 8

A	Blood (Dried)	50
	Furfuramide	5
	Synvarite Resin	5
	Wood Flour	35
	Zinc Stearate	1½

B Salt 1
Water 20
Duponol WE dry ½

Mix A in a ball mill for 10 minutes, then add B, and mix in the mill for another 20 minutes.

Transparent Moistureproof Sheeting
U.S. Patent 2,360,947

Plioform	27
Gum Dammar	55
Montan Wax	15
Benzyl Abietate	3

This mixture is applied hot to regenerated cellulose sheet.

Molded Wood Composition
Formula No. 1

Wood Flour	3
Ball Clay	2
Powdered Sodium Silicate	As required
Water	As required

The wood flour and ball clay are first mixed, and a small amount of water is added at a time until the material reaches molding consistency, when a small amount of sodium silicate is added. The sodium silicate (powdered) should be the grade having a 1 to 2 soda to silica ratio.

No. 2

Wood Flour	10
Portland Cement	10
Sodium Silicate (Powdered)	10
Water	3½

The sodium silicate should be the grade having a soda to silica ratio of 1 to 2. After thorough mixing, the compound can be molded under heavy pressure. The molds are comparatively non-shrinking, durable and water-resistant, but not refractory.

Wood and Crack Filler
Formula No. 1

A mixture of powdered wood or wood flour and resorcinol-formaldehyde resin can be used to fill scratch marks and pores before staining. The filler is brown in color and hardens or sets in a short time at room temperatures.

No. 2

A heat curing crack filler can be made by mixing wood flour and phenol-formaldehyde resin to make a paste for dark colored application. Light colored or white filler can be made by substituting urea-formaldehyde resin for the phenolic resin.

Curing can be conveniently done by using infra-red lamps.

Dental Impression Plaster
U.S. Patent 2,390,137

Calcium Carbonate	25
Calcium Sulfate (Deadburned)	40
Potassium Alginate	5
Calcium Sulfate (Hydrated)	⅕
Trisodium Phosphate	2
Add before use	
Water	28

This plaster sets in 3–6 minutes.

Denture Mold Plaster
British Patent 562,882

Plaster of Paris	150
Sodium Thiosulfate	4–14
Water	100

This compound is used for molding acrylic resins at curing temperature (90–95°C.). This mold is readily frangible without affecting the molding operation.

Synthetic Resin and Varnish

Undistilled Tall Oil	1050 lb.
Maleic Anhydride	37 lb.
Polypentek	194 lb.
Mineral Spirits	252 lb.

The tall oil and maleic anhydride are charged into a 200-gal. gas-fired, stainless steel kettle equipped with a turbine type agitator, condenser, receiver, and bottom outlet. The furnace setting is equipped with a centrifugal blower for external cooling.

The mixture is heated with good agitation to 150°C. in 2 hours and held at 150° for 1 hour to form the maleic adduct. The Polypentek is then added and the temperature increased to 275°C. in 2 hours. The charge is cooked at 275°C. for 8 hours, cooled with the help of the blower to 150°C., diluted with the mineral spirits, and discharged into suitable containers.

Thin to 50% solids with a mixture of equal parts of mineral spirits and turpentine.

When 0.05% Cobalt and 0.05% Manganese, as naphthenates, are added and a 24 hour ageing period permitted, the resulting films dry tack free on glass in 7 hours.

Flame-Resistant Resin
U.S. Patent 2,396,575
Formula No. 1

A. Vinsol Resin	100
B. Tricresyl Phosphate	25
C. Boric Acid	6–10

Melt A and B, heating them to 150°C.; then add C slowly, while stirring, until foaming ceases.

No. 2

Vinsol Resin	100
Boric Acid	6–10

Grind together, then heat slowly to 140–150°C. in shallow pans until sintered. Then remove and grind.

Electric Cable Coating
British Patent 574,252

Ethyl Cellulose (43.5–46.5%)	10
Transformer Oil	2–4
Petroleum Jelly	2–4
Castor Oil	3–6

Electrical Insulating Compound
Formula No. 1
British Patent 555,904

Polymerized Rosin	44.44
Gum Accroides	44.44
Castor Oil	8.88
Stearic Acid	2.22

No. 2
Belgian Patent 444,234

Urea Resin	70
Methylhexanone	15
Naphtha	15
Chlorinated Naphthalene	40
Ammonium Linoleate	5
Water	150

No. 3
Belgian Patent 446,384

Polyisobutylene	100
Lampblack	30–120
Graphite	225–250

Insulating Tape Impregnant
Canadian Patent 422,634

Polystyrene	51.6
Rubber	23.2
Polyisobutylene	21.4
Polyethylene	3.8
Monostyrene	15.0

Thermoplastic Shoe Stiffener
U.S. Patent 2,390,347

Rosin	37.5
Vistanex	2.5
Calcium Resinate	20.0
Gum Copal	10.0
Montan Wax	15.0
Crepe Rubber	15.0

Thermoplastic Box Toe Stiffener
Formula No. 1
U.S. Patent 2,331,095

Ethyl Cellulose	17.5
Dibutyl Phthalate	7.5
Rosin	75.0

Melt together and mix until uniform.

No. 2
U.S. Patent 2,378,674

Rosin	78.0
Candelilla Wax	2.0
Cumar	8.2
Reclaim Rubber	10.7
Anti-Oxidant	0.1

Shoe-Filler Composition
U.S. Patent 2,350,252

Ethyl Acetate	84
Petroleum Naphtha	25
Dibutyl Phthalate	10
Nitrocellulose Film Scrap	18

Dissolve by mixing slowly then add

Tall Oil (Refined)	41
Congo Gum (Raw Powdered)	12

and mix well.

Rubberless Eraser
U.S. Patent 2,404,322

Polyvinyl Acetate	88
Dibutyl Phthalate	6
Rosin	6

Artificial Straw

Cellulose Acetate	32.3
Cellulose Acetate Scrap	3.0
Tricresyl Phosphate	19.3
Dimethoxyethyl Phthalate	5.4
Titanium Dioxide	1.6
Acetone	40.0

Mix until uniform. Filter and deaerate for two days. The mixture is then fed through rayon spinnerets and coated on silk or rayon threads depending on the diameter of straw fiber desired. It is then carried through a horizontal drying chamber and wound on spools.

Artificial Bristles

Cellulose Acetate	135
Acetone	165

Mix until dissolved; filter avoiding bubble formation. Force through a 40 thread spinneret of 0.25–0.6 mm. diameter. Pass through horizontal drying chamber and wind on large (50 in.) spools and then cut into 50 in. lengths. Dry on heating plates at 40°C. Spinning speed is at 15–30 m. per minute.

Catalysts for Hardening Urea-Formaldehyde Resins
Formula No. 1
(For cold hardening)

Ammonium Chloride	15
Water	85

No. 2
(For hot hardening)

Ammonium Chloride	15
Ammonia (25%)	20
Urea	30
Water	31
Methyl Cellulose	1

Plasticizer for Polyvinyl Acetate

Dibutyl Phthalate	3
Tricresyl Phosphate	2
Polyvinyl Acetate	10

Plastic Film Coating
Styrene Organosols (Styrasols)

These are soft pastes which can be easily spread on cloth, paper, etc., and when fused for a few minutes at temperatures in excess of 320°F., produce films having the following properties:

1. Outstanding resistance towards acids, alkalies, water and alcohol.
2. Freedom from blocking at elevated temperatures.
3. Fair low temperature flexibility.
4. Excellent resistance to water vapor transmission.

These organosols are prepared in the laboratory by mixing the ingredients in a suitable container by means of a paddle and then giving the resulting paste 2 passes through a cold 3 or 5 roll mill.

Formula No. 1
Piccolastic Powder	10
Piccolastic A-5	15
Mineral Oil	5

No. 2
Piccolastic A-50	20
Piccolastic B-75	20
Piccolastic A-5	10
Piccolastic Powder	20
Mineral Oil	20

No. 3
Piccolastic A-5	10
Piccolastic A-50	20
Piccolastic Powder	20
Piccovar C-120	20
Mineral Oil	20

No. 4
Piccolastic Powder	10
Piccolastic A-5	10
Troluoil	20
Solvesso #1	35

Vinyl Organosol—Clear Base
Vinylite VYNV	100
Dioctyl Phthalate	60
Solvesso #2	30
Apcothinner	45
Calcium Stearate	12

Place all ingredients into a pebble mill. Churn for 24 hours and then discharge. A smooth white paste will result which can be easily spread on cloth, paper, etc. When fused at 350°F. for 1 minute, a clear homogenous film will be formed.

Opaque, colored films or transparent, colored films can be made by adding filler and colors to the above formulation. A convenient method for most plants is to grind the colors and fillers in plasticizers by means of an ink mill, adding them to the clear base. Some typical formulations are listed below. (Grind, using a 3-roll mill; minimum of 3 passes.) All colors have fairly good fastness in vinyl resins.

Clay Filler
Whitetex Clay	55
Dioctyl Sebacate	45

Black Paste
Lampblack	15
Dioctyl Sebacate	85

Blue Paste
Phthalocyanine Blue	35
Dioctyl Sebacate	65

White Paste
Titanium Dioxide (Rutile)	75
Dioctyl Sebacate	25

Red Paste
Toluidine Red	35
Dioctyl Sebacate	65

Yellow Paste
Lemon Chrome	25
Dioctyl Sebacate	75

Green Paste
Phthalocyanine Green	25
Dioctyl Sebacate	75

Plating with Plastics

A. Vinylite:

Vinylite Molding Powder
(About 85% vinyl chloride)	5 g.
Butyl Acetate	60 cc.
Tributyl Citrate	1.2 cc.
Absolute Ethanol	27 cc.
Butyl Acetate	9 cc.

Dissolve 5 g. Vinylite molding powder (about 85% vinyl chloride) in 60 cc. of butyl acetate. Add 1.2 cc. of tributyl citrate (less if more rigidity is desired). Then add, in several steps, a mixture of 27 cc. absolute ethanol and 9 cc. butyl acetate. Shake up the suspension after each addition until it is again uniform. Electrolyze at room temperature.

Average current density (30 minutes): 147 microamp. per square centimeter. Current yield (30 minutes): 0.071 g. Vinylite per milliampere-hour.

B. Vinyl Chloride (Koroseal)
Vinyl Chloride (Koron 101)	1 g.
Dioxane	40 cc.
Butyl Acetate	10 cc.
*Tributyl Citrate	0.5 cc.
Absolute Ethanol (99%)	13 cc.

The same procedure is followed as for A.

*The amount and type of plasticizer may be varied.

Surface Dyeing of Plastics

The following formula is satisfactory for surface coloring of methacrylate, polystyrene and vinyl plastics:
Dye	0.25
Merpentine	7.5
Methyl or Ethyl Alcohol	2.5

Dilute with 90 parts of water. Dye at 200°F. for 3–30 minutes depending on the depth of shade required.

The following dyes may be used:
Acetamine Yellow RR
Acetamine Yellow N
Celanthrene Fast Yellow GL Conc. 300%
Du Pont Oil Yellow
Acetamine Orange GR Conc. 175%
Acetamine Orange 3R Conc.
Du Pont Oil Orange
Acetamine Scarlet B
Celanthrene Brilliant Red Conc. 200%
Du Pont Oil Red
Celanthrene Pure Blue BRS 400%
Du Pont Anthraquinone Blue AB Base
Du Pont Anthraquinone Blue SKY Base
Du Pont Anthraquinone Iris R Base
Du Pont Anthraquinone Green G Base
Celanthrene Brown Y
Du Pont Oil Brown N
Celanthrene Purple Conc. 175% (Pat.)
Celanthrene Violet CB (Pat.)
Du Pont Anthraquinone Violet Base
Acetamine Black CBS

Cellulose acetate may be dyed using the following formula:

Dye 0.25
Denatured Alcohol 100.00

In addition to the dyes in the above list, the following may be used:

Du Pont Auramine SP Conc.
Luxol Fast Yellow G
Luxol Fast Orange GS
Luxol Fast Orange R
Luxol Fast Brown G
Luxol Fast Brown K
Luxol Fast Brown R
Du Pont Rhodamine B Extra
Du Pont Rhodamine 6GDN Extra
Du Pont Fuchsine Conc. Powder
Du Pont Safranine T Extra Conc. 125%
Luxol Fast Red B
Luxol Fast Red BB
Luxol Fast Scarlet C
Du Pont Crystal Violet Powder
Du Pont Methyl Violet Conc.
Du Pont Victoria Blue B Conc.
Luxol Fast Blue AR
Luxol Fast Blue G
Luxol Fast Blue MBS (Pat.)
Du Pont Brilliant Green Crystals
Du Pont Victoria Green Small Crystals
Luxol Brilliant Green BL
Luxol Fast Green B
Du Pont Nigrosine SSJ Powder
Luxol Fast Black L

Cellulose nitrate can be dyed using the following formula:

Dye 0.5
Methyl or Ethyl Alcohol 50.0
Amyl Acetate 20.0

The same dyes may be used as were recommended for use with cellulose acetate.

Acetamine, Celanthrene and Luxol are registered trade-marks of E. I. du Pont de Nemours & Co., Inc.

Fluorescent Coatings and Plastics

Formulating with fluorescent dyes differs markedly from formulating with pigments in two major respects: first, in the amount of color used; and second, in shading for color values.

The amount of dyestuff used must be carefully worked out for each specific application to obtain a maximum of fluorescent brilliance. A range from a minimum quantity to a maximum quantity generally can be worked out through which a variation in intensity and shade can be controlled, and at the same time a peak of fluorescent brilliance maintained. In producing pastel shades, the dye content may tend to run below the minimum. This may result in almost complete loss of fluorescence under ultraviolet light because the fluorescent molecules are so far apart that they produce only a faint bluish-white glow or are completely undetectable. Minute quantities of soluble dyes produce relatively large shade changes and even full strength formulations may contain only 0.4–0.5% of dye in the final dry film. In producing the heaviest shades, care must also be taken not to exceed the established maximum dye content for a particular application. The effect of an excess of dye is a marked dulling effect on fluorescent brilliance.

No hard and fast rules can be given establishing the quantity of dye to be used. However, the following table shows a series of five brilliantly fluorescent enamels (formulated for test purposes only) which may be used as a guide or starting point for dye usages when developing fluorescent top-coatings. The table also illustrates how combinations of various solvents may be used advantageously to increase dye concentrations when this is necessary.

Shade matching with fluorescent colors not only presents those problems ordinarily encountered in color work, but also is complicated by the fact that the fluorecsent color may be entirely different from that of the visible light color. Thus, a dye or compound that is colorless or faintly tinted in visible light may be blue (Calcomine Fluorescent Violet G) or green (anthracene) under ultraviolet light. However, by careful manipulation these complications can be used to advantage to produce an infinite number of different effects in shade and tone. The table shows how Calcomine Fluorescent Violet G has been given a bright blue daylight color with Calco Ultramarine Blue No. 1401. The addition of the non-fluorescent ultramarine blue appeared to actually increase the fluorescent intensity of the Calcomine Fluorescent Violet G. However, when certain non-fluorescent dyes were substituted for the ultramarine blue in the same formulation, the intensity of fluorescence of the coating was so reduced that it was of no interest.

The addition of a small amount of non-fluorescent dyestuff to a fluorescent coating may reduce its brilliance far out of proportion to the amount added. This effect is probably produced by the differences in absorption and reflectance properties of the various dyes in both the ultraviolet and visible light wave lengths. It is probable that this interference between dyes could be predicted by spectrophotometric tests, but these are not generally available and the same result can be obtained visually by actual blend tests.

This table gives examples of the use of both non-fluorescent dyes and pigments. In the case of the green, a non-fluorescent dyestuff has been added to a fluorescent yellow to produce both a green daylight effect and a green fluorescence under ultra-violet light. In this case, a slight loss of fluorescent intensity is noted. In order to give a blue color in both visible and ultra-violet light, a non-fluorescent blue pigment has been added to a coating which is colorless in visible light and fluoresces blue in ultra-violet light. In this instance, no loss of fluorescence is noted and in fact an increase in intensity has been obtained.

The toning or shading of fluorescent colors is most readily accomplished by using other fluorescent colors whenever the proper shades are available. However, it must be kept in mind that the regular color matching rules no longer apply when dealing with fluorescent colors viewed in ultra-violet light. For example, a mixture of Calco Fluorescent Yellow AB (greenish yellow fluorescence) and Calcomine Fluorescent Violet G (blue fluorescence) would be expected to fluoresce green if judged by normal color blending rules, but under

Formulations for Fluorescent Enamels

Color	Dye or resin	Dye g.	Resin g.	Butanol ml.	Xylene ml.	Carbitol ml.	Butyl cellosolve ml.	Total formula ml.	Amount of dye in dry film %	Notes
Yellow	Calco Fluorescent Yellow AB	0.5	..	..	..	7.5	17.5	25.0	0.8	Dissolve dye hot
	Beetle 227-8	..	61.8	38.6	38.6	..	..	139.0	..	
Green	Calco Fluorescent Yellow AB	0.5	..	..	..	7.5	17.5	25.0	0.4	Dissolve hot and filter
	Calcocid Milling Fast Green C.R.	0.5	..	25.0	77.2	..	..	25.0	0.4	
	Beetle 227-8	..	123.5	77.2	77.2	..	..	278.0	..	
Red	Calco Rhodamine B Stearate	6.25	140.0	20.0	25.0	..	..	50.0	4.27	
	Beetle 227-8	..	140.0	88.0	88.0	..	..	315.0	..	
Orange	Calcozine Red 6G Ex.	0.5	..	12.5	12.5	..	..	25.0	0.715	
	Beetle 227-8	..	69.4	43.4	43.4	..	..	156.0	..	
Blue	Calcomine Fluorescent Violet G	1.6	..	..	..	12.0	28.0	40.0	0.9	Dissolve hot and filter
	Ultramarine Blue 1401	8.0	17.8	12.3	14.3	..	..	50.6	4.5	Mix in ball mill for 24 hours
	Beetle 227-8	..	160.2	100.0	100.0	..	..	360.0	..	

ultra-violet light the mixture fluoresces yellow. Here, again, these complicating factors permit unusual effects to be obtained with a little ingenuity in manipulation.

Effect of Vehicles on Fluorescent Dyes

In formulating with fluorescent dyes, care must be exercised in selecting the vehicle. For example, Calco Rhodamine B Stearate formulated with Melmac 245-8 changes on baking to a deep maroon with no fluorescence. However, when formulated with Beetle 227-8 and baked, no change occurs.

Still more important in the selection of the vehicle is its effect on the permanence of fluorescence to visible and ultra-violet light exposures. Not all dyes are equally affected, but differences will be observed. When formulated in cellulose nitrate at less than 1 hour of exposure, the fluorescence had almost completely disappeared. By contrast, when formulated in the Beetle 227-8, even at 128 hours, about 50% of the original intensity was retained.

Of course, it must be remembered that organic dyestuffs do not, in general, exhibit the permanence of pigment-type fluorescent compounds. Therefore, fluorescent dyes should not be used in coatings which are to be exposed to ultraviolet, daylight, or incandescent light for extended periods of time. The usefulness of fluorescent dyes is the greatest where extreme brilliance without fastness is required.

Effect of Base to Which Coatings Are Applied

Since many fluorescent lacquers are transparent, it is obvious that the texture of the base, its color and other properties will have a marked influence on the intensity of a fluorescent coating. The greatest loss of fluorescence is caused by the color of the base. Maximum brilliance is obtained over white. Theoretically, all other factors being equal, the least fluorescence is obtained over black.

Heat Stability of Fluorescent Dyes

The application of fluorescent dyes in baked enamels, molded plastics and other finishes requires a greater or lesser degree of heat stability for their successful use. Most of these applications reach their maximum temperature below 200°C.

In general, the fluorescent dyes are stable to 200°C. In certain cases they tend to volatilize at given temperatures above which there will be a loss in color values. However, in all cases, the dyes have retained their fluorescence.

Inasmuch as specific engineering conditions are unpredictable and because most plastics vary in their chemical reactivity, the ultimate results obtained may be influenced by many factors. Therefore, only trials will determine the suitability of any coloring agent for a specific plastic.

In preparing ethyl cellulose Plastic Peel, the resins, plasticizers, mineral oil, and stabilizer, all ingredients except the ethyl cellulose are heated to 190°C. (375°F.) preferably. The ethyl cellulose is then added as rapidly as possible, so that each particle will be wet before the viscosity of the mixture is increased by the dissolving ethyl cellulose. The temperature is then

Hot Melt Coatings
Plastic Peel
(Hot melt strippable protective coatings)
Formula No. 1
Light Colored

Ingredient	Amount
Ethyl Cellulose N–50	30
Piccolastic A–50	20
Staybelite Ester No. 10	25
White Oil L–1	25
Paraffin Wax	3
Menthylphenol	1

	No.2	No.3	No.4	No.5	No.6	No.7	No.8	No.9	No.10
Ethyl Cellulose N–50	25	25	25	25	25	25	25	25	25
Lopor 45 (White Oil)	67	67	67	67	67	67	67	67	67
Paraffin Wax	3	3	3	3	3	3	3	3	3
Age-Rite Alba	1	1	1	1	1	1	1	1	1
Cumar P–10	5								
Beckosol 24		5							
Ester Gum 8L			5						
Petrex 130H				5					
Staybelite Ester No. 10					5				
Nevinol						5			
Raw Castor Oil							5		
Baker's No. 15 Oil								5	
Dioctyl Phthalate									5

No. 11

Ingredient	Amount
Ethyl Cellulose N–50	25
Baker's No. 15 Castor Oil	10
Lopor 45	62
Paraffin Wax	3
Menthylphenol	1

raised again to 190°C. (375°F.) and held there until all the ethyl cellulose is dissolved (10 to 15 minutes). Keep at 190°C. until all bubbles disappear.

No. 12

Ingredient	Amount
Ethyl Cellulose	30
Beeswax	35
SAE Lubricating Oil No. 40	15
Castor Oil	10
SAE Lubricating Oil No. 60	5

NO. 13

Ingredient	Amount
Ethyl Cellulose	25
Beeswax	30
SAE Lubricating Oil No. 40	10
Paraffin Wax	25

Melt all the ingredients and mix in the molten state.

Hot Melt Coating
Formula No. 1

A hot melt combination of 55% of Acrawax C and 45% ethyl cellulose N-100 forms an extremely hard, tough high melting point product which exhibits a minimum of shrinkage and gives a high gloss finish. This melt shows unusual tenacity in its adherence to surfaces such as glass and metal and of course also is noteworthy for water insolubility, solvent, oil and grease resistance, and is extremely hard.

No. 2

A hot melt coating composition adapted to form non-blocking coatings, consists essentially of:

Ethyl Cellulose (47–49%
  Ethoxyl, Viscosity Be-
  low 40 Centipoises)   10
12-Hydroxy Stearic Acid   44
Compatible Oil-Soluble
  Phenol-Formaldehyde
  Resin   18
Paraffin Wax   28

Aminoplast Resins

Impregnation With Aminoplast Resins

Set plaster casts can be impregnated with Durite aminoplast resins by dipping, spraying or brushing. (Resins may be diluted with up to equal parts of water.) Depending on the time the casts are immersed in the solution, penetration will range from surface coating to complete saturation with decided improvement in all the mechanical properties. Spraying or brushing will provide case hardening with subsequent improved surfaces of the casts, as well as increased resistance to weathering, water, weak acids and alkalies.

After the cast is impregnated to the desired depth, the resins are cured by heat in the same manner as previously indicated for plaster-resin mixes, care being exercised not to overcure at high temperatures or for too long a time.

Preparation and Lubrication of Molds

The addition of Durite resins to the plaster mix does not call for any changes in the preparation or lubrication of the mold over that employed for casting plain plaster or Hydrocal mixes. Many different types of lubricants or parting compounds are used.

Korogel (Goodrich special synthetic rubber) molds require no lubrication.

Natural or synthetic rubber molds should be brushed well with a soap solution, infrequent application only being needed.

Wood or plaster molds should first be shellacked and then be sanded smooth. Then apply, as frequently as may be needed, a thin coating of:

Johnson's Glo-Coat Wax or equivalent, or

a thin solution of paraffin wax, cut with kerosene, or with a mixture of about 60 parts benzol and 40 parts toluol by volume.

Glue or gelatin molds are generally of two types—

(a) A mixture of glue and water which dries hard. Considerable shrinkage occurs during the drying. This type is most generally used. Treat these molds as though they were made from wood or plaster, but omit the shellacking operation.

(b) Flexible glue, such as Wacko and other trade-named brands, does not dry hard or shrink. Follow the standard practice of first coating the mold with a water solution of potash alum, allow it to dry and then lubricate it in the same way as if it were made from wood or plaster.

Impregnation With Resorcinol Resins

Durite resorcinol type resins, setting at room temperatures of 75°F. or over have been especially developed for plaster of Paris induration where the cured cast must be able to withstand higher heats, where excellent resistance to the elements is desired and where heating to effect cure is impracticable. The colors of these resins render them unsuitable wherever pastel shades or natural white plaster finish is required.

Durite 3026P is a dark-colored resin syrup used in conjunction with a white powdered coagent 3026AP. When kept tightly closed and in a dry place at normal room temperatures both of these materials have an indefinite storage life.

Mixing Formulas

The following table shows formulas for five mixtures ranging from the heaviest to the lightest recommended consistencies.

Aminoplast Resin-Plaster Mixes

Plaster-resin mixes are prepared, using either ordinary plaster or Hydrocal, as shown below.

Formula	No. 1 Industrial White Molding Plaster Mix	No. 2 Industrial White Hydrocal Mix
Water	25	20
Aminoplast Resin Mix*	35	25
Plaster of Paris	100	..
Hydrocal	..	100
Potash Alum	1	1

* Aminoplast Resin Mix.
DURITE 2979 (Colorless syrup)   100
DURITE 2983AA (White powder)   20
DURITE 2987A   (White powder)   8

2987A is preferably ground with a little of the 2979 to prevent segregation of the former; the remaining ingredients are then added and thoroughly mixed. Working life of the prepared mixture is at least 96 hours at 70°F. Beyond 4 days, the viscosity of the mixture may be undesirably heavy. Individually, the three component materials have useful lives of between 1 and 2 years if kept sealed and at the lowest practical temperature.

Formula No.	3	4	5	6	7
Water	30	35	35	45	40
Durite 3026P	30	30	35	30	40
Durite 3026AP	6	6	7	6	8
Plaster of Paris	100	100	100	100	100
Set Time in Minutes	20	25	25	30	35

Mix 3026P with the water; 3026AP is then added to the solution, stirring for approximately 5 minutes. Following this, the plaster is slowly sifted in, allowed to soak 2 or 3 minutes, then stirred to a smooth, lumpfree mass.

NOTE: In general, the weight of the resin should not exceed the weight of water as shrinkage will occur when the resin content of the cast is too high. Where a vibrator is used in casting operations, a heavier consistency of mix is both permissible and desirable, making

After combining the water and resin, the plaster is sifted in, allowed to soak for a few minutes and then stirred until all lumps are removed. At this time the alum is added. The set time for the Hydrocal mix is 20 minutes and for the plaster of Paris 20 to 25 minutes. If a shorter setting time is desired, additional alum can be added without injury to the cast.

NOTE: In casting large pieces, a slower set time may be required. To allow 45 minutes initial set time, Durite coagent 2983AB should be substituted in identical weight for 2983AA in making the aminoplast resin; and the alum content of Formulas 1 and 2 should be reduced to 0.75 parts.

NOTE: All the above formulas, using U. S. Gypsum Co. products, provide a pH of 5. If the pH is too low, the plaster will set too quickly; if the pH is too high, the resin will not cure properly. Colorimetric indicators may be used to determine the pH of the plaster slurry.

Acidity of Durite aminoplast-plaster mix:

Too much emphasis cannot be placed on obtaining the proper pH value in the aminoplast resin-plaster slurry. When using, as recommended, either industrial white molding plaster or industrial white Hydrocal the pH of the slurry should normally not be below 4.0 or higher than 5.7. The pH may be raised or lowered, respectively, by decreasing or increasing the amount of alum used. If under 4.0 in pH, the mixture will set too rapidly with a very rapid rise in setting or crystallization temperature accompanied by an excessive sweating out from the cast of a solution of resin and water. If the mass is large and confined the setting temperature may rise to as high as 175–180°F. Conversely, if the pH is in excess of 5.7, the set time will be prolonged and the resin will not cure properly when the indurated casts are cured at the required 150–175°F. temperature.

Curing aminoplast treated casts:

The curing temperature should rise gradually at the rate of about 1°F. per minute to 150–175°F. It should be held within that range for as long as may be necessary to cure the resin at all points on the heaviest cross section, or until acceptable improvement in physical properties of the cast has been obtained. The duration of the cure at 150–175°F. will vary depending on the shape and size of the cast, oven conditions with respect to air circulation, etc., and can best be established by tests of trial samples of the particular part under consideration. If facilities permit and production volume warrants, the end point for the cure at the 150–175°F. oven temperature can definitely be determined by inserting or casting in a thermo-couple at the center point of the heaviest cross sectional area of the cast. After the thermo-couple reaches 150°F., continue the cure for 8 hours at 150°F. or for 4 hours at 175°F. after the couple reaches 175°F.

In cases where the curing cycle must arbitrarily be established, the following hours at 150–175°F. oven temperature are suggested after the oven temperature has reached 150°F. by a gradual rise from room temperature at the rate of about 1°F. a minute.

For sections up to
  1 in. thick   10 hours
For sections up to
  2 in. thick   15 hours
For sections up to
  3 in. thick   20 hours
For sections up to
  4 in. thick up 30 hours up

These figures are based on the assumption that the curing cycle is carried out in a well ventilated, circulating hot air oven of the atmospheric type.

Care should be observed when using glue molds that the cast be removed before excessive heat is generated as the glue will start to disintegrate at about 140°F.

Coloring of aminoplast-treated casts:

This may be accomplished by the addition of acid-tolerant pigments or dyeing the aminoplast resin solution or the resin-plaster slurry with aniline dyes of the acid tolerant type such as Calco Chemical or National Aniline normally furnish for use with urea-formaldehyde resins or adhesives. The hard surface of the finished cast is particularly suitable for practically all coloring or finishing mediums.

Yields:

Approximately 119 lb. of dried cast is obtained from each 100 lb. of plaster or Hydrocal in a plain water mix. This basic yield weight is increased proportionately, in the case of Durite treatments, by the solids content of the Durite resin which is incorporated in the mix, or impregnated into the dried plain plaster cast.

Resin Curing

After setting, the casts are subjected to heat to effect cure of the resin. Infra-red lamps have proven satisfactory for this operation as well as ovens or circulating hot air. The correct heat range is 150°F. to 175°F. to assure the proper qualities in the final product, and for best results, the heat should be slowly raised to the maximum temperature (from 100°F. at the rate of approximately 1° per minute). Higher temperatures tend to distort or crack the casting while room temperature drying will not produce the proper strength qualities. A certain amount of leaching and sweating will normally occur after the initial set due to the heat of crystalliaztion. This is particularly noticeable in conditions of extreme humidity or if the resin curing temperature is brought up too rapidly.

Resorcinol Impregnating Resins

The physical properties of set plain plaster cast or molded pieces are enhanced by impregnation with resorcinol resin solutions. For this purpose, however, a lower viscosity resin is recommended, containing coagent 3026ANF rather than the previously mentioned 3026AP. The suggested formula is as follows:

Formula No. 1

Durite 3026P	100
Water	50
Durite 3026ANF	9

Ingredients are mixed in the order given and stirred to secure complete homogeneity. The batch should be stirred from time to time after initial mixing. The mixed material has a working life of approximately 3 hours at 70°F., which is also the proper temperature for the mixing operation.

Impregnation is accomplished by dipping, spraying or brushing, depending on the depth of penetration desired. Drying and curing is effected at room temperature of 75°F. or higher in the same manner as resorcinol resin-plaster mixes previously stated.

Increasing Impact Strengths

Impact strength of resin indurated plaster casts does not show the marked improvement of the other mechanical properties. To increase materially such shock resistance, the addition to the resin mix of fibrous material, i.e., sisal or fiber glass, is recommended. As an example, sisal, chopped ¼ to ½ in. lengths is incorporated in the mix as follows:

Formula No.	1	2
Industrial White Hydrocal	100	..
Plaster of Paris	..	100
Water	20	30
Durite Resorcinol Resin 3026P	..	30
Durite Resorcinol Resin 3026AP	..	6
Durite Aminoplast Resin Mixture	25	..
Potash Alum	1½	..
Sisal	1½	1½

After curing standard ½ in. bars of Formula 1 for 3 hours at 175°F., average notched Izod impact strengths of 0.460 ft. lb. per inch square were obtained. Hydrocal casts, plain or impregnated with resin but without sisal exhibited impact strengths of about 0.110 ft. lb. To further demonstrate improved impact strengths, thin ornamental shapes having narrow cross sections were made, using resin indurated Hydrocal with and without sisal. Those containing sisal withstood repeated dropping on a concrete floor from a height of 5 ft., whereas those without sisal broke on the first drop. When sisal is added to a resin-plaster mix as in Formula 2, impact strength increases from 0.128 to 0.263 ft. lb. per inch square.

Formulas and Tests for Specific Industrial Application

Special formulas have been developed for specific industrial applications. In the following, Formula 1 is especially recommended for:

(a) Checking fixtures for contoured parts
(b) Molds or mandrels for low pressure laminating
(c) Master patterns and Keller duplicating models

Note: In (c), resin indurated forms must be given an additional surface coating of Durite Resorcinol impregnating resin.

Formula 2 is especially recommended for:

(d) Form blocks for forming acrylic sheets
(e) Stretch press dies and hydro press form blocks

Mixing Instructions

Formula No.	1	2
Water	25	25
Durite 3026P	20	20
Durite 3026AP	4	4
Industrial White Hydrocal	100	100
Fiberglas No. 22 (Chopped ½ in.)	..	0.5

The water and two resin ingredients are thoroughly mixed for approximately 5 minutes. After this, the Hydrocal is slowly sifted in and allowed to soak for 2 or 3 minutes. The fiber glass is added, in the case of Formula 2, and the mixture stirred to a smooth, creamy consistency. The approximate set time of each formula is 15 to 20 minutes. After the initial set, casts are removed from the molds; drying and curing are accomplished at room temperature of 75°F. and over or at oven temperatures up to a maximum of 175°F. for a time cycle dependent on the temperature used and on the size of the cast or molded piece.

Surface coating with Durite impregnating resin solution* will increase the surface hardness from 6 to 7 Moh's scale, and the water absorption is decreased from 11.9% for formula 1, and 11.4% for formula 2, to 8.15% and 8.05% respectively. In this instance, samples tested were immersed for 1 hour in resin solution and cured at 100°F.

For increased impermeability, casts may be further coated with Durite impregnating solution or treated with synthetic materials of the vinyl group or polymers of vinylidene chloride.

Plastic Packings

So-called plastic packings are continuous length or molded ring packings composed of asbestos fiber, graphite or mica, 0 to about 50% soft metal such as slit foil or wire of lead, copper or aluminum chopped into short pieces (lead sometimes is granulated), and rubber, synthetic rubber or similar binders. Rubber bonded types are made in (1) water resistant and (2) oil resistant modifications. Manufacturing operations include formation of a cement in a paddle mixer or tumbler, mixing of cement with fibrous and powdered ingredients, evaporation of excess solvent, and extrusion through a die by means of a tubing machine, either screw or hydraulic plunger type.

* Formula for the Resin Solution:
Durite 3026P	100
Durite 3026ANF	9
Water	100

Water-Resistant Plastic Packing
GR-S Cement Binder

GR–S Regular	1 lb.
Toluene	7 pt.

Give GR–S the usual preliminary mastication on a rubber mill or in a Banbury mixer. Sheet and slice, add to toluene in a paddle mixer or tumbler and run until free from lumps.

Plastic Mixture

GR–S Cement	1 gal.
Process Oil	1 lb.
Graphite or Mica	10 lb.
Chrysotile Asbestos Fiber	5 lb.

Plastic Mixture with Lead

GR–S Cement	1 gal.
Process Oil	1 lb.
Graphite or Mica	8 lb.
Lead (Chopped Foil or Wire)	10 lb.
Chrysotile Asbestos Fiber	5 lb.

The plastic mixture is made in a Z blade mixer by adding powdered and fibrous material alternately to the cement and plasticizer. When uniformly mixed (requiring 2 or more hours depending on the batch size), the material is dried to a crumb (with mixer open) which will just hold together when squeezed. The non-metallic plastic will contain from 15 to 20% solvent by weight at this stage.

Where the stock is to be tubed and molded into rings, the asbestos fiber used must be of the short form and the non-metallic composition must be employed. Where the stock is to be run off in a continuous coil, spinning grade fiber can be used.

Oil-Resistant Plastic Packing
Buna N Cement Binder
Buna N (Hycar, Perbunan, Chemigum, Butaprene)	1 lb.
Methyl Ethyl Ketone	3 pt.
Toluene	4 pt.

The Buna N is given a minimum mastication before it is added to the cement mixer.

Plastic Mixture
Buna N Cement	1 gal.
Bardol B	1 lb.
Graphite or Mica	10 lb.
Chrysotile Asbestos Fiber	5 lb.

Mixing procedure is the same as for water resistant plastic. Two pounds of graphite or mica may be replaced by 10 lb. of lead or an equivalent volume of copper or aluminum slit and chopped foil, or chopped wire.

Where the plastic is to be run in continuous length form, the fiber content may be increased at the expense of the graphite. Usually, it is necessary to apply an open braid of fine cotton thread to the plastic as it emerges from the extrusion machine to reinforce it during storage and shipping.

Gasket Composition
U. S. Patent 2,368,118
Blown Plasticized Asphalt	10–20
Vulcanized Corn Oil	15–50
Gilsonite	10–25
Short Asbestos Fibers	25–50
Paraffin Wax	1– 8

Molded Friction Material
(For brake linings, clutch-facings, etc.)
U. S. Patent 2,175,480
Asbestos Fiber	40–65
Lead Oxide	1–5
Albumin	1–4
Sulfurized Linseed Oil	1–18
Friction Material	To make 100

Silastic Gaskets
Crepe Silastic, 150, 160, 167 or 180, may be molded into heat resisting gaskets by sheeting out and molding in a platen press or by extruding and vulcanizing. Inorganic pigments must be introduced under cold roll conditions. To facilitate handling, a light dusting of talc may be employed. Press and cure at 500 lb. per square inch at 212 to 300°F. for 4 hours or longer, depending on the temperature. Place in the mold and remove from the mold at 100°F. or less.

Compressed Asbestos Sheet
What is known in the trade as compressed asbestos sheet is made in two basic types, (1) water-resistant and (2) oil-resistant sheets. Manufacture involves making a cement with the rubber binder and then kneading in the asbestos fiber in a dough mixer, as a preliminary step; the next operation of actually forming the sheet is done on a sheeter, the dough being squeezed between two gradually separating rolls so as to build up to the desired thickness on the larger of two rolls. Finishing operations include calendering to thickness and curing either in an air oven or in a platen type press.

Water-Resistant Compressed Asbestos Sheet
Rubber	35
GR–S (85 Mooney)	65

Sulfur	3
Accelerator	2
Zinc Oxide	5
Pigment (Carbon Black, etc.)	18
Barytes	60
Asbestos Fiber, Cleaned Shorts (Plus Suitable Longer Fiber)	450
Solvent Naphtha (165–235°F.)	650

The solvent must be drawn off from the sheeter with a sufficient volume of air. It can be recovered and reused.

Oil-Resistant Compressed Asbestos Sheet

This sheet is made in either a neoprene or Buna N type. A typical neoprene type formula is as follows:

Neoprene GN	100
Stearic Acid	1
Zinc Oxide	5
Magnesia	5
Pigment (Carbon Black, etc.)	15
Barytes	425
Toluene	700

Foamed (Sponge) Plastic

The process consists of whipping air into a special solution and then hardening at elevated temperatures. The solution used is prepared by condensing 200 parts of 30% formaldehyde with the pH adjusted with 60 parts of urea, dry and cold, neutralized to 8.0 pH. Fifty parts of hexanethiol are added and the solution is heated to 95°C. (203°F.) until 1 part of solution at 20°C. (68°F.) clouds with 5 parts of water. At this point, more urea is added until the ratio of formaldehyde to urea is 1.7 to 1. Condensation is continued until 1 part of solution at 20°C. clouds with 1.5 parts of water. The pH is then adjusted to 8.0.

A second solution is then prepared containing 15% phosphoric acid, 10% resorcin, 10% *Nekal BX* (sodium dibutylnapthyl sulfonate), and 65% water. This is the foaming solution.

In a 300-gal. kettle the following mixture is whipped into foam by stirring at 200 rotations per minute:

Foam Solution	0.3 l.
Water	18.0 l.
Resin Solution	10.0 l.

Dissolve the foam solution in 3 l. of water.

The liquid foam is run promptly into forms and set in 2 hours. It is then dried at 40 to 60°C. (104 to 140°F.) for 6 days, shrinking about 20% while drying. The resulting slabs (20 in. by 20 in. by 0.8 in. to 8 in.) weigh 15 kg. per cubic meter (about $\frac{1}{50}$ the weight of cork), and have a compressive strength of 0.2 kg. per square meter (2.9 psi).

The second method of foaming involves the addition of special agents designed to release gas under proper conditions and thus foam the soft plastics material. Experimental work had been done on polystyrene, the phenolics, and cellulose acetate using such foaming agents as ammonium carbonate. With this material, it had been possible to obtain some experimental moldings about 4 in. thick. Much more successful work is based on the use of the German product known as *Porophor N*. Chemically this product is azoisobutyric dinitrile. It has a melting point of 103 to 104°C. (218 to 220°F.), and when heated above 120°C. (248°F.) rapidly releases nitrogen. *Porophor N* had been used extensively with rubber and with polyvinyl chloride. Experimentally it had been used with some success with polystyrene. A typical formula for producing polyvinyl chloride type foam having a density of 0.05 (3 lb. per cubic foot) is as follows:

Polyvinyl Chloride Type P	48
Tricresylphosphate	24
Mesamoll (Mepasin Sulfophenolate)	8
Porophor N	20

The general process from either paste or plastics is to mold small pieces at 160°C. (320°F.), pressure 400 to 600 kg. per square centimeter (5800 to 8700 psi) for 5 to 7 minutes. The mold is then cooled slightly, and the molding removed and allowed to stand at room temperature for about 1 hour. During this time, the volume increases 6 times. It is then immersed in hot water at 80°C. (176°F.) for 12 to 30 minutes. During this time, the final volume is reached.

Rubber Compositions

Synthetic Sponge (Cellular) Rubber

TABLE 1

Formula	No. 1	No. 2	No. 3	No. 4
GR-S (15 Minutes Break)	100.0	100.0	100.0	100.0
MT Thermal Black	....	10.0	20.0	20.0
SRF Furnace Black	40.0	5.0	....	....
HMF Furnace Black	....	....	20.0	....
Whiting	....	....	....	40.0
Zinc Oxide	3.0	3.0	3.0	3.0
Benzothiazyl Disulfide	....	....	....	1.0
Tetramethyl Thiuram Monosulfide	....	....	....	0.1
N-Cyclohexyl 2-Benzothiazole Sulfonamide	1.5	1.5	1.5	....
Process Oil (Circo)	50.0	35.0	40.0	....
Dispersing Oil No. 10	....	....	....	45.0
Phenyl Beta Naphthylamine	1.0	1.0	1.0	1.0
Sulfur	2.5	2.5	2.5	2.5
Blowing Agent No. 15*	8.5	5.0	6.5	6.0
PHYSICAL PROPERTIES				
Press Cure at 324°F. (Minutes)	15	15	15	15
311°F. (Minutes)	20	20	20	20
% Blow	320	250	220	220
Apparent Density (Oz. Per Cubic Inch)	0.15	0.17	0.20	0.22

* Blowing Agent No. 15 is a mixture of powdered biuret and urea.

When special properties are desired, it is sometimes advisable to use more than one blowing agent in a formula. In this way the advantages of each blowing agent can be obtained in the finished product.

Table 2 illustrates the use of Blowing Agent No. 15 with diazoaminobenzene and with sodium bicarbonate in GR–S. Diazoaminobenzene is useful in producing cellular products with a very fine cell size, but cannot be used in many applications owing to its tendency to stain fabric or metal panels which come in contact with the rubber. By use of a standard amount of Blowing Agent No. 15 and a small amount of diazoaminobenzene, as in formula No. 6, it is possible to obtain a product with fine cell structure, but a reduced tendency to stain.

Buna N Cellular Products

It is more difficult to produce cellular products from Buna N than from GR–S because of the less plastic quality of the raw polymer and the greater difficulty of plasticizing on the mill. By proper compounding, however, Buna N products can be made fully as satisfactory as those from GR–S.

TABLE 2

Formula	No. 5	No. 6	No. 7
GR–S (15 Minutes Break)	100.0	100.0	100.0
MT Thermal Black	20.0	20.0	20.0
HMF Furnace Black	20.0	20.0	20.0
Zinc Oxide	3.0	3.0	3.0
Benzothiazyl Disulfide	....	....	1.0
Tetramethyl Thiuram Monosulfide	....	....	0.1
N-Cyclohexyl 2-Benzothiazole Sulfonamide	1.5	1.5	....
Process Oil (Circo)	40.0	40.0	40.0
Phenyl Beta Naphthylamine	1.0	1.0	1.0
Stearic Acid	....	....	15.0
Blowing Agent No. 15	6.5	6.0	3.0
Diazaminobenzene	....	0.2	....
Sodium Bicarbonate	....	....	15.0
PHYSICAL PROPERTIES			
Press Cure at 324°F. (Minutes)	15	15	15
% Blow	220	220	180
Apparent Density	0.20	0.20	0.23
Cell Size	Medium	Fine	Medium

TABLE 3

Formula	No. 8	No. 9	No. 10
Chemigum N–1 (15 Minutes Break)	100.0	....	....
Butaprene NM (15 Minutes Break)	....	100.0	....
Hycar OR-15 (15 Minutes Break)	....	....	100.0
SRF Furnace Black	30.0	....	....
FT Thermal Black	....	40.0	....
MT Thermal Black	....	....	60.0
Zinc Oxide	3.0	3.0	3.0
Stearic Acid	0.5	0.5	0.5
Phenyl Beta Naphthylamine	1.0	1.0	1.0
N-Cyclohexyl 2-Benzothiazole Sulfonamide	1.3	1.2	....
Benzothiazyl Disulfide	....	....	1.5
Tetramethyl Thiuram Disulfide	....	....	0.1
Dibutyl Phthalate	80.0	75.0	80.0
Sulfur	2.0	2.2	2.0
Blowing Agent No. 15	10.0	10.0	12.0
PHYSICAL PROPERTIES			
Press Cure at 311°F. (Minutes)	30	30	30
% Blow	240	230	210
Apparent Density (oz. per cubic inch)	0.19	0.20	0.22

To compensate for the nerve of Buna N it is suggested that those grades of Buna N easiest to break down on the mill be used. In addition it is necessary to incorporate rather large quantities of plasticizers.

Cellular GR–S Ebonite

When Blowing Agent No. 15 is incorporated into an ebonite (hard rubber) stock, a very interesting type of cellular product is obtained. Since Blowing Agent No. 15 produces a majority of closed cells when gas is evolved, the result is a light weight, but strong product which will float on water indefinitely since there is no way that the water can penetrate through the bulk of the material.

Production of cellular ebonite from GR–S presents a greater problem than encountered with soft GR–S because the stock must be held inflated for a longer period of time before the cure is completed. Formula No. 11, Table 4, has given fairly good results in the laboratory, but a better product can probably be obtained by experienced compounders.

Table 4

Formula	No. 11
GR-S (15 Min. Break)	100.0
SRF Furnace Black	30.0
MT Thermal Black	50.0
Zinc Oxide	3.0
Unsaturated Petroleum Softener	70.0
N-Cyclohexyl 2-Benzothiazole Sulfonamide	3.0
Phenyl Beta Naphthylamine	1.0
Sulfur	50.0
Blowing Agent No. 15	9.0
Physical Properties	
Press Cure at 324°F. (Minutes)	105.0
% Blow	280.0
Apparent Density (Oz. per cubic inch)	0.20

Table 5

Formula	No. 12	No. 13
Whole Tire Reclaim	100.0	100.0
MT Thermal Black	40.0	....
Whiting	....	40.0
Zinc Oxide	3.0	3.0
Stearic Acid	2.0	2.0
Benzothiazyl Disulfide	1.0	1.0
Tetramethyl Thiuram Monosulfide	....	0.2
N-Cyclohexyl 2-Benzothiazole Sulfonamide	0.2	....
Phenyl Beta Naphthylamine	1.0	1.0
Process Oil (Circo)	30.0	25.0
Sulfur	3.0	3.0
Blowing Agent No. 15	5.0	5.0
Physical Properties		
Press Cure at 311°F. (Minutes)	15.0	15.0
% Blow	360	450
Apparent Density (Oz. per cubic inch)	0.16	0.14

Reclaimed Rubber Products

Most compounders have had more experience in working with reclaimed rubber than with GR–S and Buna N. The formulae of Table 5 are therefore offered merely as guides in converting present formulae to the use of Blowing Agent No. 15 and are not suggested as finished commercial compounds.

Windshield Wiper Compounds

Color—Gunmetal:

Smoked Sheets	77.7
SPDX–G	0.36
Phenex	0.27
Spider Brand Sulfur	1.9
Zinc Oxide	3.0
Paraffin	0.75
Stearic Acid	0.4
Barytes	13.0
Calcene	6.0
Stabilite	1.5
Carbon Black	0.02
Tensile Strength	3900 lb.
Specific Gravity	1.11

Color—Black:

Smoked Sheets	100.0

SPDX–G	0.47
Phenex	0.35
Black Bird Sulfur	2.0
Zinc Oxide	5.0
Carbon Black	3.0
Paraffin	0.8
Calcene	10.5
Stabilite	2.0
Tensile Strength	4150 lb.
Specific Gravity	1.03

Cure for Both Compounds 8 min. at 307°F.

Synthetic Rubber Thread (Neoprene)

Neoprene GN	100
Sodium Acetate	1
Stearic Acid	2
Ceresin Wax	½
Agerite White	1
Philblack (Carbon Black)	18
Calcined Magnesia	2
Zinc Oxide	4

Rubber Eraser Stock

Factice	50
Rubber	50
Sulfur	3
Zinc Oxide	5
Captax	2
Fine Carborundum	80

Cure for 60 minutes at 270°F.

Plastic (Heat Hardening) Rubber

Rubber	100
Rosin	25
Shellac	25
Zinc Oxide	60
Mercaptobenzolthiazole	3
Sulfur	3

Cure for 60 minutes at 300°F.

Rubber-to-Brass Valve Stem Adhesion

Rubber	100.0
Captax	0.7
Agerite Powder	0.7
Zinc Oxide	15.7
Sulfur	3.4
Gastex	23.0
Carbon Black (M P B)	14.0
Stearic Acid	2.3

Press cure 15 minutes at 320°F.

Softening Rubber Articles

Practically all rubber articles which have grown hard and lost their elasticity may be softened by using glycerin.

First cleanse the article by scrubbing thoroughly with a brush dipped in warm water and place in a solution of 1 part of ammonia to 2 parts of water, allowing it to remain 1 hour or so, until the ammonia has evaporated. Then rinse with a dilute solution of glycerin and water. Wipe off and allow to dry.

To Reclaim Rubber

Formula

	No. 1	No. 2	No. 3
Scrap Rubber	100	100	
Albone C	10	10	50
Sun Reclaiming Oil		10	50
Aquarex			2

Heat Treatment:
24 hours—24 hours
at 308°F. at 308°F.

Dip or paint on article and cure 2 hours at 250°F. Refine on a mill.

No. 4

Sun Reclaiming Oil	50
Oleic Acid	16
Triethanolamine	3
Hypochlorite (Oxol)	31

Add the triethanolamine to the Oxol. Add the oleic acid to the oil. Then add the oleic acid oil mixture to the Oxol mixture with vigorous stirring.

Add to the scrap. Heat 2 hours at 250°F.

Anti-Oxidant Film for Ultra-Violet Ozone Protection of Rubber Goods

Neoprene	60
Dimethyl Trihydroxy Quinoline (Polymerized)	10
Benzene	400

Mold Wash for White Rubber Goods

Soap Bark Chips	2 oz.
Sugar	2 oz.
Water	12 gal.

Boil 5 minutes and strain.

Boring Holes in Rubber Stoppers

Bore a hole about ¼ in. or less with a cork borer wet with water. Pour a few drops of the following mixture into the cut:

Diglycol Laurate	15
Water	85

Continue using the cork borer, not pushing too hard. If the borer sticks, remove and apply more of the above fluid. After the hole is completed wash out with water.

Anti-Tack Coating for Asphalt Tiles

Ethyl Cellulose	100
Blown Castor Oil	8
Ethyl Acetate	385

Making Flexible Molds from Neoprene Latex

Small statuary, novelties, and plaques of simple or complex design may be reproduced most conveniently through the use of flexible molds made of rubber or synthetic rubber. Such molds are usually made in one piece and their flexibility minimizes difficulties with undercuts and sharp angles. Latex is the best source of rubber or synthetic rubber because relatively thick films may be deposited over the surface of an article in a very short time. Further, suitably prepared latex is readily available and expensive or complicated equipment is not required for its use.

Neoprene latex serves especially well for this purpose because films from it have very good heat, oil, and abrasion resistance. Neoprene molds reproduce fine lines and detail fully as well as does rubber. They may be used for casting at temperatures up to 300°F. and possibly higher. This permits the use of low melting alloys. The life of neoprene molds is exceptional.

Preparing the Latex

It is necessary to compound neoprene to obtain best results. The following dry basis recipes are typical:

Typical Compounds

	Formula No. 1	No. 2
Neoprene from Type 571 Conc. Latex	100.0	100.0
Zinc Oxide	25.0	25.0
Neozone D	3.0	3.0
Natural Whiting	30.0	150.0
Aquarex D	1.5	1.5
Methyl Cellulose (25 cps.)	...	1.0

These compounds may be prepared in the conventional manner. Convenient recipes for dispersing the insoluble materials in water are shown below:

Master Dispersions

Formula	55% Solids For No. 1	60% Solids For No. 2
Zinc Oxide	250.0	250.0
Neozone D	30.0	30.0
Natural Whiting	300.0	1500.0
Dispersing Agent Solution (10%)	232.0	712.0
Waterglass (10%)	15.0	45.0
Distilled or Soft Water	228.0	317.0

These dispersions may be ground in a colloid mill 20 to 30 minutes or a pebble mill 24 to 48 hours. Any convenient non-foaming dispersing agent may be used. The Aquarex D and methyl cellulose are added as a 10% aqueous solution. In preparing the compounds, the Aquarex D is added to the latex first. If methyl cellulose is used, it may be stirred into the master dispersion before adding to the latex. The latex compound may be thinned if desired by adding distilled water or the compound may be thickened by adding more methyl cellulose.

The latex compound should be allowed to stand before using until air bubbles come to the surface and can be removed. Very thick compounds may be rendered air free by allowing them to stand under vacuum.

The above compounds are but two of an almost infinite variety of mixtures which may be used. Very satisfactory ready-prepared neoprene latex mixtures may be purchased from various suppliers of compounded latex.

Preparing the Article for Duplication

After the latex compound has been prepared or purchased, the object to be duplicated is prepared for dipping by providing it with a suitable handle or small wire attachment to suspend it in the latex.

Plaster or oxychloride cement molds may be lightly waxed to prevent the latex film from adhering too tightly. Metal molds should be clean and smooth surfaced. Rough metal may be very lightly waxed. Wooden molds are usually unsatisfactory due to the effect of moisture and because latex adheres tightly to wood. Painted and lacquered wooden molds may sometimes be used. Glass or porcelain molds should be cleaned thoroughly before using.

Forming the Flexible Mold

As the first step, the article is dipped into the compounded latex and held for 1 or 2 minutes and withdrawn. The excess latex is allowed to drain and the film is air or oven dried at low temperatures until it is firm. This operation is repeated until a film thickness of 0.03 to 0.1 in., depending on the size of the mold, is produced. Film thickness may be more rapidly built up if the coagulating dip processes* are used. A typical procedure is outlined below:

1. Dip into the latex, withdraw slowly, invert the form if possible to smooth the adhering film.

2. Immediately dip into a coagulant such as the following one:

	Parts by Weight
Calcium Nitrate (Tetrahydrate)	20
Methanol	80

*Coagulating dip processes are the subject of patents held by the American Anode Company and the U. S. Rubber Company.

3. Withdraw from the coagulant and allow to dry until the surface is no longer shiny and very wet.

4. Dip again into the latex and hold until the necessary thickness is deposited (5 to 30 minutes). Alternatively, the form may be dipped into the coagulant first and then into the latex and this procedure repeated until the desired thickness is obtained.

5. Withdraw from the latex. If reinforcement of the neoprene is to be applied it should be started at this point, if not, the form should again be dipped briefly into the coagulant.

6. As the final step, the form coated with neoprene is rinsed, leached for 4 hours in tepid water, dried overnight at room temperature plus 2 to 6 hours at 170°F. in air, and vulcanized 30 minutes at 284°F. If lower vulcanization temperatures are used, longer times should be used; for instance, 90 minutes at 240°F. It is not always necessary to vulcanize molds of relatively simple design, but unvulcanized molds are more easily distorted than vulcanized ones.

Reinforcement of the Mold

After the original film on the form is made, its stiffness and thickness may be increased as desired by covering it with cotton flock which has been thoroughly wet down with either one of the typical compounds shown. The cotton flock paste may be applied by hand or by spatula during step 5 in the process as outlined above. The mold, thus reinforced, is allowed to dry for several days at room temperature before use. The cotton flock coating makes long drying necessary, particularly if the mold is to be vulcanized as outlined in step 6 above. Reinforced molds are not easily distorted but the cotton flock reinforcement may crack if bent at very sharp angles unless the cotton flock has been thoroughly wet down and mixed with latex.

Final Steps to Complete the Mold

Finally, the mold is cut off the form, using a sharp rubber cutting knife or other suitable cutting tools. Usually it is necessary to slit only one side of the mold covering. Care should be taken in slitting to see that the edges of the cut are smooth and will fit tightly together. In pouring operations the mold is held together with cord or tape. Very viscous casting mixtures should be carefully worked into the crevices and detailed parts of the mold with a stirring rod or spatula before the main pour is made.

Wax Compositions

Beeswax Substitute
Formula No. 1

Octadecane Amide	30
Mineral Oil (100 Vis. at 100°F.)	10
WW Rosin	60

No. 2

Rosin	60
Mineral Oil	10
Stearamide	30

Lanette Wax SX Replacement

Cetyl Alcohol	90
Sodium Lauryl Sulfate	10

"Lost" Wax
Carnauba Wax	8
Albacer	2
Burgundy Pitch	3
Venice Turpentine	4

Hard Wax Composition
U. S. Patent 2,374,617
Hydrogenated Castor Oil	30–60
Ester Gum	30–60
Petrolatum Wax	5–25
Petroleum Asphalt	2–25

Condenser Impregnating Wax
Ethyl Cellulose	68
Carnauba Wax	2
Piccolastic Resin	10
Acrawax C	20

Ski Wax
Czechoslovakian Patent No. 42496
Beeswax	15
Rosin	20
Paraffin	25
Coal Tar Oil	30

Melt, mix and freeze to uniform mass. Finally, some powdered aluminum is added.

Stop-Off Wax
(For Stripping Plated Coatings)
Cellosolve Ricinoleate	22
Ethyl Cellulose (N 100)	10
Acrawax C	100

Melt together and mix until uniform. Apply hot.

Wetting Agent for Wax Molds
Isopropyl Alcohol	90
Sulfonated Castor Oil	10

Vinylite Laminating Thinners

	Formula No. 1	No. 2	No. 3	No. 4
Kronisol	5	2.5	2.5	
Methyl Acetate (82%)	58			
Ethyl Acetate (85%)	10			
Butyl Acetate	10			
Methyl Ethyl Ketone		63		40
Dioxane		20		
Isophorone		2.5	2.5	
Methylene Dichloride			50	
Ethylene Dichloride			43	
Cyclohexanone				40
Propylene Oxide				20
Troluoil	15			
Acetic Acid	2	2.0	2.0	
Methanol		10		

Softening Baths for Celluloid and Cellulose Acetate Plastic

Celluloid	Alcohol 4; Acetone 1; Water 1.
Cellulose acetate	Acetone 1; Water 2; *or* Ethyl Acetate alone.
Solvents	
Cellulose nitrate	Ethyl Alcohol; Camphor.
	Acetone (Plus Water for low N content).
	Methyl Alcohol.
	Diacetone Alcohol.
	Methyl, Ethyl, Butyl and Amyl Acetates.

	Methyl Ethyl Ketone.
	Ethyl Lactate.
	Cellosolve.
	Ether-Alcohol mixture, 50:50 or 70:30
Cellulose triacetate	Methylene Chloride 9; Alcohol 1.
	Tetrachlorethane 9; Alcohol 1.
Cellulose acetate	Acetone (Plus Water for low Acetyl content).
	Methyl Acetate.
	Methyl Ethyl Ketone.
	Ethyl Lactate.
	Benzene 1; Alcohol 1, Hot.
Cellulose acetobutyrate	As Acetate.
Ethyl cellulose	Ethyl cellulose is swollen or partly dispersed by alcohols, esters, chlorinated solvents, ketones and aromatic hydrocarbons, but apart from the less usual solvents: Benzyl Alcohol, Glycol Diacetate, Butyl Lactate and Methyl Cyclohexanone, the only satisfactory dispersing agents are Benzene, Toluene or Solvent Naptha 4; Alcohol 1.
	Methylene Chloride 9; Alcohol 1.
	Methyl Acetate.
Polyvinyl chloride-acetate copolymer	Methyl Ethyl Ketone.
	Cyclohexanone.
	Methyl Cyclohexanone.
	Chlorinated Hydrocarbons such as Methylene Chloride and Chlorbenzene.
Formvar	Trichlorethylene 9; Alcohol 1.
	Benzene 7; Alcohol 3.
Methyl methacrylate	Acetone.
	Methyl Ethyl Ketone.
	Methyl and Ethyl Acetate.
	Chloroform.
	Acetic Acid.
Polystyrene	Benzene and Homologs.
	Ethyl, Butyl, Amyl and Hexyl Acetates.
	Methylene Chloride.
	Trichlorethylene.
	Carbon Tetrachloride.

Chapter XVIII

POLISHES

Automobile Polish
Formula No. 1
Diglycol Laurate	5
Turpentine	5
Sodium Fatty Acid Sulfonate (Sulfatate B)	10
Water	500

No. 2
Water	51
Light Petroleum Oil	38
Castor Oil	10
Potash Soap	1

No. 3
Paraffin Oil	24
Linseed Oil	4
Kerosene	4
Polyethylene Glycol Mono Oleate	1

Auto Polishing Cloth
Water White Gasoline	50
Paraffin Oil	50

Pieces of flannel or cheese-cloth are soaked in the mixture and allowed to dry. They are excellent for removal of dust and restoring the gloss of automotive finishes.

Metal Cleaning and Polishing Cloth

In preparing impregnated cloths for cleansing and polishing brass, copper and silver, hard soap is found a good binding agent, as well as cleaner.

Calcium Carbonate	100 g.
Kieselguhr	40 g.
Rouge	8 g.
Water	To make 1 l.

Mix the ingredients and impregnate the cloths. Press out the excess liquid and dry the cloths at 120°F. Then immerse in a hot 10% solution of hard soap. Squeeze out the excess fluid and dry again.

Metal Polish

Chip soap, 10 parts; silica dust, 20 parts; air-floated tripoli, 20 parts; pine oil, 2 parts; water, 48 parts.

Dissolve the soap in the hot water and add the previously mixed silica and tripoli without stirring; then add pine oil, with stirring, and run the hot mixture into suitable containers. The abrasives, silica and tripoli, should be able, almost 100%, to pass through a No. 325 sieve.

Furniture Polish
Formula No. 1
White Mineral Oil	70
Soya Oil Foots Acids	3
Light Blown Castor Oil	1.5
Glaurin	0.75
2–Amino–2–Methyl–2–Propanol	36.0
Water	40.0

No. 2
Mineral Oil	256
Steam Distilled Pine Oil	10
Blendene (Emulsifier)	73
Water	301

The mineral oil, pine oil and Blendene are thoroughly mixed. It is important that the liquid be clear. If not clear add Blendene in small quantities until it clears up.

Then stir in the water in small portions until entirely incorporated. Stir for 1 hour. High speed mixing is used.

Furniture Cleaner and Polish
Formula No. 1

White Beeswax	10

Melt and add slowly with stirring:

Stoddard Solvent	500
Turpentine	500

No. 2

Yellow Beeswax	100
Turpentine	175

Melt the wax and to it add slowly with stirring the turpentine. Cool slowly while mixing.

Floor Polish
Formula No. 1

Carnauba Wax	9.7
Triethanolamine	0.3
Ozokerite	7.5
Paraffin Wax	45.0
Deodorized Kerosene	187.5

Melt the carnauba wax and add the other ingredients. Warm together in a double boiler until dissolved. Stir until cool and smooth.

No. 2

V.M. and P. Naphtha	102
Paraffin Wax	80
Albacer	61
Hexalin	12
Tetralin	4

Dance Floor Wax

Carnauba Wax	12
Candelilla Wax	3
Paraffin Wax	5
Rice Flour	80

Grind to a very fine powder.

Wood Laboratory Table Polish

Dow Corning Silicone Fluid 200	95
Hard Paraffin Wax	5

Add the melted paraffin wax to the silicone heated to about 150°F. and cool.

Oil Polishes

Any of these oil polishes can be used on furniture, woodwork, and automobiles. As emulsions, they clean and polish the surface in one operation. The polish can be rubbed dry to give a glossy finish on a varnished or lacquered surface.

Oil		Amine	Oleic Acid	Water
Light Mineral Oil Sulfonated Castor Oil (50%)	48 16	Monoethanolamine 0.5	6.6	60
Light Mineral Oil Sulfonated Castor Oil (50%)	48 16	Morpholine 0.6	6.6	60
Light Mineral Oil Boiled Linseed Oil	40 8	Morpholine 1.0	4.0	60

The addition of 0.07 to 1.0 parts by weight of 10% aqueous solution of Cellosize hydroxyethyl cellulose WS–500 to the above emulsions assures stability over a longer period of time.

Dissolve the oleic acid in the oils and stir in the amine.

Stir for about 5 minutes. If the mixture is not then clear, add oleic acid a little at a time until clarity is attained.

Add the oil solution of the water with vigorous stirring to form a creamy, stable emulsion.

The clear oil solution can be marketed with directions to mix it with an equal amount of water before use, pointing out that it can be stored as an oil to be mixed with water when desired.

When these polishes are to be used on automobile or other lacquered surfaces, a small amount of a fine abrasive is frequently added as an ingredient that cleans by friction.

Wax Polishes

Wax polish emulsions require more rubbing than oil polish emulsions, but produce a harder, high luster finish. These polishes are cleansers and polishers combined and leave a bright, hard film. They are applied by rubbing well over the surface to remove dirt and streaks, and then polishing with a dry cloth. The wax mixture usually contains a hard wax, such as carnauba, and a soft wax, such as paraffin or beeswax, which acts as a plasticizer. The use of naphtha in a wax polish allows faster application without leaving a tacky film. The morpholine emulsion films become water-resistant several hours after application and will stand up under constant exposure to water fully as long as a solvent-type wax polish. The liquid cream wax polish is more easily applied than the wax paste polish and does not require as hard buffing to produce a high gloss. The liquid wax polish makes an excellent shoe cream polish and can be used with the addition of nigrosine for black shoes or, with the addition of other suitable dyes, for colored shoes.

Wax Paste Polish

Carnauba Wax	Beeswax	Naphtha	Amine		Stearic Acid	Water
30	30	50	Triethanolamine	4.3	8	65
30	30	50	Monoethanolamine	1.9	8	65
30	30	50	Morpholine	2.6	8	65

Liquid Cream Wax Polish

Carnauba Wax	Beeswax	Naphtha	Amine		Stearic Acid	Water
12	6	70	Triethanolamine	4.8	8	180
12	6	70	Monoethanolamine	2.1	8	180
14	4	25	Monoethanolamine	2.0	8	240
12	6	70	Morpholine	3.0	8	180

Automobile Polish *

Carnauba Wax	Beeswax	Naphtha	Amine		Stearic Acid	Water
9	8	75	Triethanolamine	2.7	7	75
9	8	75	Monoethanolamine	1.2	7	75
9	8	75	Morpholine	1.7	7	75

* In the automobile polish, about 25 lb. of water-absorbing abrasive such as bentonite can be added to produce a paste polish; 60 lb. of an oil-absorbing abrasive such as tripoli makes a liquid polish.

A steam- or hot water-jacketed kettle is preferred for making wax polishes, as a satisfactory temperature must be maintained to prevent caking of the wax along the sides of the kettle and to avoid discoloration by overheating the wax. A paddle-type, hand-operated stirrer or a slow speed, large-bladed propeller is also suggested for successful operation. Since morpholine has a flash point of 100°F., it should not be added to the mixture in the presence of open flames. If the wax is melted by means of a gas burner, the gas should be turned off during the addition of the morpholine.

Melt the waxes and stearic acid, add the amine, and maintain the temperature at about 90°C.

Add the naphtha slowly and stir until a clear solution is obtained and the temperature is 90° to 95°C. *Avoid the use of open flames.*

The method of adding the abrasive depends upon the type used. An oil-absorbing abrasive, such as tripoli, should be well mixed with the hot naphtha solution of waxes just before the water is added. An abrasive that absorbs water, such as bentonite, is best stirred into the finished emulsion.

Heat the water to boiling, add it to the naphtha solution, and stir vigorously until a good emulsion is obtained.

Continue stirring slowly until the emulsion has cooled to room temperature.

The proportions of waxes can be changed as desired, depending upon the ease of polishing required and the hardness of the final film. A high melting hydrocarbon wax can be used in place of all or part of the beeswax with good results. When the primary use of the automobile polish is for polishing rather than as a cleaning and polishing combination, it will be more satisfactory without an abrasive.

Rubless Polishes

Rubless Wax Floor Polish

This polish produces a glossy film that can be readily re-emulsified or removed with water. A rubless floor polish prepared in this manner should give a clear, bright film when applied to linoleum, mastic, hardwood, and other floor surfaces. The addition of dispersed shellac or casein improves its spreading and flow-out properties.

Carnauba Wax No. 1	20.0
Oleic Acid	2.25
Triethanolamine	3.30
Borax	1.50
Water	120.0

It is essential that a good grade of carnauba wax be used and that the following directions be followed closely:

A steam- or hot water-jacketed kettle is preferred for maintaining a uniform temperature and prevent overheating and caking of the wax along the sides of the kettle. A paddle-type, hand-operated stirrer or slow-speed, large-bladed propeller is recommended for successful operation.

Method 1

Melt the wax and the oleic acid, stirring occasionally to break up the wax lumps. Bring the temperature to 95°C.

Add the triethanolamine slowly,

stirring constantly until the mixture becomes *clear*.

Dissolve the borax in 20 lb. of boiling water; pour this solution into the wax mixture, stirring until a clear, viscous mixture is obtained.

Add the remaining boiling water to the mixture slowly, with steady stirring, a small portion at a time. Each portion of water should be thoroughly incorporated and the stirring continued until the mixture returns to a smooth, even consistency before the next addition is made.

The mixture will become more viscous when the water is first added and then becomes thin again. When about one-half to two-thirds of the water has been added and the mixture becomes water-thin, the rest of the boiling water can be added slowly, but continuously, with steady stirring. If the mixture becomes creamy at any time, the water is being added too rapidly and is not being thoroughly incorporated before the next addition. The final polish should be light colored and translucent, less opaque than milk.

The best results are obtained by using all of the water at boiling temperature. However, if more convenient, only about two-thirds of the water in the above formula need be heated to boiling. The rest of the water may be at room temperature when added, with constant stirring, to the hot polish. This final dilution with cold water may be made at any time.

The polish is allowed to cool with occasional stirring, covering between stirrings to prevent crusting or graining on top. If cold water can be run through the jacketed kettle, the cooling can be accomplished more quickly.

Make a dispersion of bleached, dewaxed shellac or casein and add 2 gal. of the resin dispersion for each 10 gal. of polish.

Method 2

Follow the instructions for Method 1 through the first three steps or until the borax solution has been incorporated.

Add about 6 lb. of boiling water and stir for several minutes after the mixture becomes clear.

Add the remaining water, at boiling temperature, quickly, all at one time; and stir until a smooth dispersion is obtained.

Cool as directed under Method 1.

Shellac or casein dispersion can be added, if desired, using 2 gallons of the dispersion for each 10 gallons of polish.

VARIATIONS: A cake polish can be made with about one-third the water used above. When ready for dilution, the cake is melted in a steam-jacketed kettle and the rest of the water added, half at boiling temperature and the rest at room temperature.

Triethanolamine Water-Resistant Rubless Polish

A triethanolamine polish made with a small amount of potassium hydroxide has been found to produce a film more water-resistant than the film of a polish made with borax, but not so resistant as a morpholine polish film.

The following formula is suggested as a working basis:

Carnauba Wax No. 1	40.0
Triethanolamine	4.0

Oleic Acid	8.0
Potassium Hydroxide (85% as KOH)	0.5
Water	240.0

A steam- or hot water-jacketed kettle and a paddle-type, hand-operated stirrer or slow speed large-bladed propeller are preferred for successful operation. *The wax should not be allowed to cake around the sides of the kettle at any time.*

Melt the carnauba wax and the oleic acid, stirring occasionally to break up the wax lumps, and bring the temperature to 95°C.

Dissolve the potassium hydroxide in about an equal weight of water, add this hot solution and the triethanolamine to the melted wax mixture, and stir until the mixture becomes clear.

Heat the remaining water to boiling and pour it quickly into the wax mixture all at one time. Stir continuously until the wax mixture is entirely dispersed in the water.

A shellac dispersion can be added if desired.

Morpholine Water-Resistant Rubless Polish
Method 1

This water-resistant, rubless polish is a translucent solution if prepared as directed. When spread evenly over a surface and allowed to evaporate, it dries to a hard film of high brilliance. The water resistance of the film increases for several hours after application and finally the coating is unaffected by water.

It is essential that a good grade of light-colored carnauba wax be used in the following formula:

Carnauba Wax	20.0
Oleic Acid	4.0
Morpholine	2.5
Water	120.0

For successful operation in making a rubless polish, a steam-jacketed kettle and a hand-operated paddle or slow speed, large-bladed propeller are recommended. *The wax should not be allowed to cake around the sides of the kettle at any time.*

Melt the wax in the oleic acid, stirring occasionally to break up the lumps. Bring the temperature to 95°C. and stir until well mixed.

Add the morpholine and continue stirring until the whole mixture becomes clear. *Since the flash point of morpholine is 100°F., this addition should not be made in the presence of open flames.* If the wax is melted by means of a gas burner, the gas should be turned off during the addition of the morpholine.

About 20 lbs. of water, which has been heated to the boiling point, is added and stirring is continued until a clear, viscous mixture is obtained.

Add the remainder of the boiling water, a small amount at a time, with steady stirring. Each portion should be well incorporated before another addition is made. The mixture becomes increasingly viscous and should be of the appearance of petrolatum when about one-half of the water has been added. After this stage has been reached, the mixture begins to thin out.

After about two-thirds of the water has been added, and the mixture has become definitely thinned, the remaining water can be added slowly, but continuously, with constant stirring.

The polish should be covered and stirred at intervals until cool, to prevent caking on the top.

A resin dispersion improves the spreading and flow-out of the polish.

Method 2

The technique of making a morpholine rubless polish can be simplified by the use of Tergitol wetting agent 4, and less time and effort are required to produce a more uniform product. The dried film of a polish made by this method is more even than a film of polish made by the previous method, with no impairment of its water-resistant properties. The purpose of the Tergitol wetting agent in the rubless polishes is to disperse the hot wax mixture more quickly into the hot water and to produce an emulsion with a desirable size of wax particle, which should dry to a smooth wax film of good luster. This wetting agent also produces a desirable viscosity in the emulsion at all times and permits the manufacture of a superior product.

Carnauba Wax No. 1	20.00
Tergitol Wetting Agent 4	0.75
Morpholine	2.50
Oleic Acid	4.00
Water	120.00

Less Tergitol wetting agent 4 is usually required in a polish made with carnauba wax No. 2 or 3 or carnauba wax substitutes, such as the higher melting hydrocarbon waxes, than in one made with carnauba wax No. 1. The addition of too much wetting agent produces an almost clear polish which dries with a bright, though less even, film.

The completed formula plus the resin dispersion contains about 15% total solids. About 30 lb. more water can be added to the above formula at any time to produce a polish with about 13% total solids content.

A hot water- or steam-jacketed kettle and a hand-operated paddle or slow-speed, large-bladed, motor driven propeller are recommended for successful operation in making rubless polishes.

Melt the carnauba wax in the oleic acid, stirring occasionally to break up the wax lumps. Bring the temperature of the melted wax and oleic acid to 95° to 97°C.

Stir in the morpholine and continue stirring until the mixture becomes clear. *Since the flash point of morpholine is 100°F. this addition should not be made in the presence of open flames.* If the wax is melted by means of a gas burner, the gas should be turned off during the addition of the morpholine.

Stir in the Tergitol wetting agent 4 and continue stirring for about 3 minutes after the mixture becomes clear. The temperature should be maintained at 95° to 97°C.

Heat the water to boiling temperature while the wax is melting. Add all of the water (98° to 100°C.) to the melted wax mixture and stir until a uniform dispersion is obtained. The water should all be added in about 10 to 20 seconds.

The emulsion should be stirred for 5 to 10 minutes after adding the water; it can then be covered and stirred at intervals until it has reached room temperature or it can be cooled quickly by pumping through a cooling system or by running cold water through the water jacket on the mixing kettle.

A resin dispersion, such as Manila loba resin B can be added to increase the spreading and flow-out properties.

The melted wax mixture containing the wax, morpholine, oleic acid and Tergitol wetting agent 4 at 95° to 97°C. is poured, all at one time, into the water at 98° to 100°C. with stirring; and the stirring is continued until a smooth emulsion is obtained. The polish can then be cooled as above and the Manila loba resin dispersion added. Thus a smaller wax kettle can be used, the water being heated in the larger kettle.

Resin, Shellac, Casein Dispersions
Natural Resin Dispersions

The addition of a resin dispersion increases the spreading and flow-out properties of the rubless floor polishes and improves the smoothness of the dried polish film. Shellac is preferred with the triethanolamine rubless polish, while Manila loba B resin dispersion produces better results in the morpholine rubless polish. A casein dispersion can be used with any of the rubless polishes and probably improves the smoothness of the polish film more than either a shellac or Manila loba resin dispersion. However, the water-resistant properties of the films are noticeably reduced by the casein dispersion, while neither the shellac nor Manila loba resin dispersion affects this property. Excellent leather polishes may be produced with any of the rubless polishes by incorporating the casein disperison.

	Formula No. 1	No. 2
Manila Loba B		
Resin (Powdered)	3.5	3.5
Ammonia (28%)	1.2	0.4
Triethanolamine	—	0.4
Morpholine	—	0.4
Water	32.0	32.0

Preparation of Formula No. 1:

Mix the *powdered* Manila loba B resin with the ammonia.

Stir in about 10 lb. of the water, heated to about 60°C. This produces a gummy mass, which is allowed to stand for several hours or overnight.

Add about 10 lb. more water, heated to 70°C., and stir until a uniform mixture is obtained.

Heat the mixture to about 60°C., with constant stirring, to assist in getting a smooth dispersion.

Stir in the rest of the water.

If the dispersion is not clear at any time, stir in more ammonia, a little at a time, until clarity is obtained. If a small amount of resin remains suspended, it should be removed by filtering the dispersion through a cloth.

The clear, filtered dispersion is allowed to cool and can then be added to the cold polish at any time. About 1 gal. of resin dispersion to 5 gal. of polish produces the desired results. The proportions given make sufficient resin dispersion for the amount of polish produced by any of the rubless polish formulas.

Preparation of Formula No. 2:

Dissolve the ammonia, morpholine, and triethanolamine in one-half of the water and stir in the *powdered* Manila loba B resin.

Warm to about 55°C. and hold at this temperature for about 15 minutes, with constant stirring.

The remainder of the water is then stirred in. The small amount of undispersed material can be allowed to settle and the clear liquid drawn off or the whole dispersion can be strained through several thicknesses of cheesecloth.

Shellac Dispersion

Shellac (Bleached, Dewaxed)	3.5
Ammonia (28%)	0.5
Water	32.0

Add the ammonia and about one-half of the water to the fresh shellac and warm, with constant stirring, until solution is complete. The shellac may become difficult to disperse if it is kept too long, while Manila loba B resin improves with age in this respect.

Add the rest of the water. The solution should be clear. More ammonia should be added if it is not clear.

Filter, if necessary, cool, and add to the cold polish. About 2 gal. of the shellac dispersion can be added for each 10 gal. of polish.

Casein Dispersion

Casein (Lactic Acid)	3.50
Ammonia (28%)	3.35
Phenol	0.16
Water	32.0

Soak the casein in one-half of the water for several hours or overnight.

Add the ammonia to the rest of the water, and stir this solution into the soaked casein.

Warm, with constant stirring, to about 60°C. and continue stirring at this temperature until a smooth mixture is obtained.

Add 0.16 lb. (2.5 oz.) of phenol, which acts as a preservative.

The dispersion will be slightly viscous but will become thinner as it stands. It should be aged for at least a week before it is added to the polish unless a slightly viscous polish is desired. If the polish is too viscous when the aged casein dispersion is added, 30 lb. of water can be used to thin it, and the completed polish will contain about 13% total solids. This is well within the range of most of the commercial rubless polishes, especially where the actual carnauba wax content is as high as in the suggested formulas.

The casein dispersion is easily made and keeps indefinitely when a preservative is present. Only sufficient ammonia to disperse the casein should be used, as higher amounts will increase the viscosity of the dispersion.

Polishing Paste

Kieselguhr	1	oz.
Petrolatum	½	lb.
Cottonseed Oil	1	oz.
Subcarbonate of Iron	3	oz.
Benzaldehyde	3	min.
Red Oil	27	min.

The silica is ground to a very fine powder and mixed with the iron. After melting the petrolatum, the cottonseed oil is added, and the powder stirred in. The benzaldehyde is added while cooling, and the compound is run into flat cans or other containers. The paste is applied with a soft rag.

Diamond Dust Abrasive
U.S. Patent 2,347,597

Beeswax	30
Shellac	15
Resin	5

Melt and mix in Diamond Dust 50 until uniform. Pour into molds and allow to cool.

Abrasive Cleaner

Diglycol Laurate	40
Soap	40
Fine Silica	80
Gasoline	100
Water	200

Dilute with water for use.

Lens Polishing Powder
U.S. Patent 2,383,500

Cerium Oxide	32
Barium Carbonate	8
Barium Hydroxide	1

Polishing Powder

Kieselguhr	2 oz.
Rouge	½ oz.
Prepared Chalk	½ lb.

After thorough mixing, this composition can be used in polishing silver, nickel, and other non-ferrous metals. It is used by rubbing on the metal with a damp sponge or rag which is followed by rubbing with a dry chamois or cloth.

Smoothing Compound for Lucite and Plexiglas

Silica (Powdered #200 Mesh)	40
Bentonite (Powdered Alkaline Type)	10
Vegetable Potash Soap (Anhydrous Basis)	15
Red Oxide of Iron	To color
Water	35

The bentonite is worked into a paste with a minimum amount of the water, the lumps being macerated until a smooth dispersion results. The anhydrous soap is dissolved in the remainder of the water which is then mixed with the bentonite dispersion. The powdered silica is then added, and finally sufficient red oxide of iron to bring to the desired color.

Polishing Compound for Lucite and Plexiglas

Carnauba Wax	10
V. M. and P. Naphtha	10
Santomerse #3 (Emulsifying Agent)	5
Silica Smoke (1,000 mesh)	8
Rouge (Red Oxide of Iron)	2
Boiling Water	65

The carnauba wax is melted and poured into the heated naphtha (85–95°C.) with vigorous agitation. The emulsifying agent is dissolved in the water. The wax solution is then poured, while both solutions are still quite hot, into the water solution with vigorous mechanical agitation. The mixture is continuously stirred until cooled to room temperature. While still hot, the powdered silica and the rouge are added.

NOTE: The above described *smoothing compound* and *polishing compound* are designed for manual use, the purpose of which is to remove and repolish scratched and scored surfaces of Lucite and Plexiglas bringing them back to their original transparency, gloss and surface sheen. The compositions are applied with a soft cotton rag and rubbed locally in one direction only until the scratch marks or the rough surface have attained a reasonable degree of polish with the smoothing compound, when the operation is repeated for the polishing compound. Obviously, if there are rough surfaces such as from sawing or filing or deep scratch marks these surfaces will have to be initially finished with a file, followed by #2/0 sandpaper or finer before using the smoothing solution followed by the polishing compound.

Tumbling Barrel Polish
Carnauba Wax	6
Stroba Wax	2
Diglycol Laurate S	26
Water	200

Heat and mix until emulsified.

French Polish Base
Gum Accroides	1
Shellac	3
Alcohol	To suit

Black Paste Shoe Polish
Formula No. 1
Ouricuri, Gray Carnauba or Shellac Wax	¾ lb.
Ceresin, Paraffin Wax	½ lb.
Castile or Good Curd Soap	2 oz.
Potasssium Carbonate	2 oz.
Soft Water	3 pt.
Water-soluble Nigrosine Black Dye	4–6 oz.

Melt the waxes in one container. Separately boil the water with the soap, alkali and nigrosine. Stir the black water solution into the melted waxes. Cool down quickly.

By the addition of further quantities of water, creams of various consistencies may be prepared, e.g., ½ gal. of water will give a nice firm cream.

Instead of black, the polish can be colored yellow or red with 1–2 oz. of the appropriate dye for floor or furniture polish.

No. 2
A.
Soft Water	4 gal.
Water-soluble Nigrosine Water Dye	3 lb.
Potassium Carbonate (or 8 oz. Caustic Soda)	18 oz.
Rosin (Can Be Omitted)	8 oz.

Boil the water; add the nigrosine in small quantities until quite dissolved, and then the other components. Keep just on the boiling point until mixture B is ready.

B.
Ouricuri or Gray Carnauba or Shellac Wax	13	lb.
Paraffin Wax	8¾	lb.
Ceresin or Ozokerite	7½	lb.
Oil Black	¾	lb.
White Spirit or Pool Distillate	3½	gal.
Turpentine or White Spirit	3	pt.

Melt the waxes and oil black together. Remove the flame and stir in the solvents. Add to the hot black water mixture A with stirring. Cool slightly and pour into tins or glass jars. Alternatively, some manufacturers may prefer to melt the waxes and follow with the solvents and then the black solution. A brown and toney-red boot polish can be prepared in a similar way by substituting the appropriate dyestuff for the nigrosine. The inclusion of ½ lb. triethanolamine oleate to the boiling water will improve the smoothness of the polish.

No. 3
Acrawax C	10 g.
Ceresin	3 g.
Turpentine	30 cc.

Warm together until clear; mix and pour into containers.

No. 4
Pentawax 286	22 g.
Carnauba Wax	3 g.
Stoddard Solvent	75 cc.

No. 5
A. Carnauba Wax	25 g.
Ceresin Wax	12 g.
Paraffin Wax	10 g.
Trigamine Stearate	2 g.
B. Turpentine	60 cc.
C. Distilled Water	15 cc.
Caustic Soda (50%)	½ cc.

Melt A, add B (hot), and then C (hot). Stir continuously while cooling to a cream consistency. Pour while fluid, but not too soon.

No. 6
Carnauba Wax	45.0
Solvent Naphtha	70.0
Oil-Soluble Nigrosine	0.5
Stearic Acid	15.0
Caustic Soda Solution (50%)	2.0
Hard Soap	2.5
Water	300.0
Water-Soluble Nigrosine	0.5

Dissolve the water-soluble dye in half the water. Add the caustic soda solution and, while hot, add the soap and stearic acid. Bring to boiling. Then add the remaining ingredients at about 85 to 90°C. Continue stirring until cold. Replacing the nigrosine with other suitable dyes will yield products of other requisite shades of color.

Soft Leather Polish
A. White Mineral Oil	20
Turpentine	15
Carbon Tetrachloride	5
Beeswax	3
Oleic Acid	4

Stir until dissolved, and add

B. Triethanolamine	1.5
Water	31.5

Mix A and stir until the ingredients are dissolved, then add B. Stir until a smooth emulsion is formed.

Rubber Footwear Polish
After the initial gloss of rubbers is dulled the luster can be restored by cleaning with a slightly damp rag and then applying the following solution with a clean rag, sponge or soft brush. Unlike oils and ordinary shoe polishes it will not rot or swell rubber.

Glycerin	100
Water	100
Perfume	If desired

Chapter XIX

PYROTECHNICS AND EXPLOSIVES

Waterproof Matches
U.S. Patent 2,389,552
Formula No. 1

Fifty parts of polymerized rosin of 100°C. drop melting point are dissolved in 50 parts of toluene, and to the solution are added 30 parts of red phosphorus, 20 parts of finely divided silica and 40 parts of lead dioxide. These ingredients are mixed to form a uniform paste, in which the ends of the match sticks are dipped, after which the matches are permitted to dry by evaporation of the solvent.

No. 2

Fifty parts of polymerized rosin of 115°C. drop melting point are dissolved in 75 parts of benzene, and to the solution are added 3.5 parts of potassium dichromate, 37.0 parts of potassium chlorate, 8.5 parts of glass powder, 0.5 part of zinc oxide, 4.7 parts of manganese dioxide, 3.5 parts of sulfur, 3.8 parts of iron oxide, and 1.0 part of kieselguhr. These ingredients are mixed to form a uniform paste, in which the ends of the match sticks are immersed, after which the matches are permitted to dry by evaporation of the solvent.

Smokes
White Chemical Smoke

Lubricating Oil (10 to 20 Sec. S.A.E.)	87.0
Ammonium Chloride	12.3
Sodium Stearate	0.7

The lubricating oil or a small portion of it is heated and the designated amount of sodium stearate (preferably anhydrous) is added before the oil reaches 200°F. The sodium stearate oil mixture is then heated to 380°F. and allowed to cool, and when the temperature falls to 150°F. the ammonium chloride is addded while the mass is continuously stirred. This composition produces a white smoke when injected in an oxygen-free atmosphere which is heated to 400°F. or higher, and discharged through a suitable orifice. This condition may be achieved by admitting the composition through a spray nozzle into the exhaust pipe of a standard gasoline engine.

White Military Smoke Screen

Anhydrous Ferric Chloride	53
Grained Aluminum	8
Zinc Oxide	39

Colored Smokes

For *red smoke* add 1 to 2 lb. of methyl amino anthraquinone dye or para-nitraniline-β-naphthol condensate per gallon of the above white smoke mixture.

For *yellow smoke* add 1 to 2 lb. of the dyestuff known as Auramine "O" per gallon of white smoke compound.

For *green smoke* add an aggregate of approximately 2 lb. of varying mixtures of 1,4-Di-*p*-Toluidine anthraquinone Auramine "O" to the white smoke compound.

For *blue smoke* add 1 to 3 lb. of Calco oil soluble blue "N" to the white smoke compound.

Smoke Compositions

The following compositions for colored smoke formulas are representative of mixtures in use in various munitions:

Yellow Smoke:
Formula No. 1
Auramine O	38.0
Sodium Bicarbonate	28.5
Potassium Chlorate	24.1
Sulfur	9.4

No. 2
β-Naphthaleneazodimethylaniline	50.0
Potassium Chlorate	30.0
Sugar	20.0

Red Smoke:
Formula No. 1
9-Diethylaminorosindone	48.0
Potassium Chlorate	26.0
Sugar	26.0

No. 2
1-Methylaminoanthraquinone	42.5
Potassium Chlorate	27.4
Sodium Bicarbonate	19.5
Sulfur	10.6

Orange Smoke:
Formula No. 1
1-Amino-8-Chloroanthraquinone	39.0
Auramine O	6.0
Sodium Bicarbonate	24.0
Potassium Chlorate	22.3
Sulfur	8.7

No. 2
α-Aminoanthraquinone	24.6
Auramine O	16.4
Sodium Bicarbonate	23.0
Potassium Chlorate	25.9
Sulfur	10.1

Violet Smoke:
1-Methylaminoanthraquinone	18.0
1,4-Diamino-2,3-Dihydroanthraquinone	26.0
Sodium Bicarbonate	14.0
Potassium Chlorate	30.2
Sulfur	11.8

Green Smoke:
Auramine O	11.7
1,4-Di-*p*-Toluidinoanthraquinone	28.3
Sodium Bicarbonate	24.0
Potassium Chlorate	25.9
Sulfur	10.1

Blue Smoke:
1,4-Dimethylaminoanthraquinone	50.0
Potassium Chlorate	25.0
Sugar	25.0

Explosive Type Colored Smoke Bursts

All the colored smoke mixtures described have been of the burning type. A colored smoke burst can be obtained by using a mixture of approximately equal parts of dye and EC explosive powder. The mixture is detonated with an appropriate detonator. The resulting explosive gives a large puff of colored smoke.

Black Smoke

Black dyes do not give satisfactory black smokes. In order to obtain a dense black smoke, a suitable chemical reaction is set up in which carbon is liberated. The simplest method to accomplish this has been to add an oxidizing agent to a hydrocarbon of high carbon content such as anthracene. The following formula has been found to give a heavy dense black smoke:

Anthracene	45
Potassium Perchlorate	55

Dyes for Colored Smokes

Color	Dye
Blue	2-Bromo-1-amino-4-p-toluidoanthraquinone
Bordeaux	1-Amido-4-oxyanthraquinone
Brown	1-2-4-Trioxyanthraquinone
Yellow	9-10-Dianilido anthracene
Orange	1-4-Dioxyanthraquinone
Orange	1-Aminoanthraquinone
Violet	Mixture of 1,5- and 1,8-Di-m-Toluidoanthraquinone
Whitish Yellow	Condensation product of equal parts of Phenylediamine and Phthallic anhydride
Sudan Yellow	o-Nitrodiphenylamine

Dyes for Tracer Bullets

Yellow	Quinophthalone
	o-Nitrodiphenylamine
Yellow	or 2-Nitro-1-Cyclohexylaminebenzene

Colored Military Flame
U.S. Patent 2,362,502

Barium Nitrate	40
Powdered Magnesium	28
Hexachlorethane	30
Linseed Oil	2

Ammunition Primer
Formula No. 1
U.S. Patent 2,356,211

Normal Lead Triazoacetate	10
Lead Styphnate	32
Lead Thiocyanate	8
Lead Nitrate	30
Powdered Glass	20

No. 2
(For detonating lead azide)
U.S. Patent 2,395,045

Lead Peroxide	45
Calcium Silicide	15–30
Zirconium	30–15
Sulfur	10–15

Rifle Cartridge Primer

Formula	No. 1	No. 2
Tetracene	3	2– 3
Lead Trinitroresorcinate	40	—
Barium Nitrate	42	40–45
Lead Dioxide	5	5– 8
Calcium Silicide	10	6–12
Lead Styphnate	—	30–35
Antimony Sulfide	—	6– 9

Priming Mixture for Shot-Shell Ammunition

Double Salt: Lead Trinitroresorcinate-Lead Hypophosphite	20
Barium Nitrate	35
Mercury Fulminate	25
Antimony Sulfide	15
Calcium Silicide	5

No. 2
U. S. Patent 2,005,197

Mercury Fulminate	40
Barium Nitrate	40
Potassium Dinitrophenyl Azide	5
Antimony Trisulfide	15

Priming Mixture for Rimfire Ammunition
Formula No. 1

Lead Trinitroresorcinate	38
Tetracene	2
Lead Nitrate	30
Lead Sulfocyanate	8
Ground Glass	22

No. 2
U. S. Patent 2,292,956

Double Salt: Lead Trinitroresorcinate-Lead Hypophosphite	50
Lead Nitrate	30
Glass	20

Priming Mixture for Centerfire Ammunition
Formula No. 1

Double Salt: Lead Trinitroresorcinate-Lead Hypophosphite	35
Lead Nitrate	40
Antimony Sulfide	18
Calcium Silicide	7

No. 2

Lead Trinitroresorcinate	35
Tetracene	5
Lead Nitrate	35
Antimony Sulfide	20
Calcium Silicide	5

Friction Primer for Hand Grenades

Lead Trinitroresorcinate	25
Barium Nitrate	33
Lead Dioxide	24
Silicon	15
Glass	3

Electric Blasting Cap Ignition Mixture
U.S. Patent 2,190,777

Diazonitrophenol	10–20
Barium Nitrate	10–50
Smokeless Powder	35–75

Gasless Delay Fuse Powder

This powder is used to explode detonators and is ignited by an electric match head. It consists of about 70% antimony powder and 30% potassium permanganate for slow burning, or about 46% antimony powder and 54% potassium permanganate for fast burning.

The permanganate is ground in a type of disc or plate crusher mill to approximately 80 mesh.

The antimony is ground from lumps in a vibratory ball mill. The powder is introduced by a screw feed into an air separator. The air in the separatory chamber is kept in rotation by a high speed concentric fan. The fines collected do not exceed 10 microns in particle size.

The two ingredients are blended by tumbling and the mixture is compressed into tablets in a rotary multiple punch press. The tablets are formed to give intimate contact between the ingredients.

Greater uniformity of burning time is obtained by avoiding large variations in the particle size of the powder, but the burning time is independent of the particular particle size used.

Moldable Seal for Explosive Igniter Wires
British Patent 561,198

Salicylic Acid	65–80
Asbestos Powder	35–20

Chapter XX

SOAPS AND CLEANERS

Laundry Soap
Formula No. 1

This is a smooth non-cracking 52%-fatty acid laundry soap that may be made by crutching the following ingredients into a 30% rosin kettle soap.

Kettle Soap (63% Fatty Acids)	1000
Sal Soda (Saturated Na$_2$CO$_3$Sol.)	75
Sodium Silicate (Neutral 41° Bé. diluted to 36° Bé.)	100
Borax	2
Soda Ash	4
Light Mineral Oil	8

No. 2
(White Laundry Soap)

This soap can be cut in the crutcher to a 42%-fatty acid soap, without undue cracking or separation.

Tallow	55
Coconut Oil	30
Peanut or Bean Oil	15

Thirty-five per cent of the tallow charge may be replaced by hardened and bleached whale or marine oils having an iodine value between 75 and 85.

Crutch at 170 to 180°F. adding the following:

Kettle Soap	1000
Neutral Silicate (41° Bé.)	470
Alkaline Silicate (48° Bé.)	150

The alkaline silicate may be made by dissolving 10% flake caustic in neutral sodium silicate with agitation.

No. 3

A medium fatty acid laundry soap suitable for household package use results from using the following formula:

Kettle Soap (80% Tallow, 20% Coconut Oil Base)	90
Neutral Sodium Silicate (41° Bé.)	10

Crutch together at 180 to 190°F., and run over a dryer to yield a 76% fatty acid flake.

Salt Water Soap

Tallow Soap	2
Sodium Abietate	1
Sodium Pyrophosphate	1

Improved Toilet Soap

A normal toilet soap may be improved as to feel and texture by the addition of 0.10% boric acid, 0.10% lanolin and 0.10% pearl ash.

Glossy Soap Finish
U. S. Patent 2,392,831

Dip soap cakes in

Isopropyl Alcohol	10
Water	10

at a temperature of 85°F.

Floating Castile Soap

Refined and Bleached Coconut Oil	475

Caustic Soda
  (37.4% Na_2O)   172
Water   103
Perfume   1½
Color   As desired

The oil, alkali and water are added to the crutcher and heated to approximately 140°F. to start saponification. Once started the heat of the reaction will drive the temperature up to 190 to 210°F., so that cold water circulation in the jacket may be necessary. When the saponification is complete, add the perfume and color and crutch downward until the desired specific gravity is reached. This may be anywhere between 0.60 and 1.0. The product has a fatty acid content of 60% and a superfat between 2 and 5%.

Liquid Soap Extender

Triton X–300	5.0
Tamal	0.1
Methocel (4,000 CPS)	0.4
Water	44.5
Liquid Soap	50.0

Perfuming of Soap

Soap Perfumes

In developing a successful perfume compound for soaps special attention must be paid to careful selection of the base, while top notes, etc., are generally of less importance as their effect will undoubtedly be lost in the course of storage and under the influence of the soap in general. Such bases consist primarily of the less volatile aromatics such as, in particular, the various types of musks—ambrette, ketone, and xylol—the esters of cinnamic acid and salicylic acid, all resinoids, extrols, resins, balsams, etc., as well as all aromatics and essential oils generally as far as they possess a comparatively high boiling point. In this connection the decision as to whether or not the soap is to be colored is of considerable importance as many aromatics, including a great many of the most efficacious ones from the point of view of stability and odor effectiveness, will impart a certain pigmentation upon storage. For this reason it is advisable to select a suitable color for the finished product. The question of discoloration is by no means limited to the heavier types of fancy bouquets. Lilac compositions for instance have a marked tendency to darken the soap due to their content of heliotropin, vanillin, coumarin and frequently indol. Rose compositions on the other hand are much less apt to cause discoloration. A lemon odor, even though by itself an attractive perfume note for an average type of soap, is for instance usually produced by means of citral which not only has the tendency to cause a yellowish discoloration but is moreover quite unstable.

Odor effectiveness, which is to survive long shelf life and extend into the actual use of the soap, is largely dependent upon the use of alkali-resisting combinations. Such compounds are for: lily of the valley type, linalool; rose odors, citronellal, geraniol, phenylethyl alcohol; hyacinth notes, cinnamic alcohol, bromstyrol; lilac complexes, terpineol, heliotrope, heliotropin; hawthorne, anisic aldehyde or, better, crataegin, and methyl acetophenone; violet, the various ionones; carnation complexes, eugenol and isoeugenol; orange blossom note, nerolin (bromelia), methyl anthranilate, methyl naphthyl ketone; clover types, amyl salicylate, coumarin, benzylidene acetone. Certain of these, such as heliotropin, eugenol, isoeugenol and methyl anthranilate, are known to cause discoloration.

It is obvious that the successful composition of a soap perfume necessitates very intimate knowledge of the various ingredients to be used and a thorough system of tests to be submitted before final acceptance or approval. In all, this is a specialized work which it is no small matter to undertake and requires experienced perfume chemists and the necessary equipment for practical tests. This is no doubt one of the main reasons why the creation of soap perfumes is entrusted more than any other phase of perfume making to the experienced graduate perfume chemist with the facilities of a specialized house at his command. It might perhaps be well to add here that such important work is by no means limited to what the layman would consider soaps in the usual sense of the word. It includes perfumes for all types of materials which, due to their composition, can be classified as soaps which of course includes the wide variety of creams which have gained such importance in the modern cosmetic industry.

Soap Perfuming

To perfume 100 lb. of toilet soap with a jasmine odor, the following formula is suitable.

Cinnamic Alcohol	100 g.
Styrax	100 g.
Amyl Cinnamic Aldehyde	300 g.
Orris Powder	50 g.

Linalool can be used in jasmine to the extent of 5%, in which case it can replace some of the styrax. Also 10% of hydroxy-citronellal may be added. To give the finishing touch to a jasmin, 1 gal. of 10% bromostyrol solution is sometimes used. If a good quality of bromostyrol is employed, it greatly helps to bring out the flowery odor.

A good lilac perfume which will not darken soaps is the following:

Hydroxy Citronellal Residue	25 g.
Phenyl Acetic Aldehyde (50%)	5 g.
Terpineol	40 g.
Amyl Cinnamic Aldehyde	10 g.
Cinnamic Alcohol	10 g.
Femelle Bois de Rose	9 g.
Styralyl Acetate	1 g.

Patchouli may be added as a fixative but not more than 1% should be used.

For lily of the valley or muguet the following is suitable

Linalool	40 g.
Alpha Ionone	10 g.
Isoeugenol	5 g.
Sandalwood Oil	5 g.
Algerian Geranium Oil	5 g.
Amyl Cinnamic Aldehyde	10 g.
Aldehyde C. 14	1 g.
Musk Xylol	5 g.

The musk xylol is dissolved in diethyl phthalate or benzyl benzoate.

A very fine perfume for violet soap is given by the following:

Ionone	20 g.
Methyl Heptin Carbonate	10 g.
Orris Resinoid	10 g.
Bergamot	10 g.
Cananga	10 g.
Hydroxy Citronellal	10 g.
Cinnamic Alcohol	10 g.
Amyl Cinnamic Aldehyde	10 g.
Violet Leaves Concrete	10 g.
Bourbon Vetrivert Oil	10 g.
Sandalwood	5 g.
Styrax Tincture (25%)	100 g.
Siam Benzoin (25%)	100 g.
Distilled Olibanum Oil	50 g.
Musk Tonquin (4 Oz. Per Gal.)	25 g.
Civet	35 g.

After these oils are mixed, add 10–20% of orris as an impalpable powder. The orris is poured into a pony mixer, while the oil is dissolved in 500 g. of alcohol. The latter is added to the orris powder. The pony mixer is switched on and while it is turning, 90 lb. of soap are added.

To make this perfume cheaper, some of the expensive materials may be used more sparingly or eliminated. The formula can be toned down to any desired price but at the expense of quality.

Violet Perfume

Bergamot Oil	100
Thyme Oil	100
Terpineol	100
Clove Oil	50
Turpentine	150

Perfume for Tallow Soap (Flaked)
Add 2 to 3 lb. per 1000 lb. of kettle soap.

Spike Lavender Oil	50
Java Citronella Oil	12
Geranium Bourbon Oil	25
Mace Oil	13

Perfume for Laundry Soap
Formula No. 1

Citrene	63
Sassafras Oil	30
Spike Lavender Oil	7

No. 2

Citronella Oil	30
Sassafras Oil	30
Spike Lavender Oil	20
Cinnamic Aldehyde	20

No. 3

Citrene	72
Rosemary Oil	13
Spike Lavender Oil	10
Lemongrass Oil	5

Add from 0.16 to 0.18% to the soap toward the end of the crutching period.

Perfume for Cold-Made Soap

Citrene	54
Rosemary Oil	33
Spike Lavender Oil	13

Toilet Soap Perfume
Formula No. 1

Terpineol	192
Geranium Bourbon Oil	32
Hyacinthine	16
Benzyl Acetate	16
Aubepine	10

No. 2

Geranium Bourbon Oil	50
Terpineol	40
Lavender Oil	30
Natural Bergamot Oil	15
Artificial Bergamot Oil	15
Geraniol	15
Spike Lavender Oil	10
Patchouli Oil	5
Sandalwood Oil	5
Jacinthe Styrol	3

No. 3

Jasmin Absolute or Concrete	5
Lavender Absolute Barreme	3
Ylang Ylang	3
Resinoide Olibanum	2
Resinoide Yetiver	3
Resinoide Orris Root	5
Resinoide Styrax	7
Santal E.I.	5
Patchouli Singapore	2
Cedarwood	2
Bois de Rose	7.5
Benzyl Acetate	7.5
Citronellol	10
Terpeneless Petigrain	5
Methyl Heptine Carb.	2.5
Phenyl Propyl Alcohol	5
Dimethyl Hydroquinone	2
Musk Ketone	3
Terpineol Base	21.5

Two to 6 oz. added to 100 lb. of soap base produces a soap with a lasting fragrance.

No. 4
(Lavender)

Amyl Salicylate	1
Spike Lavender Oil	16
Lavender Oil	18
Terpineol	12
Rose-Wood Oil	6
Bergamot Oil	5
Musk Essence	2
Geranium Oil	2
Vanillin	3

Mix and age 1 month.

Perfume for Cold-Made Castile Soap

Lemongrass Oil	48
Terpineol	26
Rosemary Oil	13
Cassia Oil	13

Add the perfume toward the end of the crutching period.

Color and Perfume Stabilization in Laundry Soap

Magnesium silicate added in small amounts to a laundry soap assists color and perfume preservation better than the sodium silicate frequently used. Crutch in the normal ingredients called for in the formula, then add 1 lb. of a 25% magnesium sulfate solution per 1000 lb. of lightly silicated kettle soap. Continue crutching until thoroughly dispersed.

Powdered Soaps

Household Soap Powder

A 58% fatty acid built powder, satisfactory for general household use, may be made by mechanically mixing the following ingredients:

Dried and Ground Soap Powder (85% Fatty Acids)	68
Soda Ash	22
Trisodium Phosphate	10

Dishwashing Powder
Formula No. 1

Soda Ash	58
Trisodium Phosphate	42

No. 2

Soda Ash (Monohydrate)	70
Trisodium Phosphate	30

This mixture gives an excellent free-running and non-clogging compound for restaurant use at moderate temperatures (70°F.) at a 0.3% concentration.

No. 3

Tetrasodium Pyrophosphate	100
Mercol ST	5

Persil Detergent

Sodium Perborate	10
Soap	35

Soda Ash	25
Sodium Silicate	5

Mix well and grind to a powder.

Powdered Cleaner
U. S. Patent 2,367,971

To a mixture of about 6 parts by weight of sodium hydroxide (38°), 22 parts of sodium silicate (36°), and 28 parts of borax, are added about 22 parts of hydrogen peroxide of 30% and 22 parts of sodium silicate. The mixture is stirred until it assumes gelatinous consistency, poured onto a surface and permitted to dry.

Industrial Soap Powder
Formula No. 1

Pure Kettle Soap (Dried and Ground) (85% Fatty Acids)	20
Pumice (Powdered)	30
Soda Ash	20
Perborate of Soda	15
Ammonium Carbonate	5
Perfume	1

No. 2

Coconut Oil Fatty Acids	80
Soda Ash	36
Corn Meal (Degerminated)	90
Sulphonated Castor Oil	9
Soapless Detergent (Alkyl-Aryl Sulfonated Type)	10
Di-Basic Sodium Phosphate Crystals	80
Perfume	1

The above may be crutched and mixed by adding just sufficient water to keep the mass liquid, and dropping before hydration sets in.

Industrial Hand Cleaners and Soaps
Mechanics' Soap
Formula No. 1

Wood Flour	50
Powdered Soap	40
Trisodium Phosphate	5
Borax	5

Mix the ingredients thoroughly, perfume with ¼ lb. of lemon or other perfume.

No. 2

Water	40
Kerosene	40
Oleic Acid	4
Triethanolamine	2

Mix the water and triethanolamine in one container; the kerosene and oleic acid in another container. Add the solvent solution to the water solution with stirring.

Waterless Hand Cleaner
Formula No. 1

Carbowax 1500	20
Ultrawet (Wetting Agent)	8
Sodium Pyrophosphate	4
Carboxymethyl Cellulose	6
Lanolin	5
Glycerin	5
Dioxane	20
Water	32
Perfume	To suit

The materials are mixed in a high speed stirrer in the following order: carboxymethyl cellulose, water, Carbowax, Ultrawet, sodium pyrophosphate, lanolin, glycerin and dioxane. In use the material is rubbed on the hands and removed with a paper towel.

No. 2

Sovasol #4	100
Oleic Acid	14
Triethanolamine	6
Add with stirring	
Water	60
Ammonia 28%	5

No. 3
U. S. Patent 2,383,610

Polyvinyl Alcohol	75–25
Glucarine B (Glycerin Substitute)	40–25
Sodium Pyrophosphate	1– 0.5
Trisodium Phosphate	1.5– 0.3
Nacconal NR	2.5– 1
Lysol	12– 8
Water	1100

No. 4

Tallow Soap Chips (88–92%)	12
Water	73
White Mineral Oil	5
N Brand Sodium Silicate	10

Dissolve the soap in hot water, add the silicate, mix, cool, add the mineral oil and ½ lb. of lemon perfume.

Water Softener and Cleanser

Kettle Soap (63% Fatty Acids)	600
Soda Ash	1000
Water	1100

Crutch the above ingredients at 190°F. until homogeneous, run the product over a chilling roll and allow the pasty mass to set up by standing overnight in buggies. Grind the hydrated material and package.

Saddle Soap

Palm or Tallow Soap Chips	8 lb.
Water	24 lb.
Beeswax	1 lb. 2 oz.
Neatsfoot Oil	1 lb. 2 oz.

Dissolve the soap chips in hot water, add the melted wax and neatsfoot oil and color with 5 oz. of a 2% solution of DuPont Orange #110.

Washing Fluid

Sal Soda	4	lb.
Borax	2	oz.
Sal Tartar	1	oz.
Aqua Ammonia	½	pt.
Spirits of Camphor	2	oz.
Turpentine	1	oz.
Hot Water	6	pt.

The sal soda, borax, and tartar are dissolved in the hot water and the other ingredients added. One tablespoon of this composition is recommended for each gallon of water to be used in soaking clothes overnight.

Laundry Bluing

Soluble Prussian Blue	1	oz.
Oxalic Acid	¼	oz.
Boiling Water	1	qt.

Laundry Sour
U. S. Patent 2,331,396

Sodium Acid Fluoride	98
Sodium Hexametaphosphate	2

Acid Resistant Wetting Agent (Detergent)
British Patent 573,145

Oleic Acid	26.5
Pine Oil	88.4
Potassium Hydroxide (36.5% Solution)	11.5
Tetralin	265.0
Water	20.0
Sodium Isopropyl Naphthalene Sulfonate	88.4

Let stand for 2 days and filter. This is used for removing tar, etc., from wool and for other textile cleaning.

Glove Cleaner
Formula No. 1

Soap	6
Water	2

Javelle Water 4
Aqua Ammonia ¼

The soap is first dissolved in the water and then the other ingredients are added.

No. 2

Benzine or White
 Gasoline 1 pt.
Alcohol ½ oz.
Chloroform ½ oz.
Ether ½ oz.

Cologne, lavender or another perfume may be added as desired to mask the residual odor in the cleaned articles.

Dry Cleaning Soap
(Liquid)

	Per Cent of Oleic Acid Saponified	
	80%	90%
Oleic Acid	107.0	107.0
Cleaner's Naphtha	25.0	25.0
Butyl Cellosolve	27.0	27.0
Triethanolamine	21.0	..
Monoethanolamine	..	10.5
Potassium Hydroxide (100%)	8.3	9.5
Water	13.5	13.5

Do not use near an open flame.

Heat the oleic acid, butyl Cellosolve, and naphtha to 140°F.

In a separate container dissolve the potassium hydroxide in the water and add the amine.

The water solution is then stirred into the oleic acid solution.

Stirring is continued for about 30 minutes to react all of the potassium hydroxide. The solution should be clear.

Mixed isopropanolamines may be used to replace triethanolamine in these formulas to produce greater oil solubility.

(Paste)

This dry-cleaning soap is a thin paste and is suggested for the scrubbing board. A more viscous paste may be obtained by increasing the stearic acid content and decreasing the oleic acid content by the same amount. The soap is completely soluble in naphtha.

Butyl Cellosolve	25.0
Oleic Acid	95.0
Stearic Acid	12.0
Triethanolamine	19.7
Potassium Hydroxide	8.3
Water	10.0

Heat the stearic and oleic acids and butyl Cellosolve to 140°F.

In a separate container dissolve the potassium hydroxide and triethanolamine in the water. Stir the resulting hot solution into the fatty acid solution.

Stirring is continued for about 30 minutes to complete the saponification of the potassium soap.

Rug-Cleaning Soaps

The combination of a soap and a chlorinated or hydrocarbon solvent produces an excellent rug and carpet cleaner. An emulsion of solvent, soap, and water removes grease, tar, and paint more readily than does soap and water. The amine soaps, being soluble in these solvents, allow the preparation of clear solutions of solvents, soap, and water, which can be stored indefinitely without separation. The colors in the rugs or carpets will not be harmed, but will be clarified and brightened by the cleaning process.

Formula No. 1

Oleic Acid	28
Ethylene Dichloride	13
Isopropanol (99%)	14
Butyl Cellosolve	5

Triethanolamine 16
Water 125

Adequate ventilation should be provided, and special care should be taken, to avoid inhaling the vapor and repeated contact with the skin whenever chlorinated solvents are used.

Mix the oleic acid, ethylene dichloride, isopropanol, and butyl Cellosolve; and add the amine.

Stir until thoroughly mixed and add the water. If the mixture is cloudy, add sufficient isopropanol to clear it.

An emulsion made of equal volumes of the soap and water is recommended for cleaning rugs and carpets.

No. 2
U. S. Patent 2,364,608

A	Stoddard Solvent	8 –12 gal.
	Carbon Tetrachloride	13 –19 gal.
	Deodorized Kerosene	11¼–16¼ qt.
	Diglycol Oleate	10½–15¼ lb.
B	Diglycol Stearate	10 lb.
	Water	22 gal.
	Sulfated Fatty Alcohol	61 oz.
C	Wood Flour	192 lb.

Warm B and stir until uniform. Mix A into B until emulsified. Then add with good mixing C.

Textile Tar Spot Remover
Formula No. 1

Tergitol #4 or 7	0.1
Naphtha	87.9
Diglycol Laurate S.	12.0

No. 2

Nonaethylene Glycol Monoleate S725	10
Xylol	90

Both of the above are soluble oils and may be emulsified by mixing with water.

Upholstery Cleaner

Naphtha	50
Carbon Tetrachloride	50

Cleaning Fluid
(For leather and cloth)

An excellent solution for cleaning grease stains from cloth or leather consists of the following:

Carbon Tetrachloride	80
Ligroin	16
Tertiary Amyl Alcohol	4

Disinfecting Dry Cleaning Solvent
U. S. Patent 2,348,795

Carbon Tetrachloride	96
Methanol	4
Mercuric Chloride	⅕

Dry Cleaning Fluid

Stoddard Solvent	987
Diglycol Laurate	5
Tertiary Butyl Alcohol	4
Water	4

Blending Soap with Organic Solvents

The solubility of soap in dibutyl tartrate is very high, and is greatly increased by the presence of small proportions of hydrocarbons or chlorinated hydrocarbons. Thus, 100 g. of dibutyl tartrate dissolves 41.3 g. of sodium oleate at 25°C. and an addition of about 20% of chloroform, benzene, amyl alcohol or other suitable solvents increases the solubility by 10 to 30%. A very useful property of such solutions is that they can be diluted with light petroleum fractions without precipitation of the soap. Hence, to prepare soluble oil, commercial soap, particularly castile soap, is dissolved in dibutyl tartrate containing 20% benzene or trichlorethylene. In this way, a 30% solution of soap is easily prepared, and this soap solution is diluted with vegetable or mineral oil, to make a final soap concentration of about 3 to 5%.

The oil so prepared can be thinned to a creamy emulsion with water, the emulsion thus formed being very stable. By changing the type oil, it is possible to make cutting oil, textile oil, agricultural spray, etc. The same stock solution of soap can be used for dry cleaning, since it can be diluted by dry-cleaning fluids, such as Stoddard solvent, without precipitation of soap. Such solutions have good detergent power.

Carpet Cleaner

Liquid Ammonia	4
Dilute Alcohol	3

After loosening the dirt with this cleansing liquid, the following soap solution is applied:

Soap	10
Water	20
Soda	3½
Ammonia and Dilute Alcohol	½

The carpet is wiped dry and need not be taken up.

Paint and Tar Solvent

This paint and tar remover is easily dispersed in water and makes a stable emulsion that is excellent for wool scouring.

Xylene	140.0
Trichlorethylene	47.0
Ethylene Dichloride	61.0
Isopropanol (99%)	33.0
Oleic Acid	40.0
Sulfonated Castor Oil	24.0
Triethanolamine	21.5

Mix the solvents, oleic acid, and sulfonated oil, add the amine, and stir to obtain a clear solution. Adequate ventilation should be provided, and special care should be taken to avoid inhaling vapor and repeated contact with the skin whenever chlorinated solvents are used.

Paint Spot Remover

To take paint spots out of clothing, use equal parts of turpentine mixed with ammonia.

Grease and Paint Remover

Dissolve in 1 qt. of hot water 4 oz. of castile soap cut into fine pieces. To this solution add and mix thoroughly the following:

Aqua Ammonia	4 oz.
Ether	1 oz.
Glycerin	1 oz.
Alcohol	1 oz.
Water	1 qt.

Wall-Paper Cleaner

Wheat Flour	3
Powdered Whiting	1

Mix the flour and whiting thoroughly, adding enough water to produce a dough without excess tackiness. The dough is formed into small balls by hand, and 1 qt. of the composition is considered sufficient to clean the walls of a good-sized room.

Wall Cleaner

Soda Ash	88
Ammonium Sulfate	12

The monohydrated soda ash may be used if care is taken to allow the heat of hydration to be dissipated before the ammonium sulfate is added, otherwise some ammonia will be lost.

Oil and Grease Spot Remover for Floors

A truly dry cleaner for drawing out oil and grease spots from floors and walls consists of a fine-grained Fuller's earth. This material has a remarkable affinity for oil, and is simply brushed or spread over the spot. Its effect is very rapid, although for some deep stains a few repeated applications may be advisable. After Fuller's earth has soaked up the oil stain it is brushed or wiped away.

Marble Cleaner

Muriatic Acid	2 oz.
Acetic Acid	1 oz.
Verdigris	1 dr.

This mixture is applied with a brush, and sponged off with clear water. The brushing is repeated if necessary, after which the marble can be polished with moistened pumice stone.

Telephone Mouthpiece Cleaner

Tincture of Green Soap	10
Alcohol	5
Thymol	1
Water	50
Pine Needle Oil	To suit
Peppermint Oil	To suit

Celluloid and Fabrikoid Cleaner

Stoddard Solvent	7
Acetone	1–1.5

Jewelry Cleaners
Formula No. 1

Orvus Detergent	2
Tet	5
Water	125
Perfume	To suit
Color	To suit

No. 2

Methanol	40
Ammonium Hydroxide	1
Aerosol O.T. Solution (1%)	10
Orvus Solution (1%)	10
Water	4
Color and Perfume	To suit

Cigarette (Nicotine) Stain Remover

Beeswax	10.0
Paraffin Wax	5.0
Mineral Oil	46.0
Pumice (Powdered)	8.0
Borax	0.5
Water	30.0
Perfume	0.5

Auto Cleaner

Infusorial Earth	12
Bentonite	12
Mineral Oil	6
Methyl Salicylate	1
Wood Alcohol	16
Water	90

Wash for Printing Rollers

Gasoline	122 oz.
Acetone	11 oz.
Hydroquinone	75 gr.

Denture Cleaner

Calcium Carbonate	400
Powdered Hard White Soap	120
Ammonia Water	240
Glycerin	200
Sassafras Oil	40
Saccharin	$\frac{1}{5}$

Paint Brush Cleaners

1 { Kerosene 2 pt.
    Oleic Acid 1 pt.
2 { Ammonia (Conc.) ¼ pt.
    Denatured Alcohol ¼ pt.

Stir 2 into 1 until uniform. To clean brushes, place in the mixture overnight. Wash thoroughly with warm water.

(Non-Inflammable)

Trichlorethylene	80
Benzene	20

Industrial Cleanser

Powdered Toilet Soap	30
Bentonite	30
Synthetic Detergent (Wetanol)	10
Lanolin	5
Perfume	1

Solvent Emulsion Cleaner
U. S. Patent 2,374,113

Talloil	40–60
Triethanolamine	7.6–11.4
Caustic Potash (50% Solution)	20–30
Ethylene Glycol Monobutyl Ether	12–18
Pine Oil	6.4– 9.6

Cleaning Non-Ferrous Tanks
Formula No. 1
(Copper Tanks)

Sodium Bicarbonate	1	lb.
Water	1	gal.
Sulfuric Acid (66° Bé.)		½ pt.

Dissolve the bicarbonate in water. Put the sulfuric acid in a bottle on the bottom of the tank. Remove the stopper from the bottle (by means of a string attached to the stopper).

A violent reaction occurs for a few seconds.

No. 2
(Lead Lined Tanks)

Soda Ash	18	lb.
Nitric Acid	1	gal.
Ammonium Hydroxide	2	gal.

Metal Cleaner

Sovasol #6 (Mineral Spirits)	29	gal.
Celite HSC	20	lb.
Diglycol Laurate	3	lb.
Oleic Acid	7	lb.
Triethanolamine	3	lb.

Mix well and add:

Water	43	gal.
Ammonia (28%)	1	gal.

Stir very well.

Metal Parts Cleaner
Formula No. 1

Ethylene Dichloride	31.0
Alcohol (95%)	1.5
Creosote Oil	39.0
Potassium Oleate	2.8
Sodium Chromate	0.5
Water	25.2

No. 2

Tar Acid Oil (50%)	40
o–Dichlor Benzene	20
Triton X30	1
Twitchell Base 277	30
Water	10

Aluminum Cleaner

Calgon (Sodium Hexametaphosphate)	½
Sodium Silicate (41° Bé.)	9½
Water	90

Use at 185–195°F.

Stainless Steel Cleaner

Iron Chloride	1
Conc. Hydrochloric Acid	49
Water	50

To polish, run in a tumbling barrel.

Surgical Instrument Detergent and Germicide
U. S. Patent 2,347,012

Formaldehyde	4
Alcohol	70
Water	10
Thymol	0.63–0.5
Hexamethyleneamine	0.63–0.5
Methanol	7.5
Acetone	7.5

Piston Gum and Carbon Remover
U. S. Patent 2,347,983

Dibutyl Phthalate	33–50%
Phenol or Cresol	67–50%

Motor Carbon Remover
U. S. Patent 2,367,815

Methyl Naphthalene (30–50)	} 10
Cresol (70–50)	
Refined Mineral Oil or Pine Oil	90

In the fuel supply dissolve 0.1–5% of the above mixture.

Airplane Body and Engine Cleaner

Creosote Oil	50
Orthotoluidine	20
Diethanolamine	10
Oleic Acid	10
Ethylene Glycol	10

This is commonly diluted 1 in 5 with water or paraffin oil.

Machinery Cleaner
U. S. Patent 2,356,747

Neutral Coal Tar Oil	40
Monoethanolamine	15
Oleic Acid	15
Ethylene Glycol	15
o–Toluidine	15
Ethyl Silicate	½
Phosphoric Acid	½

Cleaning Microscope Slides

Potassium Dichromate	15 g.
Sulfuric Acid	100 cc.
Water	100 cc.

Dissolve the potassium dichromate in the water and add to it very slowly the sulfuric acid. To clean the slides immerse them in the above solution for a short time and then thoroughly rinse with water. The slides should then be wiped with a soft clean towel. A drop of water placed on the cleaned slide should immediately spread to a thin layer.

Removing Carbon Deposits
(From laboratory ware)

Formula No. 1
Put enough potassium chlorate into the dried vessel, from which as much carbon has been removed by mechanical means as possible. Heat enough to just melt the potassium chlorate and slowly rotate the flask so that the molten chlorate flows over all the carbon. The reaction is spontaneous and care must be taken in applying heat. After cooling, the residue washes off readily with water.

No. 2

Trisodium Phosphate	4 oz.
Sodium Oleate	1 oz.
Soft Water	1 qt.

Allow to stand in the solution for several minutes, brush off the incrustation and rinse with water.

No. 3
First rinse the flask with acetone or carbon disulfide to remove traces of oil or tar. Add a few grams of magnesium nitrate. Heat gradually over a free Bunsen flame till the water is all expelled and the magnesium nitrate melts. Rotate the flask to distribute the melt and continue the heating till the brown

fumes of nitric oxide cease to evolve. Finally cool and dissolve the residual magnesium oxide in dilute acid by boiling.

Large deposits of carbon or tar will require a repetition of the above procedure.

No. 4

To remove thoroughly all of the carbonaceous deposit baked in the bottom of an Engler flask from gasoline distillation, place 2 or 3 g. of commercial sodium sulfate in the flask to be cleaned; apply heat directly to the flask from a Bunsen burner. Heat until all the carbon residue has been loosened. Cool, rinse and drain.

Removing Brown Stains from Burettes

Brown stains left on the inside of burettes used for $KMnO_4$ solutions may be removed by filling the burette with $FeSO_4$ solution after which the liquid is removed and completely washed out. A convenient, ready for use solution of $FeSO_4$ may be made by placing small nails in a dilute H_2SO_4 solution, keeping the flask closed except for a hydrogen vent, thus preventing oxidation of the iron.

Cleaning Fermentation Tubes and Other Glassware

Fermentation tubes (used in water testing) and other glassware difficult to clean in the ordinary way, may be cleaned as follows:

Moisten the inside of the tube with ethyl alcohol. Pour off the excess alcohol, leaving not more than 2 cc. of the liquid in the tube. Add 10 cc. of concentrated nitric acid and let it stand. Soon a vigorous reaction takes place with the elimination of large quantities of nitrogen dioxide. When the reaction stops, wash with water. As some nitric acid may be blown out of the tube, it should be placed in a sink, preferably in a hood until the reaction ceases. Do not close the tube.

Removing Films of Silicone Lubricants

Fill the apparatus with warm decahydronaphthalene (decalin) and allow to stand for 2 hours or more if necessary.

Drain and rinse once or twice with acetone and dry with a stream of filtered air.

The decahydronaphthalene can be reused several times before it becomes ineffective.

Window Cleaners
Formula No. 1

Water	200.0
Mercol ST	0.5
Triton X–30	0.1

No. 2

Aerosol O. T. (10%)	4.0
Ammonium Hydroxide	4.0
Water	492.0

No. 3

Acetone	1885
Water	1890
Kerosene	5

Automotive Glass Cleaner

Isopropyl Alcohol	15
Water	85
Methylene Blue Dye	To suit

This is an excellent glass cleaner which is harmless if accidentally spilled on lacquered surfaces.

Laboratory Glass Cleaning Solution

Trisodium Phosphate	2 oz.
Sodium Oleate	1 oz.
Distilled Water	1 pt.

Soak apparatus in the warm solution 10–15 minutes, then brush with a stiff brush.

Clearing Stopped Drains

Use 1 cup baking soda and 1 cup table salt, pouring over them a kettle of boiling water.

Cleaning Locomotive Boilers

Oil and carbon deposits are removed by boiling under 100 lb. pressure with:

Sodium Metasilicate	½ oz.
Sulfatate B (Wetting Agent)	1/10 oz.

per gallon of water.

Removal of Oils and Greases

To remove oils and greases from motor blocks, equipment, garage floors, etc., pour on Emulphor AG (oil soluble) or Diglycol laurate, rub in well, and wash off by means of a hose.

Cleaning Auto Radiators

Dissolve 2½ lb. washing soda in 3 qt. of warm water. After draining the auto radiator, pour in the soda solution and fill with water. Operate the motor for 20 minutes, then drain out the solution and refill. For very dirty radiators, the solution may be left in the radiator for several days.

Cleaning Tarnish from Silverware

Apply a saturated sodium hyposulfite solution to which a little bolted whiting has been added with a brush or cloth until the silver tarnish is removed.

Dust Cloth Emulsion

Rose (Mineral) Oil	360
Span #85	32
Tween #85	32
Tetrasan	1.5

Take 200 parts of the above mixture and 800 parts of water.

Chapter XXI

TEXTILES

Bleaching Cotton Goods
Single-Boil Hypochlorite Bleach

The conventional hypochlorite bleach method used consists of the following steps:

1. Wet with 2 to 3% malt diastase at 140°F. and pile in a bin for 2 to 3 hours. (The temperature remains near 140°F. during storage.)
2. Wash and pass through 1½% sulfuric acid solution (2° Tw.) and again wash without steeping.
3. Neutralize the excess acid with sodium carbonate in the kier and boil for 10 hours (approximately 8 hours after full pressure is reached) at 15 lb. pressure in 3.6% caustic soda (on weight of the goods).
4. Wash thoroughly in the kier and pull from the kier through a washer.
5. Bleach with sodium hypochlorite (0.75° Tw.) or 1,875 lb. of liquid chlorine per 1000 lb. of cloth and pile in a bin for about 1 hour.
6. Wash and antichlor with approximately 0.1% sodium bisulfite.
7. Wash well and transfer to the white bins.

Double-Boil Hydrogen Peroxide Bleach

This method consists of the following steps:

1. Wet with 2 to 3% malt diastase at 140°F. and pile in a bin for 2 to 3 hours. (The temperature remains near 140°F. during storage.)
2. Boil for 5 hours (approximately 3 hours after full pressure is reached) at 15 lb. pressure in 2.5% caustic soda (on the weight of the goods).
3. Drain the kier and wash the cloth thoroughly in the kier.
4. Pull the cloth from the kier through 1½% sulfuric acid (2° Tw.) and a washer into a second kier.
5. Neutralize excess acid with sodium carbonate in the kier, drain, and boil for 8 hours (approximately 6 hours after full pressure is reached) at 15 lb. pressure in 1.1% caustic soda (on the weight of the goods).
6. Wash thoroughly in the kier and bleach 5 hours (approximately 4 hours after temperature is reached) at 180°F. with ½ volume hydrogen peroxide buffered to approximately pH 11 with sodium silicate.
7. Wash thoroughly in the kier and pass the cloth through a washer into the white bins.

Non-Settling Bleach Suspension

Bleaching Powder	40
Water	60
Sugar	⅕

Textile Waterproofing

Formula No. 1
(Imprägnol)

Paraffin Wax (M. P. 48–52°C.)	17.0
Bone Glue	2.5
Alumina	3.0
Formic Acid	7.0
Rosin	0.7
Caustic Potash (50° Bé.)	0.2
Water	To suit

No. 2
(Hydrophobol)

Bleached Montan Wax	6.3
Paraffin Wax (M. P. 48–50°C.)	6.3
Stearic Acid	3.2
Formic Acid	12.0
Alumina	8.0
Zirconium Oxychloride	10.0
Water	To suit

No. 3
(Textal)

Paraffin Wax	26
Bone Glue	11½
Sodium Tetralinsulfonate	2

Add water to suit before use.

Package Dyeing of Rayon
(German Process)

(A) *Preparation of the Stock Vat*

The vatting is carried out according to the prescription for stock vats of highest and medium concentrations. The formula as chosen depends on the volume of the stock vat which is easiest to handle. For large quantities of dyestuffs, the formulas for stock vats of highest concentrations should be applied, while for small quantities of dyestuffs those for stock vats of medium concentration should be used. Ten grams of Setamol WS per liter should also be added to the stock vats.

(B) *Preparation of the Dye Bath*

To prepare satisfactory diluted acid solutions it is absolutely necessary to use distributing agents in the dye bath. They particularly serve to maintain stability and fine dispersion of the vat acid. For this purpose Setamol WS with Peregal OK has proved most suitable; 1 g. per liter of each is added to the dye bath. Exceptions are Indanthrene Blue BC, GCD and Indanthrene Brilliant Blue RCL, where the Peregal OK is to be replaced by the same quantity of Medialan A or Igepon T since Peregal OK forms precipitates which are difficult to dissolve.

The mentioned quantities of 1 g. Setamol WS, 1 g. Peregal OK and Medialan A or Igepon T respectively are not to be understepped, otherwise the dye bath may be precipitated. Generally, these quantities are sufficient to keep the dyestuff in the form of vat acid well dispersed even in the case of dye baths below the proportion 1: 20 and consequently higher concentrations. Only when using Indanthrene Blue BC Powder fine and Indanthrene Blue RCL fine for dyeing higher concentrations than 3.5 and 7 g. per liter are to be avoided.

For the vat acid procedure the following dyestuffs are not suitable:

1. Indanthrene Brilliant Pink BBL
Indanthrene Blue GC, GCN
Indanthrene Blue 3GT
Indanthrene Turkey Blue GK
Indanthrene Turkey Blue 3GK

as they cannot be vatted in the necessary concentrations, and

2. Indanthrene Blue 5G
    Indanthrene Green BB
    Indanthrene Green GT

as their vat acid solutions are precipitated.

The following dyestuffs can only be regarded suitable to a limited extent:

1. Indanthrene Yellow G

has a tendency to precipitate when being used as proper dyestuff alone, but is suitable in various combinations, e.g., with the Indanthrene Brilliant Green (Jade Green) and Indanthrene Olive Green brands with Indanthrene Olive T and Indanthrene Grey M and MG.

2. Indanthrene Brilliant Orange GR
    Indanthrene Scarlet GC
    Indanthrene Blue 3G, 3GN
    Indanthrene Brilliant Blue 3G

Of these dyestuffs the free vat acids can only be kept in solution without flocking out in concentrations of 1–2 g. per liter by increasing the quantity of Setamol WS to 2–10 g. per liter of dye bath.

The vat acid solutions prepared according to the above orders with the aid of distributing agents possess an excellent stability satisfactory for practical requirements.

The transfer of the sodium leuco compound into the free vat acid is externally recognizable by a pronounced change of color. An exception is Indanthrene Brown 3GT; with Indanthrene Brown FFR and R the change is not very pronounced.

(C) *Dyeing Process*

The caustic soda during the first half hour should be added in small portions in order to regulate the speed of exhaustion by transferring slowly and gradually the vat acid into the substantive sodium leuco compound.

The quantity of caustic soda to be added corresponds to that of a normal dyeing process with even relationship of the liquor, so that in the end the effect of the quantity of caustic soda of a normal vat is obtained. Whenever the construction of the dyeing apparatus makes it possible, it is advisable to add the caustic together with the hydrosulfite by means of a metered vessel in such a way that the concentration in the dye bath increases to 4 cc. per liter within the first half hour. Thereby the dye bath will be exhausted slowly. The remainder can now be added quickly since the zone of danger is only at the beginning of the dyeing process. If the apparatus does not permit the continuous addition of caustic this has to be added gradually. Generally, it has been found appropriate for medium and deep shades to start with $1/15$ of the total quantity of caustic, adding further $2/15$ after 10 minutes and $8/15$ ten minutes later. With very light tones, with goods to be penetrated with difficulty or with dyestuffs to be levelled with difficulty, when a continuous flowing is impossible, in the first part of the dyeing procedure the caustic has to be added in small quantities of $1/15$, $2/15$, $4/15$ or possibly even smaller up to a concentration of 4 cc. per liter. The remainder of the caustic can usually be added in one addition after the concentration has reached 4 cc. per liter.

The dyeing temperature of the vat acid procedure should be the same as usual for different dyestuffs at the beginning of the process but can be increased to a higher temperature than usual after 30 to 45 minutes in order to afford better penetration and more complete exhaustion.

Formula No. 1

Dyed on a "Krantz" machine, System Hulsenlos. A batch of 91 kg. staple fiber twining has to be dyed with a combination of

Indanthrene Blue RSN Powder	1.65%
Indanthrene Dark Blue BOA Powder f.f.d	0.38%
Indanthrene Grey M Powder f.f.d	0.06%
Indanthrene Brilliant Violet RR Powder f.f.d	0.07%

to a dark blue.

For 91 kg. of staple fiber, cross-reeled cops, use 1200 l.

Caustic Soda	2/20 1800	1/20 900	1/20 900	1/20 900	15/20 13500
Sodium Hydrosulfite (Conc.)	1/10 350	2/10 700	3/10 1200		4/10 1500

After this, the temperature is increased to 75°C. and dyed for 15 minutes.

A far reaching exhaustion of the bath takes place until the fourth addition of caustic (about 50%) and 10 minutes after the last addition of caustic the bath is completely exhausted.

No. 2
Color: Vat Navy

Stock vat:

Indanthrene Blue RSN Powder	1500 g.
Indanthrene Dark Blue BOA Powder f.f.d	300 g.
Indanthrene Grey M Powder f.f.d.	50 g.
Indanthrene Brilliant Violet RR Powder f.f.d.	60 g.
Setamol WS	2 kg.
Caustic Soda (38° Bé.)	6 l.
Hydrosulfite (Conc.)	2 kg.
Water at 65°C.	200 l.

To be vatted for 10 minutes at 60°C.

Dye-liquor 60°C.:

Setamol WS	1.2 kg.
Peregal OK	1.2 kg.

Stock vat:

Acetic Acid 40%	9 l.

The vat acid is a clear, blue solution with a red cast. Before the first addition of caustic and hydrosulfite the material is impregnated for 10 minutes with the vat acid solution.

In intervals of 10 minutes each the following agents are added:

Count: 38/2 Combed Peller Cotton Yarn

Yarn Weight: 10 lb. (Dye and chemicals are figured on the weight of the yarn).

Liquor ratio: 1 to 10

After the yarn has been wet out, add

Nekal NF	1%
Igepon T Gel	1%

Circulate 15 minutes at 100°F. before adding dye.

Make a Stock Vat Reduction of:
  Indanthrene Navy Blue
    BRP Paste   6.00%
  Indanthrene Dark Blue
    BOD Paste   3.00%
  Caustic Soda   2.00%
  Sodium Hydrosulfite   2.00%
  Nekal NF   0.80%

Vat 10 minutes at 140°F. then, before feeding the dye into the machine, bring the stock vat to a pH of 5.5 by adding 600 cc. or the required amount of 40% acetic acid.

The timer is adjusted to 12:00 o'clock at the beginning of all operations regardless of the time of the day.

When the pH of 5.5 has been obtained the dye is added as follows:

Add ⅓ of the dye at 12:00 o'clock outside-in
Add ⅓ of the dye at 12:10 o'clock inside-out
Add ⅓ of the dye at 12:20 o'clock outside-in

Make a solution of caustic soda and sodium hydrosulfite using 6% caustic soda and 6% sodium hydrosulfite.

Make a solution of 3000 cc. with water.

Start a continuous flow of caustic soda and sodium hydrosulfite solution at 12:30 o'clock to flow gradually until 1:00 o'clock.

Run until 1:45 o'clock. Check and drop.

Finish

  Use a 0.75% sodium hydrosulfite wash at 100°F. for 10 minutes
  A running wash for 10 minutes
  A cold wash for 10 minutes
  A 1% sodium perborate solution at 120°F. for 20 minutes
  A cold wash for 10 minutes
  A mixture of 1% olive soap and 0.50% Raycomine at 180°F. for 30 minutes.
  A hot wash at 140°F. for 15 minutes
  A cold wash for 10 minutes
  A 2% olive oil bath at 100°F. for 15 minutes

Dyeing temperatures:
  Start the temperature at 100°F. at 12:00 o'clock
  Raise the temperature to 140°F. by 12:45 o'clock
  Raise the temperature to 165°F. by 1:05 o'clock

No. 3

Color: Algosol Green
Count: 40/1 Combed peeler cotton yarn
Yarn weight: 10 lb. Dye and chemicals are figured on the weight of the yarn
Machine: Obermaier
Liquor ratio: 1 to 13

Procedure:

After the yarn has been wet out, add:

  Nekal NF   1.00
  Ammonia (25%)   0.5
  Acetic Acid (56%)   2.3
  Retardine   1.0

Circulate 10 minutes at 75°F. Add ½ of the dye and of the chemicals listed below:

Run 2 minutes outside-in:
  Algosol Green IBW Paste-
    Color Index 1101   0.4%
  Algosol Brown IBR of
    Powder-Prototype 118   0.32%
  Algosol Blue IBC Paste-
    Color Index 1114   0.1%
  Sulfonated Castor
    Oil (50%)   1.0%

Add the remaining ½ of the dye and chemicals. Run 2 minutes inside-out, then shift the valve at regular intervals. Circulate 11 minutes at 75°F. Raise the temperature to 95°F. and run 15 minutes.

Add: 10% sodium sulfate and run 20 minutes longer. Drop the bath. Develop with:

Sodium Nitrite	0.60%
Sulfuric Acid 66° Bé.	5.00%

Run 10 minutes cold;
circulate 2 min. outside-in
2 min. inside-out
4 min. outside-in
2 min. inside-out

Drop the developing bath and give a cold wash for 10 minutes.

Drop the bath and give a 1% soda ash wash at 110°F. for 10 minutes.

Drop the bath and give a cold wash for 10 minutes.

Finish

Olive Soap Flakes	1%
Igepon T Gel	0.5%

Use at 180°F. for 30 minutes. Give a hot wash at 140°F. for 15 minutes, then a cold wash for 10 minutes.

Algosol Dyeing Chart

Add spring at 11:50 at 75°F.
Add ½ dye at 12:00 at outside-in
Add ½ dye at 12:02 at inside-out
12:04 at outside-in
12:08 at inside-out
12:12 at outside-in

No. 4

Color: Light Green
40 cakes of 550 g.—22 kg.
Dye liquor 340 l.
Ratio 1:15

(1) *Preparation of Stock Vat:*

For this purpose 24 g. Indanthrene Yellow 5GK and 16 g. Indanthrene Turquoise Blue 3GK are stirred into 8 l. of water 50°C. with an addition of 200 cc. caustic soda (35%) and 200 g. sodium hydrosulfite. The vat is allowed to stand for 15 minutes.

(2) *Preparation of the Dye Bath:*

For the dye bath 340 l. of water of 35°C. are mixed with 1140 cc. caustic soda (35%), 1 kg. sodium hydrosulfite and 1 kg. Peregal O.

(3) *Dyeing:*

The cakes are first treated with the preparation bath (2) for 15 minutes at 35°C. The stock vat is added and the bath allowed to stand until clarified. Dyeing is done for 15 minutes at 35°C. and afterwards for 1 hour and 30 minutes at 60°C. After that, rinse cold for 30 minutes, and soap for 20 minutes at 80°C. with 1 g. olive oil soap per liter; for finishing treat for 1 hour in a bath with 2 g. Soromin SG per liter at 40°C. The cakes are removed, centrifuged for 10 minutes and dried.

No. 5

Color: Bordeaux
40 cakes of 550 g.—22 kg.
Dye liquor 340 l.
Ratio 1:15

(1) *Preparation of Stock Vat:*

To prepare the stock vat 800 g. Indanthrene Rubine B, 200 g. Indanthrene Red CC and 134 g. Indanthrene Red Brown 5 RF are pasted up with 1.8 l. sulfonated castor oil and stirred into 40 l. of water 60°C. with an addition of 2720 cc. caustic soda (35%) and 680 g. sodium hydrosulfite. The vat is allowed to stand for 30 minutes.

(2) *Preparation of the Dye Bath:*

For the dye bath 300 l. of water of 35°C. are treated with 1.36 l. caustic soda (35%) and 680 g. sodium hydrosulfite and 1.36 l. Peregal O (sharpened previously).

(3) *Dyeing:*

The cakes are treated for 15 minutes with the preparation bath (2) sharpened previously at 35°C. and removed from the bath again. While stirring, the stock vat is added to the dye bath and the bath allowed to stand for 5 minutes. The cakes are immersed into the dye bath and at first dyed for ½ hour at 35°C. and another ½ at 50°C. and then for 2 hours at 60°C. After dyeing for 1 hour and 30 minutes 340 g. sodium hydrosulfite are added. Afterwards rinse cold for 2 hours; after that 340 g. sodium perborate are added to the vat and treated for 30 minutes at 50°C. After that soaping goes on at 80°C. with 1 g. olive oil soap per l. and an aftertreatment with Soromin SG at 40°C. (2 g. Soromin SG per liter) for 15 minutes follows. The cakes are removed, hydroextracted for 10 minutes and dried.

No. 6
Color: Marine Blue
144 cakes of 550 g.—80 kg.
Dye liquor 960 l.
Ratio 1:12

(1) *Preparation of Stock Vat:*

For the stock vat 2.88 kg. Indanthrene Marine Blue R 8015 and 720 g. Indanthrene Marine Blue G 8015 are pasted up with 12 l. of sulfonated castor oil and treated with 240 l. of water at 60°C. with an addition of 7.2 l. caustic soda and 2.4 kg. sodium hydrosulfite. The vat is allowed to stand for 20 minutes.

(2) *Preparation of the Dye Bath:*

To prepare the dye bath 720 l. of water at 20°C. are treated with 2 l. caustic soda (35%), 2 kg. sodium hydrosulfite and 10 l. Peregal O (sharpened previously).

(3) *Dyeing:*

The cakes are treated for 15 minutes in the preparation bath (2) at 20°C. The stock vat is added and dyeing is carried out first at 20°C. for 1 hour and after that 1 hour at 40°C. The dye bath is dropped and a so-called blind vat is prepared by adding 9.6 l. caustic soda (35%), 2 kg. sodium hydrosulfite and 4 l. Peregal O. The goods are treated for 3 hours at 80°C., the bath dropped, and rinsed cold for 30 minutes. After that oxidation is carried out for 1 hour and 30 minutes (adding 3 l. hydrogen peroxide 35%). Rinse cold for 30 minutes and soap for 20 minutes at 80°C. (1 g. olive soap per liter), rinse 20 minutes at 60°C. The cakes are removed, hydroextracted for 10 minutes and dried.

No. 7
Color: Diazo Black
144 cakes of 550 g.—80 kg.
Dye liquor 960 l.
Ratio 1:12

(1) *Dyeing:*

For dyeing 5.6 kg. Sambesi Black V are dissolved in 20 l. of water at 80°C. and added to the stock vat while stirring well. The cakes are dyed first at 80°C. for 30 minutes, 10 kg. sodium sulfate (dissolved) are added and dyeing goes on for 1 hour at 80°C. Another 10 kg. sodium sulfate are added and dyeing is finished after 1 hour. Rinse cold for 15 minutes.

(2) *Diazotation:*

For the diazotation 3 kg. nitrite

(dissolved in cold water) and 6 l. concentrated hydrochloric acid are added. The cakes are immersed into the diazotation bath at 20°C. and diazotized for 30 minutes. After that, rinse cold for 30 minutes.

(3) *Developing:*

For this purpose 1.5 kg. Developer H are dissolved in 10 l. of water at 80°C. and added to the vat. The cakes are placed into the developing bath and developed at 20°C. for 30 minutes. Rinse cold 30 minutes. Soap 20 minutes at 50°C. (1.5 g. olive oil soap per liter) and aftertreat for 15 minutes at 40°C. with Soromin SG (2 g. per liter). The cakes are taken out, hydroextracted for 10 minutes and dried.

No. 8

Color: Naphthol Red
144 cakes of 550 g.—80 kg.
Dye liquor—960 l.
Ratio 1:12

(1) *Starting Solution:*

Six hundred grams Naphthol AS—TR are pasted up with 1.2 l. sulfonated castor oil. Three liters caustic soda (35%) are added and 60 l. boiling water are poured over the mixture. Boil until the solution is clear.

(2) *Dyeing Bath:*

For the dyeing bath 960 l. of water are sharpened previously to 55°C. with an addition of 2.5 l. caustic soda (35%). The starting solution (1) is filtered and added to the bath.

(3) *Impregnation:*

The cakes are impregnated at 50°C. for 1 hour and 30 minutes, taken out and hydroextracted for 20 minutes.

(4) *Developing:*

For developing 800 g. Fast Scarlet TR—Base are stirred into 20 l. of cold water. The solution is treated with 800 cc. concentrated hydrochloric acid. While stirring well 350 g. sodium nitrite, dissolved in 2 l. of cold water, are added slowly; the solution is allowed to stand for 30 minutes and neutralized with 600 g. sodium acetate. 150 cc. acetic acid (50%) are added. Five kilograms sodium acetate are added to the stock vat. The cakes are immersed into the developing bath at 15°C. and developed for 45 minutes. After that rinse cold for 45 minutes and soap three times, 1 hour each time, at 80° using 2 kg. Igepal C. After every soap bath, rinse cold for 20 minutes. The cakes are taken out, hydroextracted for 10 minutes and dried.

Dyeing of Nylon Hose

Previous to dyeing and after presetting, the hose are scoured to remove the size.

For 100 lb. of hosiery use

Duponol D Paste	2 lb.
Soap	2 lb.
Trisodium Phosphate	1 lb.

Make up to 50 gal. of liquor

Start scouring at 100–110°F., raise the temperature to 180–190°F. in 20 minutes and continue scouring at this temperature for ¾ of an hour. Give a rinse in warm water and dye.

12:22—inside-out
12:27—outside-in
12:32—inside-out
12:34—outside-in
12:40—inside-out

12:45—outside-in
12:55—inside-out
1:00—outside-in
Allow the temperature to rise to 95°F. by 12:30
Add salt at 12:30
Drop the bath, don't give a cold wash. Treat with:

Sodium Nitrite	
Sulfuric Acid (66° Bé.)	
Run 10 minutes cold	
Circulate 2 minutes outside-in	
Circulate 2 minutes inside-out	
Circulate 4 minutes outside-in	
Circulate 2 minutes inside-out	
Cold Wash	10 minutes
Soda Ash Wash 110°F.	10 minutes
Cold Wash	10 minutes
Soap at 180°F.	30 minutes
Hot Wash at 140°F.	15 minutes
Cold Wash	10 minutes

The following are some of the representative color formulae used for producing currently popular shades on nylon hosiery:

For 100 lb. hosiery use

Joytan

Acetamine Yellow N	0.25 lb.
Acetamine Scarlet B	0.16 lb.
Celanthrene Brilliant Blue FFS Conc. 200%	0.09 lb.

Sunniblush

Acetamine Yellow N	0.23 lb.
Acetamine Scarlet B	0.12 lb.
Celanthrene Brilliant Blue FFS Conc. 200%	0.04 lb.

Cheerglo

Acetamine Yellow N	0.30 lb.
Acetamine Scarlet B	0.16 lb.
Celanthrene Brilliant Blue FFS Conc. 200%	0.09 lb.

Brown Brandy

Acetamine Yellow N	0.50 lb.
Acetamine Scarlet B	0.30 lb.
Celanthrene Brilliant Blue FFS Conc. 200%	0.20 lb.

Since these colors are insoluble in water it is recommended that they be dispersed by pasting with the Duponol D Paste. Dilute with hot water and add through a strainer. Start dyeing at 110°F., raise the temperature to 180–190°F. in 20 minutes, continue dyeing at this temperature for ¾ of an hour. Rinse twice in warm water and once in cold water, then finish.

Snag Proofing Nylon Hose

The finish which is used for nylon hose to give a desirable appearance or hand and increase the snag resistance is carried out as follows:

Methacrol NH

Dilute 10 lb. with 2 lb. water and add this mixture to the bath the temperature of which should be 80–90°F. Raise the temperature in 15 minutes to 120°F. and keep the bath 15–20 minutes at this temperature. *Do not rinse.* Extract lightly and board.

Nylon Oxford Dyeing

For the dyeing of 1,000 yd. of 3-oz. nylon oxford the following materials and methods are recommended:

Padding Formulae for Nylon

Alizarol Orange 3R	2 lb. 12 oz.
Fast Acid Brown RG	1 lb. 11 oz.
Alizarin Cyanone Green GN Extra	1 lb. 1 oz.

Fast Wool Cyanone
3R	1 lb. 2 oz.
Glycerin	20 lb.
Shellac	2 lb.
Concentrated Ammonia	1 lb. 2 oz.

The dye is pasted with the glycerin and 6 oz. of ammonia and sufficient water is added. Bring to a boil. After boiling for a few minutes, a solution made with the shellac and the rest of the ammonia in 7 gal. of water is added, and the mixture is brought up to 25 gal. at a temperature of 180°F. The pick up is approximately 30%.

The scoured and dried 3-oz. nylon is given one dip through a three-roll padder at 180°F. and dried in a hot flue at 230°F., followed by two 5-minute and ½ minute ageings in a vat ager at 216°F. The roll is transferred to the jig for development: 90 gal. water, 11 lb. formic acid, four ends at 200–205°F. followed by a fresh bath, cold running water, one end. Fresh bath: 8 lb. soda ash, 4 lb. soap, four ends at 200–205°F. The soap is removed by two ends in hot water and finally shelled up through cold water.

Dyeing Wool With Phosphoric Acid or Phosphates

Method I—Knitting worsted yarn, 50 lb. The yarn, previously scoured with a sulfated fatty alcohol, is immersed into a Hussong machine containing in solution at 120°F., the following dyes and chemicals:

Alizarine Light Blue BGA Conc. (S)	52	g.
Xylene Light Yellow 2G (S)	32	g.
Azo Rubinole 2G 175% (S)	9½	g.
Sodium Salt of Isopropyl Naphthalene Sulfonic Acid	1	lb.
Glauber's Salt Crystals	6	lb.
Phosphoric Acid (90%)	860	g.

The dye bath is heated to the boiling point in 45 minutes, and boiling is continued for 1 hour 15 minutes. The pH of the dissolved chemicals without the wool is 2.0. On completion of the dyeing the pH rises to 4.4.

Method II—Peroxide treated knitting yarn, 108 lb. The yarn is lowered into a Hussong machine containing in solution at 140°F. the following dyes and chemicals:

Brilliant Alizarine Sky Blue BS Pat. (S)	2	oz.
Acid Violet 4BNS Conc. (S)	30	gr.
Xylene Brilliant Cyanine G (S)	140	gr.
Mono-Ammonium Phosphate	5	lb.
Isopropyl Naphthalene Sulfonic Acid	2	lb.

The pH of the dye bath without the wool is 4.7, and on completion of the dyeing 5.7. The dye bath is brought to the boiling point in 40 minutes and boiling is continued for 30 minutes. Then the yarn is reversed and boiled a further 20 minutes.

Method III—Knitting yarn, 100 lb. In a Hussong machine the water is brought to the boiling point and the steam shut off. To this is added:

Phosphoric Acid (90%)	1666	cc.

Ammonia (Sp. Gr. 0.88)   2000 cc.
Bichromate of Soda   3½ oz.
Sodium Salt of Isopropyl Naphthalene Sulfonic Acid   1 lb.

The yarn is immersed at 180°F., and circulated for 20 minutes. The yarn is then lifted out and the dissolved dye, 3½ lb. Sulphonine Blue RNC (C.I. No. 289), is added, and then the yarn is again immersed. The temperature is raised to 205°F. in 15 minutes, and kept at this temperature for 1 hour until the exhausted liquor shows a violet coloration. The pH of the liquor before adding the dye should be about pH 6.6. If preferred, the dye may be added to the chemicals at the start, which is a quicker method, but not as safe as the former.

Method IV—Worsted tops, 450 lb. The worsted tops are packed into a Callebaut De Blicquy type dyeing machine, and are wetted out with warm water and a wetting agent. Then 12 lb. phosphoric acid (90%) and 22 lb. ammonium sulfate are diluted with water and added from the overhead tank. The dye, which comprises 20½ lb. Omega Chrome Black S 160% (Color Index No. 203) is dissolved and then circulated through the wool at 110°F. The dye bath is brought to the boiling point in 45 minutes and boiling is continued for 40 minutes. The pH of the dye bath before chroming is 4.2. The dye bath is allowed to cool, and 4½ lb. sodium bichromate are dissolved and added. The boiling is continued for 30 minutes. If the overhead tank is large enough, a better method is to add the phosphoric acid, the ammonium sulfate and the dissolved dye altogether to the dry tops.

Printing on Woolen Fabrics

Proper preparation of the goods to be printed is of the utmost importance. A smooth surfaced fabric, free of lint and beard, will naturally produce the sharpest and clearest prints. Hence, any woolen fabric-bunting, shirting, flannel, sheer, etc., should be singed on the gas or plate singeing machine as the first step of preparation. Furthermore, after any bleaching, chloring or other wet treatments, care in the drying process should be exercised to insure the required smooth surface.

Following the singeing, a scouring should be given with soap and soda or ammonia at not over 120°F., and the material should then be rinsed well.

At this point, when a ground that requires no bleaching is under treatment, it would be well to subject the goods to a chlorination. In fact, as pointed out further on, chlorination of the fabrics to be printed is practically a *must* if deep shades are desired.

When bleaching is required for reasons of shade clarity or a clear white ground, such treatment usually follows the scouring but may be given, if necessary, subsequent to chlorination and may be carried out on the jig. Hydrogen peroxide is the simplest agent, 1 gal. of 100 vol. peroxide per each 50 gal. of cold water, made slightly alkaline with ammonia or silicate of soda. The goods are given two ends, then heated to 140°F. for two additional passes and stored for some hours. A thorough rinse and acidification follows.

Wool fibers possess a natural tendency, especially under influence of steam, to reduce and partially destroy certain dyestuffs. In order to counteract such a tendency and at the same time increase the affinity of fiber for dyestuff, the woolen pieces are advisedly chlorinated. The process is important to the production of satisfactory prints but should be carried out only with the greatest care since excessive treatment may well prove harmful, causing a harshened and yellowed fabric. The treatment may be given on the jig or full width on any suitable apparatus. If on the jig, the goods are given two ends in a cold chlorine solution made up with about 1 gal. of sodium hypochlorite or chloride of lime (clear) at 6° Tw. per each 100 gal. of water. At this point 3 qt. of sulfuric acid sp.g. 1.84 are slowly added as the material is given two more ends, then batched for 1 hour or so. The fabrics are rinsed well to remove all traces of the chlorine. Sulfuric acid is preferred to hydrochloric acid because of the latter's tendency to yellow wool fibers. An anti-chlor treatment in a weak bath of sodium bisulfite may follow if necessry, plus, of course, more cold rinsing.

This chloring process, if carefully applied, should give the desired results without any risk of injury to the material. However, in order to obviate any possible risk, it is suggested that consideration be given to one of the specialized products which slowly release a mild chloring action.

Careful drying at full width and examination for smoothness of surface complete the preparation.

For the printing of woolen fabrics, selected acid and direct colors are used for the most part. Certain chrome and alizarine types are applied where particularly good fastness to light and washing is sought. Where brightness of shade is stressed rather than good fastness, the basic group as well as the eosine types are given consideration. The basics, when used at all, are printed on a lightly chlored fabric in order to prevent bleeding into the white ground. A typical print paste may be made as follows:

Basic Color	1 lb.
Levelene, Cellosolve, or Other Solvent Aids	4 oz.
Acetic Acid 40%	1 gal.
Water	2 gal.
Thickener	6 gal.
Tartaric Acid	4 pt.
Tannic Acid and Acetic Acid 1:1	4 pt.

The dyestuff is first pasted with the solvent aid, the acetic acid is added and the mixture is then dissolved in the water at 160°–190°F. to which is added the thickener and, after cooling, the acids for fixing.

The eosine group is generally printed with the addition of acetic or formic acids. For this type of color, the goods, after chlorinating, are best treated in the padder or jig with a tin salt, i.e., sodium stannate of about 5° Tw., acidified with weak sulfuric acid and rinsed before drying.

Most of the chrome colors may be printed with fluoride of chrome as a fixing agent. In some instances acetate of chrome and tartaric acid are employed. The following print paste may serve as an example:

Chrome or Alizarine Color	1 lb.
Boiling Water	4 gal.
British Gum Thickening	5 gal.
Glycerin	2 lb.
Chromium Fluoride Solution *	5 pt.

Dissolve the color in the boiling water. Cool and add the other ingredients.

The following colors are some of those suitable for use in the foregoing formula:

Alizarine Yellow 2G Conc. (C.I. 36)
Alizarine Yellow 4G Conc. (C.I. 52)
Anthracene Yellow C (C.I. 343)
Alizarine Orange 2GN (C.I. 40)
Erio Chrome Orange G Conc. (C.I. 274)
Chromaven Brilliant Orange 2R
Erio Chrome Red B (C.I. 652)
Chrome Red 3B Conc. (C.I. 280)
Chrome Fast Violet B (C.I. 169)
Chromaven Brilliant Blue B (C.I. 720)
Gallocyanines (C.I. 883)
Chromaven Green G
Acid Anthracene Brown PG (Pr. 4)
Chromaven Printing Brown Paste

---

* The chromium fluoride solution is made as follows:

Chromium Fluoride Crystals	2 lb.
Water	1 gal.
Formic Acid	1 lb.
Tartaric Acid 1:1	1 pt.

Acid Anthracene Brown RH (C.I. 98)
Chrome Blue Black 2B Conc. (C.I. 202)

A great many colors of the acid and acid fast groups are suitable for wool printing. As fixing agents, acetic, oxalic and tartaric acids are most important. When a slower and milder fixing agent is required, the ammonium salts of these acids are employed. The stronger and faster acting sulfuric acid can be used in extreme cases; however, it is not recommended because it tends to attack the back greys and doctor blades.

A typical print paste for acid colors is:

Dyestuff	2 lb.
Solvent Aid	8 oz.
Glycerin	2 lb.
Water	4 gal.

Paste the color with the solvent, add the glycerin, dissolve in boiling water, cool to about 180°F. and mix in:

Thickener (Tragacanth and British Gum)	6 gal.

when cool, add

Acetic Acid 30%	5 pt.
Tartaric Acid 1:1	2 pt.

A partial list of dyestuffs suitable for use with the above formula is:

Azo Yellow G Ex. (C.I. 146)
Chinoline Yellow Conc. (C.I 801)
Naphthol Yellow S (C.I. 10)
Tartarzine (C.I. 640)
Acid Fast Yellow RS (Pr. 187)
Brilliant Milling Yellow 5G (Pr. 138)
Acid Orange Y (C.I. 151)
Acid Orange RO (C.I. 161)

Amacid Fast Orange LW
Neutral Orange G (Pr. 137)
Neutral Orange SGS (Pr. 186)
Ponceau 3RB (C.I. 280)
Amaranth Conc. (C.I. 184)
Acid Fuchsine 10B Conc. (C.I. 57)
Acid Fuchsine 4B Conc. (C.I. 30)
Amacid Brilliant Red 5B Conc. (Pr. 193)
Carmoisine B Conc. (C.I. 179)
Milling Scarlet 3R (C.I. 487)
Pontacyl Violet S4B (C.I. 698)
Acid Violet 10B (C.I. 696)
Patent Blue A (C.I. 714)
Patent Blue V (C.I. 712)
Acid Blue FG (C.I. 671)
Sulphon Cyanine 3R (C.I. 289)
Indulines (C.I. 861)
Amacid Brilliant Blue 3B (Pr. 33)
Guinea Green B (C.I. 666)
Amacid Green G (C.I. 670)
Amacid Fast Green 3G (C.I. 735)
Brilliant Milling Green B (C.I. 667)
Alizarine Green CE EX. (C.I. 1078)
Alizarine Green CG EX. (C.I. 1078)
Sulphon Cyanine Blacks (C.I. 307)

The direct and direct fast types of colors have, in some instances, a distinct advantage over the ordinary acid colors, since, in a general way, they offer better wash fastness, also water fastness as applied to wool.

Print pastes of this group are usually prepared with sodium phosphate or borax. However, a few of these require a mild acid such as acetic acid or tartaric or their slower acting ammonium salt for proper fixation on wool. In either case it is advisable to incorporate a small amount of sodium chlorate into the print paste to obtain a maximum of color fixation and as a protection against the reducing action of the wool and steam, particularly in the case of heavy shades and on non-chlorinated material. In fact, about 5 to 10 parts of chlorate per 1000 parts of paste is recommended for all types of wool printing.

A generalized print paste is:

Direct Color	2 lb.
Cellosolve, Levelene or Other Solvent	8 oz.
Thickener (British Gum)	2½ gal.
Water	5½ gal.
Glycerin	2 lb.

Paste the color with the solvent aid and dissolve in boiling water; cool and add the thickening agent and the glycerin after which are incorporated:

Sodium Phosphate or Borax	2 lb.
Water	1½ gal.
Sodium Chlorate	8 oz.

Dissolve the sodium phosphate or borax and the sodium chlorate in water before adding to the color paste.

For colors requiring an acid for fixation, the phosphate or borax is replaced by an equal amount of acetic acid 30%, tartaric acid 1:1 or the milder ammonium salt of either.

Dyes suitable for use include:
Brilliant Yellow S (C.I. 816)
Chrysophenine (C.I. 365)
Solantine Yellow 4GL

Solantine Yellow NN (C.I. 814)
Amanil Fast Yellow BX (C.I. 814)
Amanil Fast Orange GLZ Conc. (C.I. 653)
Direct Fast Orange WS (C.I. 326)
Diamine Fast Red F (C.I. 419)
Pontamine Scarlet B Conc. (C.I. 382)
Amanil Fast Scarlet 3B Conc. (C.I. 382)
Amanil Brilliant Violet 4B
Amanil Fast Violet 2RL
Diamine Sky Blue FF (C.I. 518)
Diphenyl Dark Green BN (C.I. 583)
Amanil Green B Conc. (C.I. 593)
Amanil Brown 3GS Conc. (C.I. 596)
Chloramine Brown MR (C.I. 420)
Congo Brown RLH (C.I. 598)
Amanil Catechine GS (Pr. 69)
Amanil Catechine 3G (Pr. 70)
Direct Deep Black EW (C.I. 581)

For the machine printing of woolens, the roller engravings are usually slightly deeper than those ordinarily used on cotton. After printing, drying is carried out at a moderately low temperature. The subsequent steaming is effected at low pressure and with moist steam, generally in moist wrappings. The degree of moisture is important because, if too low, fixation is apt to be incomplete and if too high, the sharpness of pattern may suffer. Time of steaming varies from 1 to 1½ hours according to the nature of the color and depth of shade.

After steaming, the goods are washed well with plenty of cold water or with warm water if necessary. Hot water should be avoided to prevent possible staining of the white or lighter ground, and feathering.

Finishing is carried out in an appropriate and convenient manner.

It is well to avoid thickeners which contain large quantities of starches since these are not easily removed in the final washing. The better soluble gums such as British gums and gum tragacanth are recommended because these produce sharper prints and are washed out with greater facility.

Screen Printing on Textiles
Vat Colors for Printing on Cotton, Linen and Viscose Process Rayon

Standard printing pastes are prepared as follows:

Vat Color Paste	20
K 19–16 Thickener *	80

K 19–16 Aqueous Solution:

Water at 170°F.	600
Potassium Carbonate	190

Dissolve, cool to 140°F. and add:

Sulfoxite C	160
Glycerin	50

To prepare a weaker printing paste the standard 20% printing pastes are to be mixed with the

---

* K 19–16 Thickener:

Water	265
Wheat Starch	30
British Gum	205

Boil with agitation for 30 minutes, cool to 180°F. and add:

Potassium Carbonate	190
Water	75

Dissolve, cool to 140°F. and add:

Sulfoxite C	160
Glycerin	50
Water	25

K 8–6 thickener* in the desired proportions.

Somewhat softer prints are obtained on rayon fabrics by using a thickener which is prepared by dissolving the potassium carbonate, Sulfoxite C and glycerin in a commercially available *textile gum* instead of using a paste prepared from starch and British gum.

After printing, the goods are aged 5–10 minutes in an air-free rapid ager, oxidized, rinsed, soaped, rinsed and dried. Cotton and linen prints are oxidized in a solution containing 0.5% sodium bichromate and 0.5% acetic acid for about 45 seconds at 130°F. Rayon fabrics are oxidized in a 2% solution of Albone C or a 2% solution of sodium perborate for about 2 minutes at 90°F. After oxidation the goods are rinsed, soaped well, rinsed and dried.

Vat colors recommended for printing by this method:

Du Pont Vat Yellow 8G Double Paste
Leucosol Yellow GC Paste
Leucosol Yellow K Paste (Pat.)
Ponsol Yellow AR Double Paste
Ponsol Yellow G Double Paste
Ponsol Yellow GGK Paste (Pat.)
Sulfanthrene Yellow R Supra
Ponsol Golden Orange G Double Paste
Ponsol Golden Orange RRT Paste
Sulfanthrene Orange R Paste
Leucosol Brown 3RN Double Paste (Pat.)
Sulfanthrene Brown G Paste
Sulfanthrene Pink FB Paste
Sulfanthrene Pink FF Paste
Leucosol Red B Paste (Pat.)
Sulfanthrene Scarlet G Paste
Sulfanthrene Scarlet 2G Paste
Sulfanthrene Red 3B Paste
Leucosol Brilliant Violet RR Paste (Pat.)
Ponsol Violet RRD Paste
Leucosol Blue G Double Paste
Leucosol Dark Blue BR Paste
Ponsol Blue GD Double Paste
Ponsol Brilliant Blue R Paste (in light shades)
Ponsol Navy Blue Double Paste
Ponsol Navy Blue RA Double Paste
Sulfanthrene Blue 2BDN Extra Paste (Pat.)
Sulfanthrene Navy Blue MR Double Paste
Ponsol Blue Green Y Double Paste (Pat.)
Ponsol Brilliant Green 2G Double Paste (Pat.)
Ponsol Brilliant Green 4G Double Paste (Pat.)
Ponsol Green 2BL Paste
Ponsol Jade Green Double Paste (Pat.)
Ponsol Jade Green Supra Double Paste (Pat.)
Leucosol Black BB Double Paste
Leucosol Black BBD Double Paste
Leucosol Black RA Double Paste (Pat.)

---

*K 8–6 Thickener:

Water	388
Wheat Starch	42
British Gum	280

Boil with agitation for 30 minutes, cool to 180°F. and add:

Potassium Carbonate	80
Water	75

Dissolve, cool to 140°F. and add:

Sulfoxite C	60
Glycerin	50
Water	25

Ponsol Direct Black 3G Double Paste (Pat.) (for grays)
Sulfanthrene Black PG Double Paste
Sulfanthrene Black PR Double Paste

Diagen Colors for Printing on Cotton, Linen and Viscose Process Rayon

Diagen Color	4
Cellosolve	4
Water (120°F.)	29
Caustic Soda (35%)	3
Dissolve and add:	
Neutral Thickener	60

In the case of the Diagen Yellows use 1½ parts caustic soda 35% instead of 3 parts.

For printing on rayon fabrics and for maximum fastness to crocking on cotton use a neutral 6% gum tragacanth paste. Greater strength but inferior fastness to crocking and inferior softness are obtained if the gum tragacanth paste is replaced by the following thickener:

Wheat Starch	80
Gum Tragacanth (6%)	360
Water	510

Boil 30 minutes, then cool and add:

Glycerin	50

Print, dry, age 4 minutes in the acid ager, rinse, soap well, rinse and dry.

If no acid ager is available the printed goods may be developed by passing them for about 10 seconds through a hot bath (225°F.) prepared as follows:

Salt	250
Sodium Acid Pyrophosphate	30
Water	715

M.P.-189 *	5

After developing in the above bath, the printed goods are rinsed, soaped, rinsed and dried as described above.

Diagen colors recommended for printing:
Diagen Golden Yellow MRS (Pat.)
Diagen Yellow AGL
Diagen Yellow AY
Diagen Yellow A2Y
Diagen Orange MG (Pat.)
Diagen Dark Brown AR (Pat.)
Diagen Bordeaux MR (Pat.)
Diagen Red AMX
Diagen Red AR
Diagen Red YN
Diagen Scarlet AR
Diagen Blue MGD (Pat.)
Diagen Blue MGR (Pat.)
Diagen Black DM (Pat.)
Diagen Black MR (Pat.)

Acid and Direct Colors for Printing on Viscose Process Rayon

Urea Method:

Acid or Direct Color	1–4
Urea	18

Mix and add:

Boiling Water	31–28

Heat, if necessary, to dissolve the color.

Textile Gum	50

Print, dry, steam 1 hour in the cottage steamer, without excess pressure, rinse in cool running water, soap lightly, rinse in warm water and dry.

Acid and direct colors recommended for printing by the urea method:
Pontamine Fast Yellow 5GL
Pontamine Fast Yellow 4GL Conc.

---

* M.P.-189 is hydrocarbon sodium sulfonate.

Du Pont Milling Yellow GN Conc. 250%
Pontamine Pure Yellow M
Pontamine Fast Yellow NNL Conc. 175%
Pontamine Fast Yellow RL
Pontamine Fast Orange EGL
Pontamine Orange R Conc.
Du Pont Milling Orange RN Conc. 125% (Pat.)
Pontamine Fast Orange ERL
Pontamine Brown BT Conc.
Pontamine Fast Scarlet 4BA
Pontamine Fast Scarlet 4BS Conc. 150%
Pontamine Scarlet B
Pontamine Scarlet 3B
Du Pont Milling Red SWB Conc. 125%
Pontamine Fast Scarlet 8BSN Conc. 125%
Pontamine Fast Pink BL Conc. 125%
Pontacyl Violet S4B
Pontamine Fast Blue RRL Conc. 175%
Du Pont Brilliant Milling Blue B Conc. 200%
Pontamine Fast Turquoise 8GL Conc. 150%
Du Pont Brilliant Milling Green B Conc.
Pontamine Fast Green 5GL
Pontamine Black E Double

Acid and Direct Colors for Printing on Silk

Acid or Direct Color	1–4
Cellosolve	2–5
Mix and add:	
Hot Water	47–41

Heat, if necessary, to dissolve the dye. Then add:

Textile Gum	50

Print, dry, steam 1 hour in the cottage steamer without excess pressure, rinse, soap lightly, rinse and dry.

Acid and direct colors recommended for printing on silk:

Du Pont Milling Yellow 5G Conc.
Du Pont Neutral Yellow GS
Du Pont Milling Yellow GN Conc. 250% (Pat.)
Du Pont Quinoline Yellow P Extra Conc.
Pontamine Yellow CH Conc.
Du Pont Milling Orange R Conc.
Du Pont Milling Orange RN Conc. 125% (Pat.)
Du Pont Silk Orange R Extra Conc. 125%
Pontamine Brown BT Conc.
Pontamine Scarlet B
Pontamine Scarlet 3B
Du Pont Milling Red SWB Conc. 125%
Du Pont Milling Red SWG Conc. 125%
Pontacyl Carmine 2G Conc. 150%
Pontacyl Fast Violet VR
Pontacyl Violet S4B
Pontacyl Brilliant Blue A Conc.
Pontacyl Brilliant Blue E
Du Pont Brilliant Milling Blue B Conc. 200%
Pontacyl Wool Blue BL Conc. 200%
Pontacyl Wool Blue GL Conc. 250%
Pontacyl Navy Blue M4B Conc. 200%
Pontacyl Brilliant Blue RR Conc. 200%
Pontacyl Green NV Extra Conc. 200%
Du Pont Brilliant Milling Green B Conc.

Du Pont Nigrosine WSB Conc. Powder
Du Pont Nigrosine WSJ Powder
Seristan Black J

Acetate Colors for Printing on Acetate Rayon

Acetate Color	1–5
Glycerin	6

Mix well and add:

Cool Water	5

Mix well and add:

Hot Water	29–33

Mix carefully and heat, if necessary, for complete dispersion. Then add:

Textile Gum	55

After printing, the fabric is dried, steamed 1 hour in the cottage steamer without excess pressure, rinsed, soaped lightly, rinsed and dried.

Acetate colors recommended for printing acetate rayons:

Acetamine Yellow RR
Acetamine Yellow N
Acetamine Orange GR Conc. 175%
Acetamine Brown SR
Acetamine Red RP Conc. 175%
Acetamine Rubine B Conc. 125%
Acetamine Scarlet B
Acetesol Fast Crimson B
Acetesol Fast Scarlet B
Celanthrene Red 3B Conc. 125%
Acetamine Violet 2R
Acetesol Fast Violet 4R
Celanthrene Red Violet R Conc. 150% (Pat.)
Celanthrene Pure Blue BRS 400%

Self-Emulsifying Base for Textile Printing
U.S. Patent 2,346,041

Ethyl Cellulose	5.00
Pine Oil and Naphtha	77.00
Sodium Oleate	12.18
Oleic Acid	0.70
Water	5.12

Natural (Buckwheat) Dye

Buckwheat hulls, cleaned by washing them with water at 40 to 45°C. and boiling them for 15 to 20 minutes in two separate portions of water, are extracted with a hot alkaline solution in the following manner:

Buckwheat Hulls	75.0 g.
Sodium Hydroxide	3.7 g.
Hot Water	To make 750.0 g.

The mixture is boiled for 1 hour, with occasional replenishment of the evaporated water. The extract is filtered and the filtrate concentrated. From a 66.6% extract the following printing paste is prepared:

Buckwheat Extract	50.00 g.
Glycerin	5.00 g.
Powdered Hyposulfite of Soda	1.50 g.
Thickener (Dextrin 2:1)	35.00 g.
Chromium Formate (16° Bé.)	7.50 g.
Calcium Acetate (15° Bé.)	1.09 g.

Fabrics printed with this paste are finished as usual. The resulting color resembles that produced by Alizarin Brown K. The color is fast to water and soap at 40°C., to perspiration, and to dry rubbing.

Sizing
Textile Sizing
Formula No. 1
(Gumminat)

Magnesium Sulfate	56
Glucose	34

Carob Bean Flour	2
Formaldehyde	½
Water	To suit

No. 2
U.S. Patent 2,176,053

Polyvinyl Acetate Resin	25
Toluol	55
Butyl Acetate	20
Water	39
Sulfonated Castor Oil	⅖
Wetting Agent	⅒

Washable Textile Sizing
British Patent 573,768

Caustic Soda	280
Water	280
Zinc Oxide	100
Mix and add:	
Water	100
1. Above Solution	120
2. Rayon (Powdered)	27
3. Water	218
4. Ice	175

Add 1 to 2, 3 and 4 and mix slowly until uniform. Filter off any lumps. Treat the cloth with the filtrate and then pass it through a 2–5% sulfuric acid solution to set, run through squeeze rolls, wash and dry.

Organdie Finish for Textiles
British Patent 573,574

Cotton lawn is impregnated with:

Polyvinyl Alcohol	25
Glycerin	5
Formaldehyde	5
Ethylene Chlorhydrin	1

Squeeze, dry at 105°C. and wash with a soap solution at 60°C.

Nylon Yarn Sizing
Formula No. 1
U.S. Patent 2,312,469

Soybean Protein	10
Soda Ash	1½
Glycerol	2
Triethanolamine Oleate	5
Water	81½

No. 2
British Patent 564,027

Peanut Oil	3
Casein (Alkaline Solution)	5
Gelatin	12
Sodium Lauryl Sulfate	5
Urea	4
Water	71

Linen Laundry Stiffening

Water	22 lb.
Starch	1 lb.
Locust Bean Gum	1 lb.

Emulsion for Textile Sizes, Softeners and Lubricants

A. Stearamide	40
Stearic Acid	10
B. Ammonia (28%)	10
Water	940

Heat the mixture A to 80°C. and add to B which has also been heated to 80°C.

Fungus- and Mildew-Proofing
Formula No. 1

Salicylanilide	3
Tricresyl Phosphate	17
Isopropyl Alcohol	80

Soak the fabric in the mixture; squeeze out and dry.

No. 2

Modified Copper Naphthenate (7%)	100
Asphalt Cut-Back	30
Emulsifier	5
Water	331

To the well mixed mixture of the copper naphthenate, the asphalt cut-back and the emulsifier, add slowly the water under vigorous mechanical agitation until an emulsion is formed. It is prefer-

No. 3
Modified Copper Naphthenate

Modified Copper Naphthenate (7%)	100
Asphalt	30
Naphtha	220

Mix the ingredients well until a homogeneous solution is obtained.

No. 4

Copapel A	10 gal.
Ammonia (28%)	1 gal.
Water	90 gal.

Mix the Copapel A with the ammonia and add the water to this mixture. Apply by padding, run through squeeze rolls, and dry.

Rotproofing Fabrics

The treatment is a two-dip process using aqueous solutions. First dip the fabrics into a 0.5% solution of dimethylglyoxime at 90° to 100°C. for 2 to 5 minutes. After draining for 1 minute dip them into a 5% solution of cupric acetate at room temperature for 2 or 3 minutes. The amount of copper fixed in the fabric may be controlled by varying the strength of the copper solution from 1% to saturation.

Rayon Tire Cord Treatment
Formula No. 1
Haemoglobin Latex

A.	Haemoglobin	3.75
	Distilled Water	12.00
	Ammonia (10%)	2.00
B.	Casein	3.75
	Hot Water (112°F.)	45.50
	Ammonia (10%)	3.00

Mix solutions A and B and allow to cool.

C.	Buna S Latex (35% solids)	64
	Distilled Water	41
	Mixture of Solutions A and B	70

No. 2
Resorcinol-Formaldehyde

Resorcinol	0.6
Water	27.78
Caustic Soda (10%)	0.24
Formaldehyde	1.38
Buna S Latex (35% solids)	11.00
Caustic Soda (10%)	0.24
Water	58.80

The fabric is passed through a tank containing one of the above mentioned solutions. It is squeezed through a pair of rubber squeeze rolls and then passed under tension over a series of drying drums heated to 260°F. The tension is sufficient to stretch the fabric to a length approximately 7% greater than the untreated length.

The fabric must be thoroughly dried, and in the case of the resorcinol-formaldehyde dip, the resin must be set up. There is a weight increase of approximately 4%. The treatment stiffens the fabrics materially, making them more difficult to handle than corresponding undipped cotton.

Delusterant for Textiles
U.S. Patent 2,376,908

Water	46
Castile Soap	1
High-Boiling Petroleum Hydrocarbons	14
Castor Oil	1

To this is aded:

Egg Albumin	1
Titanium Dioxide	1
Water	6

Flameproofing

Water-Soluble Flameproofing Formula No. 1

Borax	7 lb.
Boric Acid	3 lb.
Diammonium Phosphate	5 lb.
Water	13⅕ gal.

This formula gives very satisfactory results both in flameproofing and glowproofing. It will be found effective in weightings of 7 to 15% depending upon the fabric treated. With hand wringing the above solution gives weightings of about 10 to 12%.

No. 2

Dibasic Ammonium Phosphate	7½ lb.
Ammonium Chloride	5 lb.
Ammonium Sulfate	5 lb.
Water	12 gal.

Either the cloth may be impregnated directly with this solution, or the starch sizing may be made up with it. It has been used for curtains and for cotton fabrics in general. The ammonium chloride and, to a lesser extent, the ammonium phosphate are hygroscopic, and the use of the formula for materials used in damp locations may be inadvisable. The treatment is effective in weightings of 10 to 18% depending upon the type of fabric treated. With hand wringing a solution of the above concentration gives weightings of about 16 to 18%.

No. 3

Ammonium Sulfate	8 lb.
Ammonium Carbonate	2½ lb.
Borax	8 lb.
Boric Acid	3 lb.
Starch	2 lb.
Dextrin	6½ oz.
Water	12 gal.

The amount of water may be varied as desired. The mixture is applied at 86° to 100°F. It is useful for many purposes, particularly for laces, curtains, and aprons. It is effective in loadings of 14 to 28% depending upon the fabric. With hand wringing the above concentration deposits a loading of about 28%.

No. 4

Sodium Tungstate	20 lb. 9½ oz.
Dibasic Sodium Phosphate	9½ oz.
Water	12 gal.

Sodium tungstate has been used for flameproofing theater scenery. The addition of the sodium phosphate is recommended to prevent crystallization, resulting from the formation of an acid sodium tungstate. With hand wringing the suggested concentration gives a weighting of about 23%. The treatment is effective in preventing flaming, but afterglow continues for a considerable time, and may seriously extend the char.

Flameproofing Welding Curtains

This is especially suitable for preparing heavier duty welding curtains and has been found to provide good flame resistance and afterglow depression and to have no effect on the durability of the fabric. However, the strength of the cloth is somewhat reduced and there is some stiffening. The process is as follows: First saturate the cloth with a solution made from:

Diammonium Phosphate	2 lb.
Warm Water	1 gal.

After drying, the fabric is coated with three layers of the following mixture:

Casein	100
Water	400
Glycerin	130
Ammonia Solution (25%)	10

Water-Resistant Flameproofing Formula No. 1

Antimony Oxide	120
Vinylite VYHH	60
Methyl Ethyl Ketone	420

This amount of methyl ethyl ketone is suggested for a first trial, but it should be varied so that a 35% pickup of the antimony oxide Vinylite fraction on the cloth is produced.

Where the white color of antimony oxide is objectionable, part of it may be replaced by suitable coloring pigments matching the color of the untreated fabric. Improvement in glowproofing will be effected if 27 parts of zinc borate are added.

The Vinylite is dissolved in one half the required methyl ethyl ketone and the antimony oxide added with agitation. If equipment is available, the mix is put through a paint mill to produce a more stable dispersion. Before use, sufficient methyl ethyl ketone is added to give a suitable viscosity and pickup.

All sizing and soil must be removed from the cloth by thorough washing and rinsing before treatment. The dry cloth is passed through the impregnating mix and put through squeeze rolls to remove the excess of the impregnating mix. The setting of the squeeze rolls may be varied to help adjust the amount of the pickup. The cloth is carefully dried, then softened by treating for 30 to 60 minutes on a scouring machine using a 0.2% solution of soap or other suitable detergent at 120° to 212°F., depending upon the type of material. It is then rinsed, dried, and may be further softened by treatment on a sanding or sueding machine.

Duck treated by this process has been through 12 commercial launderings with a loss of less than 10% of an original pickup of 40.4% and with no apparent loss in flameproofing. The treated fabric is also unaffected by the Stoddard or similar dry cleaning solvents, but will not withstand cleaning with chlorinated solvents. This treatment is recommended for cotton, wool and wool-rayon fabrics.

No. 2
(Cotton Drill)

A. Urea-Formaldehyde Resin Monomer	9.62
Catalyst for Resin	0.40
Water	22.26
B. Chlorinated Paraffin (70% Chlorine Content)	14.40
Chlorinated Paraffin (42.5% Chlorine Content)	12.00
Stoddard Solvent	25.15
Oil-Modified Alkyd Resin (50% solids)	3.89
C. Antimony Oxide (300 Mesh or Finer)	12.28

Solution A is added to solution B with stirring until an emulsion is formed into which the antimony oxide is dispersed.

No. 3
(Canvas)
U.S. Patent 2,343,186

Zinc Chloride	22
Ethylene Glycol Monoethyl Ether	59
Diethylene Glycol Monoethyl Ether	29
Borax	10
Tricresyl Phosphate	120

The zinc chloride is dissolved in the ethylene glycol monoethyl ether. The diethylene glycol monoethyl ether is added with stirring after which the borax and tricresyl phosphate are added, and the mix is stirred until it becomes a clear liquid.

This liquid is then distilled at about 220°F. under reduced pressure to remove the volatile solvents.

Dissolve 30 parts of chlorinated rubber (20 C.P.) in 30 parts of toluol and 30 parts of ethylene glycol monoethyl ether, and add 40 parts of the above mentioned composition. Immerse the canvas in the bath thus prepared until saturated. Remove the excess of the mixture from the surface and dry.

The canvas treated as above is fireproof and has the flexibility desired for use as an awning, tarpaulin or the like.

Resilient Rayon Batting
U.S. Patent 2,402,532

Free-fall rayon filaments made into a batting are impregnated with:

Polyvinyl Butyral	5
Polyvinyl Alcohol	10
Butyl Acetate	10
Xylene	75

Press to remove the excess solution and dry.

Coir (Coconut Fiber) Packing Pads

Beat coir to a fluff, soak in boiled linseed oil for 10 min. and then squeeze out. Dry for 4 hours at 95°C. under slight pressure in forms of any shape.

Airplane Crash Pad
U.S. Patent 2,332,357

A crash pad for airplanes, etc., consists of a body of latex foam molded with openings extending through the body. Such a body pad is flocked by first applying an adhesive cement and then distributing fiber flock of wool, cotton, rayon, Celanese silk, etc. Thus, one or more faces of the molded foamed latex can be coated with a highly concentrated solution of rubber containing light fillers, or a polychloroprene solution cement, or an aqueous dispersion of rubber or polychloroprene. When the cement coating has dried into a tacky condition, the fiber flock, which may be of the same color as the cement, is distributed uniformly thereover and cured at about 150°F. A suitable cement may be made from 18 lb. of polychloroprene, 1 lb. each of calcined magnesia and zinc oxide, 1 lb. of rosin, and about $\frac{1}{5}$ lb. of a green pigment. Such a mixture may be dissolved in xylol to form a 25% solution. Green cement and green fiber flock will diffuse glare.

Salvaging Old Jute Fibers

Jute bags are treated with aqueous sodium carbonate or preferably with sodium hydroxide, to remove tannin-like compounds. The fabric is then kept for 24–48 hours in a bath containing 0.2 g. olein, 0.1 g. potassium nitrate, 0.1 g. magnesium sulfate, and 0.1 g. phosphoric acid per l. Owing to spontaneous fermentation the initial temperature of 30° rises to 40–45°, but should not exceed this point. The fabric is then centrifuged, and disintegrated while still wet, but not dried because of the fire hazard. The fibers are boiled for at least 1 hour in a bath containing 1% sodium hydroxide and 1% magnesium chloride. The resultant pliable material is washed with water, bleached with sodium hypochlorite, and again washed at 40° for ½ hour in a bath containing 2.5% soap, 2% petrolatum, and 0.5 sodium carbonate, or preferably ammonium hydroxide. After slight centrifuging, so that they retain 33% moisture, the fibers are ready for spinning and weaving.

Synthetic (Casein) Textile Fiber
U.S. Patent 2,169,690

Casein	24.0
Sodium Lauryl Sulfate	4.8
Calcium Hydroxide	24.0
Water	2610.0

After the above solution is homogeneous, spin in the following solution:

Phosphoric Acid	5
Calcium Acid Phosphate	10
Formaldehyde	5
Glucose	20
Water	60

Asbestos Fiber Suspension

Asbestos fibers swell in sodium carbonate solutions and shrink in aluminum chloride solutions. The sedimentation volume is increased by caustic soda, sodium chloride and sodium carbonate; it is reduced by aluminum chloride, sodium silicate and sodium oleate.

Wool Pulling Compound
(Arazym NSL)

Protease Enzyme	3–5
Kaolin	93–95
Zinc Carbonate	2

Chapter XXII

MISCELLANEOUS

Fire Extinguishing Compound
  Formula No. 1

Green Vitriol	4
Ammonium Sulfate	16
Water	100

No. 2

Alum	4½
Sodium Chloride	10
Glauber's Salt	1
Soda	1
Water Glass	1½

No. 3

Sodium Chloride	43.0
Alum	19.5
Glauber's Salt	5.1
Soda	3.5
Water Glass	6.6
Water	22.3

No. 4

Sodium Carbonate	8
Alum	4
Borax	3
Potassium Carbonate	1
Sodium Silicate	24

Mix thoroughly and add 1½ lb. of the mixture to each gal. of water in the fire extinguisher.

No. 5
U.S. Patent 2,322,781

Silica	8
Fullers' Earth	11½
Iron Oxide	10
Sodium Bicarbonate	80
Willow Charcoal	¼
Lycopodium	¾

All ingredients must be finely powdered.

No. 6
Fire Extinguishing Powder

Sodium Bicarbonate	73.25
Borax	25.00
Kieselguhr	1.10
Powdered Soap	0.65
Paraffin	Trace

U.S. Patent 2,396,275

	No. 7	No. 8
Vinsol Resin	100	100
Boric Acid	6–10	—
Borax	—	10–20

Grind together to a fine powder.

Foam Fire Extinguisher
U.S. Patent 2,355,935

Sodium Bicarbonate	47½
Ferric Sulfate (<No. 8 U.S. Standard screen and 15–18% moisture)	47½
Licorice Root (Powdered)	5

Extinguishing Phosphorus Fires
  Formula No. 1

For the more permanent quenching of phosphorus fires, a solution of a wetting agent in 5% copper sulfate solution.

Aerosol OT (25% aqueous solution)	1
Copper Sulfate Solution (5%)	99

This mixture should be shaken before use.

No. 2
Liquid Soap	1
Water	5

Chimney Soot Remover
Formula No. 1
Coarse Rock Salt	125
Zinc Dust	2
Copper Sulfate	2

Mix well in tumbling barrel.

No. 2
Flowers of Sulfur	6.25
Potassium Nitrate (Powdered)	23.75
Sodium Chloride	69.00
Ultramarine Blue	1.00

Slack Coal Briquettes
Slack Coal	48	lb.
Cement	6	lb.
Sawdust	3	pt.
Water	7½	pt.

Ram in wooden molds and air-dry for 5–6 days.

Temperature Indicating Cement
Durite Cement	900
Malachite Green	100

Cold Producing Powder
Ammonium Chloride	2 oz.
Potassium Nitrate	2 oz.

Add 2 cc. of water and mix thoroughly. Quite soon the mass begins to get cold.

Windshield Anti-Fog
Formula No. 1
1	Potassium Chlorate	4
2	Glycerin	2
3	Camphor	⅛
4	Turpentine	1

Heat 1 and 2 together on a water bath. Add 3 and remove from the water bath and add 4. The windshield must be perfectly clean before the above solution is applied. Apply the solution to both sides of the windshield and rub well with a clean cloth.

No. 2
Sulfonated Castor Oil	85
Paraffin Oil	5
Glycerin	10

No. 3
Soft Soap	63
Glycerin	32
Turpentine	5

No. 4
Fresh Tallow	6 oz.
Cocoanut Oil	3 oz.
Water	6 dr.
Alcohol	2 oz.
Brown Sugar	2 oz.
Caustic Soda Solution (40° Bé.)	5 oz.

At 167°F. melt together 6 oz. of fresh tallow and 3 flu. oz. of cocoanut oil. In another vessel, mix 6 dr. of water, 4 oz. of alcohol, 2 oz. of brown sugar, 5 fl. oz. of caustic soda solution (40° Bé.). After heating this mixture to 168°F., pour the mixtures gradually together. Pour into shallow pans and cut into sticks. Clouding is prevented by rubbing the stick on the windshield.

Anti-Freeze for Radiators (Non-Corrosive)
Formula No. 1
U.S. Patent 2,382,698

Mixture of Ethylene and Propylene Glycol (Sp. gr. 1.0775) with 2½%

Water	1 gal.
Sodium Nitrite	4 g.
Soda Ash	12 g.
Disodium Phosphate	4 g.
Linseed Meal	5 g.

Calcozine Red BX
Dye ... 0.08 g.
Sec-Dibutyl
Phthalate ... 0.50 g.

No. 2
U.S. Patent 2,386,182

Ethylene Glycol	97.350
Caustic Soda (40% Solution)	0.655
Boric Acid	5.450
Sodium Nitrite (40% Solution)	0.095
Glycol Glyceryl Monoricinoleate	1.500

No. 3
U.S. Patent 2,388,155

Ethylene Glycol	2–7
Denatured Alcohol	100
Borax	1–5
Water	1–60

No. 4

Denatured Alcohol	100
Kerosene	1–4.2
Sulfonated Castor Oil	0.03–0.3
Sodium Salicylate	0.1–0.4

Anti-Freeze Corrosion Inhibitor
U.S. Patent 2,346,635
Add 0.1–5% nitrated tallow.

Aircraft De-icing Compound
U.S. Patent 2,373,727

Ethylene Glycol	58.7
Gelatin	11.2
Gum Tragacanth	0.5
Water	19.6
Soap	1.0
Mineral Oil	9.0

Prevention of Ice Formation in Gasoline

Isopropyl Alcohol	98.9
Motor Oil	0.5
Pine Oil	0.5
Triethanolamine	0.1

Use 0.5 pt. to every 10 gal. of gasoline.

Gasoline and Hydrocarbon Liquid Thickener

A.
Coconut Oil Fatty Acids	132 lb.
Refined Naphthenic Acid	66 lb.
Oleic Acid	66 lb.
Water	225 gal.
Caustic Soda (25% Solution)	280 lb.

B.
Aluminum Sulfate Crystals	198 lb.
Water	25 gal.

Mix A well until saponified. Then add slowly B while mixing.

Filter and wash the precipitate well with water. Dry below 150°F. to a moisture content below 2%. When 12% of the above mixture is added a gel forms with gasoline, benzene, cyclohexane and other hydrocarbons.

Boiler Water Compounds
Formula No. 1

Quebracho, Chestnut Extract or a Mixture of Both	20.00
Sodium Hydroxide	36.00
Sodium Aluminate	5.00
Water	39.00

The above mixture is a thick paste which is used in conjunction with soda ash.

No. 2

Tannin (Quebracho Extract)	18.00
Sodium Carbonate	60.00
Calgon	6.00
$CuSO_4$	0.50
Borax	5.50
Water	10.00

This mixture forms a fairly dry powder.

No. 3

Tannin (Quebracho or Chestnut Extract)	11.00
Soda Ash	78.00
Copper Sulfate	1.00
Calgon	7.50
Water	2.50

This mixture also forms a dry powder.

No. 4

Sodium Carbonate	70.00
Sodium Silicate	5.00
Quebracho (Pulverized)	19.00
Calgon	4.00
Tetra Sodium Pyrophosphate	2.00

Pulverize all together and briquette into balls or cakes using about 1–2% water, then let dry.

No. 5

Soda Ash	67.50
Sodium Aluminate	5.00
Quebracho (Pulverized)	20.00
Calgon	5.00
Borax	2.50

Pulverize all together and briquette into balls or cakes using about 1–2% water, then let dry.

No. 6

Sodium Aluminate	3.00
Soda Ash	65.00
Quebracho (Pulverized)	20.00
Calgon	5.00
Dextrin	7.00

Pulverize all together and briquette into balls or cakes using 1–2% water, then let dry.

Boiler Scale Inhibitor

Ground Peat	35
Caustic Soda	2
Water	100

Boil for 6 hours.
One liter is used per cubic meter of feed water.

Bubble Fluid

Formula No. 1

1	Trihydroxyethylamine Oleate	30
	Water	1000
2	Glycerin	200

Mix slowly 1; allow to stand overnight; filter and add 2.

No. 2

Gelatin	6
Water	50
Glycerin	12
Propylene Glycol	13
Alcohol	7
Nacconol NRSF (Wetting Agent)	12

Soak gelatin in water and when swelled warm carefully and mix slowly until dissolved. Then mix in all the other ingredients except the Nacconol. Finally add the latter and mix until dissolved.

No. 3

Saponin	1
Water	1000
Glycerin	500

No. 4

Methocel (25 cps.)	2.5
Carboxymethyl Cellulose (Med. Visc.)	1.0
Aerosol OT (70%)	7.5
Glycerin	50.0
Water	250.0

Stir, using good agitation, until dissolved.

No. 5

Aerosol O.T.	2.30
Polyvinyl Alcohol (High Viscosity)	0.25
Carboxymethyl Cellulose	0.25
Glycerin	74.50
Water	22.70

Allow the second and third ingredients to soak in water and Aerosol and then warm and stir until smooth.

Foam Powder
Formula No. 1

Sodium Carbonate	41.40
Citric Acid	34.14
Wetting Agent	24.40

No. 2

Sodium Bicarbonate	30.9
Benzoic Acid	44.7
Wetting Agent	24.4

On addition of water a voluminous foam is produced.

Dyeing of Feathers for Use As Fish Lures

Most of the acid wool dyes have affinity for chicken and duck feathers. Since feathers used as fish lures are subjected to repeated wetting and drying, the milling dyes or those ordinarily dyed on wool in a weak acid bath are very useful because they are fast to wet treatments. Used alone or in admixture the following dyes make possible the production of any desired color:

Du Pont Milling Yellow GN Conc. 250% (Pat.)
Du Pont Orange II Conc.
Du Pont Resorcin Brown 3R
*Pontacyl Fast Brown CGS (Pat.)
Du Pont Milling Red SWB Conc. 125%
Du Pont Milling Red SWG Conc. 125%
Du Pont Crocein Scarlet N Extra
Pontacyl Violet C4BN
Du Pont Anthraquinone Blue SWF Conc. 150%
Pontacyl Brilliant Blue RR Conc. 200%

---

*Pontacyl is a registered trade-mark of E. I. du Pont de Nemours & Co., Inc.

Pontacyl Wool Blue BL Conc. 200%
Du Pont Anthraquinone Green GN
Pontacyl Green NV Extra Conc. 200%
Pontacyl Fast Black N2B Conc. 200%

Dyeing Method

For a small batch (2–5 g.) of feathers the dye bath is prepared by dissolving the dye in 100 cc. hot water and then adding enough cold water to total 500 cc. The amount of dye will vary with the depth of color requried. This will vary from 15 mg. to 200 mg. To the bath is added 400 mg. of Duponol * D Paste and 2 cc. of a 10% solution of sulfuric acid. The temperature of the bath is then raised to 180°F. (just under the boiling point) and the dyeing is continued for ½ hour.

Duponol D Paste is added to the bath as a wetting agent because the feathers are coated with a natural oil or wax which is difficult to penetrate. Sometimes it is preferable to treat the feathers for ½ hour at 180°F. before dyeing in a bath containing the Duponol.

Dyeing Vegetable Ivory Buttons

The following procedure and dyes are suitable:

Dye Depending Upon

Depth of Shade	2–4%
Acetic Acid (28%)	2–4%
Water	To make 100%

Run 3–5 hours at 180–200°F. Rinse and dry.

For best penetration the buttons

---

*Duponol is a registered trade-mark of E. I. du Pont de Nemours & Co., Inc.

should be entered dry into the dye bath.

Recommended dyes:
Du Pont Thioflavine TCN Conc.
Du Pont Auramine Conc.
Du Pont Basic Orange 3RN
Du Pont Chrysoidine GN
Du Pont Chrysoidine R
Du Pont Basic Brown BR
Du Pont Basic Brown GXP Conc. 150%
Du Pont Rhodamine B Extra
Du Pont Rhodamine 6GDN Extra
Du Pont Fuchsine Conc. Powder
Du Pont Safranine T Extra Conc. 125%
Du Pont Methyl Violet Conc.
Du Pont Brilliant Green Crystals
Du Pont Victoria Green Small Crystals
Du Pont Nigeria Black GX
Du Pont Nigeria Black RX

Solubilizing True Gums
U.S. Patent 2,376,656

Sodium alginate, gum tragacanth, karaya or locust bean gum is intimately mixed with 1–5 parts of 50% sodium lactate solution. This prevents the gum from balling up when water is added and gives uniform quick dispersion.

Gum Tragacanth Substitute
British Patent 120,183

1 Indian Gum	1	lb.
2 Water	2½	lb.
3 Sodium Peroxide	2	g.
4 Sodium Peroxide	2	g.
5 Water	100	cc.

Add 1 to 2 in small portions stirring slowly until uniform. Then add 3 to it. Stir for 1½ hours at 30°C. and add 4 and 5. Boil for 3 hours. Cool and dilute with:
Water   To make 1 gal.

Electrical Resistance
U.S. Patent 2,340,506

Magnesium Ferrite	80
Varnish	5
Soft Soap	5
Iron Oxide	10

This is molded by pressure extrusion and then fired.

X-Ray Contrast Composition (Medical)
U.S. Patent 2,307,189

Tetraiodophenolphthalein	300
Water	1600

Warm to 40°C. and stir until dissolved, then add

Sodium Hydroxide	5

Heat to 80°C. while stirring and allow to stand overnight.

X-Ray Opaque Cream

Lead Peroxide	30
Petroleum Jelly	60

Non-Foaming Drilling Mud Additive
U.S. Patent 2,349,585

Methyl Violet	3.33
Bentonite	1.67
Water	95.00
Diglycol Laurate	0.05–1

Sealing Porous Formations in Oil Wells
U.S. Patent 2,398,347

A mud of the following composition is used:

Water	75
Barytes	21
Bentonite	3½
Feathers	½

Heat Absorbing Glass Batch
U.S. Patent 2,397,195

Sand	1000
Limestone	280
Soda Ash	260
Salt Cake	50
Salt	30
Borax	20
Fluorspar	30
Iron Scale	8
Charcoal (Powdered)	5

This is mixed and heated in a glass melting furnace.

Infra-Red Crystal Lens

Thallium Iodide	420
Thallium Bromide	580

This is carefully fused and mixed then cooled slowly to encourage the formation of a single large crystal. The large crystal is cut and polished in accordance with optical methods. It has a refractive index of 2.2–2.5.

Synthetic Perspiration
Formula No. 1
Acid Solution

Sodium Chloride	10
Disodium Orthophosphate (Anhydrous)	1
Water	To make 1000

No. 2
Alkaline Solution

Sodium Chloride	10
Ammonium Carbonate (U.S.P.)	4
Disodium Orthophosphate (Anhydrous)	1
Water	To make 1000

Sea Water Imitation

Magnesium Chloride	11.0 g.
Anhydrous Calcium Chloride	1.2 g.
Anhydrous Sodium Sulfate	4.0 g.
Sodium Chloride	25.0 g.
Water	To make 1 l.

The above chemicals are dissolved in slightly less than 1 liter of distilled water and made up to the mark with distilled water, to make exactly 1 liter of solution. This sea water imitation is satisfactory for tests, especially corrosion tests.

Standard Soil Mixture (Testing)

Ethyl Cellulose (Viscosity 8–12, dissolved in 62 Toluene and 38 Ethyl Alcohol)	1.0
Naphtha	14.0
Butanol	0.5
Lampblack	2.0
Hydrogenated Vegetable Oil	2.5
Mineral Oil	20.0
Sodium Alginate	0.8
Cold Water	57.1
Starch	1.3
Acetic Acid	0.5
Morpholine	0.3

Delaying Setting of Bleaching Powder Slurries

Sucrose (0.3–1%) is added to bleaching powder slurries with water to prevent setting.

Hydrogen Sulfide Generation

Sulfur	3
Paraffin	1
Asbestos	Sufficient

Mix and heat slightly 3 parts of sulfur with 1 part of paraffin. Then mix with sufficient shredded asbestos to make a porous mass. Partly fill an 8 inch pyrex test tube, connect with a delivery tube and safety bottle. Heat. This furnishes a good supply of H_2S without leakage into the room as the generation of H_2S ceases as soon as the heat is removed. This mixture keeps well. The test tube may be heated over again until the reactants are used up.

Nitrogen Gas Generating Composition
U.S. Patent 2,371,707

Magnesium Oxide	8
Ammonium Nitrite	25
Water	70

This compound is used for the manufacture of sponge rubber. It is mixed with the rubber batch and "blows" during curing to form the sponge rubber.

Doctor (Plumbite) Solution
(For removal of sulfur from petroleum)
U.S. Patent 2,178,742

Litharge	10
Caustic Soda (10% Solution)	100
Sugar	¼

Heat at 100°C. until dissolved.

Non-Gelling Starch
U.S. Patent 2,400,402

Acid Modified Dry Starch	99
Triethanolamine Stearate	1

Grind and mix until uniform.

Factory Deodorant Spray

Boric Acid	6 oz.
Chloral Hydrate	4 oz.
Potassium Chlorate	6 oz.
Sodium Nitrate	6 oz.
Methanol	16 oz.

Add enough water to make 2½ gal. of solution.

Preservation of Gross Specimens
Solution No. 1

Formaldehyde	200 cc.
Water	1000 cc.
Potassium Nitrate	15 g.
Potassium Acetate	30 g.

Solution No. 2

Potassium Acetate	200 g.
Glycerin	400 cc.
Water	2000 cc.

Place the specimen in solution No. 1 for one to five to ten days, depending upon the size. The position of the specimen should be changed from day to day. There must be at least five times as much fluid as specimen. Drain and transfer the specimen to 80% ethyl alcohol for a few hours, then into 95% alcohol until the color is just restored. Finally place the specimen in solution No. 2 for preservation. It is advisable to keep the preserved specimen in the dark as light destroys the color.

Killing and Preserving Fluid
(Larvicide)
Formula No. 1

Xylene	10
Alcohol (95%)	10

No. 2

Xylene	4
Isopropyl Alcohol	6
Glacial Acetic Acid	5
Dioxan	4

Tissue Embedding and Sectioning Compositions
Formula No. 1

Paraffin Wax	90
Rosin	10

No. 2

Paraffin Wax	90
Rubber	2
Cumar	10

No. 3
Paraffin Wax	90
Clear Nevillite Resin	10

No. 4
Diglycol Stearate	50
Polyethylene Glycol 4000	50

Disperse in boiling water with good mixing.

No. 5
Diethylene Glycol Distearate	82
Ethyl Cellulose (Low Viscosity)	4
Stearin	5
Ricinoleic Diacetate (or Castor Oil)	9

This substance is soluble in most organic solvents and natural oils, including dioxan, ethylene glycol monoethyl ether (Cellosolve) and cedarwood oil, which are also suitable as clearing agents. Its chief physical properties are given below:

Melting point	48°C.
Section range	4–20 μ at room temp. 66°F.
Ribbon range	4–15 μ at room temp. 66°F.
Compression after flattening	7.6% at 10 μ

Specimens should be immersed in a bath of solvent and ester wax before being placed in the embedding medium. Since the latter is harder than paraffin wax, thinner cuts are necessary in trimming to prevent chipping, and sections should be cut more slowly.

The main advantage of this substance is the ease of *ribbon staining;* unlike sections in paraffin wax, ester wax sections may be flattened on stain solutions which penetrate the wax and stain the sections. After draining and washing, the slides are dried for about an hour at approximately 40°C.; lower temperatures cause wrinkling; the wax is then dissolved preferably by a mixed solvent consisting of:

Wax Solvent
Cellosolve	10
Ethyl Acetate	45
Xylol	45

When required, sections are generally overstained or counterstained, and differentiated in a 20% Cellosolve solution which simultaneously removes the wax. For the initial staining, methylene blue has proved most satisfactory, followed by erythrosin or eosin for counterstaining. When staining is satisfactory, the slide is transferred to the 10% Cellosolve solution to prevent further extraction of methylene blue, then to pure xylol, and finally mounted in balsam.

Dialyzers

Cellulose trinitrate membranes are good substitutes for parchment and other natural membranes. Parlodion (Du Pont) may be used, dissolving one part of the nitrate in 2 parts each of ethanol and ethyl ether. The water adhering to the Parlodion should first be removed (the shreds are preserved by covering with water), otherwise a clear solution will not be obtained. Cut off the round end of a ¾ in. or 1 in. test tube, and dip the flared end of the tube into the alcohol-ether solution of Parlodion. Upon removing the tube from the solution a film will be formed around the tube and after evaporation of the solvents the film will be found to be of sufficient strength to meet the purposes of a dialyzer. The liquid to be dialyzed is poured into the tube and contents are then set in a beaker of water. In a short time the working of the semipermeable membrane will be shown by the rise of the level of the liquid inside the tube.

Bacteria Culture Media
Formula No. 1
Plain Nutrient Agar (Used for aerobic dilution plate counts)
Beef Extract (Difco)   3 g.
Peptone (Bacto)   10 g.
Sodium Chloride   10 g.
Agar   15 g.
Distilled Water   To make 1 l.
Adjust to pH 7.0 with 1.0 N NaOH.

No. 2
Gelatin High-Salt Agar (Used for both aerobic and anaerobic dilution plate counts of halophilic or salt-tolerant bacteria)
Yeast Extract   3.0 g.
Peptone (Bacto-Tryptone)   10.0 g.
Gelatin (Bacto)   60.0 g.
Sodium Thioglycollate   0.1 g.
Sodium Chloride   175.0 g.
Agar   20.0 g.
Distilled Water   To make 1.0 l.
Adjust to pH 8.0 with 1.0 N NaOH.

No. 3
High-Salt Calcium Lactate Broth (Used for dilution tube counts of anaerobic halophilic or salt-tolerant bacteria)
Yeast Water   500 cc.
Peptone (Bacto-Tryptone)   10 g.
Gelatin (Bacto)   60 g.
Glucose   1 g.
Calcium Lactate   5 g.
Magnesium Sulfate   1 g.
Sodium Chloride   175 g.
Distilled Water   To make 1 l.
Adjust to pH 8.0 with 1.0 N NaOH.

No. 4
Beef Heart Infusion Broth (Used for dilution tube counts of anaerobic bacteria)
Extracted Ground Beef Heart *   2 g. per tube
Beef Heart Extract Glucose Broth **   10 cc. per tube

No. 5
Ammonium Nitrate   2.0 g.
Potassium Acid Phosphate   1.0 g.
Magnesium Sulfate   0.5 g.
Agar U.S.P.   10–20 g.
Distilled Water   1000 cc.

Treatment of Glass Wool Air Filters
A. Paraffin Oil   10
   Stearic Acid   4
   Bentonite   3
Stir until dissolved, add solution of
B. Triethanolamine   1.5
   Water   81.5

* Add 1 pound of ground beef heart to 1,000 cc. of distilled water; digest on a steam bath for 2 hours; press out extract; recover pressed residue.
** To the beef heart extract (see above), add 10 g. of glucose, 10 g. of NaCl, and enough distilled water to bring volume to 1 liter; adjust to pH 8.0 with 1.0 N NaOH.

Mix A and add B. Stir until well emulsified.

A warm treatment of glass wool with this emulsion will improve its efficiency and durability when used as a filter medium, as in air conditioning.

Glycerin Substitute

Magnesium Chloride	33.3
Urea	34.5
Water	32.2

One part of this mixture with 2 parts of water lowers the freezing point to $-12°C$.

TABLES

Weights and Measures
Troy Weight
24 grains = 1 pwt.
20 pwts. = 1 ounce
12 ounces = 1 pound

Apothecaries' Weight
20 grains = 1 scruple
3 scruples = 1 dram
8 drams = 1 ounce
12 ounces = 1 pound
The ounce and pound are the same as in Troy Weight.

Avoirdupois Weight
27 11/32 grains = 1 dram
16 drams = 1 ounce
16 ounces = 1 pound
2000 lbs. = 1 short ton
2240 lbs. = 1 long ton

Dry Measure
2 pints = 1 quart
8 quarts = 1 peck
4 pecks = 1 bushel
36 bushels = 1 chaldron

Liquid Measure
4 gills = 1 pint
2 pints = 1 quart
4 quarts = 1 gallon
31½ gals. = 1 barrel
2 barrels = 1 hogshead
1 teaspoonful = 1/6 oz.
1 tablespoonful = ½ oz.
16 fluid oz. = 1 pint

Circular Measure
60 seconds = 1 minute
60 minutes = 1 degree
360 degrees = 1 circle

Long Measure
12 inches = 1 foot
3 feet = 1 yard
5½ yards = 1 rod
5280 feet = 1 stat. mile
320 rods = 1 stat. mile

Square Measure
144 sq. in. = 1 sq. ft.
9 sq. ft. = 1 sq. yard
30¼ sq. yds. = 1 sq. rod
43,560 sq. ft. = 1 acre
40 sq. rods = 1 rood
4 roods = 1 acre
640 acres = 1 sq. mile

Metric Equivalents
Length
1 inch = 2.54 centimeters
1 foot = 0.305 meter
1 yard = 0.914 meter
1 mile = 1.609 kilometers
1 centimeter = 0.394 in.
1 meter = 3.281 ft.
1 meter = 1.094 yd.
1 kilometer = 0.621 mile

Capacity
1 U. S. fluid oz. = 29.573 milliliters
1 U. S. liquid qt. = 0.946 liter
1 U. S. dry qt. = 1.101 liters
1 U. S. gallon = 3.785 liters
1 U. S. bushel = 0.3524 hectoliter
1 cu. in. = 16.4 cu. centimeters
1 milliliter = 0.034 U. S. fluid ounce
1 liter = 1.057 U. S. liquid qt.
1 liter = 0.908 U. S. dry qt.
1 liter = 0.264 U. S. gallon
1 hectoliter = 2.838 U. S. bu.
1 cu. centimeter = .061 cu. in.
1 liter = 1000 milliliters or 100 cu. c.

Weight
1 grain = 0.065 gram
1 apoth. scruple = 1.296 grams
1 av. oz. = 28.350 grams
1 troy oz. = 31.103 grams
1 av. lb. = 0.454 kilogram
1 troy lb. = 0.373 kilogram
1 gram = 15.432 grains
1 gram = 0.772 apoth. scruple
1 gram = 0.035 av. oz.
1 gram = 0.032 troy oz.
1 kilogram = 2.205 av. lbs.
1 kilogram = 2.679 troy lbs.

Approximate pH Values

The following tables give approximate pH values for a number of substances such as acids, bases, foods, biological fluids, etc. All values are rounded off to the nearest tenth and are based on measurements made at 25° C.

pH Values of Acids

Hydrochloric, N	0.1
Hydrochloric, 0.1N	1.1
Hydrochloric, 0.01N	2.0
Sulphuric, N	0.3
Sulphuric, 0.1N	1.2
Sulphuric, 0.01N	2.1
Orthophosphoric, 0.1N	1.5
Sulphurous, 0.1N	1.5
Oxalic, 0.1N	1.6
Tartaric, 0.1N	2.2
Malic, 0.1N	2.2
Citric, 0.1N	2.2
Formic, 0.1N	2.3
Lactic, 0.1N	2.4
Acetic, N	2.4
Acetic, 0.1N	2.9
Acetic, 0.01N	3.4
Benzoic, 0.1N	3.1
Alum, 0.1N	3.2
Carbonic (saturated)	3.8
Hydrogen Sulphide, 0.1N	4.1
Arsenious (saturated)	5.0
Hydrocyanic, 0.1N	5.1
Boric, 0.1N	5.2

pH Values of Bases

Sodium Hydroxide, N	14.0
Sodium Hydroxide, 0.1N	13.0
Sodium Hydroxide, 0.01N	12.0
Potassium Hydroxide, N	14.0
Potassium Hydroxide, 0.1N	13.0
Potassium Hydroxide, 0.01N	12.0
Lime (saturated)	12.4
Sodium Metasilicate, 0.1N	12.6
Trisodium Phosphate, 0.1N	12.0
Sodium Carbonate, 0.1N	11.6
Ammonia, N	11.6
Ammonia, 0.1N	11.1
Ammonia, 0.01N	10.6
Potassium Cyanide, 0.1N	11.0
Magnesia (saturated)	10.5
Sodium Sesquicarbonate, 0.1N	10.1
Ferrous Hydroxide (saturated)	9.5
Calcium Carbonate (saturated)	9.4
Borax, 0.1N	9.2
Sodium Bicarbonate, 0.1N	8.4

pH Values of Foods

Apples	2.9–3.3
Apricots	3.6–4.0
Asparagus	5.4–5.8
Bananas	4.5–4.7
Beans	5.0–6.0
Beers	4.0–5.0
Beets	4.9–5.5
Blackberries	3.2–3.6
Bread, white	5.0–6.0
Butter	6.1–6.4
Cabbage	5.2–5.4
Carrots	4.9–5.3
Cheese	4.8–6.4
Cherries	3.2–4.0
Cider	2.9–3.3
Corn	6.0–6.5
Crackers	6.5–8.5
Dates	6.2–6.4
Eggs, fresh white	7.6–8.0
Flour, wheat	5.5–6.5
Gooseberries	2.8–3.0
Grapefruit	3.0–3.3
Grapes	3.5–4.5
Hominy (rye)	6.8–8.0
Jams, fruit	3.5–4.0
Jellies, fruit	2.8–3.4
Lemons	2.2–2.4
Limes	1.8–2.0
Maple Syrup	6.5–7.0
Milk, cows	6.3–6.6
Olives	3.6–3.8
Oranges	3.0–4.0
Oysters	6.1–6.6
Peaches	3.4–3.6
Pears	3.6–4.0
Peas	5.8–6.4
Pickles, dill	3.2–3.6
Pickles, sour	3.0–3.4
Pimento	4.6–5.2
Plums	2.8–3.0
Potatoes	5.6–6.0
Pumpkin	4.8–5.2
Raspberries	3.2–3.6
Rhubarb	3.1–3.2
Salmon	6.1–6.3
Sauerkraut	3.4–3.6
Shrimp	6.8–7.0
Soft Drinks	2.0–4.0
Spinach	5.1–5.7
Squash	5.0–5.4
Strawberries	3.0–3.5
Sweet Potatoes	5.3–5.6
Tomatoes	4.0–4.4
Tuna	5.9–6.1
Turnips	5.2–5.6
Vinegar	2.4–3.4
Water, drinking	6.5–8.0
Wines	2.8–3.8

pH Values of Biologic Materials

Blood, plasma, human	7.3–7.5
Spinal Fluid, human	7.3–7.5
Blood, whole, dog	6.9–7.2
Saliva, human	6.5–7.5
Gastric Contents, human	1.0–3.0
Duodenal Contents, human	4.8–8.2
Feces, human	4.6–8.4
Urine, human	4.8–8.4
Milk, human	6.6–7.6
Bile, human	6.8–7.0

Interconversion Tables and Chart
for Units of Volume and Weight, and Energy

MULTIPLY BY

TO CONVERT FROM	To Cu. in.	To Cu. ft.	To Cu. yd.	To Fl. Oz.	To Pint	To Quart	To Gallon	To Grain	To Oz. Troy	To Oz. Av.	To Lb. Troy	To Lb. Av.	To Cc. or Gr.	To Ltr. or Kg.	To Cu. M.
Cu. in.	1.00000	0.03787	0.42143	.554112	.034632	.017316	.004329	252.891	.526857	.578037	.043905	.036127	16.3871	.016387	0.41639
Cu. Ft.	1728.00	1.00000	.037037	957.505	59.8442	29.9221	7.48052	436996	910.408	998.848	75.8674	62.4280	28316.9	28.3169	.028317
Cu. Yd.	46656.0	27.0000	1.00000	25852.6	1615.79	807.896	201.974	117990.	24581.0	26968.9	2048.42	1685.56	764556	764.556	.764556
Fl. Oz.	1.80469	.001044	0.33868	1.00000	.062500	.031250	.007813	456.390	.950813	1.04316	.079234	.065199	29.5736	.029573	0.42957
Pint	28.8750	.016710	0.46189	16.0000	1.00000	.500000	.125000	7302.23	15.2130	16.6908	1.26735	1.04318	473.177	.473177	0.44732
Quart	57.7500	.033420	.001238	32.0000	2.00000	1.00000	.250000	1460.45	30.4260	33.3816	2.53550	2.08635	946.354	.946354	0.49463
Gallon	231.000	.133681	.004951	128.000	8.00000	4.00000	1.00000	58417.9	121.704	133.527	10.1420	8.34541	3785.42	3.78542	.003785
Grain	.003954	0.42288	0.88475	.002191	0.41369	0.46850	0.41712	1.00000	.002083	.002286	0.41736	0.41428	.064799	.064799	0.46479
Oz. Troy	1.89805	.001098	0.44068	1.05173	.065733	.032867	.008217	480.000	1.00000	1.09714	.083333	.068571	31.1035	.031104	0.43110
Oz. Av.	1.72999	.001001	0.33708	.958608	.059913	.029957	.007489	437.500	.911457	1.00000	.075955	.062500	28.3495	.028350	0.42835
Lb. Troy	22.7766	.013181	0.44882	12.6208	.788800	.394400	.098600	5760.00	12.0000	13.1657	1.00000	.822857	373.242	.373242	0.43732
Lb. Av.	27.6799	.016018	0.35933	15.3378	.958611	.479306	.119826	7000.00	14.5833	16.0000	1.21528	1.00000	453.593	.453593	0.44536
CC or Gram	.061024	0.33531	0.41308	.033814	.002113	.001057	0.42642	15.4323	.032151	.035274	.002679	.002205	1.00000	.001000	.000001
Liter or Kg.	61.0237	.035315	.001308	33.8140	2.11337	1.05669	.264172	15432.3	32.1507	35.2739	2.67923	2.20462	1000.00	1.00000	.001000
Cu. M.	61023.7	35.3146	1.30795	33814.0	2113.37	1056.69	264.172	154320.	32150.7	35273.9	2679.23	2204.62	1000000	1000.00	1.00000

Note. The small subnumeral following a zero indicates that the zero is to be taken that number of times; thus, .0₄1428 is equivalent to .0001428.

Values used in constructing table:
1 inch = 2.540001 cm.
1 cu. in. = 16.387083 cc. = 16.387083₄ g H₂O at 4°C; ∴ 1 lb. av. = 27.679886 cu. in. H₂O at 4°C.
4°C. = 39°F.

1 lb. av. = 453.5926 g.
1 gal. = 8.34541 lb.
∴ 1 lb. av. = 7000 grains.
∴ 1 gallon = 58417.87 grains.
231 cu. in. = 1 gallon = 3785.4162 g.

MULTIPLY BY

TO CONVERT FROM	B.T.U.	P.C.U.	Cal.	Ft. Lbs.	Ft. Tons.	Kg. M.	HP Hrs.	KW Hrs.	Joules	Lbs. C	Lbs. H₂O
B.T.U.	1.00000	.555556	.251996	778.000	.389001	107.563	.0₃3929	.0₂2931	1055.20	.0₆6876	.001031
P.C.U.	1.80000	1.00000	45.3593	1400.40	.700202	193.613	.0₂7072	.0₂5276	1899.36	.0₁1238	.001855
Calories	3.96832	2.20462	1.00000	3091.36	1.54368	426.844	.001559	.001163	4187.37	.0₂2729	.004089
Ft. Lbs.	.001285	.0₇7141	.0₃3239	1.00000	.000500	.13255	.0₆5050	.0₃3767	1.35625	.0₈8840	.0₁1325
Ft. Tons	2.57069	1.42816	.647804	2000.00	1.00000	276.511	.001010	.0₂7535	2712.59	.0₁1768	.002649
Kg. M.	.00927	.005165	.002343	7.23301	.003617	1.00000	.0₅3653	.0₂2725	9.81009	.0₆6394	.0₄9580
HP Hrs	2544.99	.141388	641.327	1980000	990.004	273747	1.00000	.746000	2685473	.175044	2.62261
KW Hrs.	3411.57	1895.32	859.702	2654200	1327.10	366959	1.34041	1.00000	3599889	.234648	3.51562
Joules	.09477	.0₅265	.0₂388	.737311	.0₃687	.101937	.0₃3724	.0₂2778	1.00000	.0₁6518	.0₄9766
Lbs. C.	14544.0	8080.00	3665.03	113150₂	5657.63	1564396	5.71434	4.26285	153470₂	1.00000	14.9876
Lbs. H₂O	970.400	539.111	244.537	754971	377.487	104379	.381270	.284424	1023966	.066744	1.00000

"P. C. U." refers to the "pound-centigrade unit." The ton used is 2000 pounds. "Lbs. C" refers to pounds of carbon oxidized, 100% efficiency equivalent to the corresponding number of heat units. "Lbs H₂O" refers to pounds of water evaporated at 100°C. = 212°F. at 100% efficiency

By the use of the foregoing table² about 330 interconversions among twenty-six of the standard engineering units of measure can be directly estimated from the alignment chart to three significant figures or calculated by simple multiplication to six figures. The multiplier factor given in the table is located on the center scale "A" giving the point which when aligned with any number point on "C1" determines the product on "C". Imperfections in the scale due to lack of precision in printing should be checked at intervals along "A" scale by actual division of "C" by "C1", the lines being left out so that the reader can do this. A line scratched on a transparent celluloid triangle gives the best medium for making alignments.

When volume and weight interconversions are given, water is the medium the calculations are based upon. By the introduction of specific gravity factors the me-

CONVERSION OF THERMOMETER READINGS

F°	C°	F°	C°	F°	C°	F°	C°	F°	C°	F°	C°
−40	−40.00	30	−1.11	80	26.67	250	121.11	500	260.00	900	482.22
−38	−38.89	31	−0.56	81	27.22	255	123.89	505	262.78	910	487.78
−36	−37.78	32	0.00	82	27.78	260	126.67	510	265.56	920	493.33
−34	−36.67	33	0.56	83	28.33	265	129.44	515	268.33	930	498.89
−32	−35.56	34	1.11	84	28.89	270	132.22	520	271.11	940	504.44
−30	−34.44	35	1.67	85	29.44	275	135.00	525	273.89	950	510.00
−28	−33.33	36	2.22	86	30.00	280	137.78	530	276.67	960	515.56
−26	−32.22	37	2.78	87	30.56	285	140.55	535	279.44	970	521.11
−24	−31.11	38	3.33	88	31.11	290	143.33	540	282.22	980	526.67
−22	−30.00	39	3.89	89	31.67	295	146.11	545	285.00	990	532.22
−20	−28.89	40	4.44	90	32.22	300	148.89	550	287.78	1000	537.78
−18	−27.78	41	5.00	91	32.78	305	151.67	555	290.55	1050	565.56
−16	−26.67	42	5.56	92	33.33	310	154.44	560	293.33	1100	593.33
−14	−25.56	43	6.11	93	33.89	315	157.22	565	296.11	1150	621.11
−12	−24.44	44	6.67	94	39.44	320	160.00	570	298.89	1200	648.89
−10	−23.33	45	7.22	95	35.00	325	162.78	575	301.67	1250	676.67
−8	−22.22	46	7.78	96	35.56	330	165.56	580	304.44	1300	704.44
−6	−21.11	47	8.33	97	36.11	335	168.33	585	307.22	1350	732.22
−4	−20.00	48	8.89	98	36.67	340	171.11	590	310.00	1400	760.00
−2	−18.89	49	9.44	99	37.22	345	173.89	595	312.78	1450	787.78
0	−17.78	50	10.00	100	37.78	350	176.67	600	315.56	1500	815.56
1	−17.22	51	10.56	105	40.55	355	179.44	610	321.11	1550	843.33
2	−16.67	52	11.11	110	43.33	360	182.22	620	326.67	1600	871.11
3	−16.11	53	11.67	115	46.11	365	185.00	630	332.22	1650	898.89
4	−15.56	54	12.22	120	48.89	370	187.78	640	337.78	1700	926.67
5	−15.00	55	12.78	125	51.67	375	190.55	650	343.33	1750	954.44
6	−14.44	56	13.33	130	54.44	380	193.33	660	348.89	1800	982.22
7	−13.89	57	13.89	135	57.22	385	196.11	670	354.44	1850	1010.00
8	−13.33	58	14.44	140	60.00	390	198.89	680	360.00	1900	1037.78
9	−12.78	59	15.00	145	62.78	395	201.67	690	365.56	1950	1065.56
10	−12.22	60	15.56	150	65.56	400	204.44	700	371.11	2000	1093.33
11	−11.67	61	16.11	155	68.33	405	207.22	710	376.67	2050	1121.11
12	−11.11	62	16.67	160	71.11	410	210.00	720	382.22	2100	1148.89
13	−10.56	63	17.22	165	73.89	415	212.78	730	387.78	2150	1176.67
14	−10.00	64	17.78	170	76.67	420	215.56	740	393.33	2200	1204.44
15	−9.44	65	18.33	175	79.44	425	218.33	750	398.89	2250	1232.22
16	−8.89	66	18.89	180	82.22	430	221.11	760	404.44	2300	1260.00
17	−8.33	67	19.44	185	85.00	435	223.89	770	410.00	2350	1287.78
18	−7.78	68	20.00	190	87.78	440	226.67	780	415.56	2400	1315.56
19	−7.22	69	20.56	195	90.55	445	229.44	790	421.11	2450	1343.33
20	−6.67	70	21.11	200	93.33	450	232.22	800	426.67	2500	1371.11
21	−6.11	71	21.67	205	96.11	455	235.00	810	432.22	2550	1398.89
22	−5.56	72	22.22	210	98.89	460	237.78	820	437.78	2600	1426.67
23	−5.00	73	22.78	215	101.67	465	240.55	830	443.33	2650	1454.44
24	−4.44	74	23.33	220	104.44	470	243.33	840	448.89	2700	1482.22
25	−3.89	75	23.89	225	107.22	475	246.11	850	454.44	2750	1510.00
26	−3.33	76	24.44	230	110.00	480	248.89	860	460.00	2800	1537.78
27	−2.78	77	25.00	235	112.78	485	251.67	870	465.56	2850	1565.56
28	−2.22	78	25.56	240	115.56	490	254.44	880	471.11	2900	1593.33
29	−1.67	79	26.11	245	118.33	495	257.22	890	476.67	2950	1621.11

ALCOHOL PROOF AND PERCENTAGE TABLE

U.S. Proof at 60° F.	Per cent Alcohol by Volume at 60° F.	Per cent Alcohol by Weight	U.S. Proof at 60° F.	Per cent Alcohol by Volume at 60° F.	Per cent Alcohol by Weight
0	0.0	0.00	57	28.5	—
1	0.5	—	58	29.0	23.82
2	1.0	0.80	59	29.5	—
3	1.5	—	60	30.0	24.67
4	2.0	1.59	61	30.5	—
5	2.5	—	62	31.0	25.52
6	3.0	2.39	63	31.5	—
7	3.5	—	64	32.0	26.38
8	4.0	3.19	65	32.5	—
9	4.5	—	66	33.0	27.24
10	5.0	4.00	67	33.5	—
11	5.5	—	68	34.0	28.10
12	6.0	4.80	69	34.5	—
13	6.5	—	70	35.0	28.97
14	7.0	5.61	71	35.5	—
15	7.5	—	72	36.0	29.84
16	8.0	6.42	73	36.5	—
17	8.5	—	74	37.0	30.72
18	9.0	7.23	75	37.5	—
19	9.5	—	76	38.0	31.60
20	10.0	8.05	77	38.5	—
21	10.5	—	78	39.0	32.48
22	11.0	8.86	79	39.5	—
23	11.5	—	80	40.0	33.36
24	12.0	9.68	81	40.5	—
25	12.5	—	82	41.0	34.25
26	13.0	10.50	83	41.5	—
27	13.5	—	84	42.0	35.15
28	14.0	11.32	85	42.5	—
29	14.5	—	86	43.0	36.05
30	15.0	12.14	87	43.5	—
31	15.5	—	88	44.0	36.96
32	16.0	12.96	89	44.5	—
33	16.5	—	90	45.0	37.86
34	17.0	13.79	91	45.5	—
35	17.5	—	92	46.0	38.78
36	18.0	14.61	93	46.5	—
37	18.5	—	94	47.0	39.70
38	19.0	15.44	95	47.5	—
39	19.5	—	96	48.0	40.62
40	20.0	16.27	97	48.5	—
41	20.5	—	98	49.0	41.55
42	21.0	17.10	99	49.5	—
43	21.5	—	100	50.0	42.49
44	22.0	17.93	101	50.5	—
45	22.5	—	102	51.0	43.43
46	23.0	18.77	103	51.5	—
47	23.5	—	104	52.0	44.37
48	24.0	19.60	105	52.5	—
49	24.5	—	106	53.0	45.33
50	25.0	20.44	107	53.5	—
51	25.5	—	108	54.0	46.28
52	26.0	21.28	109	54.5	—
53	26.5	—	110	55.0	47.24
54	27.0	22.13	111	55.5	—
55	27.5	—	112	56.0	48.21
56	28.0	22.97	113	56.5	—

TABLES

U. S. Proof at 60° F.	Per cent Alcohol by Volume at 60° F.	Per cent Alcohol by Weight	U. S. Proof at 60° F.	Per cent Alcohol by Volume at 60° F.	Per cent Alcohol by Weight
114	57.0	49.19	158	79.0	72.38
115	57.5	——	159	79.5	——
116	58.0	50.17	160	80.0	73.53
117	58.5	——	161	80.5	——
118	59.0	51.15	162	81.0	74.69
119	59.5	——	163	81.5	——
120	60.0	52.15	164	82.0	75.86
121	60.5	——	165	82.5	——
122	61.0	53.15	166	83.0	77.04
123	61.5	——	167	83.5	——
124	62.0	54.15	168	84.0	78.23
125	62.5	——	169	84.5	——
126	63.0	55.16	170	85.0	79.44
127	63.5	——	171	85.5	——
128	64.0	56.18	172	86.0	80.62
129	64.5	——	173	86.5	——
130	65.0	57.21	174	87.0	81.90
131	65.5	——	175	87.5	——
132	66.0	58.24	176	88.0	83.14
133	66.5	——	177	88.5	——
134	67.0	59.28	178	89.0	84.41
135	67.5	——	179	89.5	——
136	68.0	60.32	180	90.0	85.69
137	68.5	——	181	90.5	——
138	69.0	61.38	182	91.0	86.99
139	69.5	——	183	91.5	——
140	70.0	62.44	184	92.0	88.31
141	70.5	——	185	92.5	——
142	71.0	63.51	186	93.0	89.65
143	71.5	——	187	93.5	——
144	72.0	64.59	188	94.0	91.02
145	72.5	——	189	94.5	——
146	73.0	65.67	190	95.0	92.42
147	73.5	——	191	95.5	——
148	74.0	66.77	192	96.0	93.85
149	74.5	——	193	96.5	——
150	75.0	67.87	194	97.0	95.32
151	75.5	——	195	97.5	——
152	76.0	68.92	196	98.0	96.82
153	76.5	——	197	98.5	——
154	77.0	70.10	198	99.0	98.38
155	77.5	——	199	99.5	——
156	78.0	71.23	200	100.0	100.00
157	78.5	——			

Buffer Systems

The following table gives some common buffer systems and the approximate pH of maximum buffer capacity. The zone of effective buffer action will vary with concentration but the general average will be ± 1.0 pH from the value given, for concentrations approximately 0.1 molar.

Glycocoll - Sodium Chloride - Hydrochloric Acid	2.0
Potassium Acid Phthalate-Hydrochloric Acid	2.8
Primary Potassium Citrate	3.7
Acetic Acid-Sodium Acetate ...	4.6
Potassium Acid Phthalate-Sodium Hydroxide	5.0
Secondary Sodium Citrate	5.0
Carbonic Acid-Bicarbonate	6.5
Primary Phosphate-Secondary Phosphate	6.8
Primary Phosphate-Sodium Hydroxide	6.8
Boric Acid-Borax	8.5
Borax	9.2
Boric Acid-Sodium Hydroxide ..	9.2
Bicarbonate-Carbonate	10.2
Secondary Phosphate-Sodium Hydroxide	11.5

Courtesy of W. A. Taylor & Company

REFERENCES AND ACKNOWLEDGMENTS

Agricultural Chemicals
American Dye Reporter
American Electrochemical Society
American Journal of Public Health
American Leather Belt Association
American Paint Journal
American Perfumer
American Photography
Archives of Dermatology and Syphilology

Baker's Weekly
Brewers' Technical Review

Calco Technical Bulletins
Camera
Canadian Textile Journal
Ceramic Age
Chemical Abstracts
Chemical Industries
Chemist Analyst
Chemist and Druggist
Combustion
Confectioner's Journal

Dairy World
Dental Items
Drug and Cosmetic Industry
Druggists' Circular

Eastern Regional Research Laboratory
Electric Journal
Electrochemical Society Journal

Flavours
Food Industries
Food Materials and Equipment

Glass Industry

Hide and Leather

Ice Cream Review
India Rubber World
Indian Lac Research Institute
Industrial Chemist
Industrial Engineering Chemistry
Industrial Finishing
Industry and Power
Instruments
International Tin Research and Development Council
Iowa State College Bulletin

Journal of Chemical Education
Journal of the American Dental Association
Journal of the American Leather Chemists Association
Journal of the American Medical Association
Journal of the American Pharmaceutical Association
Journal of the Society of Chemical Industry
Journal of the Society of Leather Trades

Manufacturing Chemist
Manufacturing Confectioner
Materials and Methods
Meat
Melliand Textile Review
Metal Industry
Metal Progress
Military Surgeon
Mineralogist
Modern Medicine
Modern Plastics

National County Agent
National Provisioner
Nickelsworth
N. Y. State Pharmacist

Oil and Soap
Oil, Paint and Drug Reporter

Paint, Oil and Chemical Review
Paper Trade Journal
Pharmaceutical Journal
Physics
Plant Physiology
Plastics
Power
Practical Druggist
Products Finishing

Rayon Textile Monthly
Rock Products
Rubber Age

Schimmel Briefs
Science
Soap
Soap, Perfumery and Cosmetics
Steel

U. S. Bureau of Mines
U. S. Bureau of Ships
U. S. Chemical Warfare Service
U. S. Department of Agriculture
U. S. National Bureau of Standards

TRADE-NAME CHEMICALS

During the past few years, the practice of marketing raw materials, under names which in themselves are not descriptive chemically of the products they represent, has become very prevalent. No modern book of formulae could justify its claims either to completeness or modernity without numerous formulae containing these so-called "Trade Names."

Without wishing to enter into any discussion regarding the justification of "Trade Names," the Editors recognize the tremendous service rendered to commercial chemistry by manufacturers of "Trade Name" products, both in the physical data supplied and the formulation suggested.

Deprived of the protection afforded their products by this system of nomenclature, these manufacturers would have been forced to stand helplessly by while the fruits of their labor were being filched from them by competitors who, unhampered by expenses of research, experimentation and promotion, would be able to produce something "just as good" at prices far below those of the original producers.

That these competitive products were "just as good" solely in the minds of the imitators would only be evidenced in costly experimental work on the part of the purchaser and, in the meantime, irreparable damage would have been done to the truly ethical product. It is obvious, of course, that under these circumstances, there would be no incentive for manufacturers to develop new materials.

Because of this, and also because the "Chemical Formulary" is primarily concerned with the physical results of compounding rather than with the chemistry involved, the Editors felt that the inclusion of formulae containing various trade name products would be of definite value to the producer of finished chemical materials. If they had been left out many ideas and processes would have been automatically eliminated.

As a further service the better known "trade name" products are included with the list of chemicals and supplies.

CHEMICALS AND SUPPLIES: WHERE TO BUY THEM*

Numbers on right refer to list of suppliers on pages directly following this list. Thus to find out who supplies borax look in left hand column, alongside borax, on page 402. The number there is 67. Now turn to page 418 and find number 67. Alongside is the supplier, American Potash & Chemical Corp., New York, N. Y.

Product	No.	Product	No.
A		Albone C	393
A. A. P. Naphthols	29	Albron	21
A-Syrup	855	Albumen	476
Abalyn	557	Albusol	697
Abietic Acid	557	Alcohol, Denatured	301
Abopon	509	Alcohol, Pure	301
Absorption Base	605	Aldehol	633
Accelerator 808	393	Aldehyde C_{14}	1139
Accelerator 833	393	Aldol	793
Accelerators, Vulcanization	779	Alframine	743
Accroides, Gum	941	Alginic Acid	16
Acetamide	123	Alizarin	769
Acetanilide	792	Alkalies	795
Acetic Acid	793	Alkaloids	602
Acetic Anhydride	215	Alkanet	577
Acetoin	679	Alkanol	393
Acetone	978	Alkyd Resins	489
Acetphenetidine	731	Alloxan	411
Acetyl Cellulose..See Cellulose Acetate		Almond Oil	693
Acetyl Salicylic Acid	565	Aloes	870
Acidolene	707	Aloin	585
Acids, Fatty	382	Aloxite	221
Acimul	509	Alperox	679
Acrawax	509	Alpha Naphthol	479
Acrawax B	509	Alphanaphthylthiourea........See Antu	
Acrawax C	509	Alphasol	45
Acriflavine	1	Altax	1141
Acrylic Resins	895	Alumina	587
Acryloid	895	Aluminum	887
Activated Charcoal	341	Aluminum Acetate	793
Adeps Lanae..............See Lanolin		Aluminum Bronze Powder	48
Adheso Wax	509	Aluminum Chloride	798
Adipic Acid	393	Aluminum Hydrate	79
A. D. M. No. 100 Oil	87	Aluminum Hydroxide	
Advawet 33	11	See Aluminum Hydrate	
Aerogel	749	Aluminum Nitrate	545
Aerosol	45	Aluminum Oleate	979
Agar	721	Aluminum Silicate	500
Agene	807	Aluminum Stearate	1193
AgeRite Alba	1141	Aluminum Sulfate	483
AgeRite Powder	1141	Aluminum Sulfocarbolate	901
Akcocene	45	Aluminum Tristearate	697
Aktivin	15	Alundum	803
Albacer	509	Alvar	977
Albalith	785	Amaranth	721
Albasol	773	Amberette	1065
Albatex	277	Amberlac	895
Albolith	785	Amberol	895

* Chemicals not listed here may be located by communicating with Chemical Industries, 522 Fifth Avenue, New York 18, N. Y., or consulting their Annual Buyers' Guidebook Number.
Dyestuffs included in the various formulae can be obtained from General Dyestuff Corporation, National Aniline Division of Allied Chemical & Dye Corp., or Calco Chemical Division of American Cyanamid Co.

CHEMICALS AND SUPPLIES: WHERE TO BUY THEM

Product	No.
Amerine	515
Ameripol	515
Amerith	231
Amidine	209
Amidol	407
Aminoacetic Acid	149
m-Aminobenzoic Acid	149
Amino Glycol	301
Aminomethylpropanediol	301
p-Aminophenol	1151
Aminostearin	509
Aminox	779
Ammonia	135
Ammoniac Resin	45
Ammonium Alginate	16
Ammonium Bichromate	759
Ammonium Bifluoride	849
Ammonium Carbonate	1193
Ammonium Chloride	917
Ammonium Citrate	853
Ammonium Fluoborate	483
Ammonium Laurate	509
Ammonium Linoleate	509
Ammonium Nitrate	331
Ammonium Oleate	509
Ammonium Persulfate	187
Ammonium Phosphate	266
Ammonium Stearate	509
Ammonium Sulfamate	393
Ammonium Sulfate	886
Ammonium Sulfite	697
Ammonium Sulfocyanide	See Ammonium Thiocyanate
Ammonium Sulforicinoleate	923
Ammonium Thiocyanate	647
Ammonium Thioglycollate	711
Amorphous Wax	850
Amsco Solvent	63
Amyl Acetate	1054
Amyl Alcohol, Tertiary	978
Amyl Mercaptan	975
Amyl Salicylate	994
Amyl Valerianate	1139
Anethol	92
Anhydrone	123
Aniline Chloride	See Aniline Hydrochloride
Aniline Dyes	881
Aniline Hydrochloride	379
Aniline Oil	379
Anise Oil	681
Anisyl Acetate	92
Ansol	1127
Anthraquinone	201
Antidolorin	461
Antimony	1162
Antimony Chloride	545
Antimony Oxide	734
Antimony Sulfide	885a
Antimony Trichloride	See Antimony Chloride
Antimony Trioxide	See Antimony Oxide

Product	No.
Anti-Oxidants	393
Antox	393
Antu	393
Apco	77
Apocthinner	77
Appramine	1169
APS-202	77
Aquadag	5
Aqualube	509
Aquapel	825
Aquaplex	895
Aquaresin	509
Aquarex	393
Aquarome	439
Aquasol	45
Arachis Oil	See Peanut Oil
Araskleen	749
Aratone	99
Archer-Daniels No. 635	87
Archer-Daniels-Midland Oil	87
Arctic Syntex	293
Areskap	749
Aresklene	749
Aridex	393
Arlacel	103
Arlex	103
Arnica, Tincture of	946
Arochlor	1063
Aroflex	1059
Arolite	1059
Arosol	485
Arsenic	69
Arsenic Oxide	48
Arsenious Oxide	See White Arsenic
Artisil	939
Asbestine	607
Asbestos	867
Asbestos Fiber	621
Ascarite	1089
l-Ascorbic Acid	853
Aseptex	509
Asetoform	523
Asphalt	129
Asphalt, Blown	194
Asphaltum	17
Astrinite	499
Astrulan	45
Atabrin	1192
Atrapol	821
Aubepine	800
Aurosal	115
Avenex	883
Avirol	769
Avitex	393
Avocado Oil	383
Avonac	769
Azo	71

B

Badex	1051
Bakelite	111
Bakers' Plasticizer	113

Product	No.
Ball Clay	536
Balsams	583
Barak	393
Bardal	135
Bardex	135
Bardol B	135
Barite	See Barytes
Barium Carbonate	131
Barium Nitrate	207
Barium Peroxide	131
Barium Silico Fluoride	49
Barium Sulfate	131
Barium Sulfide	349
Baroid	775
Barretan	135
Bartyl	503
Barytes	775
Basic Colors	29
Batavia Dammar	45
Bay Oil	681
Bay Rum	917
Bayberry Wax	171
Beckacite	891
Beckamine	891
Beckolin	891
Beckophen	891
Beckosol	891
Beechwood Creosote	697
Beeswax	387
Beetle Resin	45
Bellaphan	631
Belro	557
Bensapol	1197
Bentonite	43
Benzaldehyde	565
Benzidine	479
Benzidine Sulfate	769
Benzine	63
Benzocaine	1
Benzoic Acid	225
Benzol	135
Benzotriazole	407
Benzoyl Peroxide	679
Benzyl Acetate	633
Benzyl Alcohol	633
Benzyl Benzoate	750
Benzyl Butyrate	503
Benzyl Cellulose	11
Benzyl Valerianate	92
Bergamot Oil	823
Beryllium	145
Beryllium Carbonate	185
Beryllium Salts	453
Beta Naphthol	201
Beutene	779
Bicarbonate of Soda	275
Bismuth	247
Bismuth Subnitrate	792
Bitter Almond Oil	800
Bitumen	See Asphaltum
Black Leaf	1099
Blanc Fixe	589

Product	No.
Blandol	1007
Blankit	487
Bleaching Powder	423
Blendene	509
Blood Albumen	753
Blood, Dried	91
Bludtan	249
Boea	603
Bole	1181
Bonderite	831
Bondogen	1141
Bone Ash	355
Bone Black	987
Bone Glue	343
Bone Meal	119
Bone Oil	1085
Borax	67
Bordeaux Mixture	727
Bordow	379
Boric Acid	167
Borol	389
Botanical Products	837
Bresin	559
Brilliant Green	477
Brisgo	559
Bromelia	837
Bromine	365
Bromo Acid	881
Bromo-Fluorescein	509
Bronze Powder	381
Brosco	953
BR-—Resin	111
B. R. T. No. 7	135
Burgundy Pitch	671
Butacite	393
Butalyde	301
Butaprene	446
Butex	813
Butoben	731
Butyl Acetate	301
Butyl Acetyl Ricinoleate	303
Butyl Alcohol (Normal)	877
Butyl Aldehyde	303
Butyl Amine	303
N-Butyl-p-Aminobenzoate	471
Butyl Butyrate	1054
Butyl Carbitol	215
Butyl Cellosolve	215
Butyl Lactate	303
n-Butyl Mesityl Oxide	301
Butyl Parahydroxy Benzoate	565
Butyl Propionate	330
Butyl Stearate	641
1,3-Butylene Glycol Phthalate	749
Butyric Ether	801
Butyric Ester	See Ethyl Butyrate

C

Cadalyte	519
Cadmium	1133
Cadmium Fluoborate	483

CHEMICALS AND SUPPLIES: WHERE TO BUY THEM

Product	No.
Cadmolith	259
Cajuput Oil	591
Calagum	509
Calamine	946
Calcene	299
Calcium Acetate	521
Calcium Arsenate	173
Calcium Carbonate	669
Calcium Carbonate (Precipitated)	731
Calcium Chloride	745
Calcium Chloride (Anhydrous)	433
Calcium Citrate	853
Calcium Cyanamid	45
Calcium Fluoride	483
Calcium Hydroxide	See Lime Hydrate
Calcium Lactate	91a
Calcium Ligno-Sulfonate	702
Calcium Myristate	142
Calcium Nitrate	1068
Calcium Oleate	773
Calcium Oxide	See Lime
Calcium Phosphate	875
Calcium Polysulfide	199
Calcium Propionate	509
Calcium Resinate	739
Calcium Silicate	299
Calcium Silicofluoride	28
Calcium Stearate	1069
Calcium Sulfate	See Plaster of Paris
Calcium Sulfide (Luminous)	57
Calcocid	201
Calcolac	201
Calcoloid	201
Calcozine	201
Calgon	191
Calgonite	203
Calomel	1201
Calorite	1029
Camphor	917
Camphor Oil	693
Canaga Oil	837
Candelilla Wax	605
Cantharides, Tincture of	946
Cantharidin	837
Capillary Syrup	See Glucose Syrup
Caproic Acid	215
Captax	1149
Caramel Coloring	645
Caraway Oil	681
Carbamide	393
Carbitol	215
Carbitol Acetate	215
Carbofrax	221
Carbolac	197
Carbolic Acid	See Phenol
Carbolic Oil	893
Carbolineum	219
Carbon, Activated	619
Carbon Bisulfide	123
Carbon Black	1115
Carbon, Decolorizing	341
Carbon Tetrachloride	799

Product	No.
Carbonex	135
Carboraffin	25
Carborundum	221
Carboseal	215
Carbowax	215
Carboxide	215
Cardamom Seed	789
Carmine	901
Carnauba Wax	919
Carob Bean Flour	1093
Carragheen	See Irish Moss
Casco	227
Casein	227
Cassia Oil	391
Castile Soap	315
Castor Oil	113
Castor Oil, Blown	113
Castor Oil, Sulfonated	1197
Castrolite	923
Castung	113
Catalin	33
Catalpo	751
Catechol	841
Catylon	547
Caustic Soda	713
CCH	713
Cedar Oil	383
Celascour	29
Celeron	317
Celito	621
Cellit	See Cellulose Acetate
Cellosize	215
Cellosolve	215
Cellosolve Acetate	215
Cellosolve Ricinoleate	509
Celluloid	231
Celluloid Scrap	973
Cellulose Acetate	231
Cellulose Acetate-Butyrate	1083
Cellulose Ether of Sodium Glycollate	393
Cellulose Fiber Pulp	184
Cellulose Nitrate	733
Cement	843
Censteric	303
Ceraflux	509
Cercon	309
Cerelose	327
Cereps	1171
Ceresalt	409
Ceresin Wax	983
Cerium Oxide	715
Cerol	939
Cetamin	509
Cetec	489
Cetyl Alcohol	587
Cetyl Trimethyl Ammonium Bromide	821
Chalk, Precipitated	335
Charcoal	889
Chemigum	517
Chestnut Extract	894

Product	No.
China Clay	1071
China Wood Oil	125
Chloral Hydrate	792
Chloramine	1
Chlorasol	215
Chlorex	215
Chlorinated Lime	713
Chlorinated Naphthalene	531
Chlorinated Paraffin	780
Chlorinated Rubber	559
Chlorine (Liquid)	423
Chlorocresol	135
p-Chloro-m-Cresol	749
Chloroform	379
Chlorophyll	41
Chloropicrin	81
Chlorosol	215
Chlorthymol	823
Cholesterin	731
Cholesterol	See Cholesterin
Chremnitz White	441
Chrome Alum	1151
Chrome Green	639
Chrome Yellow	79
Chromic Acid	759
Chromic Oxide	See Chromium Oxide
Chromium Oxide	573
Chromium Sulfate	731
Cicoil	177
Cinchona Tincture	946
Cinchophen	201
Cinelin	279
Cinnamic Aldehyde	994
Cinnamon Bark Oil	391
Cinnamon Oil	391
Cinnamyl Alcohol	779
Citral	503
Citrene	503
Citric Acid	853
Citronella Oil	931
Citronellol	439
Clarinol	559
Clay	1117
Clay No. 33	751
Clove Oil	272
Clovel	509
Coal Tar	333
Coal Tar Colors	645
Coal Tar Oil	135
Cobalt Acetate	183
Cobalt Chloride	719
Cobalt Driers	719
Cobalt Linoleate	719
Cobalt Naphthenate	811
Coblac	159
Cocoa Butter	19
Cocoa Butter Oxanhydride	107
Coconut Butter	873
Coconut Oil	117
Coconut Oil Fatty Acid	9
Cod Liver Oil	917
Codite	317

Product	No.
Collodion	323
Collodion Wool	See Pyroxylin
Colloidal Clay	1117
Colloresin	487
Colophony	See Rosin
Color L34	509
Colors, Certified	645
Colors, Dry	571
Colors, Oil Soluble	305
Colza Oil	See Rapeseed Oil
Cominol	305
Concretes, Perfume	1139
Condensite	111
Congo Resin	See Resins, Natural
Conoco	319
Convertit	809
Copaiba Resin	681
Copal	45
Copal, Esterified	459
Copapel A	487
Copper Carbonate	323
Copper Chloride	123
Copper Chloride, Basic	483
Copper Cyanide	541
Copper Fluoborate	483
Copper Nitrate	123
Copper Oxides	573
Copper Sulfate	127
Coppercide	605
Coresin	1148
Corn Oil	59
Corn Oil, Vulcanized	1021
Corn Sugar	1019
Corn Syrup	327
Cosmic Black	1109
Cotton Seed Oil	141
Cotton Seed Oil, Hydrogenated	873
Cottonseed Meal	421
Coumarin	715
Coumarone Resin	135
Cream of Tartar	545
Creosol Sulfonic Acid	749
Creosote	647
Cresols	325
Cresophan	523
Cresylic Acid	135
Cromodine	37
Cryolite	1159
Cryptone	785
Crysalba	749
Crystal Violet	150
Cumar	135
Cupric Chloride	See Copper Chloride
Curbay Binder	1127
Cuttle Fish Bone	463
Cyclamal	485
Cycline	749
Cyclohexanol	393
Cyclohexylamine	393
Cyclohexyl Butyrate	800
Cyclohexylcyclohexanol	135
Cyclonol	393

CHEMICALS AND SUPPLIES: WHERE TO BUY THEM

Product	No.
Cyclo Rubber	393
Cymanol	601

D

Product	No.
2-4D	646
Dammar Gum	671
Dapol	501
Darco	341
Darvan	1141
Daxad	359
DDT	646
Decresol	45
Degras	53
Dehydrol	945
Deo-base	1007
Deramin	509
Derris Extract	965
Derris Root	147
Dextrins	753
Dextrose	327
Diacetin	641
Diafoam	895
Diakonn	599
Diamond K Linseed Oil	1017
Diamylamine Phosphate	975
Diamyl Phthalate	1127
Diastafor	1027
Diastase	1073
Diatol	1127
Diatomaceous Earth	1193
Dibutyl Cellosolve Phthalate	749
Dibutyl Phthalate	641
Dibutyl Sebacate	540
Dibutyl Tartrate	633
Dicalite	363
Dichloramine T	749
o-Dichlorbenzene	575
Dichlorbenzol	575
Dichlorethylether	215
Dichlorodifluoromethane	643
Dichloroethyl Ether	215
2-4 Dichlorophenoxyacetic Acid	646
Diethanolamine Lactate	509
Diethylcarbonate	1127
Diethyl Oxalate	1127
Diethyl Phthalate	1143
Diethylene Glycol	215
Diethylene Glycol Distearate	509
Diethylene Glycol Monoethyl Ether	215
Diethylene Glycol Monostearate	509
Diethylenetriamine	215
Digestase	699
Diglycol Laurate	509
Diglycol Oleate	509
Diglycol Stearate	509
Dilecto	317
Dimethoxyethylphthalate	813
Dimethylglyoxime	407
Dimethylolurea	393
Dimethylphthalate	1054
Dinitrophenol	769
Dioctyl Phthalate	1054

Product	No.
Diolin	393
Dionin	731
Dioxan	215
Dioxane	215
Dipentene	559
Diphenyl	1063
Diphenyl, Chlorinated	749
Diphenylamine	393
p-Diphenyl Disulfonic Acid	749
Diphenyl Oxide	503
Diphenyl Phosphate	749
Diphenyl Phthalate	749
Dipropylene Glycol	379
Discolite	923
Disodium Ethylene Bisdithiocarbamate	393
Disodium Hydrogen Phosphate	1153
Disodium Phosphate	1153
Disperso	1193
Distoline	1195
Dow Plasticizers	379
Dowicide	379
Dowmetal	379
D. P. G.	749
Dreft	873
Drierite	537
Driers (Paint and Oil)	811
Driers, Varnish	811
Drop Black	1185
Dry Ice	713
Dulux	393
Duolith	649
Duponol	393
Duphax	1021
Duphonol	393
DuPont Rubber Red	393
Duprene	393
Duraplex	895
Durez	497
Durite	1053
Durocer	509
Dutox	393
Dyal	982
Dyestuffs	769
Dynax	393
Dyphene	982

E

Product	No.
East-India Gum	45
Eastman Products	407
Egg, Dried	869
Egg Yolk	1051
Elaine	427
Elemi	1127
Emulgor A	509
Emulphor	487
Emulphor DDT	479
Emulsifier A3076A	509
Emulsifier S489	509
Emulsone	509
Emulsone B	509
Eosin	891

Product	No.
Epersol-Y	419
Ephedrine	731
Ephedrine Hydrochloride	523
Ephedrine Sulfate	523
Epinephrine	731
Epsom Salts	483
Erinoid	65
Erio Chrome Dyes	477
Erythrosin	881
Escolite	329
Essences, Flavoring	595
Essential Oils	431a
Esso	1033
Ester Gum	667
Esterol	827
Estersol	1127
Ethanol	See Ethyl Alcohol
Ethavan	749
Ether	215
Ethocel	379
Ethox	1145
Ethyl Abietate	557
Ethyl Acetate	733
Ethylamine	151
Ethyl Anisate	1139
Ethyl Butyrate	432
Ethyl Cellulose	559
Ethyl Cinnamate	432
Ethyl Formate	439
3-Ethyl-1,3-Hexanediol See Rutgers	612
Ethyl Lactate	45
Ethyl Mercury Phosphate	393
Ethyl Parasept	565
Ethyl Propionate	383
Ethyl Protol	377
Ethyl Ricinoleate	445a
Ethyl Silicate	215
Ethylene Chlorhydrin	215
Ethylene Diamine	151
Ethylene Dichloride	379
Ethyleneglycol	215
Ethylene Glycol Ethers	215
Ethylene Glycol Monobenzyl Ether	215
Eucalyptus Oil	447
Eugenol	503
Eugenol Acetate	800
Eulan	487
Euresol	157
Exton	393

F

Product	No.
Fabroil	489
Factice	1021
Factolac	577
Falba Absorption Base	851
Falkene	435
Falkide	435
Falkomast	435
Falkote	435
Falkovar	435
Falkyd	435

Product	No.
Feectol	925
Feldspar	313
Fermate	393
Fer-ox	1193
Ferric Chloride	See Iron Chloride
Ferrisul	749
Ferrite Yellow	159
Ferro Chrome	425
Ferro Manganese	425
Ferro Silicon	425
Ferrous Chloride	See Iron Chloride
Ferrox	1047
Fiberglas	824a
Fiberlon	443
Filac	581
Fillers	1189
Film Scrap	581
Filter-Cel	621
Filtrol	445
Fireclay	499a
Fish Glue	563
Fish Liver Extract	430
Fish Oil	435
Fish Oil Soap	336
Fixalt	749
Fixtan	519
Flavine	1205
Flavors	595
Flaxseed	161
Flectol	925
Flexalyn	559
Flexol DOP	215
Flexoresin	509
Flexowax C	509
Fluid Extracts	946
Fluoboric Acid	849
Fluorspar	569
Foamapin	509
Foamex	509
Formaldehyde	565
Formalin	See Formaldehyde
Formamide	393
Formic Acid	1153
Formica	455
Formvar	977
Freon	643
Freon-12	643
Friction Black Pulp	403
Fuller's Earth	935
Furfuramide	883
Fusel Oil	1127
Fyrex	1153

G

Product	No.
G Protein	509
Gallagum	509
Gallic Acid	1207
Gamboge	919
Gamma Valero Lactone	749
Gammexane	599
Gardinol	393
Gastex	481

CHEMICALS AND SUPPLIES: WHERE TO BUY THEM

Product	No.
Gelatin	95
Gelloid	1197
Gelowax	509
Gelozone	1183
Gelva	977
Gelva Resin	977
Gentian Violet	881
Geraniol	633
Geranium Bourbon Oil	800
Geranium Lake	611
Geranium Oil	947
Geranyl Acetate	969
Geranyl Formate	1139
Gilsonite	671
Ginseng	665
Glandular Products	1191
Glassheen	509
Glauber's Salt	613
Glaurin	509
Glucarine	509
Glucarine B	509
Glucose	941
Glue	337
Glumide	509
Glutrin	909
Glycerawax	509
Glycerin	291
Glycerol	See Glycerin
Glyceryl Mcnooleate	509
Glyceryl Mono Stearate	509
Glyceryl Monostearate (S-928)	509
Glyceryl Phthalate	509
Glyceryl Sebacate	915
Glyceryl Stearate	509
Glyceryl Tristearate	509
Glyco Wax A	509
Glyco Wax B	509
Glycol	See Ethylene Glycol
Glycol Bori-Borate	509
Glycol Distearate	509
Glycol Oleate	509
Glycol Phthalate	509
Glycol Stearate	509
Glycolac	509, 651
Glycomel	509
Glycomine	509
Glycopon	509
Glycoride	509
Glycosterin	509
Glycox 1300-1400	509
Glyptal	491
Gold Chloride	697
Grapeseed Oil	339
Graphite	589
Graphite, Colloidal	7
Green Soap	721
Green Vitriol	971
Ground Nut Oil	See Peanut Oil
Guai-a-phene	295
Guantal	925
Guiacol	565
Gum Arabic	391
Gum Balata	677
Gum Batu	1093
Gum Benzoin	835
Gum Copal	671
Gum Dammar	1093
Gum Elemi	459
Gum Karaya	463
Gum, Locust Bean	605
Gum Manila	603
Gum Mastic	941
Gum Myrrh	577
Gum Sandarac	671
Gum Tragacanth	729
Gums, Varnish	603
Gutta Percha	385
Gypsum	1129

H

Product	No.
Haemoglobin	1
Halex	181
Halowax	531
Harcol	509
Harveg	551
Harvite	987
Haskelite	549
Heliotropin	715
Heliozone	393
Hemlock Bark	1077
Henna Leaves	837
Herbs	281
Hercolyn	559
Hercusol	559
Herkolite	489
Hevealac	509
Heveatex	561
Hexachlorethane	575
Hexamethylene Diisocyanate	395
Hexamethylenetetramine	565
Hexamine	565
Hexone	215
Hexyl Alcohol	215
Hormones	943
Hyamine	915
Hycar	411
Hydralite C	487
Hydrated Lime	See Lime
Hydristear	1187
Hydrocal	1123
Hydrochloric Acid	483
Hydrofluoric Acid	507
Hydrogen Peroxide	1167
Hydrogenated Castor Oil	393
Hydrogenated Cottonseed Oil	873
Hydrogenated Fish Oil	1003
Hydromalin	509
Hydromel	509
Hydronol	393
Hydropel	305
Hydroquinone	407
Hydroresin	509
Hydrowax	509
Hydroxyethylcellulose	215

Product	No.
Hydroxylamine Sulfate	301
12-Hydroxy Stearic Acid	509
Hygropon	509

I

Product	No.
Ichthyol	731
Idalol	555
IG Wax	487
IG Wax O	487
Igepon	487
Ignex	509
Indalone	1127
Indanthrene Dyes	487
Indian Red	159
Indigisols	213
Indigo	885
Indium	145
Indolebutyric Acid	37
Indolylacetic Acid	731
Indur	893
Indusoil	601
Infusorial Earth	
See Diatomaceous Earth	
Insect Wax, Chinese	917
Invadine	277
Invert Sugar	809
Iodine	792
Iodoform	853
Ionone	1151
Iridium	115
Irish Moss	837
Iron Ammonium Citrate	957
Iron Blue	79
Iron Chloride	269
Iron Oxide	159
Iron Sulfate	483
Isco	605
Isoamyl Acetate	1054
Isoamyl Butyrate	92
Isoamyl Isovalerate	92
Isobornyl Thiocyanate	557
Isobutyl Acetate	521
Isobutyl Ketone	439
Isobutyl Salicylate	439
Isocholesterol	See Cholesterin
Isohol	509
Isolene	777
Isoline	1195
Isomerpin	393
Isophan	233
Isophorone	215
Isopropanol	See Isopropyl Alcohol
Isopropanolamine	215
Isopropyl Acetate	535
Isopropyl Alcohol	1025
Isopropyl Ether	215
Isopropylnaphthalene Sulfonic Acid	393
Ivory Black	159

J

Product	No.
Japan Drier	739
Japan, Liquid	570

Product	No.
Japan Wax	1001
Jasmine, Artificial	383
Jasmogene	1149
Javelle Water	1170
Jointex	509

K

Product	No.
K. D. Gum	671
Kainite	467
Kalite	1141
Kaolin	951
Kapsol	1145
Karo	327
Kellin	1017
Kellogg KVO	1017
Kellogg Varnish Oil	1017
Kellsey	1017
Keltone	635
Kerol	185
Kerosene	297
Kerosene, Deodorized	983
Ketanol	393
Ketonone	45
Kilfoam	37
Kolineum	647
Kopol	143
Koreon	759
Korogel	515
Kronisol	1145
Kronitex	1145
Kryocide	849

L

Product	No.
Laboratory Equipment, 239, 261, 267, 417, 449, 783, 961	
Lacquer Blue	79
Lacquers	685
Lactac	393
Lactic Acid	85
Lactoid	181
Lactol Spirits	259
Lactose	206
Laevo-Ascorbic Acid	853
Lake Colors	639
Lamp Black	159
Lanette Wax	393
Lanolin	731
Lanthanum Oxide Hydrate	672
Lard Oil	431
Latac	393
Latex	561, 779
Laurex	779
Lauric-Ol	505
Lauryl Alcohol	393
Laurylamine	773
Lauryl Pyridinium Chloride	509
Lauryl Sulfonate	393
Lavekol	487
Lavender Oil	1139
Lavene	503
Lead	401
Lead Acetate	483

CHEMICALS AND SUPPLIES: WHERE TO BUY THEM

Product	No.
Lead Arsenate	483
Lead Borate	719
Lead Carbonate, Basic	982
Lead Fluoborate	401
Lead Oleate	979
Lead Oxides	401
Lead Stearate	735
Lead Titanate	1097
Lecithin	505
Lecivit	627
Lemon Chrome	675
Lemon Juice, Concentrated	763
Lemon Oil	591
Lemonal	1139
Lemonone	509
Leonil	45
Le Page's Cement	929
Leptyne	859
Lerbinia	439
Lethane	915
Leucosol	393
Leucotrope W	1197
Leukanol	915
Leukonin	545
Levelene	29
Lewisol	667
Licorice	687
Licorice Root	687
Lignin	702
Lime	843
Lime, Hydrated	605
Limestone	955
Linalöol	503
Linalyl Acetate	503
Linalyl Formate	503
Lindol	235
Lingasan	557
Linoleic Acid	509
Linseed	See Flaxseed
Linseed Oil	161
Linseed Oil, Sulfurized	1021
Liquid Paraffin	See Mineral Oil
Literite	675
Litex	1079
Litharge	401
Lithium Carbonate	715
Lithium Chloride	715
Lithium Salts	715
Lithoform	37
Lithopone	649
Locust Bean Powder	391
Lognite	1205
Logwood Extract	47
Lopor 45	432
Lorinol	163
Lorol	393
Lubrisol	1007
Lucidol	679
Lucite	393
Lufax	915
Lumarith	235
Luminol	407
Luxene	111
Luxol	393
Lycopodium	721
Lynsol	557
Lysol	661

M

Product	No.
Mabelite	45
Mace Oil	681
Mafos	713
Magnesia	223
Magnesia, Light Calcined	493
Magnesite	493
Magnesium	379
Magnesium Ammonium Phosphate	749
Magnesium Carbonate	731
Magnesium Chloride	1193
Magnesium Hydroxide	705
Magnesium Peroxide	187
Magnesium Powder	145
Magnesium Silicate	1181
Magnesium Silicofluoride	345
Magnesium Stearate	1193
Mahogany Soaps	1007
Maize Starch	See Corn Starch
Makalot	285
Malachite Green	881
Maleic Acid	769
Manganar	393
Manganese	13
Manganese Dioxide	381
Manganese Resinate	979
Manila Gum	603
Manjak	1209
Mannitan Monostearate	103
Mannitol	103
Manol	983
Manox	701
Mapico	159
Mapromin	905
Mapromol	905
Marble Dust	533
Marcol	1023
Marlite	873
Marseilles Soap	See Castile Soap
Mazola	327
Melamac	45
Mellittis	503
Menhaden Oil	109
Menthol	585
Menthone	503
Menthyl Salicylate	523
Mercapto-Acetic Acid	711
Mercaptobenzothiozole	779
Mercol ST	1088
Mercurous Nitrate	1201
Mercury	585
Mercury Compounds	1201
Mercury Oxycyanide	149a
Merpentine	393
Merpol	393
Mersol	749

Product	No.
Mertanol	749
Merusol	1031
Mesityl Oxide	303
Metaldehyde	793
Methacrylic Acid	393
Methalate C	473
Methanol	521
Methasol	599
Methenamine	565
Methicol	501
Methocel	379
Methox	1145
Methyl Abietate	559
Methyl Acetate	215
Methyl Acetone	353
Methyl Alphanapthylacetic Acid ..	37
Methyl Amyl Ketone	215
Methylanol	393
Methyl Anthranilate	451
Methylate	473
Methyl Bromide	379
Methyl Caproate	800
Methyl Cellosolve	215
Methyl Cellosolve Acetate	461
Methyl Cellulose	379
Methylcyclohexanone	135
Methyl Ethyl Ketone	215
Methyl Hexalin	135
Methyl Hexanone	779
Methyl p-Hydroxybenzoate	565
Methyl Ionone	503
Methyl Isobutyl Ketone	523
Methyl Naphthalene	1148
Methyl α-Naphthylacetic Acid	37
Methyl Orange	289
Methyl Parasept	565
Methyl Salicylate	379
Methyl Violet	881
Methylene Blue	881
Methylene Chloride	1157
Methylene Dichloride	1157
Metol	407
Metro-Nite	737
Metso	855
Mica	1009
Micoid	741
Migasol	277
Milcol	509
Milk Gloss	509
Milk Sugar	697
Mineral Black	675
Mineral Jelly	See Petrolatum
Mineral Oil	See Paraffin Oil
Mineral Rubber	129
Mineral Seal Oil	1033
Mineral Spirits	63
Minium	See Lead Oxide
Mirasol	824
Mischmetal	145
Molasses	64
Moldex	509
Moldslip	509

Product	No.
Molybdenum	285
Molybdenum Compounds	285
Molybdenum Oxide	285
Molybdic Acid	123
Monex	779
Monoamylamine Oleate	509
Monoethanolamine	215
Monoethanolamine Lactate	509
Monolite	489
Monostearin	509
Monostyrene	379
Monosulph	773
Montan Color	509
Montan Wax	1055
Morpholine	215
Mowilith	11
Mulsene	509
Multifex	593
Muriatic Acid	See Hydrochloric Acid
Musk Xylol	503
Mustard Oil	1171
Mycoban	393
Myristic Acid	91

N

Naccolene	769
Naccon	769
Nacconol	769
Naphtha	347
Naphthalene	135
Naphthalene, Chlorinated	531
Napthenic Acid	509
α-Naphthol	479
β-Naphthol	201
Naphthylacetic Acid	37
α-Naphthylthiourea	See Antu
Napoleum Spirits	347
Narobin	509
National Oil Red	769
NDGA	525
Neatsfoot Oil	773
Nekal	487
Nelgin	509
Neolan	277
Neomerpin	393
Neoprene	599
Neozone D	393
Neroli Oil	800
Neutroleum	473
Neville Resin	781
Nevindene	545
Nevinol	781
Nevtex	781
N-Glo-5	791
Nickel Chloride	323
Nickel Sulfate	545
Nicotine	1099
Nicotine Sulfate	655
Nicotinic Acid	201
Nigrosine	201
Nipabenzyl	937
Nipagen	513

CHEMICALS AND SUPPLIES: WHERE TO BUY THEM

Product	No.
Nipasol	937
Nitramon	393
Nitrated Cotton	See Pyroxylin
Nitre Cake	1103
Nitric Acid	749
Nitrobenzol	201
Nitrocellulose	395
Nitrocotton	See Nitrocellulose
Nitroethane	303
p-Nitrophenol	749
Nitropropane	301
Nonaethylene Glycol	215
Nonaethylene Glycol Laurate	509
Nonaethylene Glycol Stearate S	509
Nopco	773
Novolak	111
Nuad	811
Nuba Resin	781
Nu-Char	601
Nulomoline	809
Nuodex	811
Nuodex Copper, Cobalt	811
Nusoap	811

O

Product	No.
Ocenol	395
Ochres	999
Octadecane Amide	91
Octyl Acetate	215
Octyl Alcohol	215
Oenanthic Ether	681
Oil, Citronella	591
Oil, Mineral	1029
Oil Root Beer C	969
Oilate	811
Oildag	5
Oilsolate	895
Oiticica Oil	615
Olate	873
Olein	241
Oleoresins	969
Oleyl Alcohol	393
Olive Oil	659, 1057
Olive Oil Substitute	509
Olive Oil, Sulfonated	923
Ondulum	509
Opal Wax	393
Orange Oil	923
Oroco	921
Orvus	873
Ortho Dichlorbenzene	575
Ortho-Phenylphenate	379
Orthosil	849
Osmo-Kaolin	457
Ouricuri Wax	375
Oxalic Acid	759
Oxgall	781
Oxycholesterol	42
Oxygen	265
Oxynone	925

Product	No.
Oxyquinoline Sulfate	149
Ozokerite Wax	1057

P

Product	No.
Palm Kernel Oil	117
Palm Oil	1193
Palm Oil Fatty Acids	1171
Palmitic Acid	91
Pancreas	91
Panoline	1033
Para Aminophenol	1151
Parachlormetacresol	749
Parachol	509
Paracide	575
Para-dor	379
Paradura	827
Paraffin, Chlorinated	780
Paraffin Oils	959
Paraffin Wax	815
Para-flux	529
Paralac	599
Paraldehyde	565
Paramet	827
Paranitrophenol	749
Paranol	827
Para-Phenylenediamine	73
Parapont	393
Parasept	565
Paratoluene Sulfone Chloride	749
Paris Black	159
Paris Green	357
Paris White	1013
Parlon	559
Paroil	23
Peachol	509
Peanut Oil	421
Pearl Essence	725
Pectin	205
Peerless Clay	1141
Pegopren	511
Pentacetate	975
Pentachloronitrobenzene	749
Pentachlorphenol	379
Pentaerythritol	565
Pentaerythritol Abietate	559
Pentaerythritol Esters	565
Pentalyn	559
Pentasol	975
Pentawax	565
Penicillin	301
Pennyroyal Oil	681
Pentrol	633
Peppermint Oil	693
Peptone	369
Perchloric Acid	731
Perchloroethylene	393
Perchloron	849
Peregal	479
Perfume Bases	595
Perilla Oil	625
Permosalt	509
Permosalt A	509

Product	No.	Product	No.
Perone	393	Plexigum	915
Perrol	1139	Plexite	915
Perspex	599	Plioform	517
Peru Balsam	377	Pliolite	517
Petitgrain Oil	681	Podophyllin Resin	870
Petrex	557	Pliowax	517
Petrobenzol	77	Pollopas	865
Petrohol	1023	Polyacrylic Esters	393
Petrolatum	847	Polycol	509
Petrolene	77	Polyethylene	215
Petroleum Ether	997	Polyethylene Glycol Laurate	509
Petroleum Jelly	1007	Polyglycerol Monostearate	509
Petroleum Spirits	1061	Polyglycol 400 Monostearate	509
Petromix	1007	Polyisobutylene	1033
Petropol 2138	63	Polymerized Glycol Stearate	509
Pharmagel B	1064	Polypale Resin	557
Pharmasol	213	Polypentek	565
Phenac	45	Polyrin	509
Phenex	529	Polystyrene	749
Phenol	31	Polyvinyl Acetal	793
Phenol-Formaldehyde Resins	397	Polyvinyl Acetate	215
Phenolic Resin	891	Polyvinyl Alcohol	793
Phenothiazine	201	Polyvinyl Butyral	749
Phenyl Acetaldehyde	800	Polyvinyl Butyral Resin	749
Phenyl Cellosolve	215	Polyvinyl Chloride	749
Phenyl Chloride	575	Polyzime	1073
Phenyl Cyclohexanol	503	Ponolith	649
Phenyl Ethyl Butyrate	503	Pontalite	393
Phenyl Mercuric Nitrate	411	Pontianak Resin	739
Phenyl Mercuric Stearate	149a	Pontol	393
o-Phenylphenate	379	Potash, Caustic	795
Phenyl Salicylate	379	Potash Water Glass	855
p-Phenylenediamine	73	Potassium Abietate	509
Phi-Sol	821	Potassium Alginate	16
Phloroglucinol	411	Potassium Borofluoride	849
Phobophene	1179	Potassium Bromate	379
Phosphoric Acid	1153	Potassium Bromide	565
Phosphorus	609	Potassium Carbonate	1105
Phosphotex	749	Potassium Chlorate	1105
Phthalic Anhydride	749	Potassium Chloride	67
Phthalocyanine Blue	393	Potassium Citrate	853
Piccolastic	845	Potassium Dichromate	871
Piccolyte	845	Potassium Dihydrogen Phosphate	266
Piccolyte Resin	845	Potassium Hydroxide	731
Piccovar	845	Potassium Iodine	792
Picric Acid	731	Potassium Metabisulfite	697
Pigment 725	785	Potassium Nitrate	See Saltpeter
Pigment Colors	79	Potassium Oleate	509
Pilocarpine Nitrate	602	Potassium Palmolate	509
Pimento Oil	693	Potassium Perchlorate	1105
Pine Needle Oil	383	Potassium Permanganate	225
Pine Oil	495	Potassium Rosin Soap	335
Pine Tar	1011	Potassium Silicate	855
Pinene	557	Potassium Silicofluoride	1060
Pitch	1209	Potassium Soap	648
Plaskon	861	Potassium Tartrate	731
Plastacele	393	Prague Powder	525
Plaster of Paris	1181	Prepared Chalk	601
Plastogen	1141	Prestabit	487
Plastopal	11	Proflavine	697
Platinum Chloride	514	Proofit	509
Plexiglas	915	Proflex	509

CHEMICALS AND SUPPLIES: WHERE TO BUY THEM

Product	No.
Propionic Acid	393
Propyl Alcohol	1054
Propyl p-Hydroxybenzoate	565
Propylene Dichloride	215
Propylene Glycol	379
Propylene Glycol Monostearate	509
Propylene Glycol Stearate	509
Propylene Oxide	215
Protease Enzyme	1165
Protectoid	235
Protoflex	509
Protovac	227
Provatol	413
Proxate	673
Prussian Blue	639
Prystal	33
Psyllium Seeds	657
Puerine	707
Pumice	335
Pylam Red	881
Pyrax	1141
Pyrefume	837
Pyrethol	717
Pyrethrum	837
Pyrethrum Extract	723
Pyridin	647
Pyro	See Pyrogallic Acid
Pyrogallic Acid	1207
Pyrolusite	381
Pyrophyllite	1181
Pyroxylin	557
Pyroxylin Solutions	415

Q

Product	No.
Quakersol	839
Quartz Sand	273
Quassia Extract	946
Quebracho	47
Quince Seed	577
Quinine Bisulfate	523
Quinine Hydrochloride	731
Quinine Salicylate	792
Quinoline	135

R

Product	No.
Raisin Seed Oil	901
Rancidex	509
Rapeseed Oil	125
Rapidase	1165
Rauzene	1127
Rayox	1141
Red Oil	241
Red Oil Soap	648
Red Squill	577
Redmanol	111
Reogen	1141
Resin DA 1	509
Resin R-H-35	393
Resinox	897
Resins, Natural	45
Resins, Synthetic	891

Product	No.
Resipon	89
Resoglaz	11
Resorcin	841
Resorcinol-Formaldehyde Resin	845
Revertex	899
Rezidel	509
Rezinel	509
Rezyl	45
Rheolan	509
Rhodium	115
Rhonite	915
Rhoplex	915
Rhotex	915
Rice Starch	86
Ricinoleic Acid	113
Rochelle Salts	853
Rodo	1141
Rose Water	681
Rosemary Oil	823
Roseol	693
Rosin	495
Rosin Oil	777
Rosin, Polymerized	559
Rosin, Polypale	557
Rosoap A	509
Rotenone	1091
Rouge	273
Rubber	405
Rubber Hydrochloride	703
Rubber Latex	677
Rubber Resin	509
Rubber, Synthetic	393, 515, 517
Rubidium Salts	689
Rum Ether	469
Rutgers 612	215
Rutile	1097

S

Product	No.
"S" Syrup	855
Saccharine	565
Sal Soda	275
Salicylanilide	479
Salicylic Acid	379
Salt	755
Salt Cake	45
Saltpeter	331
Santicizers	749
Santobane	749
Santocel	749
Santolite	749
Santomask	749
Santomerse	749
Santowax	749
Santox	749
Sapamine	277
Saponin	627
Sardine Oil	675
Savolin	509
Schultz Silica	251
Selenium	61
Sellatan A	477
Sepia	See Cuttle Fish Bone

Product	No.
Serinol	629
Serrasol	263
Sesame Oil	1171
Shellac	1207
Shellac Wax	589
Sherpetco	983
Sicapon	509
Siennas	441
Silastic	380
Silex	1181
Silica	133
Silica Aerogel	345
Silica Black	617
Silica Smoke	935
Silicic Acid	483
Silicon	145
Silicon Fluoride	1060
Silicone Fluid	491
Silvatol	277
Silver	539
Silver Chloride	514
Silver Cyanide	323
Silver Nitrate	407
Silver Oxide	514
Slag Wool	955
Slaked Lime	See Lime
Slate Powder	1193
Soap	1041
Soap Bark	870
Soap Base	648
Soapstone	935
Soda Ash	361
Soda, Caustic	713
Soda, Sal	311
Sodium Abietate	557
Sodium Acetate	1105
Sodium Acid Fluoride	1060
Sodium Alginate	635
Sodium Aluminate	765
Sodium Antimonate	245
Sodium Aluminum Silicofluoride	1060
Sodium Aluminum Sulfate	123
Sodium Arsenite	543
Sodium Benzoate	575
Sodium Bicarbonate	275
Sodium Bichromate	871
Sodium Bisulfite	519
Sodium Borate	See Borax
Sodium Borophosphate	509
Sodium Carbonate	1005
Sodium Chlorate	817
Sodium Chlorite	713
Sodium Choleate	369
Sodium Citrate	853
Sodium Cyanide	393
Sodium Fluoride	45
Sodium Hexametaphosphate	191
Sodium Hydrosulfite	923
Sodium Hydroxide	731
Sodium Hypochlorite	351
Sodium Hypochlorite Liquid	907
Sodium Hyposulfite	519
Sodium Lactate	85
Sodium Lauryl Sulfate	509
Sodium Metaphosphate	191
Sodium Metasilicate	855
Sodium Naphthalene Sulfonate	547
Sodium Nitrate	139
Sodium Nitrite	1005
Sodium Oleate	509
Sodium Orthophenylphenate	379
Sodium Oxalate	1153
Sodium Pectate	205
Sodium Perborate	393
Sodium Phosphate	1063
Sodium Propionate	509
Sodium Pyrophosphate	1153
Sodium Silicate	855
Sodium Silico Fluoride	519
Sodium Stannate	545
Sodium Stannite	719
Sodium Stearate	773
Sodium Sulfacetamide	429a
Sodium Sulfate	483
Sodium Sulfite	727
Sodium Sulfocyanide	971
Sodium Taurocholate	91
Sodium Tetradecyl Sulfate	215
Sodium Thiocyanate	911
Sodium Thioglycollate	1043
Sodium Titanium Fluoride	1060
Sodium Tungstate	123
Soft Soap	648
Soligen	11
Solox	1127
Solozone	393
Soluble Castor Oil	547
Solux	393
Solvent Naphtha	135
Solvesso	1033
Solwax	893
Sorbitan Monooleate	103
Sorbitol	103
Sorbitol Laurate	103
Sorbitol Oleate	103
Sorbitol Stearate	103
Sovasol	1004
Soya Protein	505
Soybean Flour	1019
Soybean Oil	1017
Soyflake	1017
Span	103
Sperm Oil	321
Spermaceti	1055
Speron	197
Spindle Oil	1033
Squill	837
SRA Black	29
Stabilite	529
Stacol	509
Stago CS	169
Standard Viscous Oil	1029
Standoil	1017
Stannic Chloride	See Tin Chloride

CHEMICALS AND SUPPLIES: WHERE TO BUY THEM

Product	No.
Stannous Chloride	See Tin Chloride
Stannous Fluoborate	483
Starch	1045
Staybelite	559
Staybelite Ester	557
Stearacol	509
Stearamide	91
Stearic Acid	241
Stearin	1175
Stearin Pitch	527
Stearite	1193
Stearol	863
Stearoricinol	509
Stearyl Alcohol	901
Stetsol	593
Stoddard Solvent	347
Storax	693
Stramonium	837
Stripolite	923
Stripper, T. S.	89
Stroba Wax	509
Strontium Hydrate	130
Strontium Nitrate	519
Strontium Sulfate	130
Strychnine	853
Styrax	See Storax
Styrene	379
Styrene Dibromide	379
Styrex	379
Subcarbonate of Iron	697
Sublan	509
Succinic Acid	769
Sucrose Octoacetate	793
Sulfadiazine	201
Sulfanilamide	201
Sulfanole	1169
Sulfatate	509
Sulfated Castor Oil	547
Sulfathiazole	201
Sulfite Liquor	724
Sulfite Lye	724
Sulfo Turk A	509
Sulfo Turk B	509
Sulfo Turk C	509
Sulfogene	393
Sulfonamide	201
Sulfonated Castor Oil	189
Sulfonated Coconut Oil	923
Sulfonated Fatty Alcohol	509
Sulfonated Hydrogenated Castor Oil	773
Sulfonated Liquid Petrolatum	1007
Sulfonated Mineral Oil	509
Sulfonated Olive Oil	1197
Sulforicinol	155
Sulfur	1047
Sulfur Chloride	1047
Sulfur Dioxide	1157
Sulfuric Acid	733
Sulfurized Oils	1021
Sulfurized Sperm Oil	773
Sunoco Spirits	1061
Sunsoy	1017

Product	No.
Superba Black	159
Superphosphate	1155
Surfex	879
Suspendite	509
Suspensone	509
Syncrolite	627
Syntex	293
Synthane	1067
Synthenol	1017
Synvarite Resin	1070

T

Product	No.
T-Tree Oil	193
Talc	335
Talcum	See Talc
Tall Oil	249, 601
Tallow	1171
Tamal	915
Tamol	915
Tanax	45
Tannic Acid	1207
Tannin Extract	47
Tantalum	437
Tar Acid Oil	135
Tartar Emetic	85
Tartaric Acid	523
Tartrazine	881
Tea Seed Oil	683
Tegin	513
Teglac	45
Tegofan	513
Telloy	1141
Tellurium	1141
Tellurium Oxide	69
Tenex	495
Tenite	1083
Tepidone	393
Tergitol	215
Terpesol	559
Terpineol	591
Tetrachlorethane	379
Tetrachlorethylene	393
Tetradecanol	215
Tetralin	393
Tetrasodium Pyrophosphate	266
Tetrone	393
Textac	559
Textone	713
Thallium Bromide	417
Thallium Iodide	417
Thallium Sulfate	627
Thanite	557
Theop	509
Thermoplex	895
Thiamin Hydrochloride	731
Thiocarbamilid	749
Thioglycollic Acid	1043
Thionex	393
Thiourea	627
Thorium Salts	1173
Thyme Oil	823
Thymol	981

Product	No.
Thyroprotein	1
Ti-Cal	649
Tidolith	1119
Timonex	1087
Tin	1113
Tin Chloride	971
Tin Oxide	719
Tinctures	829
Tintite	509
Ti-Sil	649
Titanium Dioxide	373
Titanium Tetrachloride	1047
Titanox	1097
Ti-Tone	649
Ti-Tree Oil	101
Tocopherol	943
Tollac	781
Tolu Balsam Oil	377
Toluene	135
p-Toluene Sulfone Chloride	749
o-Toluidine	639
Toluol	623
Toluquinone	731
Toners	1039
Tonsil	935
Tornesit	557
Triacetin	793
Triamylamine	975
Tributyl Citrate	303
Tributyl Phosphite	301
Tributyrin	1083
Trichlorethylene	393
Trichlorobutyl Alcohol	1139
Triclene	393
Tricresol	749
Tricresyl Phosphate	523
Triethanolamine	215
Triethanolamine Lactate	509
Triethanolamine Naphthenate	509
Triethanolamine Oleate	509
Triethanolamine Phthalate	509
Triethanolamine Stearate	509
Triethylene Glycol Di-2-Ethyl Hexoate	215
Triethylene Glycol Ester of Hydrogenated Rosin	557
Trigamine	509
Trigamine Stearate	509
Trihydroxyethylamine See Triethanolamine	
Trikalin	509
Trimethylene Glycol	873
Triphenylguanidine	393
Triphenylphosphate	749
Tripoli	1075
Trisodium Phosphate	1153
Tritolyl Phosphate	750
Triton	915
Triton NE	915
Troluoil	77
Truline Binder	557
Tuads	1141

Product	No.
Tung Oil	See China Wood Oil
Tungsten	437
Tunguran A	11
Turkelene	509
Turkerol	509
Turkey Red Oil	773
Turmeric	837
Turpentine	83
Turpentine Substitute	77
Turpentine (Venice)	777
Turtle Oil	967
Tween	103
Twitchell Base	427
Tysenite	1141

U

Product	No.
Uformite	895
Ultramarine Blue	1039
Ultranate	97
Ultrasene	97
Ultravon	277
Umbers	441
Undecalactone	1139
Undecylenic Acid	779
Unilith	1119
Union Solvent	1111
Unyte	861
Uranium Nitrate	545
Uranyl Nitrate	697
Urea	981
Urea-Formaldehyde Resin	45
Ureka C	925
Ursulin	45
Uversol	545

V

Product	No.
Valex	193
Vanadium Pentoxide	1136
Van Dyke Brown	1181
Vandex	1141
Vanilla Beans	1093
Vanillal	993
Vanillin	969
Vanzyme	1141
Varcrex	211
Varcum	1147
Varnish	757
Varnish Gums and Resins	45
Varnolene	1033
Varsol	1033
Vaso	1157
Vat Colors	29
Vatsol	45
Veegum	1141
Vegetable Colors	885
Vegetable F Wax	1055
Velsicol 1068	1148
Verdigris	381
Vermiculite	567
Vermilion	441
Victron	1131

CHEMICALS AND SUPPLIES: WHERE TO BUY THEM

Product	No.
Vinapas	11
Vinsol	557
Vinyl Acetate	793
Vinyl Chloride	215
Vinylite	215
Virifoam	509
Viscogum	509
Viscoloid	393
Vistac	11
Vistanex	11
Vitamins	731
Vitroil	See Sulfuric Acid
V. M. P. Naphtha	1033
Volclay	43
Vultex	1161

W

Product	No.
Water Glass	See Sodium Silicate
Wax L33	509
Wax, Microcrystalline	1206
Wax, Synthetic	509
Wetanol	509
Wetting Out Agents	509
Whale Oil	1171
White Arsenic	857
White Lead	771
White Oil	See Mineral Oil
White Wax	See Beeswax
Whitex Clay	751
Whiting	299
Witch Hazel Extract	365
Witco #1	1193
Witco Yellow	1193
Wood Flour	675
Wood Oil	See China Wood Oil
Woodruff Essence	595
Wool Fat	See Lanolin
Wool Wax	165
Wool Wax Alcohols	42
Wyo-Jel	1203

X

Product	No.
X-13	483
Xerol	471
Xylene	See Xylol
Xylerol	509
Xylol	135
Xynomine	821

Y

Product	No.
Yeast	1027
Yeast Extract	1027
Yelkin	921
Yellow Wax	See Beeswax
Ylang Ylang	451
Yumidol	509

Z

Product	No.
Zein	59
Zelan	393
Zenite	393
Zikol	791
Zimate	1141
Zinc	553
Zinc Carbonate	1193
Zinc Chloride	1193
Zinc Chromate	1163
Zinc Fluoborate	483
Zinc Fluosilicate	1060
Zinc Lactate	85
Zinc Oleate	545
Zinc Oxide	731
Zinc Peroxide	731
Zinc Phosphide	817
Zinc Resinate	791
Zinc Silicofluoride	545
Zinc Stearate	731
Zinc Sulfate	927
Zinc Sulfide	785
Zinc Undecylenate	142
Zinol	791
Zirconium	1095
Zirconium Hydrate	245
Zirconium Oxide	453
Zirconium Oxychloride	1095
Zirex	791
Zopaque	259

SELLERS OF CHEMICALS AND SUPPLIES*

No.	Name	Address
1.	Abbott Laboratories	North Chicago, Ill.
5.	Acheson Colloids Corp.	Port Huron, Mich.
7.	Acheson Graphite Corp.	Niagara Falls, N. Y.
9.	Acme Oil Corp.	Chicago, Ill.
11.	Advance Solvents & Chem. Corp.	Jersey City, N. J.
13.	Ajax Metal Co.	Philadelphia, Pa.
15.	Aktivin Corp.	New York, N. Y.
16.	Algin Corp. of America	New York, N. Y.
17.	Allied Asphalt & Mineral Corp.	New York, N. Y.
21.	Aluminum Co. of America	Pittsburgh, Pa.
23.	Amecco Chemicals, Inc.	New York, N. Y.
25.	American Active Carbon Co.	Columbus, O.
27.	American Agar Co., Inc.	San Diego, Calif.
28.	American Agricultural Chemical Co.	New York, N. Y.
29.	American Aniline Products, Inc.	New York, N. Y.
31.	American-Brit. Chem. Supplies, Inc.	New York, N. Y.
33.	American Catalin Corp.	New York, N. Y.
35.	American Cellulose Co.	Indianapolis, Ind.
37.	American Chemical Paint Co.	Ambler, Pa.
41.	American Chlorophyll, Inc.	New York, N. Y.
42.	American Cholesterol Products, Inc.	Milltown, N. J.
43.	American Colloid Co.	Chicago, Ill.
45.	American Cyanamid & Chem. Co.	New York, N. Y.
47.	American Dyewood Co.	New York, N. Y.
48.	American Firstoline Corp.	New York, N. Y.
49.	American Fluoride Corp.	New York, N. Y.
51.	American Insulator Corp.	New Freedom, Pa.
53.	American Lanolin Corp.	Lawrence, Mass.
57.	American Luminous Products Co.	Huntington Park, Calif.
59.	American Maize Products Co.	New York, N. Y.
61.	American Metal Co.	New York, N. Y.
63.	American Mineral Spirit Co.	New York, N. Y.
64.	American Molasses Co.	New York, N. Y.
65.	American Plastics Corp.	New York, N. Y.
67.	American Potash & Chem. Corp.	New York, N. Y.
69.	American Smelting & Refining Co.	New York, N. Y.
71.	American Zinc Co.	New York, N. Y.
73.	Amido Products Co.	New York, N. Y.
77.	Anderson Prichard Oil Corp.	Oklahoma City, Okla.
79.	Ansbacher-Siegle Corp.	Rosebank, N. Y.
81.	Ansul Chemical Co.	Marinette, Wis.
83.	Antwerp Naval Stores Co., Inc.	Boston, Mass.
85.	Apex Chemical Co.	New York, N. Y.
86.	Arabol Mfg. Co.	New York, N. Y.
87.	Archer-Daniels-Midland Co.	Minneapolis, Minn.
89.	Arkansas Co.	Newark, N. J.
91.	Armour & Co.	Chicago, Ill.
91a.	Armstrong Inc., C. M.	New York, N. Y.
92.	Aromatic Products, Inc.	New York, N. Y.
93.	Asbury Graphite Mills	Asbury Park, N. J.

*Just when this list was ready to go to press, it was carefully checked and it was found that some of the firms listed went out of business. Their names have been cancelled, and this accounts for the discontinuity in numbering.

SELLERS OF CHEMICALS AND SUPPLIES

No.	Name	Address
95.	Atlantic Gelatine Co.	Woburn, Mass.
97.	Atlantic Refining Co.	Philadelphia, Pa.
99.	Atlantic Research Associates	Newtonville, Mass.
103.	Atlas Powder Co.	Wilmington, Del.
109.	Badcock, Robert & Co.	New York, N. Y.
111.	Bakelite Corp.	New York, N. Y.
113.	Baker Castor Oil Corp.	Jersey City, N. J.
115.	Baker & Co., Inc.	Newark, N. J.
117.	Baker, Franklin Co.	Hoboken, N. J.
119.	Baker, H. J. & Bro.	New York, N. Y.
121.	Baker, J. E., Co.	York, Pa.
123.	Baker, J. T. Chem. Co.	Philipsburg, N. J.
125.	Balfour, Guthrie & Co., Ltd.	New York, N. Y.
127.	Barada & Page, Inc.	Kansas City, Mo.
129.	Barber Asphalt Co.	Philadelphia, Pa.
130.	Barium Chemicals, Inc.	Willoughby, O.
131.	Barium Reduction Corp.	Charleston, W. Va.
133.	Barnsdall Tripoli Corp.	Seneca, Mo.
135.	Barrett Co.	New York, N. Y.
139.	Battelle & Renwick	New York, N. C.
141.	Battleboro Oil Co.	Battleboro, N. C.
142.	Beacon Co.	Boston, Mass.
143.	Beck, Koller & Co.	Detroit, Mich.
145.	Belmont Smelting & Refining Wks.	Brooklyn, N. Y.
149.	Benzol Products Co.	Newark, N. J.
149a.	Berk & Co. Inc., F. W.	Wood Ridge, N. J.
150.	Bernard Color & Chem. Co.	New York, N. Y.
151.	Bersworth Labs., F. C.	Framingham, Mass.
153.	Beryllium Corp. of America	New York, N. Y.
155.	Bick & Co., Inc.	Reading, Pa.
157.	Bilhuber-Knoll Corp.	New York, N. Y.
159.	Binney & Smith	New York, N. Y.
161.	Bisbee Linseed Co.	Philadelphia, Pa.
165.	Bopf-Whittam Corp.	Linden, N. J.
167.	Borax Union, Inc.	San Francisco, Calif.
169.	Borne Scrymser Co.	New York, N. Y.
171.	Bowdlear Co., W. H.	Syracuse, N. Y.
173.	Bowker Chemical Corp.	New York, N. Y.
175.	Bradley & Baker	New York, N. Y.
177.	Brazil Oiticica, Inc.	New York, N. Y.
179.	British Drug Houses, Ltd.	London, England
181.	British Xylonite Co.	London, England
183.	Brooke, Fred L. Co.	Chicago, Ill.
184.	Brown Co.	Portland, Me.
185.	Brush Beryllium Co.	Cleveland, O.
187.	Buffalo Electro Chem. Co.	Buffalo, N. Y.
189.	Burkard-Schier Chem. Co.	Chattanooga, Tenn.
191.	Buromin Corp.	Pittsburgh, Pa.
193.	Bush, W. J. & Co., Inc.	New York, N. Y.
194.	Byerlyte Corp.	Cleveland, O.
197.	Cabot, Godfrey L., Inc.	Boston, Mass.
199.	Calcium Sulfide Corp.	Damascus, Va.
201.	Calco Chemical Co.	Bound Brook, N. J.
203.	Calgon, Inc.	Pittsburgh, Pa.
205.	Calif. Fruit Growers Exchange	Ontario, Calif.
206.	Calif. Milk Products Co.	Philadelphia, Pa.
207.	Campbell, C. W. Co., Inc.	New York, N. Y.
209.	Campbell, John & Co.	New York, N. Y.
211.	Campbell Rex & Co.	London, England
213.	Carbic Color & Chemical Co.	New York, N. Y.
215.	Carbide & Carbon Chem. Corp.	New York, N. Y.
219.	Carbolineum Wood Preserving Co.	Milwaukee, Wis.

No.	Name	Address
221.	Carborundum Co.	Niagara Falls, N. Y.
223.	Carey, Philip Co.	Lockland, Ohio
225.	Carus Chem. Co., Inc.	La Salle, Ill.
229.	The Casein Mfg. Co. of Amer., Inc.	New York, N. Y.
231.	Celanese Corp.	New York, N. Y.
239.	Central Scientific Co.	Chicago, Ill.
241.	Century Stearic Acid & Candle Wks.	New York, N. Y.
245.	Ceramic Color & Chem. Mfg. Co.	New Brighton, Pa.
247.	Cerro de Pasco Copper Corp.	New York, N. Y.
249.	Champion Paper & Fiber Co.	Canton, N. C.
251.	Chaplin-Bibbo	New York, N. Y.
253.	Chazy Marble Lime Co., Inc.	Chazy, N. Y.
255.	Chesebrough Mfg. Co.	New York, N. Y.
257.	Chemical & Pigment Co.	Baltimore, Md.
263.	Chemical Solvents, Inc.	New York, N. Y.
265.	Cheney Chem. Co.	Cleveland, Ohio
266.	Chew, John A., Inc.	New York, N. Y.
267.	Chicago Apparatus Co.	Chicago, Ill.
269.	Chicago Copper & Chem. Co.	Blue Island, Ill.
271.	Chipman Chem. Co., Inc.	Bound Brook, N. J.
272.	Chiris, Antoine Co.	New York, N. Y.
273.	Chrystal, Charles B. Co., Inc.	New York, N. Y.
275.	Church & Dwight Co., Inc.	New York, N. Y.
277.	Ciba Co., Inc.	New York, N. Y.
279.	Cinelin Co.	Indianapolis, Ind.
281.	Clarke, John & Co.	New York, N. Y.
283.	The Cleveland-Cliffs Iron Co.	Cleveland, Ohio
285.	Climax Molybdenum Co.	New York, N. Y.
287.	Clinton Co.	Clinton, Ia.
289.	Coleman & Bell Co.	Norwood, Ohio
293.	Colgate-Palmolive-Peet Co.	Jersey City, N. J.
295.	Colledge, E. W., Inc.	Cleveland, Ohio
297.	Colonial Beacon Oil Co.	Everett, Mass.
299.	Columbia Alkali Corp.	New York, N. Y.
301.	Commercial Solvents Corp.	New York, N. Y.
305.	Commonwealth Color & Chem. Co.	Brooklyn, N. Y.
307.	Compagnie Duval.	New York, N. Y.
309.	Conewango Refining Co.	Warren, Pa.
311.	Consolidated Chem. Sales Corp.	Newark, N. J.
313.	Consolidated Feldspar Corp.	Trenton, N. J.
315.	Conti Products Corp.	New York, N. Y.
317.	Continental Diamond Fiber Co.	Bridgeport, Pa.
319.	Continental Oil Co.	Ponca City, Okla.
321.	Cook Swan Co., Inc.	New York, N. Y.
325.	Coopers Creek Chem. Co.	W. Conshohocken, Pa.
327.	Corn Products Refining Co.	New York, N. Y.
329.	Cowles Detergent Co.	Cleveland, Ohio
330.	C. P. Chemical Solvents, Inc.	New York, N. Y.
331.	Croton Chem. Corp.	Brooklyn, N. Y.
333.	Crowley Tar Products Co.	New York, N. Y.
335.	Crystal, Charles B. Co., Inc.	New York, N. Y.
336.	Crystal Soap & Chem. Co.	Philadelphia, Pa.
337.	Cudahy Packing Co.	Chicago, Ill.
339.	Danco, Gerard J.	New York, N. Y.
341.	Darco Corp.	New York, N. Y.
343.	Darling & Co.	Chicago, Ill.
345.	Davison Chem. Corp.	Baltimore, Md.
347.	Deep Rock Oil Corp.	Chicago, Ill.
349.	De Lore, C. P. Co.	St. Louis, Mo.
351.	Delta Chem. Mfg. Co.	Baltimore, Md.
353.	Delta Chem. & Iron Co.	Wells, Mich.
355.	Denver Fire Clay Co.	Denver, Colo.

SELLERS OF CHEMICALS AND SUPPLIES 421

No.	Name	Address
357.	Devoe & Reynolds Co.	New York, N. Y.
359.	Dewey & Almy Chem. Co.	Boston, Mass.
363.	Diamond Alkali Co.	Pittsburgh, Pa.
364.	Dicalite Co.	New York, N. Y.
365.	Dickinson, E. E. Co.	Essex, Conn.
367.	Dickinson, J. Q. & Co.	Malden, W. Va.
369.	Difco Laboratories, Inc.	Detroit, Mich.
371.	Digestive Ferments Co.	Detroit, Mich.
373.	Marshall Dill.	San Francisco, Calif.
375.	Distributing & Trading Co.	New York, N. Y.
377.	Dodge & Olcott Co.	New York, N. Y.
379.	Dow Chemical Co.	Midland, Mich.
380.	Dow Corning Corp.	Midland, Mich.
381.	Drakenfeld, B. F. & Co.	New York, N. Y.
382.	Drew, E. F. Co.	New York, N. Y.
383.	Dreyer, P. R. Co.	New York, N. Y.
385.	Dreyfus Co., L. A.	Rosebank, N. Y.
387.	Drury, A. C. & Co., Inc.	Chicago, Ill.
389.	Ducas, B. P. Co.	New York, N. Y.
391.	Duche, T. M. & Sons.	New York, N. Y.
393.	DuPont, E. I., de Nemours & Co., Inc.	Wilmington, Del.
397.	Durite Plastics.	Philadelphia, Pa.
401.	Eagle-Picher Lead Co.	Cincinnati, Ohio
403.	Eakins, J. S. & W. R., Inc.	Brooklyn, N. Y.
405.	Earle Bros.	New York, N. Y.
407.	Eastman Kodak Co.	Rochester, N. Y.
409.	Economic Materials Co.	Chicago, Ill.
411.	Edwal Labs.	Chicago, Ill.
413.	Eff Laboratories Inc.	Cleveland, Ohio
415.	Egyptian Lacquer Co.	Kearney, N. J.
417.	Eimer & Amend.	New York, N. Y.
421.	Elbert & Co.	New York, N. Y.
423.	Electro Bleaching Gas Co.	New York, N. Y.
425.	Electro-Metallurgical Co.	New York, N. Y.
427.	Emery Industries, Inc.	Cincinnati, Ohio
429.	Empire Distilling Corp.	New York, N. Y.
429a.	Emulsol Corp.	Chicago, Ill.
430.	Enco Chemical Corp.	New York, N. Y.
431.	Enterprise Animal Oil Co.	Philadelphia, Pa.
431a.	Essential Aromatics Corp.	New York, N. Y.
432.	Esso Marketers.	New York, N. Y.
432a.	Fairmount Chemical Co.	Newark, N. J.
433.	Fales Chem. Co., Inc.	Cornwall Landing, N. Y.
435.	Falk & Co.	Pittsburgh, Pa.
437.	Fansteel Metallurgical Corp.	No. Chicago, Ill.
439.	Felton Chemical Co.	Brooklyn, N. Y.
441.	Fezandie & Sperrle, Inc.	New York, N. Y.
443.	Fiberloid Corp.	Indian Orchard, Mass.
445.	Filtrol Co.	Los Angeles, Calif.
445a.	Fine Organics, Inc.	New York, N. Y.
446.	Firestone Tire & Rubber Co.	Akron, Ohio
447.	Fishbeck, Chas. Co.	New York, N. Y.
449.	Fisher Scientific Co.	Pittsburgh, Pa.
451.	Florasynth Laboratories	New York, N. Y.
453.	Foote Mineral Co.	Philadelphia, Pa.
455.	Formica Insulation Co.	Cincinnati, Ohio
457.	Fougera, E. & Co.	New York, N. Y.
459.	France, Campbell & Darling.	Kenilworth, N. J.
461.	Franco-American Chemical Wks.	Carlstadt, N. J.
463.	Frank-Vliet Co.	New York, N. Y.
465.	Franks Chem. Products Co., Inc.	Brooklyn, N. Y.
467.	French Potash Co.	New York, N. Y.

No.	Name	Address
469.	Fries, Alex. & Bro.	Cincinnati, Ohio
471.	Fries Bros.	New York, N. Y.
473.	Fritzchie Bros.	New York, N. Y.
475.	Garrigues, Stewart & Davies, Inc.	New York, N. Y.
476.	Gartenberg, H. & Co., Inc.	Chicago, Ill.
477.	Geigy Co., Inc.	New York, N. Y.
479.	General Aniline Works, Inc.	New York, N. Y.
481.	General Atlas Carbon Co.	New York, N. Y.
483.	General Chemical Co.	New York, N. Y.
485.	General Drug Co.	New York, N. Y.
487.	General Dyestuffs Corp.	New York, N. Y.
489.	General Electric Co.	Pittsfield, Mass.
491.	General Electric Co.	Schenectady, N. Y.
493.	General Magnesite & Magnesia Co.	Philadelphia, Pa.
495.	General Naval Stores Co.	New York, N. Y.
497.	General Plastics Corp.	London, England
499.	General Plastics, Inc.	No. Tonawanda, N. Y.
499a.	General Refractories Co.	Philadelphia, Pa.
500.	Georgia Kaolin Co.	Elizabeth, N. J.
501.	Girdler Corp.	Louisville, Ky.
503.	Givaudan-Delawanna, Inc.	New York, N. Y.
505.	Glidden Co.	Cleveland, Ohio
507.	Globe Chem. Co.	Cincinnati, Ohio
509.	Glyco Products Co., Inc.	Brooklyn, N. Y.
513.	Goldschmidt Corp.	New York, N. Y.
514.	Goldsmith Bros. Smelt. & Refining Co.	New York, N. Y.
515.	Goodrich, B. F. Co.	Akron, Ohio
517.	Goodyear Tire & Rubber Co.	Akron, Ohio
519.	Grasselli Chemical Co.	Cleveland, Ohio
521.	Gray, W. S. Co.	New York, N. Y.
523.	Greeff, R. W. & Co.	New York, N. Y.
525.	Griffith Laboratories.	Chicago, Ill.
527.	Gross, A. & Co.	New York, N. Y.
529.	Hall, C. P. & Co.	Akron, Ohio
531.	Halowax Corp.	New York, N. Y.
533.	Hammill & Gillespie, Inc.	New York, N. Y.
535.	Hamilton, A. K.	New York, N. Y.
537.	Hammond Drierite Co.	Yellow Springs, Ohio
539.	Handy & Harman.	New York, N. Y.
540.	Hardesty Chemical Co.	New York, N. Y.
541.	Hardy, Charles, Inc.	New York, N. Y.
543.	Harrison Mfg. Co.	Rahway, N. J.
545.	Harshaw Chemical Co.	Cleveland, Ohio
547.	Hart Products Corp.	New York, N. Y.
549.	Haskelite Mfg. Corp.	Chicago, Ill.
551.	Haveg Corp.	Newark, Del.
553.	Hegeler Zinc Co.	Danville, Ill.
555.	Heine & Co.	New York, N. Y.
557.	Hercules Powder Co.	New York, N. Y.
561.	Heveatex Corp.	Melrose, Mass.
563.	Hewitt, C. B. & Bro.	New York, N. Y.
565.	Heyden Chemical Corp.	New York, N. Y.
567.	Hill Bros. Chemical Co.	Los Angeles, Calif.
569.	Hillside Fluor Spar Mines.	Chicago, Ill.
570.	Hilo Varnish Co.	Brooklyn, N. Y.
571.	Holland Aniline Dye Co.	Holland, Mich.
573.	Hommel, O. Co.	Pittsburgh, Pa.
575.	Hooker Electro-Chemical Co.	New York, N. Y.
577.	Hopkins, J. L. & Co.	New York, N. Y.
579.	Hord Color Products.	Sandusky, Ohio
581.	Horn Jefferys & Co.	Burbank, Calif.
583.	Horner, James B., Inc.	New York, N. Y.

SELLERS OF CHEMICALS AND SUPPLIES

No.	Name	Address
585.	Huisking, Chas. L. & Co., Inc.	New York, N. Y.
587.	Hummel Chemical Co., Inc.	New York, N. Y.
589.	Hurst, Adolph & Co., Inc.	New York, N. Y.
591.	Hutchinson, D. W. & Co., Inc.	New York, N. Y.
593.	Hycar Corp.	Akron, Ohio
595.	Hymes, Lewis Associates.	New York, N. Y.
599.	Imperial Chem. Industries.	London, England
601.	Industrial Chem. Sales Co.	New York, N. Y.
602.	Inland Alkaloid Co.	Tipton, Ind.
603.	Innes, O. G. Corp.	New York, N. Y.
605.	Innis Speiden Co.	New York, N. Y.
607.	International Pulp Corp.	New York, N. Y.
609.	International Selling Corp.	New York, N. Y.
611.	Interstate Color Co., Inc.	New York, N. Y.
613.	Iowa Soda Products Co.	Council Bluffs, Ia.
615.	Jackson, L. N. & Co.	New York, N. Y.
617.	Jacobson, C. A., W. Va. Univ.	Morgantown, W. Va.
619.	Jennison-Wright Co.	Toledo, Ohio
621.	Johns-Manville Corp.	New York, N. Y.
623.	Jones & Laughlin Steel Corp.	Pittsburgh, Pa.
625.	Jones, S. L. & Co.	San Francisco, Calif.
629.	Kali Mfg. Co.	Philadelphia, Pa.
633.	Kay Fries Chem., Inc.	New York, N. Y.
635.	Kelco Co.	San Diego, Calif.
637.	Kentucky Clay Mining Co.	Mayfield, Ky.
639.	Kentucky Color & Chem. Co.	Louisville, Ky.
641.	Kessler Chem. Corp.	Philadelphia, Pa.
643.	Kinetic Chem., Inc.	Wilmington, Del.
645.	Kohnstamm, H. & Co.	New York, N. Y.
646.	Kolker Chem. Works.	Newark, N. J.
647.	Koppers Products Co.	Pittsburgh, Pa.
648.	Kranich Soap Co.	Brooklyn, N. Y.
649.	Krebs Pigment & Color Corp.	Newark, N. J.
651.	Kuhlman, Establs.	Paris, France
655.	Lattimer-Goodwin Chem. Co.	Grand Junction, Ohio
657.	Laxseed Co.	New York, N. Y.
659.	Leghorn Trading Co., Inc.	New York, N. Y.
661.	Lehn & Fink Corp.	New York, N. Y.
663.	Leonhard Wax, Theo. Co., Inc.	Haledon, Paterson, N. J.
665.	Lewis, C. H. & Co.	New York, N. Y.
667.	Lewis, John D., Inc.	Providence, R. I.
669.	Limestone Products Corp. of America.	Newton, N. J.
671.	Lincks, Geo. H.	New York, N. Y.
672.	Lindsay Light & Chem. Co.	Chicago, Ill.
673.	Liquid Carbonic Corp.	Chicago, Ill.
675.	Litter, D. H., Co.	New York, N. Y.
677.	Littlejohn & Co., Inc.	New York, N. Y.
679.	Lucidol Corp.	Buffalo, N. Y.
681.	Lueders, Geo. & Co.	New York, N. Y.
683.	Lundt & Co.	New York, N. Y.
685.	Maas & Waldstein.	Newark, N. J.
687.	MacAndrews & Forbes Co.	New York, N. Y.
689.	Mackay, A. D.	New York, N. Y.
691.	Magnetic Pigment Co.	New York, N. Y.
693.	Magnus, Mabee & Reynard, Inc.	New York, N. Y.
695.	Makalot Corp.	Boston, Mass.
697.	Mallinckrodt Chemical Works.	St. Louis, Mo.
699.	Malt Diatase Co.	Brooklyn, N. Y.
701.	Manchester Oxide Co.	Manchester, England
702.	Marathon Corp.	Rotschild, Wis.
703.	Marbon Corp.	Gary, Ind.
705.	Marine Magnesium Prod. Corp.	S. San Francisco, Calif.

No.	Name	Address
707.	Martin Dennis Co.	Newark, N. J.
709.	Martin, L. Co.	New York, N. Y.
711.	Martin Laboratories	New York, N. Y.
713.	Mathieson Alkali Co.	New York, N. Y.
715.	Maywood Chemical Works	Maywood, N. J.
717.	McCormick & Co.	Baltimore, Md.
719.	McGean Chem. Co.	Cleveland, Ohio
721.	McKesson & Robbins, Inc.	New York, N. Y.
723.	McLaughlin, Gormley, King & Co.	Minneapolis, Minn.
724.	Mead Corp.	Chillicothe, Ohio
725.	Mearl Corp.	New York, N. Y.
727.	Mechling Bros. Chem Co.	Camden, N. J.
731.	Merck & Co.	Rahway, N. J.
734.	Metal & Thermit Corp.	New York, N. Y.
735.	Metasap Chem. Co.	Harrison, N. J.
737.	Metro-Nite Co.	Milwaukee, Wis.
739.	Meyer & Sons, J.	Philadelphia, Pa.
741.	Mica Insulator Co.	New York, N. Y.
743.	Michel Export Co.	New York, N. Y.
745.	Michigan Alkali Co.	New York, N. Y.
747.	Miller, Carl F. Co.	Seattle, Wash.
749.	Monsanto Chem. Works	St. Louis, Mo.
750.	Montrose Chemical Corp.	Newark, N. J.
751.	Moore-Munger, Inc.	New York, N. Y.
753.	Morningstar, Nicol, Inc.	New York, N. Y.
755.	Morton Salt Co.	Chicago, Ill.
757.	Murphy Varnish Co.	Newark, N. J.
759.	Mutual Chem. Co. of America	New York, N. Y.
763.	Mutual Citrus Products Co.	Anaheim, Calif.
765.	National Aluminate Corp.	Chicago, Ill.
767.	National Ammonia Co., Inc.	Philadelphia, Pa.
769.	National Aniline & Chem. Wks.	New York, N. Y.
771.	National Lead Co.	New York, N. Y.
773.	National Oil Products Co.	Harrison, N. J.
775.	National Pigments & Chem. Co.	St. Louis, Mo.
777.	National Rosin Oil & Size Co.	New York, N. Y.
779.	Naugatuck Chem. Co.	Naugatuck, Conn.
780.	Naylee Chemical Co.	Philadelphia, Pa.
781.	Neville Co.	Pittsburgh, Pa.
783.	N. J. Laboratory Supply Co.	Newark, N. J.
785.	N. J. Zinc Co.	New York, N. Y.
789.	Newmann-Buslee & Wolfe, Inc.	Chicago, Ill.
791.	Newport Industries, Inc.	New York, N. Y.
792.	New York Quinine & Chem. Wks., Inc.	Brooklyn, N. Y.
793.	Niacet Chem. Co.	Niagara Falls, N. Y.
795.	Niagara Alkali Co.	New York, N. Y.
797.	Niagara Chemicals Corp.	Niagara Falls, N. Y.
798.	Niagara Chlorine Products Co.	Lockport, N. Y.
799.	Niagara Smelting Corp.	Niagara Falls, N. Y.
800.	Norda Essential Oil & Chem. Co.	New York, N. Y.
801.	Northwestern Chem. Co.	Wauwatosa, Wis.
803.	Norton Co.	Worcester, Mass.
805.	Norwich Pharmacal Co.	Norwich, N. Y.
807.	Novadel-Agene Corp.	Newark, N. J.
809.	Nulomoline Co.	New York, N. Y.
811.	Nuodex Products, Inc.	Elizabeth, N. J.
813.	Ohio-Apex, Inc.	Nitro, W. Va.
815.	Oil States Petroleum Co.	New York, N. Y.
817.	Oldbury Electro-Chem. Co.	New York, N. Y.
819.	Olive Branch Minerals Co.	Cairo, Ill.
821.	Onyx Oil & Chem. Co.	Jersey City, N. J.
823.	Orbis Products Corp.	New York, N. Y.

SELLERS OF CHEMICALS AND SUPPLIES 425

No.	Name	Address
824.	Osborn, C. J. Co.	New York, N. Y.
824a.	Owens-Corning Fiberglas Corp.	New York, N. Y.
825.	Papermakers' Chem. Corp.	Wilmington, Del.
827.	Paramet Chem. Corp.	Long Island City, N. Y.
829.	Parke, Davis & Co.	Detroit, Mich.
831.	Parker Rust Proof Co.	Detroit, Mich.
833.	Patent Chemicals, Inc.	New York, N. Y.
835.	Peek & Velsor, Inc.	New York, N. Y.
837.	Penick, S. B. & Co.	New York, N. Y.
839.	Penn. Alcohol Corp.	Philadelphia, Pa.
841.	Penn. Coal Products Co.	Petrolia, Pa.
843.	Penn.-Dixie Cement Corp.	New York, N. Y.
845.	Penn. Industrial Chem. Corp.	Clairton, Pa.
847.	Penn. Refining Co.	Butler, Pa.
849.	Penn. Salt Mfg. Co.	Philadelphia, Pa.
850.	Petrolite Corp.	New York, N. Y.
851.	Pfaltz-Bauer, Inc.	New York, N. Y.
853.	Pfizer, Chas. & Co., Inc.	New York, N. Y.
855.	Phila. Quartz Co.	Philadelphia, Pa.
857.	Philipp Bros.	New York, N. Y.
859.	Pittsburgh Plate Glass Co.	Pittsburgh, Pa.
861.	Plaskon Corp.	Toledo, Ohio
863.	Plymouth Organic Labs.	New York, N. Y.
867.	Powhatan Mining Corp.	Woodlawn, Baltimore, Md.
869.	Pray, W. P.	New York, N. Y.
870.	Prentiss, R. J. Co.	New York, N. Y.
871.	Prior Chem. Corp.	New York, N. Y.
873.	Procter & Gamble Co.	Cincinnati, Ohio
875.	Provident Chem. Wks.	St. Louis, Mo.
877.	Publicker, Inc.	Philadelphia, Pa.
879.	Pure Calcium Products Co.	Painesville, Ohio
881.	Pylam Products Co.	New York, N. Y.
883.	Quaker Oats Co.	Chicago, Ill.
885.	Ransom, L. E. Co.	New York, N. Y.
885a.	Rare Metal Products Co.	Belleville, N. J.
886.	Republic Chemical Corp.	New York, N. Y.
887.	Reynolds Metals Co., Inc.	New York, N. Y.
889.	Read, Chas. L. & Co., Inc.	New York, N. Y.
891.	Reichhold Chemicals, Inc.	Detroit, Mich.
893.	Reilly Tar & Chem. Corp.	Indianapolis, Ind.
894.	Republic Chem. Corp.	New York, N. Y.
895.	Resinous Prod. & Chem. Co.	Philadelphia, Pa.
897.	Resinox Corp.	New York, N. Y.
899.	Revertex Corp.	Brooklyn, N. Y.
901.	Revson, R. F. Co.	New York, N. Y.
903.	Rhone-Poulenc, Inc.	Paris, France
905.	Richards Chem. Works	Jersey City, N. J.
907.	Riverside Chem. Co.	No. Tonawanda, N. Y.
909.	Robeson Process Co.	New York, N. Y.
911.	Rochester Gas & Elec. Corp.	Rochester, N. Y.
913.	Rogers & McClellan	Boston, Mass.
915.	Rohm & Haas	Philadelphia, Pa.
917.	Rosenthal-Bercow Co.	New York, N. Y.
919.	Ross, Frank B. Co., Inc.	New York, N. Y.
921.	Ross-Rowe, Inc.	New York, N. Y.
923.	Royce Chem. Co.	Carlton Hill, N. J.
925.	Rubber Service Labs. Co.	Akron, Ohio
927.	Russell, W. R. & Co.	New York, N. Y.
929.	Russia Cement Co.	Gloucester, Mass.
931.	Ryland, H. C., Inc.	New York, N. Y.
933.	Saginaw Salt Products Co.	Saginaw, Mich.
935.	Salomon, L. A. & Bro.	New York, N. Y.

No.	Name	Address
939.	Sandoz Chem. Works	New York, N. Y.
941.	Scheel, Wm. H.	New York, N. Y.
943.	Schering Corp.	Bloomfield, N. J.
946.	Schieffelin & Co.	New York, N. Y.
947.	Schimmel & Co.	New York, N. Y.
951.	Schofield-Daniel Co.	New York, N. Y.
953.	Scholler Bros., Inc.	Philadelphia, Pa.
955.	Schundler, F. E. & Co.	Joliet, Ill.
957.	Schuylkill Chem. Co.	Philadelphia, Pa.
959.	Schwabacher, S. & Co., Inc.	New York, N. Y.
961.	Scientific Glass Apparatus Co.	Bloomfield, N. J.
965.	Seacoast Laboratories	New York, N. Y.
967.	Edwin Seebach Co.	New York, N. Y.
969.	Seeley & Co., Inc.	New York, N. Y.
971.	Seldner & Enequist, Inc.	Brooklyn, N. Y.
973.	Serinsky, Moses Co.	Indianapolis, Ind.
975.	Sharples Chemicals, Inc.	Philadelphia, Pa.
977.	Shawinigan, Ltd.	New York, N. Y.
978.	Shell Chem. Co.	New York, N. Y.
979.	Shepherd Chem. Co.	Norwood, Cincinnati, O.
981.	Sherka Chem. Co., Inc.	Bloomfield, N. J.
982.	Sherwin-Williams Co.	Cleveland, Ohio
983.	Sherwood Petroleum Co.	Englewood, N. J.
985.	Shields, Thomas J. Co.	New York, N. Y.
987.	Siemon Colors, Inc.	Newark, N. J.
989.	Siemon & Co.	Bridgeport, Conn.
991.	Silica Products Co.	Kansas City, Mo.
993.	Silver, Geo., Import Co.	New York, N. Y.
994.	Simons, Harold L. Co.	Long Island City, N. Y.
995.	Sinclair Refining Co.	Olmstead, Ill.
997.	Skelly Oil Co.	Chicago, Ill.
999.	Smith Chem. & Color Co.	Brooklyn, N. Y.
1001.	Smith & Nichols, Inc.	New York, N. Y.
1003.	Smith, Werner G. Co.	Cleveland, Ohio
1004.	Socony-Vacuum Co.	New York, N. Y.
1005.	Solvay Sales Corp.	New York, N. Y.
1007.	Sonneborn, L., Sons.	New York, N. Y.
1009.	Southern Mica Co.	Franklin, N. C.
1011.	Southern Pine Chem. Co.	Jacksonville, Fla.
1013.	Southwark Mfg. Co.	Camden, N. J.
1015.	Sparhawk Co.	Sparkhill, N. Y.
1017.	Spencer Kellogg & Sons Sales Corp.	Buffalo, N. Y.
1019.	Staley, A. E. Mfg. Co.	Decatur, Ill.
1021.	Stamford Rubber Supply Co.	Stamford, Conn.
1023.	Stanco Distributors.	New York, N. Y.
1025.	Standard Alcohol Co.	New York, N. Y.
1027.	Standard Brands, Inc.	New York, N. Y.
1027.	Standard Oil Co. of Calif.	San Francisco, Cal.
1031.	Standard Oil Co. of Indiana	Chicago, Ill.
1033.	Standard Oil Co. of N. J.	New York, N. Y.
1035.	Standard Oil Co. of N. Y.	New York, N. Y.
1037.	Standard Silicate Co.	Pittsburgh, Pa.
1039.	Standard Ultramarine Co.	Huntington, W. Va.
1041.	Stanley, John T. Co.	New York, N. Y.
1043.	Stanton Lab.	Wyncote, Pa.
1045.	Starch Products Co.	New York, N. Y.
1047.	Stauffer Chem. Co.	New York, N. Y.
1049.	Stauffer Chem. Co. of Texas	Freeport, Texas
1051.	Stein, Hall & Co.	New York, N. Y.
1053.	Stokes & Smith Co.	Philadelphia, Pa.
1054.	Stoney-Mueller, Inc.	Lyndhurst, N. J.
1055.	Strahl & Pitsch.	New York, N. Y.

SELLERS OF CHEMICALS AND SUPPLIES 427

No.	Name	Address
1059.	Strohmeyer & Arpe Co.	New York, N. Y.
1059.	Stroock & Wittenberg Corp.	New York, N. Y.
1060.	Sundheimer, Henry, Inc.	New York, N. Y.
1061.	Sun Oil Co.	Philadelphia, Pa.
1064.	Swift & Co.	Chicago, Ill.
1065.	Synfleur Scientific Labs.	Monticello, N. Y.
1067.	Synthane Corp.	Oaks, Pa.
1068.	Synthetic Nitrogen Products Co.	New York, N. Y.
1069.	Synthetic Products Co.	Cleveland, Ohio
1070.	Synvar Corp.	Wilmington, Del.
1071.	Tainton Trading Co.	New York, N. Y.
1073.	Takamine Laboratory, Inc.	Clifton, N. J.
1075.	Tamms Silica Co.	Chicago, Ill.
1077.	Tanners Supply Co.	Grand Rapids, Mich.
1079.	Tannin Corp.	New York, N. Y.
1081.	Tennant & Sons, C. Co. of N. Y.	New York, N. Y.
1083.	Tenn. Eastman Corp.	Kingsport, Tenn.
1085.	Texas Chem. Co.	Houston, Texas
1087.	Texas Mining & Smelting Co.	Laredo, Texas
1088.	Theobold Industries, Inc.	Kearney, N. J.
1089.	Thomas, Arthur H. Co.	Philadelphia, Pa.
1091.	Thorocide, Inc.	St. Louis, Mo.
1093.	Thurston & Braidich.	New York, N. Y.
1095.	Titanium Alloy Mfg. Co.	Niagara Falls, N. Y.
1097.	Titanium Pigment Corp.	New York, N. Y.
1099.	Tobacco By-Products & Chem. Corp.	Louisville, Ky.
1101.	Trask, Arthur C. Co.	Chicago, Ill.
1103.	Trojan Powder Co.	Allentown, Pa.
1105.	Turner, Joseph & Co.	Ridgefield, N. J.
1107.	Uhe, George Co.	New York, N. Y.
1109.	Uhlich, Paul Co.	New York, N. Y.
1111.	Union Oil Co.	Los Angeles, Calif.
1113.	Union Smelting & Refining Co., Inc.	Newark, N. J.
1115.	United Carbon Co.	Charleston, W. Va.
1117.	United Clay Mines Corp.	Trenton, N. J.
1119.	United Color & Pigment Co.	Newark, N. J.
1121.	U. S. Bronze Powder Works, Inc.	New York, N. Y.
1123.	U. S. Gypsum Co.	Chicago, Ill.
1125.	U. S. Industrial Alcohol Co.	New York, N. Y.
1127.	U. S. Industrial Chem. Co.	New York, N. Y.
1129.	U. S. Phosphoric Prod. Corp.	New York, N. Y.
1131.	U. S. Rubber Products, Inc.	New York, N. Y.
1133.	U. S. Smelting, Refining & Mining Co.	New York, N. Y.
1135.	Utah Gilsonite Co.	St. Louis, Mo.
1136.	Vanadium Corp. of America	New York, N. Y.
1137.	Van Allen, L. R. & Co.	Chicago, Ill.
1139.	Van-Ameringen Haebler, Inc.	New York, N. Y.
1141.	Vanderbilt, R. T. Co.	New York, N. Y.
1143.	Van Dyk & Co., Inc.	Jersey City, N. J.
1145.	Van Schaack Bros. Chem. Co.	Chicago, Ill.
1147.	Varcum Chem. Corp.	Niagara Falls, N. Y.
1148.	Velsicol Corp.	Chicago, Ill.
1149.	Verley, Albert & Co.	Chicago, Ill.
1151.	Verona Chem. Co.	Newark, N. J.
1153.	Victor Chem. Works.	Chicago, Ill.
1155.	Virginia-Carolina Chem. Corp.	Richmond, Va.
1157.	Virginia Smelting Works.	W. Norfolk, Va.
1159.	Vitro Mfg. Co.	Pittsburgh, Pa.
1161.	Vultex Chem. Co.	Cambridge, Mass.
1162.	Wah-Chang Trading Corp.	New York, N. Y.
1163.	Waldo, E. M. & F., Inc.	Muirkirk, Md.
1165.	Wallerstein Co., Inc.	New York, N. Y.

No.	Name	Address
1167.	The Warner Chem. Co.	New York, N. Y.
1169.	Warwick Chem. Co.	West Warwick, R. I.
1170.	Washine National Sands, Inc.	Long Island City, N. Y.
1171.	Welch, Holme & Clark Co.	New York, N. Y.
1173.	Welsbach & Co.	Gloucester, N. J.
1175.	Werk, M. Co.	Cincinnati, Ohio
1177.	Western Charcoal Co.	Chicago, Ill.
1179.	Westinghouse Elec. & Mfg. Co.	E. Pittsburgh, Pa.
1181.	Whittaker, Clark & Daniels	New York, N. Y.
1183.	Wiffen & Co., Sons, Ltd.	London, England
1185.	Wilckes-Martin-Wilckes Co.	New York, N. Y.
1187.	Will & Baumer Candle Co.	New York, N. Y.
1189.	Williams, C. K. & Co.	Easton, Pa.
1191.	Wilson Laboratories:	Chicago, Ill.
1192.	Winthrop Chemical Corp.	New York, N. Y.
1193.	Witco Chem. Co.	New York, N. Y.
1195.	Woburn Chem. Corp.	Harrison, N. J.
1197.	Wolf, Jacques & Co.	Passaic, N. J.
1199.	Wood Flour, Inc.	Manchester, N. H.
1201.	Wood Ridge Mfg. Co.	Wood Ridge, N. J.
1203.	Wyodak Chem. Co.	Cleveland, Ohio
1205.	Young, J. S. & Co.	Hanover, Pa.
1206.	Ziegler, G. S. & Co.	New York, N. Y.
1207.	Zinsser, Wm. & Co.	New York, N. Y.
1209.	Zophar Mills, Inc.	Brooklyn, N. Y.

INDEX

A

Abrasive, Diamond Dust 333
Absorption Base 39, 102
Acidproof Cement 21, 22, 24
Acidproofing Laboratory
    Tables 286
Acids, pH Value of 392
Acknowledgments 399
Adhesion, Increasing Asphalt . 19
Adhesive. See also Bonding
    Cement, Fly Paper, Glue,
    Laminating Lute, Mucilage, Paste, Putty, Seal.
    Benzene Resistant 30
    Cellophane 17
    for Cellophane, Laminating.. 28
    Cellulose Acetate 27
    Cellulose Ether 19
    Cement 25
    Denture 79
    for Glassine, Laminating ... 28
    Heat Sensitive 29
    Laboratory 29
    Laminating 28
    Leather 29
    Lucite 28
    Paper to Metal Foil 17
    Paper, Thermoplastic Coating
    for 18
    Paper to Tin 17
    Plexiglas 28
    Plywood 19
    Pressure Sensitive 29
    Rubber to Brass 318
    Rubber Latex 19
    Starch (High-Strength) 16
    Tape, Coating for 18
    Tape, Pressure Sensitive ... 18
    Tape, Thermoplastic 18
    Tin 25
    Vinylite Coated Cloth 27
    Vinylite Sheeting 26
    Waxed Paper 17

Adhesives 11
Advice 5
Aerosol Insectide 205
Agitation 33
Air Filter Treatment 390
Alcohol Anti-Freeze 15
    Proof and Percentage Table
                        396, 397
Aldo 28. See Glyceryl Monostearate S.
Aldo 33. See Monostearin.
Alkalies, pH Value of 392
Alloy, Ceremic-Metal Sealing .. 31
    Glass-Metal Sealing 31
Alloys 251
Aluminum Cleaner 246, 351
Amino Acid Elixir 92
Aminoplasts 307
Ammunition Primer 338, 339
Anodizing 250
Ant Control 203
    Poison 8
Anti-Acid, Stomach 92
Anti-Carburizing Composition.252
Anti-Chap Lipstick 56
Anti-Foam. See Defoamer.
Anti-Fog, Windshield 381
Anti-Freeze 15
    Radiator 381
Anti-Friction Alloy 251
Anti-Oxidant for Rubber 319
Anti-Perspirant. See Cosmetics, Deodorant.
Anti-Pyorrhea Powder 82
Antiseptic, Cattle Wound 137
    Powder, Ear 93
    Sticks 109
    Surgical 109
Antiseptics 109
Anti-Stick Coating, Adhesive
    Roll 18
Anti-Tack Coating for Asphalt.319
Anti-Vesicant Ointment 102
Apparatus 3

Apple Pie Filling 151
Aquarium Cement 12, 25
Aromatics. See also Perfumes.
   Stability of 86
   Tenacity of 87
Artificial Leather Dope 220
Asbestos Cement Sheet
   Coating 230
   Fiber Suspension 379
   Sheet, Compressed 313
Asphalt Tile, Anti-Tack Coating for 319
"Asplit" Cement 23
Asthma Inhalant 93
Atabrine Tablets 93
Athlete's Foot Remedies 93, 94
Attractant. See also Lure.
   Mosquito 207
Auto Polish. See Polish, Auto.
Automobile Radiator Leak Seal 19

B

Baby Oil, Antiseptic 92
Bacterial Slime, Inhibiting....290
Bactericide 290
   Plant 208
Baked Goods, Improved
   Quality 168
Bakers' Pan Grease 173
Baking Powder 8, 147
   Powder Biscuit 158
Barley Smut Control 129
Bases, pH Value of 392
Bate, Oropon 219
Bath Preparations 68
Bedbug Exterminator 8
Beeswax Substitute 321
Belt Dressing 226
Belts, Cleaning Leather 220
Benzene Hexachloride Insecticide. See Gammexane.
Biologic Materials, pH Value
   of 392
Bird Food 9
Bleach, Cotton Goods 355
   Hair 65

Bleach—Continued
   Hydrogen Peroxide 355
   Laundry 13
   Non-Settling 355
   Stable 386
   Wood 229
   Wood Floor 12
Blueberry Pie Filling 152
Bluing, Laundry 13, 346
Body Deodorant 77
Boil Leak, Stopping 20
Boiler Compound 15
   Scale Inhibitor 383
   Water Compound 382
Bonding Glass to Aluminum .. 20
   Vinylite to Cloth 27
Bookbinding Preservative 219
Borating Solution 31
Borax, Quality of 32
Bottle Cap 19
   Cork Seal 29
   Seal 19
Brake Lining, Molded 313
Bread Baking, Improved 171
   Dough Improver 173
   Fresh-Keeping 171
   Irish Soda 173
Brick, Building 229
Brilliantine 64, 65
Briquettes, Slack Coal 381
Bristles, Artificial 299
Bubble Bath Powder 68
   Fluid 383
Buckshot, Non-Poisonous 251
Building Board 230
Bull Semen Preservative 137
Burn Treatments 95
Bursting Disc Alloy 251
Butterscotch Pie Filling 155
   Topping 176

C

Cabbage Maggot Control 212
Cable Coating, Electric.......298
Cakes, Improved 169
Calamine Lotion 96

INDEX

Calcium Arsenate, Fatted 207
Calculating Costs 5
Calculations 2
Camphor Solution 96
Canary Bird Food 9
Candied Fruit Peel 176
Candles 11
Candy 174-177
    Coatings 176
Canvas, Waterproofing 14
Cap, Bottle 19
    Gelatin Sealing 19
Carbide Tips, Removing 256
Casein Dispersion 332
    Manufacture 139
    Spreader 210
Catalyst, Urea-Formaldehyde
    Resin 299
Cattle Feed, Dairy 130
    Feed, Molasses 133
    Grub Control 210
    Lice Control 210
    Spray, DDT 211
    Tick Control 211
    Wound Antiseptic 137
    Wound Dressing 137
Caulking Cement, Metal Joint. 23
    Composition 23
    Compound 285
Cautions 4
Cedar Rust Control 209
Celery Blight, Preventing 124
Cement, Acidproof 21, 22, 24
    Aquarium 12, 25
    "Asplit" 23
    Brick 24
    Caulking 23
    for "Celluloid" 25
    for Cellulose Acetate 25, 28
    Coating, Refractory 231
    Coke-Oven 21
    Concrete Tank 24
    De Khotinsky 30
    Dry Furnace 21
    for Ethyl Cellulose 25
    Floor Hardener 11
    for Formvar 25

Cement—Continued
    Furnace 21
    High Temperature 21
    High Temperature
        Conduction 233
    Improved 230
    Iron Pipe Leak 23
    Jointing 24
    Light Weight 230
    Litharge-Glycerin 20
    Metal to Glass 20
    for Methylmethacrylate 25
    Paper Pad End Binding 17
    Pipe Joint 23
    for Plastics 25
    Plugging 23
    for Polystyrene 25
    for Polyvinyl Chloride 25
    Refractory 21
    Rubber (Self-Curing) 30
    Sealing 23
    Solvent Resistant 24
    Spark Plug 25
    Synthetic Resin 24
    Temperature Indicating ...381
    for Thermoplastics 25
    for Thermosetting Plastics .. 28
    Tile 24
    Tin Adhesive 25
    Tungsten 233
    for Vinylite 25
    Waterproofing 14
    Weatherproof 230
Ceramic Finish 230
Chapped Lip Preventative 56
Cheese Cake, French 163
Chemical Heating Pad 63
Chemicals, Quality of 32
    Variation in 32
    Where to Buy 4, 400
Cherry Pie Filling 152
    Spread 191
Chest-Rub Salve 7
Chick Coccidiosis Control 137
Chicken Feed 135
    Louse Powder 211
    Roost Spray 211

Chinch Bug Control 212
Chlorine Water. See Javelle Water.
Chocolate Bars, Sugarless 177
    Cream Pie Filling 155
Chrome Pickling Bath 249
    Tan 219
Cinnamon Oil, Artificial 181
    Sugar, Artificial 181
Citrus Brown Rot Control 210
Clarification, Liquid 4
Cleaner (see also Remover) ... 12
    Abrasive 12, 333
    Airplane 352
    Aluminum 246, 351
    Auto Radiator 354
    Beer Pipe 5
    Carbon Deposit 352
    Carpet 349
    Celluloid 350
    Dental 79
    Denture 79, 350
    Dish 344
    Drain 354
    Engine 352
    Fabrikoid 350
    Fruit 121
    Furniture 325
    Glass. See Cleaner, Window.
    Glove 346
    Household 13
    Industrial 351
    Industrial Hand 345
    Iron Tank 351
    Jewelry 350
    Laboratory Glass .352, 353, 354
    Lead-Lined Tank 351
    Locomotive Boiler 354
    Machinery 352
    Magnesium Weld 246
    Marble 350
    Metal 246, 247, 351
    Metal Cloth 324
    Microscope Slide 352
    Milk Pipe 5
    Paint 349
    Paint Brush 13, 351

Cleaner—Continued
    Powdered 345
    Printing Roller 350
    Radiator 247
    Silicone 353
    Silver Tarnish 354
    Skin 50
    Solvent Emulsion 351
    Stainless Steel 351
    Steel Tool 246
    Straw Hat 13
    Surgical Instrument 352
    Telephone Mouthpiece 350
    Upholstery 348
    Wall Paper 13, 349
    Waterless Hand 345
    Window 13, 353
    Windshield 353
Cleaning Fluid 348
Cleanser. See Cleaner.
Coagulant, Latex 320
Cocoa Malt Powder 8
    Powder, Sweet 8
Coconut Bars 158
Coffee Cake Dough 158
Cold Cream. See Cream, Cold.
    Producing Powder 381
Color, Certified 83
    Stabilization 344
Coloring (see also Dyeing) .33, 83
    Anodized Aluminum 235
    Cast Resins 310
    Copper Antique 9
    Cosmetics 83
    Organosols 300
    Plastics 301, 305
    Silver Black 235
    Smoke 336
    Stainless Steel Black 236
    Steel 9
    Steel Black 235
    Wood 228
Colors, Blended Oil Soluble ... 86
    Blended Water Soluble 86
    Insoluble 85
    Treating and Handling 197
Compact Powder, Face 78

INDEX

Compounding Vessels 3
Concrete, Dark Gray 229
    Floor Treatment 286
    Hardener 230
Condiments 192, 193
Confections. See Candy.
Consultation 5
Containers, Mixing 3
Copper, Coloring 9
Core Binder 253
    Casting Foundry 253
    Crack Filler 253
    Wash, Foundry 252
Cork Substitute 229
Corn Borer Dust 213
Corrosion Inhibitor, Anti-
    Freeze 382
    Prevention. See Corrosion
    Proofing and Rust Prevention.
    Proofing Aluminum 248
    Proofing Grease 227
    Proofing Magnesium 248
Cosmetic Emulsions 34
Cosmetics 32
Costs, Calculating 5
Crab Lice Treatment 96
Crash Pad, Airplane 378
Crayon, Marking (Green) 9
Cream, Absorption Base 39
    All-Purpose 44
    Antiseptic 102
    Barrier 106
    Brittle-Nail 72
    Brushless Shaving 6, 69
    Camphor-Peppermint 44
    Cleansing 5, 44
    Cleansing Liquefying 5
    Cold 5
    Cold (Non-Greasy) 6
    Dry Skin 38
    Filling for Pies 166
    Finishing 44
    Flash-Burn 94
    Fondant, Chocolate Dipping.176
    Fortified Whipped 178
    Foundation 47

Cream—Continued
    Greaseless 41
    Hair 63
    Hand 48
    Imitation Whipped 178
    Industrial Skin Protective ..106
    Lathering Shaving 71
    Liquid (Cosmetic) 40
    Medicated 100
    Modern Cold 36
    Neutral Cleansing 2
    Night 39
    Penicillin 103
    Permanent Wave 62
    Powder 47
    Pre-Shave 72
    Puff Paste 167
    Sports 44
    Stabilized Whipped 177
    Sulfathiazol 103
    Sun Protective 52
    Sunscreen 52
    Tropical Cold 38
    Tropical Powder 48
    Vanishing 6, 41
Crucible Lining 231
Culture Media, Bacteria 389
Custard Powder 179
Cutting Oil (see also Soluble
    Oil) 119, 223

D

DDT Cattle Spray 211
    Emulsion 113
    Powder 205
    Powder, Wettable 205
    Solvent 205
    Spray 204
Decolorizing 4
Defeathering Compound, Poultry 138
Defoamer, Paper Pulp 290
    Yeast Fermentation 173
Defoaming Oil 226
Dehydrated Sweet Potatoes,
    Improving 196

Dehydrating Meat 195
  Vegetables 195
De-Icing Compound, Airplane. 382
Delousing Preparation 205
Delusterant, Textile 375
Dental Amalgam 251
  Casting Wetting Agent 253
  Cleanser 79
  Grinding Coolant 223
  Gum Hardener 79
  Impression Plaster 297
  Paste. See Tooth Paste.
  Plate Adhesive 79
  Powder. See Tooth Powder.
  Preparations 79
Dentifrice. See Dental Preparations, Mouth Wash, Tooth Powder.
Denture Adhesive 79
  Cleaner 79
Deodorant, Cosmetic ... 75, 76, 77
  Spray 8
Deodorizer, Plaster Dressing.. 97
Depilatories 68, 69
Depilatory, Hide 219
Dermatitis Treatment 107
Derris Diluents 208
Detergent. See also Cleaner, Soap.
  Persil 344
  Textile 346
Developer, Fine Grain 293
  Low Temperature 292
  Photographic 15, 292
Developing Old Printing Paper. 293
Devil's Food Cake Dough..... 162
Dialyzers 388
Dielectric, Ceramic 232
Diglycol Laurate 36
  Stearate S. 36
Dishwashing Powder 344
Disinfectant. See also Fungicide.
  Citrus 216
  Cresol 8
  Seed 216
  Wheat Seed 216

Disinfecting Seed 128
Dissolving 4
Distemper. See Paint, Emulsion.
Doctor Solution 387
Dog Flea Powder 211
  Repellent 139
  Shampoo 139
Dough, Danish 158
  Soya Flour Sweet 157
Doughnut Dough 163
  Glaze 164
  Icing 165
Drain Root Destroyer 217
Dressing, Salad 189
Drilling Mud, Non-Foaming .. 385
Drug Products 92
Dry Cleaning Fluid 12
  Cleaning Fluid, Disinfecting. 348
  Cleaning Soap 347
Dubbing, Leather 219
Dust Cloth Impregnant 354
Dyeing. See also Coloring.
  Feathers 384
  Gloves 221
  Natural 373
  Nylon Hose 362
  Plastics 301
  Rayon, Package 356
  Vegetable Ivory Buttons ... 384
  Vehicles 305
  Wool 364
Dyes for Coloring Smoke 338
  Cosmetic 33

E

Ear Antiseptic Powder 93
  Oil 93
Egg Preserving Coating 138
  Production Increaser 138
  White Substitute 168
Eggs, Detecting Cold Storage .. 196
Electro-Plating. See Plating.
Elixir, Amino Acid 92
Emulsifier. See Emulsifying Agent.

Emulsifying Agent 34
Emulsion, Acrawax 116
  Acrawax C 116
  Almond Oil 181
  Animal Oil 118
  Base, Salad 190
  Beeswax 114
  Benzyl Benzoate 96, 113
  Carnauba Wax ..114, 289, 327,
                328, 329, 330
  Castor Oil 112
  Changes 35
  Chlorinated Paraffin 26
  Chlorinated Rubber 26
  Chlorinated Solvent 111
  Cinnamon Oil 182
  Cleaning 354
  Cocoa Butter 113
  Cod-Liver Oil 120
  Cosmetic 34
  Cottonseed Oil 112
  Creosote Oil 111
  Cutting Oil 119
  DDT 113, 204
  Dichloroethyl Ether 111
  Ester Gum 115
  Ethylene Dichloride 111
  Fatty Amide 116
  Flavor 181
  "Flexalyn" 113
  Glycol Resinate 113
  Grease 114
  Hydrocarbon 111
  Inversion 34
  Irish Peat Wax 116
  Japan Wax 114
  Kerosene 111
  Lanolin 114
  Lard Oil 112
  Lemon Oil 181, 182
  Lethane 209
  Light Mineral Oil 117
  Linseed Oil 112
  Lubricating Oil 112
  Mace Oil 182
  Manila Resin 331
  Mineral Oil 119, 325

Emulsion—Continued
  Naphtha 111
  Neatsfoot Oil 112
  Oil 112
  Oil-in-Water 34
  Olive Oil 112
  Orange Oil 182
  Paint 266
  Paraffin Wax 114
  Petrolatum 114
  Pine Oil 112
  "Polypale" Ester 113
  Polyvinyl Copolymer 266
  Propylene Dichloride 111
  Shellac 332
  Solvent 111
  Stability 34
  "Staybelite" 113
  Stearamide 374
  Sulfathiazole 113
  Testing Type of 34
  Turpentine 97
  Type 34
  Vegetable Oil 118
  Velsicol 208
  Water-in-Oil 34
  Wax 114, 281, 326, 334
  White Mineral Oil 112, 117
Emulsions 111
  Making 111
  with Pectin 120
Enamel. See also Paint.
  Black Lusterless 261
  Black Machinery 260
  Camouflage 274
  Crinkle Finish 230
  Floor 274
  Fluorescent 304
  Gloss Gray 260
  Gloss White 260
  Gray Machinery 261
  Marine Interior 265
  Non-Scumming 231
  Olive Drab Gloss 261
  Olive Drab Lusterless 261
  Opaque 231
  Quick Drying 272

Enamel—Continued

Vehicle, Floor	275
Energy Conversion Tables	394
Ephedrine Preparations	97
Equipment	3
Manufacturing	33
Eradicator, Ink	201
Eraser, Rubber	318
Rubberless	299
Essence. See Flavor.	
Essential Oils, Stability of	87
Oils, Tenacity of	87
Etchant, Glass	9
Etching Magnesium	241
Metallographic	241
Ethylene Dichloride, Soluble	119
Glycol Anti-Freeze	15
Extract, Flavor. See Flavor.	
Eye Antiseptic Ointment	75
Drops	75
Preparations	72
Shadow	74

F

Face Lotion, Vinegar	3
Powder	78
Powder, Tropical	78
False Teeth Preparations. See Denture Preparations.	
Feathers, Dyeing	384
Fiber, Synthetic Textile	379
Fiberboard, Greaseproofing	14
Figuring	2
Files, Chemical Sharpening of	257
Filler, Clay	300
Crack	297
Wood	297
Filling Containers	33
Filter, Improved Glass Wool	389
Light	294
Lubricating Oil	227
Porous Metal	256
Filtering	4
Finish. See Enamel, Lacquer, Paint, Varnish, Sizing.	

Fire Extinguisher	380
Extinguisher, Dry	14
Extinguisher, Liquid	14
Kindler	15
Fireproofing. See also Flameproofing.	
Canvas	14
Light Fabrics	14
Paint	264, 265
Paper	14
Fires, Extinguishing Phosphorus	380
Fixatives. See Fixing Agents, Perfume.	
Fixing Agents, Perfume	88
Solution, Succinic Acid	293
Flame, Colored Military	338
Flameproofing. See also Fireproofing.	
Textiles	376
Water Resistant	377
Flannel Cakes	173
Flavor, Celery Vinegar	183
Emulsions	181
Fruit Vinegar	183
Ginger Ale	180
Imitation Apricot	180
Imitation Banana	180
Imitation Peach	180
Imitation Pineapple	180
Imitation Raspberry	180
Imitation Rum	180
Malt Vinegar	183
Mustard Vinegar	183
Rock and Rye Whisky	181
Spice	182
Spice Curry	182
Spice Fish	182
Spice Meat	182
Spice Mixed Pickle	182
Spice Pickle	182
Spice Sausage	182
Spice Smoked Meat	182
Spice Tarragon	182
Spice Vinegar	183
Tarragon Vinegar	183
Tooth Paste	80, 81

Flavor—Continued
  Vanilla 8
  Wine Vinegar 183
Flavors, Imitation 180
  Treating and Handling 196
Flea Beetle Control 212
Flint, Lighter 252
Floor Composition 230
  Hardener, Cement 11
  Hardener, Concrete 230
  Oil 10
  Sanitary 230
  Treatments. See Wood Treatments.
  Wax 10
Flotation Reagent 257
Flour Enrichment Mix 173
Flower Preservative, Cut 122
Fluorescent Coatings 302
Flux, Aluminum Melting 254
  Aluminum Welding 254
  Arc Welding 254
  Electric Welding 254
  Magnesium Welding 254
  Non-Corrosive 253
  Silver Soldering 254
  Soft Soldering 254
  Soldering 15, 253
  Stainless Steel Soldering 253
  Tin-Bronze Melting 254
Fly Paper 8
  Paper, Sticky 203
  Spray 8
Foam Powder 384
  Killer. See Defoamer.
Food Products 147
Foods, pH Value of 392
Foot Balm 78
  Powder 7, 78
  Preparations 78
Formulation, Cosmetic 31
Foundry Core Wash 252
Friction Material, Molded 313
Fruit Bars 175
  Cake Dough 161
  Cleaning and Disinfecting ..122
Fruits, Treating and Handling.196

Fudge Bars 174
Fumigant, Corn Crib 216
  Dairy 216
  Factory 216
  Insect 216
  Warehouse 215
Fumigating Cone 110
  Sickrooms 110
Fungicide (see also Disinfectant) 209
  Apple 209
  Leather 219
Fungus Proofing 374
Furnace Cement 21
  Lining 232
Furniture Polish. See Polish, Furniture.
  Stain, Walnut 287
Fuse Powder, Delayed 339

G

Galvanizing Alloy 251
Gammexane Insecticide 209
Gasket Composition 313
Gasoline, Preventing Ice in ...382
  Solidified 15
  Thickener 382
Gasolineproof Coating 281
Gel, Aluminum Hydroxide 92
Gelatin, Dissolving 19
  Seal (Cap) 19
Germicide, Surgical Instrument
                          109, 352
Germicides 109
Glass Cleaner. See Window Cleaner.
  Etchant 9
  Marking Ink 2
"Glassine" Paper 13
Glassware, Where to Buy 4
Glaurin 36
Glaze, Candy 176
  Doughnut 164
  Waterproof Label 281
Glue 16
  Casein 289

Glue—Continued
 Dissolving 16
 Flexible 2
 Perfume 17
Glycerin Anti-Freeze 15
 Substitute 390
Glyceryl Monostearate 35
Glycol Anti-Freeze, Ethylene.. 15
Goat Louse Dip 211
Grafting Wax. See Wax, Grafting.
Grain Smut Control 129
Graphite Lubricant. See Lubricant, Graphite.
Grease. See also Lubricant.
 Bakers' Pan 173
 Graphite 11
 Paint, Theatrical 48
 Remover. See Remover.
Greaseproofing (see also Oil-proofing) 290
Grinding 4
 Coolant, Dental 223
 Oil, Glass 223
Gross Specimens, Preserving ..387
Gum Rosin Yield, Increasing..122
 Tragacanth Substitute 385
Gun Lubricant. See Lubricant, Gun.
Gypsum Board 230

H

Hair Bleach 65
 Cream 63
 Dressing, Liquid 64
 Pomade 64
 Preparations 57
 Setting Mix 63
 Straightener 63
 Tonic 57
 Wave Lotion, Permanent... 61
Hand Cleaner, Industrial 345
 Grenade Primer 339
 Jelly, Glycerin 51
 Lotion. See Lotion, Hand.
Hard Rubber, Cellular 317

Hardener, Cement Floor 11
Heartwood, Detecting 229
Heat Absorbing Glass 386
 Pad Mix, Hair Waving..... 63
Heating 3
Hectograph Composition 200
Herbicide 216, 217
Hog Feed 134
Honey, Artificial 189
Horse Feed 134
Hot Melt Coatings 306
Hydraulic Brake Fluid 226
 Pressure Fluid 226
Hydrogen Peroxide Stabilizer.110
 Sulfide Generation 386
Hydroponic Plant Food 121

I

Ice Cream 183
 Cream Mix, Powdered .185, 186
 Cream Stabilizer 186
 Milk 186
 Rhubarb 186
Icing Cream Baking 166, 172
 Doughnut 165
 Improved 172
 Improved Cake 166
 Meringue 167
 Non-Sticky 172
Ignition Mixture, Blasting Cap.339
Incense Cones 110
Infra-Red Crystal Lens 386
Inhibitor, Pickling 246
Ink, Bakelite Stamping 201
 Ceramic Stencil 200
 Eradicator 201
 Glass Etching 200
 Glass Marking 2
 Hectograph 200
 Indelible 9
 Indestructible 200
 Laundry Marking (Indelible) 9
 Marking 201
 Metal Etching 201
 Printing 201
 Remover. See Remover.

Ink—Continued
  Rubber-Stamp Pad 200
  Spirit 199
  Steam-Setting Printing 201
  Thermofluid Printing 201
  Thermoplastic Printing 201
  Typewriter Ribbon 200
  from Used Mimeograph
    Paper 199
  Waterproof Ruling 199
  Writing (Blue-Black) 9
Insect Bite Lotion 207
  Control. See Insecticide.
  Repellent 7
Insecticide. See also Ant Poison, Bedbug Exterminator, Fly Spray, Fumigant, Mothproofing, Repellent, Spray.
  Aerosol 205
  Airplane 206
  Animal 210
  Ant 203
  Blueberry Thrip 214
  Body Louse 203
  Bulb 211
  Chigger 215
  Cockroach 203
  Codling Moth 213
  Colorado Beetle 214
  Corn 211
  Corn Borer 213
  Corn Worm 213
  Cotton Flea Hopper 213
  DDT 204
  Emulsion 204
  Fire Ant 215
  Fly 204
  Gammexane 209
  Grape-Bud Beetle 213
  Grasshopper 213
  Horticultural 214
  Lacebug 214
  Lethane 209
  Louse 205
  Lucerne Snout Beetle 215
  Mealybug 214

Insecticide—Continued
  Mexican Bean Beetle 213
  Mound Ant 215
  Nicotine 209
  Ornamental Plant 214
  Pea Aphid 213
  Peach-Tree Borer 214
  Pine Sawfly 215
  Plant 211
  Red Spider 214
  Rotenone 207
  Scale Insect 214
  Seed 214
  Silverfish 203
  Slug 215
  Velsicol 208
  Vine Moth 214
  Wheat 215
Insulating Tape Impregnant ..298
Insulation, Building 234
  Ceramic 232, 233, 234
  Mortar 234
  Thermal 234
Intensifier, Photographic 294
Interconversion Tables 393
Itch Remedy 96

J

Javelle Water 13
Jelly, Apple-Raspberry 187
  Coagulant 187
  Cranberry 188
  Imitation 188
  Low Cost 187
  Orange 188
Jute Fibers, Salvaging 378

K

Kalsomine 286
Kerosene, Soluble 119
Killing and Preserving Fluid...387
Kindler, Fire 15

L

Label, Peelable Adhesive 29
Lacquer, Cellolyn 279

Lacquer—Continued
 Cellulose Acetate 278, 280
 Cellulose Ether 278
 Clear 277, 280
 Cloth 277
 Engine Gray 277
 Furniture 279
 Metal 277
 Primer 277
 Remover. See Remover.
 Thinner 280
 Wood 276
 Wood Polishing 277
Lamb Feed 135
Laminating Paper, Adhesive for 28
 Thinner, Vinylite 322
Larvicide 387
Laundry Bleach. See Bleach, Laundry.
 Blue 13
 Ink 9
 Linen Stiffening 374
 Sour 346
Layer Cake Dough 159
 Cake Dough, Yellow 161
Leaf Spot Treatment, Beet 124
Leak Cement, Iron Pipe 23
 Stopping Water 19, 20
Leather Conditioning 220
 Durability, Increasing 219
 Preservative 10
 Soap. See Saddle Soap.
 Treating Preparations 219
 Waterproofing. See Waterproofing, Shoe.
Leg Make-Up 54, 55
Lemon Chiffon Pie Filling.152, 156
 Extract, Pure 8
 Pie Filling 153
 Pie Filling, Sugarless 156
 Powder for Pie Filling 165
Lens, Seal for 20
Liniment 7, 96
 Sore Muscle 7
 Turpentine 97
Linseed Oil, Soluble 118
Lip Pomade 56

Lip—Continued
 Rouge, Liquid 56
 Stain, Liquid 56
Lipstick 55
Litharge-Glycerin Cement 20
Lithographic Plate Dampener.201
Loss, Preventing 4
"Lost" Wax 322
Lotion, Antiseptic Skin 109
 Calamine 96
 Dandruff 60
 Face 49
 Hair 57
 Hand 6, 50
 Hand (Milky) 6
 Insect Bite 207
 Milky Sun 51
 Sun Tan 52
 Vinegar Face 3
Lubricant 11
 Antifoaming 226
 Bolt Drawing 223
 Clock 225
 Cold Rolling 223
 Die Cutting 290
 Door 225
 Drawer 225
 Drill 224
 Extreme Pressure 223
 Fine Instrument 225
 Gasoline Insoluble 224
 Gear 226
 Goldbeater's 225
 Gun 11
 Mastic 225
 Medicinal 106
 Mold Parting 224
 Nitration Resistant 224
 Optical Instrument 225
 Packing 224
 Paper Cap 290
 Pipe Thread 223
 Plug Valve 224
 Rayon 226
 Rope 225
 Rubber 224
 Rubber Molding 224

INDEX

Lubricant—Continued
  Slide Rule 225
  Solvent Insoluble 224
  Split Die 224
  Stuffing Box 224
  Textile 226, 374
  Tire Rim 223
  Valve 223
  Water Soluble 224, 226
  Waterproof 225
  Window 225
  Wire Drawing 223
  Wire Rope 225
Lubricating Jelly, Medicinal ..106
Lure, Japanese Beetle 215

M

Macaroon Dough, Almond 159
Make-Up, Liquid 48
Malted Milk, Frosted Chocolate 179
  Milk Powder 8
Manufacturing Operations 32
Mapeleine Syrup, Sugarless ...177
Marshmallow 167, 168
Mascara 72
Masking Coating 243
Matches, Waterproof 336
Measuring 4
  Temperature 3
Meringue Icing 167
  Pie 154
  Sugarless 167
Metal, Coloring 235
  Oil Absorbing 251
  Polish 9
  Porous 251, 252
  Powder, Spherical 252
  Stripping 243
  Substitute, Porous 252
Methyl Methacrylate Protective Coating 282
Mexican Bean Beetle Control .213
Mica, Synthetic 231
Mildew Control 209
  Control, Leather 219

Mildewproofing 374
Military Smokes ...336, 337, 338
Milk, Beauty 51
  Ice 186
Mixing 4
Moistureproofing. See Waterproofing.
Mold Growth, Preventing 196
  Parting Lubricant 224
  Plaster, Dental 297
  Plaster-of-Paris—Resin 307
  Wash, Rubber 319
  Wetting Agent for Wax 322
Molding Material 11
  Sand, Magnesium 253
Molds, Flexible 319
Monostearin Emulsifier 35
Mortar, Insulating 234
Mosquito Preparations. See Insecticides, Repellents.
  Repelling Oil 7
Mothproofing 203
  Stainless 8
Motor Fuel. See Gasoline.
Mouth Wash 7
Mucilage, Casein 16
  Cosmetic 49
  Gum Arabic 16
Muffin Dough, Banana 159
  Dough, Honey 158
Mule Feed 134
Mushroom Bubbles, Preventing 124
Mustard, French 192
  Table 192

N

Nail Lacquer Remover 72
  Polish Drier 72
Nails, Cream for Brittle 72
Narcissus Rot, Preventing 125
Nasal Preparations 97
Neatsfoot Oil, Soluble 118
Neoprene Latex Dispersion ...319
Nirosan Dust 214
Nitrogen Generation 387

Non-Block. See Anti-Stick.
Noodle Dough 173
Nougat Fruit-Nut Bars 174
   Santo Domingo 174
   Short 175
Nuts, Treating and Handling ..197
Nylon Dyeing 362, 363
   Hose, Snagproofing 363

O

Oat Smut Control 129
Oil. See also Lubricant.
   Cutting 119, 223
   Ear 93
   Emulsions of 112
   Floor. See Floor Oil.
   Glass Grinding 223
   Penetrating 11, 226
   Polish. See Polish, Oil.
   Polishing 325
   Remover. See Remover.
   Soluble 117, 223
   Soluble Animal 118
   Soluble Vegetable 118
   Sulfonated Castor 221
   Sun Tan 52
   Torpedo 224
Oilproofing Paper 14
Oils, Essential 87
Ointment. See also Cream, Salve.
   Antiseptic 2
   Base 100
   Base, Vanishing 101, 102
   Base, Washable 102
   Coal Tar 102
Olive Oil 194
   Oil, Soluble 118
   Pimento Spread 191
Olives, Stuffed 194
Onion Smut, Preventing 124
Orchid Preservative, Cut 122
Organdie Finish 374
Organosols 300
Oropon Bate 219
Overheating, Preventing 3
Oxidation Inhibitor, Fat-Oil ..226

P

Packing Pads, Coir 378
Paste, Mechanical 25
Plastic 312
Paint. See also Enamel.
   Anti-Fouling 263, 264
   Awning 265
   Black Hull 264
   Brick 258
   Brush Cleaner 13
   Chemically Resistant 284
   Cold Water 286
   Crankcase Insulating 287
   Deck 275
   Emulsion 266, 267
   Exterior House 258
   Exterior Metal 261
   Exterior Trim 259
   Fire Resistant Canvas 264
   Flat Wall 270
   Fungus Treating 285
   Gasoline Proof Tank 284
   Gloss Emulsion 266
   Implement 260
   Interior Emulsion 266
   Interior Gloss 271
   Interior Semi-Gloss 271
   Marine Interior Flat 265
   Mosquitoproof 282
   Oleoresinous Water 267
   Plastic Nail 284
   Plastics 284
   Polyvinyl Emulsion 266
   Primer, Ferrous Metal 261
   Primer, Rust Inhibiting 261
   Primer, Wood 287
   Red Barn 285
   Red Lead 262
   Remover (see also Remover, Paint) 12, 281
   Sealer 270
   Shipbottom 263, 264
   Solvent 349
   Stucco 258
   Temperature Sensitive 282
   Thermocolor 282

INDEX

Paint—Continued
- Tractor 260
- Trellis 259
- Undercoat 270
- Utility Exterior 259
- Varnish Emulsion 266
- Vehicle, Aluminum 276
- Wagon 260
- Wall Emulsion 266
- Water. See Paint, Emulsion.
- White Stencil 281
- Wood Emulsion 266

Paper Finish 289
- "Glassine" 13
- Greaseproofing 14
- Improved Drawing 290
- Oilproofing 14
- Surface Finish for 290
- Waterproofing 13

Parting Compound, Foundry ..253
Paste, Mechanical Packing 25
- Paperhanger's 11
- Wall Paper 11
Peach Chiffon Pie Filling 153
- Pie Filling 153
Peanut Butter, Flavored..191, 192
Pecan Wafer Dough, Crisp ...158
Peelable Coating 243
Penetrating Oil 226
Penicillin Cream 103
- Glycerin Jelly 103
- Pastilles 97
- Suppositories 99
Perfume, Glue 17
- Jasmine 90, 342
- Lavender 344
- Lilac 90, 342
- Lily of the Valley........342
- Muguet 342
- Rose 90
- Soap 341, 342
- Stabilization 344
- Violet 90, 343
Perfumes, Stability of 86
Perfuming 33
Permanent Wave Lotion 61
Persil Detergent 344

Perspiration, Synthetic 386
Petrol. See Gasoline.
Petroleum Tank Coating 284
pH values 392
Photographic Acid Hardening
- Fixing Bath 15
- Chemicals 290
- Developer 15
Pickle Spread 191
Pickling Bath, Chrome.......249
- Bath Inhibitor 246
- Solution 246
Pie Crust Mix 157
- Dough 157
- Filling 148
- Filling Base 165
Pig Feed 134
Pigeon Repellent 138
Pigment, Fire Retardant 286
- Yellow Fluorescent 285
Pill Coating, Enteric 96
- Excipient 97
Pine Oil, Soluble 120
Plant Food, Hydroponic 121
- Growth Regulator 121
- Hormone. See Plant Growth Regulator.
Plaster Dressing Deodorizer ..97
- of Paris Molds 307
- Patching (Wall) 11
- Wall 230
Plastic, Black Polyvinyl 296
- Cheap 296, 297
- Films 300
- Foamed 314
- Peelable Coatings 306
- Rosin 296
- Rubber, Heat Hardening ...318
- Sheet, Moistureproof 297
- Wood Dough 12
- Wood Flour 297
Plasticizer, Polyvinyl 299
Plastics. See also Molding Material.
- Coloring 301
- Fluorescent Coating of ...302
- Identifying 295

Plastics—Continued
- Paint for 284
- Plating on 238
- Plating with 301
- Softening 322

Plating, Black Nickel 236
- Copper 237
- Indium 236
- Lead 237
- Nickel 236
- on Plastics 238
- Silver 236
- Tin 237
- with Plastics 301
- Zinc 238

Platinum Substitute 251
Plywood Adhesive 19
Poison, Insect. See Insecticide.
- Ivy Control 216
- Oak Control 216

Poisons 4
Polish 10, 324
- Auto 324
- Auto (Clear Oil) 10
- Bright-Drying 327
- Floor 325
- French 334
- Furniture 324
- Furniture (Oil and Wax) ... 10
- Laboratory Table 325
- Leather 335
- Lens 333
- Lucite 333
- Metal 9, 324
- Oil 325
- Paste 332
- Plastics 333
- Plexiglas 333
- Powder 333
- Rubber Footwear 335
- Rubless Floor 327
- Shoe 334
- Steel Tool 246
- Tumbling Barrel 334
- Wax 326
- White Shoes. See Shoe Dressing, White.

Polishing Cloth, Auto 324
Popcorn, Revivifying Dried ...197
Potato Psyllid Control 212
- Sprouting, Preventing 121
Poultry Feed 135
- Increasing Weight of 137
- Inhalant 138
Pound Cake Dough 159
Powdering 4
Preparations, Simple 5
Preservative, Leather 10
- Wood 228
Preshaving Cream 72
Prices, Chemical 5
Prickly Heat Preparations 98
Primer. See also Paint Primer.
- Undercoat, White 258
Printing Textiles, Screen 369
- Type Alloy 251
Protective Grease, Metal 248
Pudding Powder 180
Pulverizing 4
Puppies, Shampoo for 139
Putty 12
- Plastic 25
Pyorrhea Powder 82
Pyroxylin. See Celluloid.

R

Rabbit Deterrent 138
Radiator Anti-Freeze 15
- Cleaner 247
- Leak Seal 19
Rash Lotion 98
Rat Control 217
Rayon Tire Cord 375
Recovering Silver from Negatives 294
References 399
Refractory Cement 21
- Coating 232
- Lining 231, 232
Relish, Pepper-Onion 193
Remover. See also Eradicator.
- Chimney Soot 381
- Cigarette Stain 350

Remover—Continued
　Cylinder Carbon 352
　Grease 349
　Ink 13
　Nail Lacquer 72
　Nicotine Stain 350
　Oil 350, 354
　Paint 281, 349
　Piston Gum 352
　Rust 13
　Spot 13, 348, 349
　Tar 346, 348, 349
　Varnish 281
Repellent. See also Insecticide.
　Midge 207
　Mosquito 206
Resilient Rayon Batting 378
Resin, Aminoplast 307
　Casting 307
　Curing 310
　Flame Resistant 298
　Impregnating 307
　Synthetic 298
Resistance, Electrical 385
Rhubarb Rot, Preventing 124
Rifle Cartridge Primer 338
Road Marker 230
Rodenticide. See Rat Control.
Root Destroyer, Drain 217
Rotenone Insecticides 207
Rotproofing. See also Mildew Control.
　Textiles 375
Rouge, Face 78
Rubber, Cellular 315
　Footwear Polish 335
　Mold Wash 319
　Reclaiming 318
　Softening 318
　Stoppers, Boring Holes in .. 319
　Thread, Synthetic 318
Rust Inhibitor 247
　Prevention 9
　Remover. See also Remover.
　Remover, Radiator 247
　Removing Solution 246
Rustproof Coating 263

Rustproofing. See also Corrosionproofing.
　Steel 247

S

Saddle Soap 12
Salad Dressing 189
　Emulsion Base 190
Salt Water Protective Grease..227
Sandwich Spread, Olive 192
Sapwood, Detecting 229
Saran Coating 285
Sauce, Barbecue 193
　Chili 193
　Mustard 193
　Sea Food 192
Sausage, Beef-Soya 194
　Pork-Soya 195
Scabies Lotion 98
Scent. See Perfume.
Sea Water, Synthetic 386
Seal, Can Seam 25
　Cork 29
　for Explosive Igniter Wires..339
　Glass to Brass 20
　Glass High Vacuum 31
　Knot (Varnish) 282
　Metal-Glass 31
　Oil Well Wall 385
　Pipe Thread 223
　Threaded Joint 23
Sealer, Asphalt Clear 282
Sealing Alloy, Ceramic-Metal.. 31
　Alloy, Glass-Metal 31
Seasoning, Sausage 195
Secret Writing Detector 202
Seed Damping-Off, Preventing.122
　Disinfection 122, 216
　Germination, Testing 130
　Potato Treatment 125
Sellers of Chemicals 418
Shampoo, Coconut Oil 66
　Cream 67
　Dandruff 68
　Dog 139
　Golden 67
　Hair 65

Shampoo—Continued

Jelly	67
for Puppies	139
Silver	67
Soapless	68
Shaving Cream. See also Cream, Shaving.	
Cream, Brushless	69
Cream Improver	72
Cream, Lathering	71
Stick, Brushless	71
Sheep Worm Control	137
Sheepskin, Preserving Wool	222
Sherbet. See Ice.	
Shingle Stain	287
Shoe Dressing, White	10
Filler	299
Polish	334
Polish, Black	10
Stiffener, Thermoplastic	299
Shortening, Improved	177
Saving	169
Short-Stop, Succinic Acid	293
Silastic Gasket	313
Sizing, Emulsion	374
Nylon Yarn	374
Paper	289
Textile	373
Washable	374
Slushing Grease	248
Smoke, Chemical	336, 337
Colored	336
Screen	336
Smut Control, Grain	129
Soap	12
Dry Cleaning	347
Extender, Liquid	341
Finish, Glossy	340
Floating	340
Improved Toilet	340
Industrial Powdered	345
Laundry	340
Leather. See Saddle Soap.	
Liquid	12
Mechanic's	345
Mechanic's Hand (Paste)	12
Paste	12

Soap—Continued

Perfuming	341, 342
Powdered	344
Rug Cleaning	347
Saddle	12, 346
Salt Water	340
Solvent	348
Soapstone, Molded	234
Softener, Skin	50
Soil Mixture, Standard	386
Solder, Aluminum	255
Aluminum Foil	255
Berzelit	256
Brazing	255
Low Melting	255
Silver	256
Soft	256
Wiping	255
Zinc	255
Soldering Flux. See also Flux.	
Flux, Non-Corrosive	15
Solubilizing Gums	385
Soluble Oil	223
Oils	117, 118, 119, 120
Solutions, Making	4
Solvent, Soap	348
Sorghum Silage, Improved	134
Soybean Milk	179
Spoilage, Preventing	4
Sponge, Plastic	314
Rubber	315
Starch Surgical	109
Spot Remover. See Remover, Spot.	
Spray. See also Insecticide.	
Agricultural. See Insecticide.	
Citrus	204
Factory Deodorant	387
Fly	204
Insecticide. See Fly Spray.	
Nicotine	209
Pea Aphid	213
Spreader	210
Sticker	210
Stabilizer, Ice Cream	186
Stain Remover. See Remover.	
Stainless Steel Pretreatment	236

INDEX

Starch, Non-Gelling 387
  Paste, Edible 189
Steatite Insulation 232
Steel, Blue-Black Finish on ... 9
Sticking Tape. See Adhesive Tape.
Stocking, Liquid 54, 55
Stoke's Liniment 97
Straining 4
Straw, Artificial 299
  Hat Cleaner 13
Strawberry Pie Filling 153
  Whip Pie Filling 154
Strip Coating 243
Stripping, Metal Plate 243
  Oxide Films 246
Stuffing, Fowl 196
Styptic Cotton 99
Sulfa Drug Ointment 103
Sulfadiazine Tablets 98
Sulfonamide Chewing Wafers.. 99
  Ointment 104
Sulfonated Castor Oil 221
Sun Burn Ointment 53
  Protective Cosmetics 51
  Tan Powder 54
  Tan Products 32
Supplies, Where to Buy 400
Suppositories, Vaginal 99, 100
Synthetic Fiber, Casein 379
  Flavor. See Flavor, Imitation.
Syrup, Artificial Lemon 183
  Compound Table 177

T

Tables 391-397
Tank Lining, Acidproof 282
Tanning Liquor, Chrome 219
  Liquor, Vegetable 219
Tarnishproof Silver 251
Technical Magazines 5
Temperature 3
  Conversion Table 395
Tempering Steel 252
Termite Control 212

Textile Treatments 355
Thermometers 3
Thickener, Hydrocarbon 382
Thinner, Lacquer 280
Tile, Ceramic 234
Tinplate, Protective for 248
Tints for Paint 258
Tire Leak Seal 19
Tissue Embedding Composition 387
Tobacco Spot, Preventing 125
Tomato Culture, Seedless 121
  Seed Disinfectant 216
  Yield, Improving 121
Toners 85
Tooth Cavity Filling 79
  Paste 79
  Powder 7, 81
Toothache Oil 79
Trade Name Chemicals 398
Transparentizing Paper 289
Tree Wound Dressing 122
Tropical Ulcer Treatment 105
Turkey Anthelmintic 137
  Feed 136
  Worm Control 137

V

Vaginal Douche 99
Vanilla Concentrate 181
  Cream Pie Filling.....155, 156
  Flavor, Artificial 8
  Powder 165
  Sugar 181
Vanishing Cream. See Cream, Vanishing.
Varnish. See also Enamel, Paint.
  Chemically Resistant 269
  Clear 269
  Electrical Insulating 276
  Exterior 269
  Fiber 276
  Impregnation 276
  Interior 275
  Label 281

Varnish—Continued
  Lithographic (Improved) ...287
  Mixing 275
  Paper 276
  Phenolic Mixing 276
  Pulp 276
  Quick Drying 275
  Remover 281
  Spar 269
  Synthetic Resin 298
  Tall Oil 276
Venereal Prophylactic 100
Vinegar Essence, Pickling ...193
  Flavor 183
  Tablets 193
Volume-Weight Tables 393

W

Wall Paper Cleaner 13
  Paper Remover 284
  Patching Plaster 11
War Gas Ointment 102
Wart Removal 99
Washing Fluid 346
Water Freezing Protection ... 15
  Purity of 32
  Softener 346
  Treatment. See Boiler Compound.
Waterproofing Canvas 14
  Cement 14
  Compound 14
  Fiberboard 13
  Paper 13, 28, 290
  Shoes 10
  Textile 356
Wax Coating, Fungusproof 285
  Coating, Waterproof 281
  Compositions 321
  Condenser Impregnating 322
  Dance Floor 325
  Grafting 11
  Hard 322
  Liquid 10
  "Lost" 322
  Polish 10, 326

Wax—Continued
  Poultry Defeathering 138
  Ski 322
  Stop-Off 322
Weed Killer. See Herbicide.
Weighing 4
Welding Composition, White
  Metal 255
  Flux. See Flux, Welding.
  Spatter Composition 282
Wetting Agent, Acid-Resisting.346
  Agent, Dental Casting 253
  Agent, Wax Mold 322
Wheat Cakes 173
Where to Buy Chemicals 400
  to Buy Supplies 400
Whitewash 286
  Antiseptic 286
Window Cleaner 13
Wine, Rhubarb 198
Wire Rope Preservative 225
Wood Bleach 229
  Coloring 228
  Floor Bleach 12
  Preservative 212, 228
  Stain, Light-Fast 287
  Treatments. See Floor, Furniture.
Wool Fulling Compound 379
Wound Ointment, Tropical ...105
Wrinkle Finish 285

X

X-Ray Barrier Coating 284
  Contrast Composition 385
  Opaque Cream 385

Y

Yeast Cake Improver 168
  Fermentation Defoamer 173

Z

Zein Dispersion 120